Environmental Management of Concentrated Animal Feeding Operations (CAFOs)

Environmental Management of Concentrated Animal Feeding Operations (CAFOs)

Frank R. Spellman
Nancy E. Whiting

CRC Press
Taylor & Francis Group
Boca Raton London New York

CRC Press is an imprint of the
Taylor & Francis Group, an **informa** business

CRC Press
Taylor & Francis Group
6000 Broken Sound Parkway NW, Suite 300
Boca Raton, FL 33487-2742

© 2007 by Taylor & Francis Group, LLC
CRC Press is an imprint of Taylor & Francis Group, an Informa business

First issued in paperback 2019

No claim to original U.S. Government works

ISBN 13: 978-0-367-45305-3 (pbk)
ISBN 13: 978-0-8493-7098-4 (hbk)

Library of Congress Cataloging-in-Publication Data

Spellman, Frank R.
   Environmental management of concentrated animal feeding operations (CAFOs) / Frank R. Spellman and Nancy E. Whiting.
      p. cm.
   Includes bibliographical references and index.
   ISBN-13: 978-0-8493-7098-4 (alk. paper)
   ISBN-10: 0-8493-7098-1 (alk. paper)
   1. Agricultural pollution--United States. 2. Livestock factories--Environmental aspects--United States. 3. Manure handling. 4. Livestock--Housing--Waste disposal. 5. Agricultural laws and legislation--United States. I. Whiting, Nancy E. II. Title.

TD195.A34S64 2007
363.72'88--dc22   2006038866

**Visit the Taylor & Francis Web site at
http://www.taylorandfrancis.com**

**and the CRC Press Web site at
http://www.crcpress.com**

Contents

Preface

"We don't think of hogs as 'animals,' Bob, not in the same way as cats and dogs and deer and squirrels. We say 'pork units'—a crop, like corn or beans."

(Proulx, 2002, p. 302)

Concentrated animal feeding operations (CAFOs) are part of a recent agricultural trend toward large-scale, corporate-owned and managed livestock production. A far cry from the traditional family farm, CAFOs have lowered prices and increased production—but at a cost. While some of the problems CAFOs can cause are economic and related to small farm and farming community culture, or societal with a focus of the animal rights activism groups, other problems are ecological and environmental and will only be solved by intelligently crafted and enforced legislation.

In many states, lawsuits against CAFOs for unsound environmental practices demonstrate that CAFOs are creating problems. In short, regulations and legislation have fallen behind CAFOs creation and operation, enforcement of existing regulations is spotty, and problems associated with CAFOs are still being identified—although you can be sure those who neighbor CAFOs can identify some big issues. Chief among these are problems associated with massive quantities of animal waste, including air pollution and odor problems, as well as soil and water pollution problems.

These big-agribusiness businesses don't use traditional pasturing and feeding practices. Typically, manure is removed from livestock buildings or feedlots and stored in stockpiles or lagoon systems until it can be spread on farm fields, sold to other farmers as fertilizer, or composted. When properly designed, constructed, and managed, CAFO-produced manure is an agronomically important and environmentally safe source of nutrients and organic matter necessary for the production of food, fiber, and good soil health. Experience has demonstrated that, when properly applied to land at proper levels, manure will not cause water quality problems. However, cleanly—environmentally soundly—disposing of animal waste in the quantities that CAFOs produce can only be described as a challenge, and in many areas, the corporations that own the CAFOs have not been required to meet that challenge.

PURPOSE OF TEXT

This book is designed to provide practical information on the concepts and practices involved in the operation and maintenance of CAFOs, paying particular attention to regulatory requirements and compliance and best available technology. Both students and educators can use this practical text. However, *Environmental Management of Concentrated Animal Feeding Operations (CAFOs)* is also tailored to the information needs of practicing agronomists, rural community authorities, and other personnel involved in animal feeding operations.

Frank R. Spellman
Norfolk, Virginia

Nancy E. Whiting
Columbia, Pennsylvania

1 Introduction

[D]on't let the folks down there know that you are looking for sites for hog facilities or they will prevaricate and try to take us to the cleaners, they will carry on to various editors, every kind of meanness and so forth, as they have been brainwashed by the Sierra Club to think that hog facilities are bad, even the folks who love baby back ribs, even the ones hunting jobs... The panhandle region is perfect for hog operations—plenty of room, low population, nice long dry seasons, good water. There is no reason why the Texas panhandle can't produce seventy-five percent of the world's pork.

(Proulx, 2002, p. 6)

1.1 SETTING THE STAGE

The debate over the future of agriculture and agricultural policy is not new. It can be found in the history of such agrarian protest movements as the Grange (1870s), populism (1890s), the Farm Holiday Movement (1930s), the National Farmers Organization (1950s), and the immigration and environment groups of the last two decades. Every movement or trend in agriculture has had some farmers and rural residents who lauded progress, others who lamented the loss of the agrarian nature of America that allowed more to provide for themselves, and many more who were ambivalent (Schwab, 1998).

A current debate in agriculture over concentrated animal feeding operations (CAFOs) is gaining steam. The debate is fueled by the serious impact of CAFOs on the environment and on the social fabric of rural living. Because livestock factories produce and store large quantities of animal waste in leak-prone lagoons, America's water is at risk. Fish and wildlife suffer from manure spills. Livestock factories pose a threat to our air quality. The huge corporations that run the livestock factories edge out family farmers, who often use more environmentally friendly techniques. Owning a home might be the American dream, but a hog operation in the backyard is a nightmare for property values and for those who live on adjoining properties.

The use of CAFOs is a big issue now and we expect it to get bigger. Public awareness is rising. Hog farms in North Carolina, for example, have been in the national news because of water pollution and odor problems. Poultry farms and beef feedlots are drawing attention in other parts of the United States. The urban-rural interface is also garnering attention. Consider, for example, the various aspects of urban-wildlife interfaces, urban sprawl and its impact on forestry, compared to the impact of a family farm. Other issues include air pollution, antibiotics, hormones, and other chemical contaminants that are deposited into the environment via the manure environmental medium interface. The fact is that the popular myth behind the common image of traditional, idyllic, self-contained and self-operated farm operations stands in stark contrast to the modern reality of factory farming, based on CAFOs.

Siting of CAFOs is a divisive issue, pitting neighbors against neighbors in rural communities, amid angry debates about odors, environmental degradation, increasing concentration among producers, and a sense of injustice among those who feel unheard and disenfranchised (Barrette, 1996; Keller & Miller, 1997). Finances are of grave concern, too, for those in rural farming communities impacted by a CAFO presence.

CAFOs are farming operations where large numbers of livestock or poultry are housed inside buildings or in confined feedlots. How many animals? The U.S. Environmental Protection Agency (USEPA) defines a CAFO or industrial operation as a concentrated animal feeding operation where animals are confined for more than 45 days per year. To classify as a CAFO, such an operation must also have 1000 animal units more than a standardized number based on the amount of waste each species produces, basically 1000 lb of animal weight. Thus dairy cattle count as 1.4 animal units each. A CAFO could house more than 750 mature dairy cattle (milking or dry cows) or 500 horses and discharge into navigable water through a man-made ditch or a similarly man-made device. CAFO classification sets numbers for various species per 1000 animal units—for example:

- 2500 hogs
- 700 dairy cattle
- 1000 beef cattle
- 100,000 broiler chickens
- 82,000 layer hens

Unless you've seen such an operation, getting a grasp on the scope of the problem can be difficult. By using comparison, we quantify the issue: How does the amount of CAFO-generated animal manure compare to human waste production? Let's take a look at it.

Here's a small-scale number: One hog excretes 2.5 per day times more waste than an adult human—nearly 3 gallons (Cantrell, Perry, & Sturtz, 2004).

Here's a medium-scale number: A 10,000-hog operation produces as much waste in a single day as a town of 25,000 people (Sierra Club, 2004)—but the town has a treatment plant.

Here's a big picture approach: The USEPA estimates that human uses generate about 150 million tons (wet weight) of human sanitary waste annually in the United States, assuming a U.S. population of 285 million and an average waste generation of about 0.518 tons per person per year. The U.S. Department of Agriculture (USDA) estimates that operations that confine livestock and poultry animals generate about 500 million tons of excreted manure annually. The USEPA estimates there to be over 450,000 CAFOs in the United States today, producing 575 billion pounds of manure annually (USEPA, 2003).

Here's the bottom line: By these estimates, all confined animals generate well over *3 times* more raw waste than is generated by humans in the United States. Much of this waste undergoes no—or very little—waste treatment. Waste-handling is a major business concern and expense for any CAFO. Unless regulation and legislation support sound environmental practices for these operations, CAFO owners have little incentive to improve their waste-handling practices.

1.2   WATER SUPPLY, USE, AND WASTEWATER TREATMENT

Because wastewater treatment is so interconnected with other uses of water, it's considered a water use. Much of the water used by homes, industries, and businesses is treated prior to its release back into the environment, becoming part of the endless water cycle.

The scope of treatment processes inclusive in the term "wastewater treatment" is unknown to most people, who think of it solely in terms of sewage treatment. Nature has an amazing ability to cope with small amounts of water waste and pollution through its self-purification process. However, nature would be overwhelmed if the billions of gallons of wastewater and sewage produced every day were not treated before release back into the environment. Wastewater treatment plants reduce pollutants in wastewater to a level nature can handle.

We treat used water for many reasons. Principal among them is the matter of caring for our environment and our own public health. We treat used water because clean water is critical to our water supply, as well as to plants and animals that live in water. Human health, environmental health, and many commercial interests (for example, the fishing industry and sport fishing enthusiasts) depend on biota that can survive in only clean, healthy water systems—and we, of course, as today's responsible adults, hold our water system in its entirety for future generations.

Our rivers and ocean waters teem with life that depends on healthy shorelines, beaches, and marshes, which provide critical habitats for hundreds of species of aquatic and semiaquatic life. Migratory water birds use these areas for feeding and resting. Species of both flora and fauna are adapted to live in these zones that lie between or on the verge of water. These areas are extremely vulnerable to certain types of pollution.

Water is one of our most used playgrounds. The scenic and recreational values of our waters are serious factors for many people in deciding where to live. Tourists are drawn to water activities, such as swimming, fishing, boating, hunting, and picnicking. Improper treatment of wastes, of course, impacts these activities as well.

In short, if used water is not properly cleaned, it carries waterborne disease. Because we live, work, and play so close to water, harmful pathogenic organisms must be removed or made harmless to make water safe, regardless of the aesthetic factors involved in untreated wastes in the water system. So we treat our wastewater before releasing it to the environment. The major aim of used water treatment is to remove as much of the suspended solids and other contaminants as possible before the remaining water, called *effluent,* is discharged to the environment.

Treatment involves several interrelated steps. *Primary treatment* removes about 60% of suspended solids from used water. This treatment also involves aerating (mixing up) the used water to put oxygen back in, essential because as solid material biodegrades, it uses up oxygen needed by the plants and animals living in the water. *Secondary treatment* removes more than 90% of suspended solids. In some cases, *tertiary treatment* takes the waste removal further or addresses the removal of specific waste elements (nutrients such as nitrogen and phosphorus, for example) not removed by other means. After treatment, used water is returned to the water cycle as treated effluent. Whether from consumer or industrial sources, the returned treated water should be returned at least as clean as—if not cleaner than—the receiving body of water.

What does wastewater contain and where does it come from? Used water (wastewater) carries such substances as human waste, food scraps, oils, soaps, and chemicals. In homes (consumer use), this includes water from showers, sinks, toilets, bathtubs, washing machines, and dishwashers. Businesses and industries also contribute their share of used water (industrial waste) that must be cleaned and thus recycled. The treatment of both household wastewater and industrial wastewater is regulated and monitored. These are point-source pollutants; the sources are identifiable and limited in scope—we know what is in this wastewater, where the sources are and, in a general way, how much will be produced (the quantity can be predicted to fall within the capacity of the system's treatment capability).

Other wastes that enter our water supply are more difficult to define, quantify, and control. Stormwater or storm runoff, for example, is also a major contributor to the endless wastewater stream. The average person might assume that rain that runs off a home, into a yard, and then down the street during a storm event is fairly clean, but it is not. Harmful substances wash off roads, fields, lawns, parking lots, and rooftops and can harm our still waters (lakes, ponds, etc.) and running waters (rivers, streams, etc.). Stormwater also, obviously, is incident-related. If rain is absent, stormwater doesn't enter the system. Stormwater provides non-point-source pollutants to the wastwater stream. We know in a general way what they will carry, and we put systems in place to channel and control the stream. But complex programs and modeling are used—and essential—to evaluating how to handle stormwater to avoid serious problems with treatment and control, specifically to handle many different levels of force in storm incidents of differing durations.

Agricultural sources also contribute several problems to wastewater treatment, problems that have been difficult to evaluate, identify, and control. Historically, especially before the chemical industry provided manufactured fertilizers and pesticides for crop farm production and growth hormones and antibiotics for livestock production, the techniques individual farmers used were reasonably environmentally friendly. The size of the farm dictated the limits of production: Any individual farmer has limits that include his own physical and financial ability to work and effective natural limits on how many bushels of grain or animals per acre the land itself support because, in general, overloading those capacities provides negative results. As modern practices evolved, though, our water systems suffered. Fertilizers, pesticides, hormones, and antibiotics send non-point-source pollution directly into local water systems, which create downstream problems in the water system. A number of both crop and livestock farming practices contribute to soil erosion which, of course, affects both soil and water quality. However, the most recent changes in farming practices in the last several decades (perhaps best defined as the switch from the small farmer to "agribusiness," or factory farms) and the increasing demand for inexpensive meat products have created a new set of problems to address. While factory farming of crops presents its own set of issues, we address problems created by the factory farming of livestock, which creates agricultural point-source pollution of extreme scope.

1.3   ANIMAL FEEDING OPERATIONS AND ANIMAL WASTE TREATMENT

Waste problems, and innovative solutions to these problems, are not new, even in mythical terms. For example, the ancient Greek gods had animal housing. Hercules, the first environmental engineer, first formulated the solution to an animal waste pile-up in ancient Greek animal housing. The "Hercules" engineering principle states that: "The solution to pollution is … dilution." Hercules applied this principle when he cleaned up the royal Aegean Stables, which had not been cleaned for at least 30 years. He cleaned the stables by diverting the flow of two upstream rivers and directing the combined flow through the stables. Today we call this practice "flushing." This flushing idea of Hercules worked so well in cleaning the stables that the same idea was later applied in the design of human toilets and sewer systems. As the world became industrialized, the Hercules idea was applied just as successfully in the dispersion of air pollutants through tall chimneys. These chimneys are not significantly different from the sewer pipes that take waste away. Note that the Hercules principle has ample scientific base. Also note that, although treating today's massive quantities of animal waste from CAFOs is certainly a Herculean task, unfortunately he is no longer around to solve current problems.

Fast-forwarding from mythical times back to the present, the fact is that other animal waste problems and solutions are also not new. For example, field spreading of human and animal wastes is accomplished naturally under nomadic and pasture social systems. The early Chinese practiced intentional manure conservation and reuse. In Iceland, slotted floors (allowing waste material to drop below the floor surface) go back at least 200 years. From at least the 19th century, dairy barns with wastes from scores of animals were contained in one building. Huge poultry centers with wastes concentrated in a small area have been around for decades, as have some very large swine and beef units.

Figuratively speaking, and in general, animal manure deposited by animals managed by standard grazing livestock methods does not create serious environmental problems, especially if a farmer limits herd size to numbers the acreage can support without environmental damage, restricts livestock access to stream beds, and applies practices that include soil erosion prevention methods, such as greenbelts for waterways and shoreline planting. Accidentally stepping into such deposits is an occupational hazard, of course.

It is important to understand that the manure deposited by a large herd of animals that is not assimilated through the soil surface and is carried off by storm runoff into local streams or other water bodies being an obvious "isolated incident" problem, small-farm animal manure waste is not the problem we are addressing. Agribusiness and large-scale, factory-farming practices have created a different farm category in CAFOs, the livestock version of factory crop farming, which produce a massive quantity of manure. In the 1920s, no one was capable of spilling millions of gallons of manure into a local stream in a single event. Ikerd (1998) points out that such an event is possible today because of the piling up of too much manure in one place. Simply, the piling up is the result of greater concentration and reduced diversity in farm operations.

Agribusinesses don't use traditional pastures and feeding practices. Typically, manure is removed from livestock buildings or feedlots and stored in stockpiles or lagoon or pond systems until it can be spread on farm fields, sold to other farmers as fertilizer, or composted. When properly designed, constructed, and managed, CAFO-produced manure is an agronomically important and environmentally safe source of nutrients and organic matter necessary for the production of food, fiber and good soil health. Experience has demonstrated that, when properly applied to land at proper levels, manure will not cause water quality problems. When properly stored or deposited in holding lagoons or ponds, properly conveyed to the disposal outlet, and properly applied to the appropriate end-use, potential CAFO waste environmental problems can be mitigated.

But CAFOs must be monitored and controlled. CAFOs are inherently potential sources of contaminants (pollutants) to the three environmental mediums: air, water, and soil. Let's take a look at manure handling and storage practices recommended by USDA and USEPA (1998) that should be employed to prevent water pollution from CAFOs. In addition to water pollution prevention, it should be noted that manure and wastewater handling, storage, and subsequent application and treatment practices should also consider odor and other environmental and public health problems.

- Divert clean water—Siting and management practices should divert clean water from contact with feedlots, holding pens, animal manure, and manure storage systems. Clean water can include rainfall falling on roofs of facilities, runoff from adjacent lands, or other sources.
- Prevent leakage—Construction and maintenance of buildings, collection systems, conveyance systems, and permanent and temporary storage facilities should prevent leakage of organic matter, nutrients, and pathogens to groundwater or surface water.
- Provide adequate storage—Liquid manure storage systems should safely store the quantity and contents of animal manure and wastewater produced, contaminated runoff from the facility, and rainfall. Dry manure, such as that produced in certain poultry and beef operations, should be stored in production buildings or storage facilities, or otherwise stored in such a way so as to prevent polluted runoff. Location of manure storage systems should consider proximity to water bodies, floodplains, and other environmentally sensitive areas.
- Manure treatment—Manure should be handled and treated to reduce the loss of nutrients to the atmosphere during storage, to make the material a more stable fertilizer when land-applied, and to reduce pathogens, vector attraction, and odors, as appropriate.
- Management of dead animals—Dead animals should be disposed of in a way that does not adversely affect groundwater or surface water or create public health concerns. Composting, rendering, and other practices are common methods used to dispose of dead animals.

1.3.1 MANURE TREATMENT

According to Sutton and Humenik (2003), "Advanced technologies are being developed for the biological, physical, and chemical treatment of manure and wastewaters. Some of these greatly reduce constituents in the treated solids and liquids that must be managed on the farm. By-product recovery processes are being developed that transform waste into value-added products that can be marketed off the farm" (p. 2).

1.3.1.2 An Example: Animal Waste Treatment: Lagoons

Primarily because it is an economical means of treating highly concentrated wastes from confined livestock operations, the most widespread and common treatment technique for managing animal waste is the use of lagoons. In the late 1960s, considerable attention was paid to the impact of lagoons on surface water quality; since the 1970s, that attention has shifted to the potential impacts on groundwater quality. Unfortunately, lagoons are prone to leaks and breakage. Groundwater has been contaminated with bacteria from them. Floods push the wastes into streams, lakes, and oceans when they overrun the lagoons. North Carolina, with its concentration of factory farms, has been the focus of massive water contamination because of its waste lagoons. The storage lagoons for factory farms are often stinking manure lakes the size of several football fields, containing millions of gallons of liquefied manure. Remember, a single animal factory can generate waste that is equivalent to the waste of a small town.

In the past 30 years, several studies on the effectiveness of factory farm lagoons, specifically on lagoon liners, in preventing environmental damage have been conducted. Consider the following review of studies on effective lagoon construction versus defective construction.

Sewell et al. (1975) found on an anaerobic dairy lagoon that the lagoon bottom seals within 2 months of start up, and little or no pollutants were found in the groundwater after this time. Ritter et al. (1984) studied a two-stage anaerobic swine lagoon for 4 years and determined that the contaminant concentration increased in wells (50 m from the lagoon) the first year and then steadily decreased afterward. Their data led them to speculate that biological sealing takes place over a period of time depending on the loading rate to a lagoon. Collins et al. (1975) studied three swine lagoons, each in high water table areas. They found no significant effect on groundwater beyond 3 m from the lagoon edge. Miller et al. (1985) studied the performance of beef lagoons in sandy soil and found that the lagoons had effectively sealed to infiltration within 12 weeks of addition of manure. Humenik et al. (1980) summarized research conducted by others on the subject of lagoon sealing and concluded that the studies indicated that lagoon sealing may be expected within about 6 months, after which the area of seepage impact becomes restricted to approximately 10 m.

On the other hand, Hegg et al. (1978; 1981) collected data from a dairy lagoon and from newly established swine lagoons and found that some of the monitoring wells became contaminated while others did not. This led them to conclude that seepage does not occur uniformly over the entire wetted perimeter of a lagoon, but at specific unpredictable sites where sealing has not taken place. Similarly, Ritter et al. (1980) monitored an anaerobic two-stage swine lagoon for 2 years and found that one of the wells showed contamination that indicated localized seepage, whereas the other monitoring well indicated that the lagoon system produced a minimum impact on groundwater quality and that sealing had gradually taken place. Note, however, that these studies do not take flooding into account. The problem remains that lagoons have the unpredictable potential to affect both groundwater and surface water.

In many states, notwithstanding the USEPA's and USDA's manure handling, storage, and treatment recommendations, lawsuits against CAFOs for unsound environmental practices demonstrate that CAFOs are still creating problems. In short, regulations and legislation have fallen behind CAFO creation and operation, enforcement of existing regulations is spotty, and problems associated with CAFOs are still being identified—although you can be sure those who neighbor CAFOs can identify some big issues, both environmental and social. In this book, we discuss these "big issues" in detail.

1.4   "GET BIG OR GET OUT"

Corporate livestock factory owners and management tout themselves as "saviors" to the rural communities they target. Everyone is promised salvation: job creation for local inhabitants, increased tax revenues for local coffers, expanded markets for family farmers, and increased purchasing power for hometown businesses, with high-tech production for consumers. Simply, the mantra the "Big Boys" profess is "Get Big or Get Out." However, the facts of the industry paint a different picture. Corporate livestock factories actually disable community development with self-serving contracts and tax breaks, market-monopolizing strategy, and few local purchases (Cantrell et al., 1999).

But do bigger farms really benefit the overall economy of the community? Consider the effects of hog mega-barns on communities, the environment, and independent hog producers in Canada (National Farmers Union, 2000):

- Large corporate hog producers and attendant vertical integration threaten family farm hog production by pushing down prices, closing markets for family farmers, and obscuring price signals. In effect, the domination of the hog production and packing sectors by a handful of large, vertically integrated corporations destroys the open market in hogs.
- Vertical integration and the transfer of hog production from family farms to large corporate packers and processors is a policy decision, not an inevitable result of economic forces. Governments, at all levels, can make choices that will either turn agriculture over to distant corporations or retain it in the hands of local families.
- Although the corporate proponents of large hog barns promise jobs, economic development, and markets for feed grains, these corporate barns provide significantly fewer of these benefits than do the family farm hog producers these corporations displace. Corporations employ fewer people per hog and spend less in their communities than family-farm hog producers.
- While large hog barns do not deliver promised economic benefits, they do pose real environmental threats to surface water and groundwater.

Large hog farms also give off objectionable odors, increase fly populations, destroy the quality of life for surrounding residents, and lower property values.

While communities naturally want to attract jobs, wealth, and capital for investment, transferring hog production from local families to corporations facilitates and accelerates the extraction of wealth and capital from rural areas. In short, after considering the pros and cons of corporate factory farms versus the traditional family farm, the family farmer may conclude that it might be better to "stay small and stay put."

1.5 CAFOS: CURRENT TREND

The trend toward CAFOs has accentuated a long-term shift with profound implications for future livestock production. Table 1.1 illustrates the historic shift to CAFO production. Table 1.1 also shows the decrease in small family farm units. In the past, small family farms (nearly an anachronism today) raised livestock, such as hogs, to diversify output that typically included grains and other crops and cattle, because of their profitability. Barrette (1996) calls this practice "mortgage lifting."

Note that the geography of CAFOs is not only different but also in a state of flux. For example, Schwab (1998) points out that chicken production has historically been concentrated in the South and Mid-Atlantic states, whereas cattle and hog production have been located largely in the Midwest and, to some degree for cattle, on the High Plains. Change in geographical location is most apparent in the swine industry. For example, North Carolina has emerged as a major player. The Rocky Mountain states have also come into play (McBride, 1997). These choices, according to some industry critics, appear to be motivated by lax regulations in these locations.

TABLE 1.1

Trend in U.S. Livestock Production, 1969–1992

	Number of farms		Average number of heads per farm	
	1969	**1992**	**1969**	**1992**
Hogs	644,882	186,627	138	588
Milk cows	567,786	153,945	20	61
Beef cows	845,514	767,919	41	40
Broilers	33,221	21,777	70,798	237,622
Layer hens	470,832	70,623	632	2,985

Source: Adapted from McBride (1997).

1.6   DEFINITIONS OF KEY TERMS

Every topic studied, including CAFOs, has its own language for communication. To work at even the edge of CAFOs and the scientific and socioeconomic disciplines closely related to CAFOs, one must acquire a familiarity with the vocabulary used in the industry and in this text.

While providing a glossary of key terms and acronyms at the end of technical publications is useful, for this text, including many of these key terms and acronyms early in the text facilitates more orderly, logical, step-by-step learning. Simply, we feel key word definitions and acronyms should be right up front and readily accessible. Terms not defined here will be defined when used in the text. Many of the terms defined below will be highlighted and redefined, for emphasis, in the text.

Absorption: the physical integration of a liquid into the pore spaces of a solid (for example, water absorption into a sponge).

Accuracy: the closeness of an individual measurement or of the average of a number of measurements to the true value. Deviation from the true value is a measure of bias in the individual measurement or averaged value.

ACWF: America's Clean Water Foundation.

Adsorption: electrochemical attraction of positively or negatively charged molecules onto solids with an opposite charge.

Advection: the process by which solutes are transported by the bulk motion of flowing groundwater.

AED: aerodynamic equivalent particle diameter.

AER: allowable emission rate.

Aeration: creation of contact between air and a liquid by spraying the liquid in the air, bubbling air through the liquid, or agitating the liquid to promote surface absorption.

Aerobic: achieving solids reduction in manure mixtures using microorganisms that require oxygen. The breakdown of organic material tends to be odor free.

Aerobic bacteria: bacteria that require free elemental oxygen for their growth. Oxygen in chemical combination will not support aerobic organisms.

AFO: animal feeding operation. As defined by the USEPA (40 CFR 122.23), a "lot or facility" where animals "have been, are, or will be stabled or confined and fed or maintained for a total of 45 days or more in any 12 month period and crops, vegetation, forage growth, or post-harvest residues are not sustained in the normal growing season over any portion of the lot or facility."

Agitation: the remixing of liquid and settled solids.

Agriculture, animal: the use of land for animal feedlots or animal waste areas.

Agriculture, crop: the use of land for the production of row crops, field crops, tree crops, timber, bees, apiary products, or fur-bearing animals.

Agricultural storage: facilities for the warehousing of agricultural products. Example: grain elevators.

Agricultural wastes: waste normally associated with the production and processing of food and fiber on farms, feedlots, ranches, ranges, and forests, which may include animal manure, crop- and food-processing residues, agricultural chemicals, and animal carcasses.

Agricultural waste management system: a combination of conservation practices formulated to appropriately manage a waste product that, when implemented, will recycle waste constituents to the fullest extent possible and protect the resource base in a nonpolluting manner.

Alluvial: pertaining to or composed of alluvium or deposited by a stream or running water.

Alluvium: a general term for clay, silt, sand, gravel, or similar unconsolidated material deposited during a comparatively recent geologic time by a stream or other body of running water as a sorted or semisorted sediment in the bed of the stream or in its flood plain or delta, or as a cone or fan at the base of a mountain slope.

Alternate animal feedlot: a lot or building, or combination of lots and buildings, intended for the confined feeding, breeding, raising, or holding of animals and specifically designed as a confinement area where manure can accumulate or where the concentration of animals is such that vegetative cover cannot be maintained within the enclosure during the months of May, June, July, and August. Open lots used for the feeding and rearing of poultry (poultry ranges) are considered animal feedlots; pastures are not considered animal feedlots.

Alternate conservation practice: alternative method or field-specific condition that provides pollutant reductions equivalent to or better than the reductions that would be achieved by a 100-foot setback.

Ambient air quality: quality of the outdoor air to which humans are exposed during the course of their normal lives.

Amendment: an ingredient, such as waste hay, cotton gin trash, or peanut hulls, that is added to corral surfaces to improve dust and odor control or to enhance the composting process.

Amino acids: organic nitrogen compounds that form the building blocks of proteins.

Ammonia volatilization: loss of ammonia (NH_3) to the atmosphere.

Anaerobic: the absence of molecular oxygen, or growth in the absence of oxygen, such as anaerobic bacteria.

Anaerobic bacteria: bacteria that do not require the presence of free or dissolved oxygen.

Anaerobic digester: a heated, air-tight apparatus that facilitates anaerobic digestion.

Anaerobic digestion: conversion of organic matter in the absence of oxygen under controlled conditions to such gases as methane and carbon dioxide.

Anaerobic lagoon: a facility to treat animal waste by predominantly anaerobic biological action using anaerobic organisms, in the absence of oxygen, for the purpose of reducing the strength of the waste.

Animal feeding operation (AFO): any facility that feeds livestock or poultry in confinement such that the animals are not sustained on forages growing in the confinement area or that relies on imported feed. As defined by the USEPA (40 CFR 122.23): a "lot or facility" where animals "have been, are, or will be stabled or confined and fed or maintained for a total of 45 days or more in any 12 month period and crops, vegetation, forage growth, or post-harvest residues are not sustained in the normal growing season over any portion of the lot or facility."

Animal feedlot: a lot or building, or combination of contiguous lots and buildings, intended for the confined feeding, breeding, raising, or holding of animals and specifically designed as a confinement area where manure can accumulate or where the concentration of animals is such that vegetative cover cannot be maintained within the enclosure. Open lots used for the feeding and rearing of poultry (poultry ranges) are considered animal feedlots; pastures are not considered animal feedlots.

Animal unit (AU): a unit of measure used to compare different animal species:

1. USEPA (66FR 2960-3138): 1 cattle excluding mature dairy and veal cattle; 0.7 mature dairy cattle; 2.5 swine weighing over 55 pounds; 10 swine weighing 55 pounds or less; 55 turkeys; 100 chickens; and 1 veal calf.
2. USDA: 1,000 pounds of live animal weight.

Animal waste area: a holding area or lagoon used or intended to be used for the storage or treatment of animal manure and other waste products associated with an animal feedlot.

Anion: negatively charged ion that can adsorb to negatively charged particles. Common soil anions are nitrates and orthophosphates.

Anion exchange: ion exchange process in which anions in solution are exchanged for other anions from an ion exchanger.

Anthropogenic: caused by humans.

Aquitard: a geologic formation, group of formations, or part of a formation through which virtually no water moves.

ARS: Agricultural Research Service, a division of the USDA.

Artesian well: a well deriving its water from a confined aquifer in which the water level stands above the ground surface; synonymous with flowing well.

ASM: aerosol mass spectrometer.

ASTM: American Society for Testing and Materials.

Atmospheric stability: a property that depends on inversion strength—how rapidly air temperature rises with altitude (in units of degrees Celsius per 100 m). Strong inversions near the ground tend to stabilize the atmosphere, trap emissions, and result in higher pollutant concentrations. For a discussion of the meteorological effects on carbon monoxide concentrations, see NRC (2002).

Available nitrogen: form of nitrogen that is immediately available for plant growth.

Available nutrient: a nutrient molecule that can be adsorbed and assimilated by growing plants.

Available phosphorus: form of phosphorus that can be immediately used for plant growth.

Available water capacity (available moisture capacity): the capacity of soils to hold water available for use by most plants, commonly defined as the difference between the amount of soil water at field capacity and the amount at wilting point and commonly expressed as inches of water per inch of soil. The capacity, in inches, in a 60-inch profile is expressed as:

Very low	0 to 3 inches
Low	3 to 6 inches
Moderate	6 to 9 inches
High	9 to 12 inches
Very high	> 12 inches

BACT: best achievable control technology.

Bacteria: a group of essentially one-cell microscopic organisms lacking chlorophyll that are usually regarded as plants.

bar: a unit of pressure equal to one atmosphere (14.7 psi).

Basalt: a general term for dark-colored, iron- and magnesium-rich igneous rocks, commonly extrusive, but locally intrusive. It is the principal rock type making up the ocean floor.

Baseflow: water that, having infiltrated the soil surface, percolates to the groundwater table and moves laterally to reappear as surface runoff.

BAT: best available technology (economically achievable).

Bedrock: the solid rock that underlies soil and other unconsolidated material or that is exposed at the surface.

Best management practice (BMP): defined by the National Pollutant Discharge Elimination System (NPDES) regulations as defined by a schedule of activities, prohibitions of practices, maintenance procedures, and other management practices to prevent or reduce the pollution of waters in the United States. BMPs also include treatment requirements, operating procedures, and practices to control plant site runoff, spillage or leaks, sludge or waste disposal, or drainage from raw material storage (40 CFR 122.2).

Bioaerosol: particulate matter in the atmosphere containing materials of biological origin that can cause disease, such as toxins, allergens, viruses, bacteria, and fungi.

Biochemical oxygen demand (BOD): an indirect measure of the concentration of biodegradable substances present in an aqueous solution. Determined by the amount of dissolved oxygen required for the aerobic degradation of the organic matter at 20° C.

Biodegradation: the destruction or mineralization of natural or synthetic organic materials by microorganisms.

Biological wastewater treatment: forms of wastewater treatment in which bacterial or biochemical action is intensified to stabilize or oxidize unstable organic matter. Oxidation ditches, aerated lagoons, anaerobic lagoons, and anaerobic digesters are examples.

Biomagnification: the process by which toxic substances become concentrated in animal and plant tissues.

Biomass: organic plant materials, such as cornstalks, small grain straw, and other plant fibers. Total amount of living material, plants, and animals above and below ground in a particular area.

BMP: best management practice.

BOD: biochemical oxygen demand.

BW: body weight.

C: carbon.

CAA: Clean Air Act.

Candidate measure (CM): a practice with the potential to reduce pollutant loading and, thereby, the potential to improve water quality.

Capillary fringe: the zone at the bottom of the vadose zone, where groundwater is drawn upward by capillary force.

Carbonate: a sediment formed by the organic or inorganic precipitation from aqueous solution of carbonates of calcium, magnesium, or iron.

Carbon-nitrogen ratio (C/N): the weight ratio of carbon to nitrogen.

Cation: positively charged ion; can adsorb to a soil particle.

Cation exchange: ion exchange process in which cations in solution are exchanged for other cations on the surface of a surface-active (ion exchanger) material, such as a clay colloid or organic colloid.

Cation-exchange capacity: the total amount of exchangeable cations that can be adsorbed by a soil or a soil constituent, expressed in terms of milliequivalents per 100 grams of soil at neutrality (pH 7.0) or at some other stated pH value.

Center point irrigation system: automated irrigation system consisting of a sprinkler line rotating about a pivot point and supported by a number of self-propelled towers. Water is supplied at the pivot point and flows outward through the line, supplying the individual outlets.

CERCLA: Comprehensive Environmental Response, Compensation, and Liability Act.

CFR: Code of Federal Regulations.

CH_4: methane.

Character descriptors: terms used by trained odor panelists to describe an odor's character (e.g., minty, citrusy, or earthy).

Chemical oxygen demand (COD): an indirect measure of the biochemical load exerted on the oxygen content of a body of water when organic wastes are introduced into the water. If the wastes contain only readily available organic bacterial food and no toxic matter, the COD values can be correlated with BOD values obtained from the wastes.

Chlorinated hydrocarbons: a class of synthetic organic compounds used by industry, farms, and households for a variety of purposes, including pest control. These organic compounds can also be produced by chlorinating sewage effluent, which is done to aid oxidation and kill pathogens contained in untreated effluent.

Clay: as a soil separate, the mineral soil particles less than 0.002 millimeters in diameter. As a soil textural class, soil material that is 40% or more clay, less than 45% sand, and less than 40% silt.

Coarse textured soil: sand or loamy soil.

COD: chemical oxygen demand.

Coliform bacteria: a group of bacteria predominately found in soil. Fecal coliform species inhabit the intestines of humans and animals. Coliform bacteria include all aerobic and facultative anaerobic, gram negative, non-spore-forming bacilli that ferment lactose with production of gas. This group of "total" coliforms includes *Escherichia coli*, which is considered to be a typical coliform of fecal origin.

Complexation (chelation): the reaction between a metallic ion and a complexing organic agent that forms a complex chemical ring structure and the effective removal of the metallic ion from the system.

Comprehensive nutrient management plan (CNMP): A conservation plan unique to animal feeding operations that incorporates practices to use animal manure and organic by-products as a beneficial resource. A CNMP addresses nature resource concerns dealing with soil erosion, manure, and organic by-products and their potential impacts on water quality that may derive from a CAFO. A CNMP is developed to assist a CAFO owner or operator in meeting all applicable local, tribal, state, and federal water quality goals or regulations. For nutrient-impaired stream segments or water bodies, additional management activities or conservation practices may be required to meet local, tribal, state, or federal water quality goals or regulations.

CNMPs must meet National Resource Conservation Service (NRCS) technical standards. For those elements included by the owner or operator in a CNMP for which NRCS currently does not maintain technical standards (i.e., feed management, vector control, air quality), producers should meet criteria established by land grant universities, industry, or other technically qualified entities. Within each state, the NRCS State Conservationist has the authority to approve non-NRCS criteria established for use in the planning and implementation of CNMP elements.

Composting: controlled microbial degradation of organic waste yielding an environmentally safe and nuisance-free soil conditioner and fertilizer.

Cone of depression: a depression in the groundwater table or potentiometric surface with the shape of an inverted cone that develops around a well from which water is being withdraw. It defines the area of influence of a well.

Confined animal feeding operation (CAFO): the statutory basis for federal regulation of air pollution, which is revised and reauthorized every 5 years.

Confined aquifer: a formation in which groundwater is isolated from the atmosphere at the point of discharge by impermeable geologic formations. Confined groundwater is generally subject to pressure greater than atmospheric pressure.

Conservation cropping sequence: an adapted sequence of crops designed to provide adequate organic residue for maintenance or improvement of soil tilth and for other conservation purposes.

Conservative pollutants: pollutants that are not altered as they are transported from their source to the receiving water.

Conservation practice: a specific structural, managerial, or cultural treatment of natural resources commonly used to meet a specific need in planning and carrying out soil and water conservation programs.

Contamination: the degradation of water quality as a result of natural processes or human activity. No specific limits are established because the degree of permissible contamination depends on the intended end use or uses of the water.

Conventional tillage: those primary and secondary tillage operations considered standard for the specific location and crop.

CO_2 equivalent: the mass of carbon dioxide with the same climate change potential as the mass of the greenhouse gas in question.

Cover crop: a close-growing crop whose main purpose is to protect and improve soil and use excess nutrients or soil moisture during the absence of the regular crop or in nonvegetated areas of orchards and vineyards.

Cultural eutrophication: the process of nutrient enrichment artificially accelerated by some action(s) of human society.

CWA: Clean Water Act.

Darcy's law: a derived equation for the flow of fluids on the assumption that the flow is laminar and that inertia can be neglected.

Deep percolation: downward movement of water through soil to below the root zone.

Demineralization: total removal of all ions.

Denitrification: the reduction of nitrates, with nitrogen gas evolved as an end product.

Desorption: the release of sorbed ions or compounds from solid surfaces.

Detection threshold: volume of nonodorous air needed to dilute a unit volume of odorous sample air to the point where trained panelists can detect no difference between the sample and the non-odorous air.

Digestion: commonly, the anaerobic breakdown of organic matter in water solution or suspension into simpler or more biologically stable compounds, or both. Organic matter can be decomposed to soluble organic acids or alcohols and then to gases such as methane and carbon dioxide. Bacterial action alone cannot complete destruction of organic solid materials.

Direct runoff: both surface flow and the interflow component of subsurface flow.

Discharge criteria: in this text, the conditions established in CAFO regulations to describe the circumstances under which a medium-size AFO is defined as a CAFO or a small-size AFO can be designated as a CAFO:

1. Pollutants are discharged into waters of the United States through a man-made ditch, flushing system, or other similar man-made device.
2. Pollutants are discharged directly into waters of the United States that originate outside of and pass over, across, or through the facility or otherwise come into direct contact with the animals confined in the operation (40 CFR 122.23(b)(6)(ii)).

Dispersion: the spreading and mixing of chemical constituents in groundwater caused by diffusion and mixing because of microscopic variations in velocities within and between pores.

Dissolved oxygen (DO): the molecular oxygen dissolved in water, wastewater, or another liquid; generally expressed in milligrams per liter, parts per million, or percent of saturation.

DM: dry matter.

Dry-weight percentage: the ratio of the weight of any constituent to the oven-dry weight of the whole substance.

Effluent: liquid discharge of a manure treatment process.

Effluent limitations: defined by the NPDES regulations as any restriction imposed by the Director on quantities, discharge rates, and concentrations of pollutants that are discharged from point sources into waters of the United States, the contiguous zone, or the ocean (940 CFR 122.2).

Effluent limitations guideline (ELG): a technical EPA document that sets effluent limits for a given industry and its pollutants; available at http://www.epa.gov/OCEPAterms.

Effluent standard: designated limit in the amount of any constituent within an effluent.

Electronic nose: device that detects a select number of individual chemical compounds to measure an odor.

Electronic particle counter: device that reports the number of dust particles per volume of air that a filter collects in and near animal facilities.

Electronic sensor: device that measures gas concentrations.

Emission flux: the rate of mass emission per unit of area (tons per hour per hectare, for example), typically from an area such as a waste lagoon or field.

Emission rate: the rate of mass emissions (tons per hour, for example).

Emissions inventory: the list of all applicable regulated pollutants and their expected annual emissions. In the case of a cattle feedlot, the emissions inventory has generally been limited to emissions from flaker cyclones, hay grinding, grain unloading, and feed loading.

Enrichment ratio: the ratio of pollutant concentration in runoff or sediment to its concentration in the soil or soil water, respectively.

Equipotential line: a contour line on the water table or potentiometric surface along which the pressure head of groundwater in an aquifer is the same. Fluid flow is normal to these lines in the direction of decreasing fluid potential.

Erosion: the wearing away of land surface by water, wind, ice, or other geologic agents and by such processes as gravitational creep.

Escherichia coli (E. coli): a species of bacteria that lives in the intestinal tracts of warm-blooded animals and whose presence is considered indicative of fecal contamination.

Eutrophication: a natural or artificial process of nutrient enrichment whereby a water body becomes abundant in plant nutrients and low in oxygen content.

Evapotranspiration: the loss of water from an area by evaporation from the soil or snow cover and transpiration by plants.

Exchange capacity: the abundance of sites (within the soil sample) with the potential for being actively engaged in ion adsorption.

Facultative bacteria: bacteria that can grow in the presence, as well as in the absence, of oxygen.

Fault: a fracture or a zone of fractures along which displacement of the sides has occurred, relative to one another and parallel to the fracture.

Federal Clean Air Act Amendments (CAAA): the statutory basis for federal regulation of air pollution, which is revised and reauthorized every 5 years.

Federal operating permit (FOP): an operating permit obtained under the auspices of the Clean Air Act that outlines the maximum emissions rates and abatement measures required of all sources under the permit's purview.

Feedlot: an animal feeding operation where beef cattle are finished to slaughter weight; it consists of fenced earthen or concrete paddocks, with cattle having little or no access to pasture.

Feed option: pavement extending 8 to 15 ft from a feed bunk to prevent erosion or potholing from hoof action or other cattle activity.

Fertilizer value: the potential worth of the plant nutrients in wastes that are available to plants when applied to soil; the cost of obtaining the same nutrients commercially.

FID: flame ionization detector.

Field (moisture) capacity: the moisture content of a soil, expressed as a percentage of the oven-dry weight, after the gravitational, or free, water has drained away.

Field sniffer: trained panelist who determines odor intensity in the field.

Filtration: the process of passing a liquid though a filter to remove suspended matter.

Fine textured soil: sandy clay, silty clay, and clay.

Flocculation: the process of using chemical compounds that will react with nutrients to produce a precipitant or complex that can easily be separated from a waste stream.

Flow lines: lines indicating the direction followed by groundwater toward points of discharge. Flow lines are perpendicular to equipotential lines.

Flushing system: a system that collects and transports or moves waste material with the use of water, such as in washing of pens and flushing confinement livestock facilities.

Fugitive emissions: emissions identified with a discrete process but not traceable to a single emission point, such as the end of a stack (grain unloading, for example). Fugitive emissions from a cattle feedlot or an open lot dairy include dust resulting from cattle activity on the feedlot surface or from vehicle traffic on unpaved roads. Analogous to nonpoint source water pollution.

Gas chromatograph/mass spectrometer: research laboratory device that both identifies and measures gas concentrations by analyzing very small samples of air injected into a carrier (nitrogen or helium) gas stream. This gas stream is passed through a column that adsorbs and desorbs the chemicals in the air at different rates, plus a detector, which identifies individual chemicals and the amount in the sample.

GC: gas chromatography.

Grassed infiltration area: an area with vegetative cover where runoff water infiltrates into the soil.

Groundwater: water filling all the unblocked pores of underlying material below the water table.

Groundwater table: the surface between the zone of saturation and the zone of aeration; the surface of an unconfined aquifer.

H_2S: hydrogen sulfide.

ha: hectare; an area 100 meters square, or about 2.5 acres.

Half-life: time required for half of a specified substance to be transformed to another substance.

HAP: hazardous air pollutant.

Head: energy contained in a water mass; expressed in elevation (feet) or pressure (pounds per square feet).

Head loss: part of head energy that is lost because of friction as water flows.

Hedonic tone: scale used to describe an odor; ranges from –10, which is unpleasant, to +10, which is pleasant.

Holding pond: a storage area, usually earthen, where lot runoff, lagoon effluent, and other dilute wastes are stored before final disposal. It is not designed for treatment.

Horizon, soil: a layer of soil, approximately parallel to the surface, having distinct characteristics produced during soil-forming processes.

Humus: dark, high-carbon residue resulting from plant decomposition. Similar residues are in composted manure and well-digested sludge.

Hydraulic conductivity: the rate of flow of water (in gallons per day) through a cross section of one square foot under a unit hydraulic gradient at the prevailing temperature (gpd/ft^2). In the SI system, the units are $m^3/day/m^2$ or m/day.

Hydraulic gradient: the rate of change in total head per unit of distance of flow in a given direction.

Hydrologic condition: description of the moisture present in a soil by amount, location, and configuration.

Hydrologic soil groups: a classification system used by the Natural Resources Conservation Service to group soils according to their runoff-producing characteristics. The chief consideration is the inherent capacity of soil to permit infiltration when bare of vegetation. The slope and the kind of plant cover are not considered but are separate factors in predicting runoff. Soils are assigned to four groups. Group A soils have high infiltration rates when thoroughly wet and low runoff potentials. They are mainly deep, well drained, and sandy or gravelly. At the other extreme, group D soils have very slow infiltration rates and, thus, high runoff potentials. They have a claypan or clay layer at or near the surface, have a permanent high water table, or are shallow over nearly impervious bedrock or other material.

Igneous rock: rock that solidified from molten or partly molten material, that is, from magma.

Indicator tube: glass tube with both ends sealed that measures a wide range of gases.

Infiltration: the process of water entering soil through the surface.

Infiltration rate: the rate at which water enters soil under a given condition; expressed as depth of water per unit time, usually inches per hour.

Influent: a liquid entering a container or process.

Intensity: the strength of an odor sample.

Interflow: water that enters the soil surface and moves laterally through the soil layers to reappear as surface flow. Flow takes place above groundwater level.

Ion: a charged element or compound that has gained or lost electrons so that it is no longer electrically neutral.

Jerome® meter: portable electronic device that measures hydrogen sulfide concentrations by sampling the air for several seconds and providing a nearly instantaneous reading.

Karst topography: topography formed in limestone, gypsum, and other similar type rock by dissolution, characterized by sinkholes, caves, and rapid underground water movement.

kg: kilogram, or 1,000 grams (about 2.2 pounds).

km: kilometer, or 1,000 meters.

Labile: readily coming into equilibrium.

Lagoon: a treatment structure for agricultural wastes that can be aerobic, anaerobic, or facultative, depending on their loading and design, and can be used in series to produce a higher quality effluent.

Land application: application of manure, sewage sludge, municipal wastewater, and industrial wastes to land for reuse of the nutrients and organic matter for their fertilizer and soil-conditioning values.

Land application area: defined by CAFO regulations as land under the control of an AFO owner or operator, whether it is owned, rented, or leased, to which manure, litter, or process wastewater from a production area is or may be applied (40 CFR 122.23(b)(3)).

Landscape: the environment, both natural and built, that surrounds us.

Landscape character: a measure of an apparent harmony or unity among all landscape elements, built and natural, that can be intensified or preserved to make a memorable scene.

Landscape quality: a composite of landscape conditions and perceived values that provide diverse and pleasant surroundings for human use and appreciation. Recognized components of landscape quality include visual resource, landscape use, viewscape, and visibility.

Leaching: the removal of soluble constituents from soils or other material by water.

Lidar: a device similar to radar that emits pulsed laser light rather than microwaves.

Limiting nutrient: a nutrient that restricts plant growth.

Liquid manure: a mixture of water and manure that behaves more like a liquid than a solid and is generally less than 5% solids.

Livestock wastes: manure with added bedding, rain, or other water, soil, etc. It also includes wastes, such as milkhouse or washing wastes, not particularly associated with manure, such as hair, feathers, and other debris.

LOAEL: lowest observed adverse effect level.

Lot: any paved or unpaved outdoor animal area, such as a feedlot, handling area, or resting areas.

LU: live unit; 500 kg of body weight.

Macronutrient: a chemical element required in relatively large amounts for proper plant growth.

Manure: the fecal and urinary defecations of livestock and poultry, which does not include spilled feed, bedding, or additional water or runoff.

Mechanical solids separation: the process of separating suspended solids from a liquid-carrying medium by trapping the particles on a mechanical screen or sieve or by centrifugation.

MDA single-point monitor: units used to monitor ambient air concentrations of individual gases, such as hydrogen sulfide, over extended periods.

Microclimate: climate as experienced at the scale of a particular site. Includes such elements as solar orientation, wind direction, temperature, and precipitation.

Micronutrient: a chemical element required in relatively small amounts for proper plant growth.

Milkhouse waste: wastewater from milkhouse operations.

Mineralization: the microbial conversion of an element from an organic to an inorganic state.

Molecular diffusion: dispersion of a chemical caused by the kinetic activity of the ionic or molecular constituents.

Morphology, soil: constitution of soil, including the texture, structure, consistency, color, and other physical, chemical, and biological properties of the various soil horizons that make up the soil profile.

MS: mass spectrometer.

N: nitrogen.

N_2: dinitrogen molecule.

National Ambient Air Quality Standards (NAAQS): a list of maximum concentrations, or pollutant thresholds, above which human exposure can result in adverse health effects. Serves as an administrative benchmark for clean air.

National Pollutant Discharge Elimination System (NPDES): the NPDES program operates under the Clean Water Act to prohibit the discharge of pollutants in waters of the United States unless a special permit is issued by the EPA; a state; or, where delegated, a tribal government on an Indian reservation (http://www.epa.gov/OCEPAterms/).

NH_3: ammonia.

Nitrate nitrogen: the nitrogen component of the final decomposition product of the organic nitrogen compounds; expressed in terms of the nitrogen part of the compound.

Nitrification: oxidation of an ammonia compound to nitric acid, nitrous acid, or any nitrate or nitrite, especially by the action of nitrobacteria.

Nitrogen: a chemical element commonly used as a nutrient in fertilizer that is also a component of animal wastes. As one of the major nutrients required for plant growth, nitrogen can promote algal blooms that cause water body eutrophication if water runs off or leaches out of surface soil. Nitrogen is immediately usable for plant growth in available forms.

Nitrogen cycle: the succession of biochemical reactions that nitrogen undergoes as it is converted to organic or available nitrogen from the elemental form. Organic nitrogen in waste is oxidized by bacteria into ammonia. If oxygen is present, ammonia is bacterially oxidized first into nitrite and then into nitrate. If oxygen is not present, nitrite and nitrate are bacterially reduced to nitrogen gas, completing the cycle.

Nitrogen fixation: the biological process by which elemental nitrogen is converted to organic or available nitrogen.

nm: nanometer; 10^{-9} m.

N_2O: nitrous oxide.

NO: nitric oxide.

NO_x: nitric oxide and nitrogen dioxide rapidly interconverted in the atmosphere.

Nonattainment area (NAA): area found to exceed the National Ambient Air Quality Standards for any one or more regulated pollutant and subsequently required to implement a regional plan to reduce emissions and bring the area into attainment.

Nonconforming use: the use of any land, building, or structure that does not comply with the use regulations of the zoning district in which such use is located but which complied with the use regulations in effect at the time the use was established.

Nonconformity: a nonconforming use, noncomplying structure, or other development situation that does not comply with currently applicable zoning regulations but which complied with zoning regulations in effect at the time the use or development was established.

Nonpoint source (NPS): entry of effluent into a water body in a diffuse manner, with no definite point of entry.

Notice of intent (NOI): a notification submitted to a permitting authority to indicate that a discharger intends to be covered under a general permit and will comply with the permit conditions. For CAFOs, a notice of intent to be covered under a general permit must include the information specified in 40 CFR 122.21(i)(1) and any other information specified by the permitting authority in the general permit.

No-till: a planting procedure that requires no tillage except that done by a coulter in the immediate area of the crop row.

NPDES permit: an authorization, license, or equivalent control document issued by the USEPA or an approved state agency to implement the requirements of the NPDES regulations; for example, a permit to operate a CAFO.

NRCS: National Resource Conservation Service.

Nuisance: any condition that inhibits the reasonable use or enjoyment of property.

Nutrient assimilation: the conversion or incorporation of plant nutrients into plant cells and tissue.

Nutrient excretion factor: an estimate of an element (for example, nitrogen) excreted by an animal, usually reported in kilograms per day (or year) per animal (animal unit or kilogram of body weight).

Nutrients: elements required for plant or animal growth, including macronutrients (nitrogen, phosphorus, and potassium) and micronutrients (a number of other elements that are essential but needed in lesser amounts).

Nutrient transformation: the changing of form of a plant element that can affect the stability, availability, or mobility of the compound. An example is the changing of ammonium nitrogen to nitrate nitrogen.

Odor patch: single-use piece of cardboard or plastic coated with a chemical that changes color when exposed to the gas being measured.

Odor plume: A downwind air mass containing odorous gases from an odor source, such as an animal production building or a manure storage facility.

Odor unit: volume of diluted air divided by the volume of odorous sample air at either detection or recognition.

OH: hydroxyl radical.

Olfactometer: device that delivers known concentrations of an odorous air sample to a sniffing port for evaluation by trained human panelists who determine the odor detection or recognition thresholds that are reported in odor units.

Olfactometry: means of measuring odor by using the highly sensitive human sense of smell.

Organic matter: chemical substances of animal or vegetable origin that contain carbon.

Orthographic: relating to the physical geography of mountains and mountain ranges.

Ozonation: water and odor control technology using ozone, either in the air or dissolved in water, to oxidize pathogens and odorous compounds.

PAN: peroxyacetyl nitrate.

Pathogens: disease-causing microorganisms; generally associated with viruses or bacteria.

Perched water: unconfined groundwater separated from an underlying main body of groundwater by an unsaturated zone (generally an aquaclude).

Percolation: the downward movement of water through soil.

Permanent wilting point: the moisture content of soil, on an oven-dry basis, at which a plant (specifically a sunflower) wilts so much that it does not recover when placed in a humid, dark chamber.

Permeability: the quality of soil that enables water to move downward through the profile; measured as the number of inches per hour that water moves downward through saturated soil. Terms describing permeability are:

Very slow	Less than 0.06 inches/hr
Slow	0.06 to 0.2 inches/hr
Moderately slow	0.2 to 0.6 inches/hr
Moderate	0.6 to 2.0 inches/hr
Moderately rapid	2.0 to 6.0 inches/hr
Rapid	6.0 to 20 inches/hr
Very rapid	More than 20 inches/hr

Permitting authority: a state agency (or governmental entity, such as a tribal government) that has received authority from the EPA to administer the NPDES program. For states that have not received authority from the USEPA to administer the NPDES program, the USEPA Regional Administrator is the permitting authority.

Persistence: calculated value that relates to the time required for a gas to degrade into a harmless compound.

pH: a measure of hydrogen-ion concentration. 7 is neutral; 0 to 7 is acidic; 7 to 14 is alkaline.

Phosphate: phosphate ion exists in water as H^2PO^{4-}. Otherwise, an ester or salt of phosphoric acid, such as calcium phosphate rock.

Phosphorus: one of the primary nutrients required for the growth of plants; often the limiting nutrient for the growth of aquatic plants and algae.

Phytase: enzyme that, when added to rations of nonruminant animals, makes the phosphorus in grains and other feed ingredients more available during digestion.

Plenum: duct that transports ventilation air to or from a building.

PM: particulate matter.

PM2.5: particulate matter with an aerodynamic diameter of 2.5 micrometers or less.

PM10: particulate matter with an aerodynamic diameter of 10 micrometers or less.

Point source: "Any discernible, confined, and discrete conveyance, including but not limited to any pipe, ditch, channel, … concentrated animal feeding operation … from which pollutants are or may be discharged. This term does not include agricultural stormwater discharges and return flows from irrigated agriculture" (33 US § 1362(14)).

Pollutant: a resource out of place; a resource is anything useful.

Pollutant delivery ratio (PDR): the fraction of a pollutant leaving an area that actually enters a body of water.

Pollutant threshold: maximum value that relates to the time required for a gas to degrade into a harmless compound.

Ponding: standing water on soils in closed depressions. Unless the soils are artificially drained, the water can be removed only by percolation or evapotranspiration.

Porous dam: a runoff control structure that reduces the rate of runoff so that solids settle out in the settling terrace or basin. The structure may be constructed of rock, expanded metal, or timber arranged with narrow slots.

Potassium: one of the primary nutrients required for the growth of plants.

Potential to emit: amount of emitted pollutant that would be expected from a facility operating year-round at full capacity.

ppb: parts per billion by volume.

ppm: parts per million by volume.

Precision: agreement among individual measurements of the same property, under prescribed similar conditions.

Process wastewater: defined by CAFO regulations as water directly or indirectly used in the operation of an AFO for any or all of the following: spillage or overflow from animal or poultry watering systems; washing, cleaning, or flushing pens, barns, manure pits, or other AFO facilities; direct contact swimming, washing, or spray cooling of animals; or dust control. Process wastewater also includes any water that comes into contact with raw materials, products, or by-products, including manure, litter, feed, milk, eggs, or bedding [40 CFR 122.23(b)(7)].

Production area: defined by CAFO regulations as the part of an AFO that includes the animal confinement area, the manure storage area, the raw materials storage area, and the waste containment areas. The animal confinement area includes but is not limited to open lots, housed lots, feedlots, confinement houses, stall barns, free stall barns, milkrooms, milking centers, cowyards, barnyards, medication pens, walkers, animal walkways, and stables. The manure storage area includes but is not limited to lagoons, runoff ponds, storage sheds, stockpiles, underhouse or pit storages, liquid impoundments, static piles, and composting piles. The raw materials storage area includes but is not limited to feed silos, silage burners, and bedding materials. The waste containment area includes but is not limited to settling basins and areas within berms and diversions that separate uncontaminated stormwater. Also included in the definition of production areas is any egg washing or egg process facility and any area used in the storage, handling, treatment, or disposal of mortalities [40 CFR 122.23(b)(8)].

Pumping test: a test conducted to determine aquifer yield or well characteristics.

Recharge water: water placed in either gravity or pull-plug manure drain systems to assist in the transport of manure out of an animal building.

Recognition threshold: volume of nonodorous air needed to dilute a unit volume of odorous sample air to the point where trained panelists can correctly recognize the odorous air.

Reduced tillage: a management practice whereby the use of secondary tillage operations is significantly reduced.

Residence time: amount of time ventilation air is in contact with biofilter media.

Resource base: the combination of soil, air, water, plants, and animals that makes up the natural environment.

Resource management system (RMS): a combination of conservation practices and management identified by the primary use of land or water that, when installed, will at a minimum protect the resource base.

Respirable dust level: measurement of the dust particles small enough to enter the human respiratory system that a filter collects in and near animal facilities.

Ridge planting: the practice of growing a row crop on the ridges between the furrows.

Root zone: the part of soil that can be penetrated by plant roots.

Runoff: the part of precipitation or irrigation water that appears in surface streams or water bodies; expressed as volume (acre-inches) or rate of flow (gallons per minute, cubic feet per second).

Run-on: the water moving by surface flow onto a designated area. Run-on occurs when surface water from an area at a higher elevation flows down onto an area of concern, such as a feedlot, vegetated filter strip, or riparian zone.

S: sulfur.

Salt: a compound made up of the positive ion of a base and the negative ion of an acid.

Sampling: collection of a small part of an entity, such as a water body, that can be tested and from which conclusions can be drawn about the entity as a whole.

Scentometer: handheld device used to measure ambient odor levels in the field.

Sediment delivery: sediment arriving at a specific location.

Sediment delivery ratio (SDR): fraction of eroded soil that actually reaches a water body.

Sediment yield: quantity of sediment leaving a specified land area.

Sedimentation tank: a unit in which water or wastewater containing settleable solids is retained to remove by gravity a part of the suspended matter. Also called *sedimentation basin, settling basin, settling tank,* or *settling terrace.*

Septage: the mixed liquid and solid contents pumped from septic tanks and dry wells used for receiving domestic type sewage.

Sequencing batch reactor: an in-vessel, self-contained system with alternate aerobic and anaerobic stages of biological treatment to biodegrade organic matter, convert nitrogen to nitrogen gas, and precipitate phosphorus in settled sludge.

Setback: specified distance from surface waters or potential conduits to surface waters where manure, litter, and process wastewater may not be applied.

Settleable solids: matter in wastewater that either settles to the bottom or floats to the top during a preselected settling period.

Sewage sludge: settled sewage solids combined with varying amounts of water and dissolved materials that are removed from sewage by screening, sedimentation, chemical precipitation, or bacterial digestion.

Sheet erosion: soil erosion occurring from a thin, relatively uniform layer of soil particles on the soil surface. Also called *interrill erosion.*

Shelterbelts: extended windbreak of living trees and shrubs established and maintained to direct, diffuse, and filter odor plumes.

Slope: the inclination of a land surface from the horizontal. Percentage of slope is the vertical distance divided by the horizontal distance, multiplied by 100. Thus, a slope of 20% is a drop of 20 ft in 100 ft of horizontal distance.

Slotted flooring: a flooring surface in a building that has open spaces or grooves to allow material to drop below the floor surface.

Sodicity: the degree to which a soil is affected by exchangeable sodium, expressed as a sodium adsorption ratio (SAR) of a saturation extract.

Soil amendment: any material, such as lime, gypsum, sawdust, or synthetic conditioner, that is worked into the soil to make it more amendable to plant growth. Amendments may contain important fertilizer elements, but the term commonly refers to added materials other than fertilizer.

Soil and water conservation practices (SWCPs): the manipulation of such variables as crops, rotation, tillage, management, and structures to reduce the loss of soil and water.

Soil organic matter: the organic fraction of soil that includes plant and animal residue at various stages of decomposition, exclusive of undecayed plant and animal residue. Often used synonymously with *humus*.

Soil profile: a section of soil viewed on a vertical plane extending through all its horizons and into the parent material.

Soil solution: the liquid phase of soil, including dissolved organic and inorganic materials.

Solid manure storage: a storage unit in which accumulations of bedded manure or solid manure are stacked before subsequent handling and field spreading. The liquid part, including urine and precipitation, may or may not be drained from the unit.

Solids content: the residue remaining after water is evaporated from a sample at a specified temperature, usually about 215°F (103°C).

Sorbed: adsorbed or absorbed.

Spatial: the occupied space relationship between a soil or soil map unit to the landscape or geomorphic surface on which the soil or map unit is located.

State air pollution regulatory agency (SAPRA): a state agency that administers ambient monitoring programs, issues operating permits, conducts compliance inspection, and compels federally mandated emissions-reduction programs for nonattainment areas. SAPRAs also develop and submit for EPA approval implementation plans that bring nonattainment areas into compliance with the National Ambient Air Quality Standards within a reasonable time.

State implementation plan (SIP): a state's plan for bringing its federally designated nonattainment areas into compliance with the National Ambient Air Quality Standards.

Stocking density: number of cattle per unit corral area. Increased density can modestly reduce downwind dust concentrations, but it reduces the linear bunk space available to each animal and may result in behavioral changes that increase stress and reduce production.

Stream classification: the identification of specific water uses for water courses.

Struvite: a colorless to yellow or pale-brown mineral that can build up as crystals on pump impellers and in pipes conveying wastewater.

Subsurface runoff: water that infiltrates soil and then moves laterally or vertically below the surface; includes baseflow and interflow.

Supernatant: the liquid fraction in a lagoon.

Surface water: where this text says "surface water," it refers to "waters of the United States."

Suspended solids: (1) undissolved solids that are in water, wastewater, or other liquids and are largely removable by filtering or centrifuging. (2) The quantity of material filtered from wastewater in a lab test, as prescribed in *APHA Standard Methods for the Examination of Water and Wastewater* or a similar reference.

Symbiotic: of or relating to a relationship in which two organisms live together in close association in which neither are harmed and both benefit.

Synoptic: of or relating to data obtained nearly simultaneously over a large area of the atmosphere.

Synthetic organic compounds: organic compounds created by industry either inadvertently as a part of a chemical process or for use in a wide array of applications for modern day life. Some that have been created are persistent in the environment (slow to decompose) because oxidizers, such as soil microbes, cannot readily use them as an energy source.

Texture, soil: the relative proportions of sand, silt, and clay particles in a mass of soil.

Tg: teragram, 1×10^{12} g.

Tilth, soil: the physical condition of soil as related to tillage, seedbed preparation, seedling emergence, and root penetration.

Total dust concentration: measurement of all of the dust particles that a filter collects in and near an animal facility.

Total solids: the sum of dissolved and undissolved solids in water or wastewater, usually stated in milligrams per liter (mg/L).

Toxicity: degree of harmful effect an element or compound has on a living organism, plant, or animal. Excessive amounts of toxic substances, such as sodium or sulfur, can severely hinder establishment of vegetation or severely restrict plant growth.

TSP: total suspended particulates.

Uncertainty: the degree of confidence that can be assigned to a numerical measurement in terms of both accuracy and precision.

Unconfined aquifer: an aquifer in which the water table is exposed to the atmosphere through openings in the overlying materials.

Universal Soil Loss Equation (USLE): an empirical equation estimating the amount of soil loss that is used for the evaluation of a resource management system for water erosion control.

Urban expansion zone: a boundary or mapped area surrounding a municipality and officially designated by the governing body of the municipality as the area in which future urban development will be allowed to occur as the municipality grows.

USDA: U.S. Department of Agriculture.

USEPA: U.S. Environmental Protection Agency.

Vadose zone: the zone containing water under less pressure than that of the atmosphere, including soil water, intermediate vadose water, and capillary water. This zone is limited above by the land surface and below the surface of the zone of saturation, that is, the water table.

Vector: a bearer or carrier, such as an organism (often an insect), that carries and transmits disease-causing microorganisms.

Vegetative practices: candidate measure that includes vegetation as the principal method of pollution control.

VOC: volatile organic compound.

Volatile compounds: by-products of animal manure decomposition that readily become vapors (e.g., ammonia, carbon dioxide, and methane).

Volatile organic compounds (VOCs): organic molecules, usually arising from the decomposition of manure, that tend to move from liquid into the air above animal facilities.

Volatile solids (VS): weight lost upon ignition at 550°C (using Method 2540 E of the American Public Health Association). Volatile solids provide an approximation of moisture and organic matter present.

Volatilization: the loss of gaseous components, such as ammonium nitrogen, from animal manure.

Waste storage pond: an impoundment made by excavation or earthfill for temporary storage of animal or other agricultural waste.

Waste treatment lagoon: an impoundment made by excavation or earthfill for biological treatment of animal or other agricultural wastes. Lagoons can be aerobic, anaerobic, or facultative, depending on their loading and design.

Water management system: a planned system in which the available water supply is effectively used by managing and controlling the moisture environment of crops to promote the desired crop response, to minimize soil erosion and loss of plant nutrients, to control undesirable water loss, and to protect water quality.

Water quality: the excellence of water in comparison with its intended use or uses.

Water table: the surface between the vadose zone and groundwater; the surface of a body of unconfined groundwater at which the pressure is equal to that of the temperature.

Waters of the United States: according to the Code of Federal Regulations, defined as:

a. All waters that are currently used, were used in the past, or may be susceptible to use in interstate or foreign commerce, including all waters that are subject to the ebb and flow of the tide

b. All interstate waters, including interstate wetlands

c. All other waters, such as interstate lakes, rivers, streams (including intermittent streams), mudflats, sandflats, wetlands, sloughs, prairie potholes, wet meadows, playa lakes, or natural ponds, the use, degradation, or destruction of which would affect or could affect interstate or foreign commerce, including any such waters:
   1. that are or could be used by interstate or foreign travelers for recreational or other purposes
   2. from which fish or shellfish are or could be taken and sold in interstate or foreign commerce
   3. that are used or could be used for industrial purposes by industries in interstate commerce
d. All impoundments of waters otherwise defined as waters of the United States under this definition
e. Tributaries of waters identified in paragraphs (a) through (d) of this definition
f. The territorial sea
g. Wetlands adjacent to waters (other than waters that are themselves wetlands) identified in paragraphs (a) through (f) of this definition.

Waste treatment systems, including treatment ponds or lagoons designed to meet the requirements of the Clean Water Act (other than cooling ponds as defined in 40 CFR 423.11(m), which also meet the criteria of this definition), are not waters of the United States. This exclusion applies only to man-made bodies of water, which neither were originally created in waters of the Unites States (such as disposal area in wetlands) nor resulted from the impoundment of waters of the United States. [At 45 FR 48620, July 21, 1980, the USEPA suspended until further notice in § 122.2, the last sentence, beginning "This exclusion applies ..." in the definition of "Waters of the United States."] Waters of the United States do not include prior converted cropland. Notwithstanding the determination of an area's status as prior converted cropland by any other federal agency, for the purposes of the Clean Water Act, the final authority regarding Clean Water Act jurisdiction remains with the EPA [40 CFR 122.2].

Wet basis: the fraction of a given constituent in a moist mixture as a proportion of the total weight of dry matter plus incorporated water; always numerically less than the corresponding "dry basis" proportion.

Wet scrubber: air treatment technology that uses wet chemicals or water to remove dust and gas compounds from an air stream.

Wet-weight percentage: the ratio of the weight of any constituent to the typical hydrated weight of the whole plant part as harvested.

1.7 SUMMARY

Although CAFOs can cause problems, factory farming in general, and the techniques and products that make factory-farming possible, have done much for the easy availability of inexpensive, high-quality food. While regulations, controls, management, and limits are necessary to protect the environment, CAFOs are not a fad and, despite the problems they create, this industry is going to grow.

CHAPTER REVIEW QUESTIONS

1. Describe and define the term "CAFO."
2. What threat do CAFOs pose to air quality? Water quality? Fish and wildlife? General health? Local farming community economics?
3. What's the difference between a CAFO and an AFO?
4. What's an animal unit, and what's the purpose of creating the concept?

5. How many animals (hogs, dairy cattle, beef cattle, laying hens, broiler chickens) does it take to equal 1000 animal units?
6. Compare U.S. human waste output to U.S. confined animal waste output, in small-, medium-, and large-scale cases.
7. Discuss the purpose of and the processes of water and wastewater treatment.
8. What wastes are common in domestic wastewater? In industrial wastewater? In agricultural wastewater? What treatment processes are used for the three different types of wastes?
9. What are the advantages and disadvantages of land application of manures?
10. What are the differences between land-applying family farm animal wastes and land-applying CAFO wastes?
11. Describe and discuss storage and handling considerations for manures.
12. Describe and discuss the use of lagoons for treating animal wastes. What effects can lagoons have on water systems, surface waters, wells, and groundwater?
13. What types of CAFOs are common to what geographic regions? Why?
14. Describe, define, and discuss the similarities and differences between feedlots, animal feedlots, and alternate animal feedlots.
15. What is an NPDES permit? Why is it important?
16. What are discharge criteria, and why are they important?
17. What's important about hydrologic soil groups and available water capacity in terms of agricultural wastes and CAFOs?
18. What do the following abbreviations stand for, and why is each important: CNMP, RMS, SWCP, NAAQs?
19. What are the "Waters of the United States"? What doesn't that term include?

THOUGHT-PROVOKING QUESTIONS

1. Many agricultural practices are designed to protect the soil and water. Just as agriculture is sometimes said to be a major source of water pollution, it is also a major source of practices to protect the environment. No area of activity does more to protect water than agriculture. Do you agree with this statement? Explain.
2. Examine the history of agricultural activism in the United States. How have these movements affected American society?
3. Are urban sprawl and the growth of factory farming related issues? If so, how? If not, why not?
4. Animal rights activists are often strongly critical of CAFO animal treatment. What do they object to and why? What is your reaction to the descriptions of CAFO animal treatment?
5. Some animal activist groups are more extreme than others. How does their extremity help to forward their cause? How does it hurt their cause?
6. What roles do marketing and public image play in CAFO acquisition and operations? Examine state and local government roles in attracting industry, and look at the reactions of local groups and governments to big-box commercial enterprises that wish to enter a community—Wal-Mart, for example. What public, commercial, and governmental economic factors are at work?

REFERENCES

Barrette, M., 1996. Hog-tied by feedlots. *Zoning News* (October): 1–4.
Cantrell, P., Perry, R., & Sturtz, P., 1999. Economic disaster: Boom or bust? http://www.inmotionmagazine.com/hwdisas.html. Accessed December 20, 2004.
Cantrell, P., Perry, R., & Sturtz, P., 2004. The environment (…and factory farms). @ http://www.inmotionmagazine.com/hwdisas.html. Accessed December 15, 2004.

Collins, E.R. et al., 1975. Effect of anaerobic swine lagoons on groundwater quality in high water table soils. *The Proceedings of the Third International Symposium on Livestock Wastes*, pp. 303–305.

Hegg, R.O., King, T.G., & Ianzen, I.I., 1981. Four year study of the effect on groundwater from a dairy lagoon in the piedmont. *American Society of Agricultural Engineers.*

Hegg, R.O., King, T.G., & Wilson, T.V., 1978. The effect on groundwater from seepage of livestock manure lagoons. *Water Resources Research Institute Tech., Report No. 78.*

Humenick, F.J., Overcash, M.R., Baker, J.C. & Western, P.W., 1980. Lagoons: State of the Art. *Proceedings of the International Symposium on Livestock Wastes*, pp. 211–216.

Ikerd, J., 1998. Large scale, corporate hog operations: Why rural communities are concerned and what they should do. http://www.ssu.missouri.edu/faculty/jikerd/papers/top-toh.html. Accessed December 21, 2004.

Keller, D. & Miller, D., 1997. Neighbor against neighbor. *Progressive Farmer*, 112, no. 11 (October): 16–18.

McBride, W.D., 1997. *Change in U.S. Livestock Production, 1969–92*. Agricultural Economic Report No. 754. Washington, D.C.: Economic Research Service, U.S. Department of Agriculture.

Miller, H.H., Robinson, J.B., & Gillam, W., 1985. Self-sealing of earthen liquid manure storage ponds: I. A case study. *Journal of Environmental Quality*, 14: 533–538.

Proulx, A., 2002. *That Old Ace in the Hole*. New York: Scribner, p. 6.

National Farmers Union, 2000. The effects of hog mega-barn on communities, the environment, and independent hog producers. http://www.hogwatchmanitoba.org/hardisty.html. Accessed December 21, 2004.

Ritter, W.R., Walpole, E.W., & Eastburn, R.P., 1980. An anaerobic lagoon for swine manure and its effect of the groundwater quality in sandy-loan soils. *Proceedings of the Fourth International Symposium on Livestock Wastes*, pp. 244–246.

Ritter, W.R., Walpole, E.W., & Eastburn, R.P., 1984. Effect of an anaerobic swine lagoon on groundwater quality in Sussex Country Delaware. *Agricultural Wastes*, 10: 267–284.

Schwab, J., 1998. *Planning and Zoning for Concentrated Animal Feeding Operations*. Chicago, Illinois: American Planning Association: Planning Advisory Service Report Number 482.

Sewell, J.I., Mulling, J.A., & Vaigneur, H.O., 1975. Dairy lagoon system and groundwater quality. *The Proceedings of the Third International Symposium on Livestock Wastes*, pp. 286–288.

Sierra Club, 2004. *Clean Water & Factory Farms*. http://www.sierraclub.org/factoryfarms/faq.asp. Accessed December 2, 2004.

Sutton, A., & Humenik, F., 2003. *CAFO Fact Sheet #24: Technology Options to Comply with Land Application Rules*. Ames, Iowa: MidWest Plan Service, Iowa State University.

USDA & USEPA, 1998. Comprehensive nutrient management plan components. www.epa.gov/fedrgstr/EPA-water/1998/September/Day-20/w25138.htm. Accessed December 12, 2004.

USEPA, 2003. What's the problem? www.epa.gov/region09/crosspr/animalwaste/problem.html. Accessed December 21, 2004.

2 CAFO Regulations and CNMPs

"But hog farms make jobs for local people. I mean, this is a region where there aren't many jobs, so that's something. Helping the economy and all. Mr. Skin there had a job from them."

Why Bob, you are innocent to the facts of life. One hog farm site makes a very few jobs at minimum wage. They run three shifts but everything's automated and computer controlled. The corporations don't buy locally. They buy bulk supplies in the world market, truck it in. Good business. The hog farms come in, they look like they're bringin money into the region so some a the locals just lap it up. Give them tax breaks. Then, where there was eight thousand pigs all of a sudden there is fifty thousand. They polluted Tulsa's water supply. They poisoned the rivers in North Caroline. They run wild in Oklahoma until very recent when Oklahoma begin to lay out some rules. That's when they commenced a come down here into the Texas panhandle.

(Proulx, 2002, p. 113)

2.1 INTRODUCTION

In the chapter opening, the statement "they run wild in Oklahoma until very recent when Oklahoma begin to lay out some rules" sets the stage for this chapter: Concentrated animal feeding operation (CAFO) regulations (rules). Regulations at the federal, state, and local levels affect the operations of CAFOs. Federal and state regulations set an important context for local regulations and, in some cases, preempt certain categories of local legislation entirely. For example, many states have exemptions for agriculture from county zoning laws in their county zoning enabling legislation. The purpose of this chapter is to establish the federal, state, and local context within which regulation of CAFOs occurs. *Note:* Much of the information in this chapter is adapted from the Environmental Protection Agency's (USEPA's) *Producers' Compliance Guide for CAFOs*, last accessed December 12, 2004 @ [www.epa.gov/npdes/pubs/cafo_prod_guide_entire_doc.pdf].

In addition to presenting an overview of federal, state and local regulations, we also discuss U.S. Department of Agriculture's (the Natural Resources Conservation Service [NRCS]) *Comprehensive Nutrient Management Plan (CNMP)* in this chapter. CNMPs are developed in accordance with NRCS conservation planning policy and rely on the planning process and established conservation practice standards.

2.2 FEDERAL REGULATIONS: CLEAN WATER ACT

The primary legislative basis for environmental regulation of CAFOs is Section 402 of the Clean Water Act (CWA), which specifically cites CAFOs as point sources. The regulatory compliance guidelines (working documents), revised in February 2003, for CAFOs are:

- The National Pollutant Discharge Elimination System (NPDES) Permit Regulation for CAFOs (40 CFR Part 122)
- The Effluent Limitations Guidelines and Standards for CAFOs (40 CFR Part 412)

The USEPA issues, enforces, and occasionally updates its regulations. Both of the previously mentioned regulations have requirements for CAFOs, so the USEPA revises them at the same time to make sure that their requirements are consistent.

2.2.1   NPDES Program

The NPDES Program was created under the federal CWA to protect and improve water quality by regulating point source dischargers, operations that discharge pollutants from discrete conveyances directly into waters of the United States.

Key term: A *discharge*, in general, is the flow of treated or untreated wastewater from a facility to surface water.

Key term: The umbrella term *pollutant* covers a wide variety of materials that might contaminate waters of the United States. Pollutants from CAFOs could include nutrients, suspended solids, oxygen-demanding substances, or pathogens.

Key term: A *discrete conveyance*, in general, is any single, identifiable way for pollutants to be carried or transferred to waters, such as a pipe, ditch, or channel.

Key term: In this text, the term *waters of the United States* is defined as in 40 CFR 122.2 and in Chapter 1 of this text. Where this text says "surface waters," it means "waters of the United States," which includes, but is not limited to:

- Waters used for interstate or foreign commerce (e.g., the Mississippi River or the Gulf of Mexico).
- All interstate waters, including wetlands (any river, stream, lake, or other water body that crosses state borders).
- Waters used for recreation by interstate or foreign travelers (e.g., a lake in one state that attracts fishermen from neighboring states).
- Waters from which fish or shellfish are taken to be sold in other states or countries.
- Waters used for industrial purposes by industries involved in interstate commerce.
- Tributaries and impoundments or dams of any waters described above.
- Territorial seas.
- Wetlands adjacent to any waters described above.

Key term: The term *waters of the United States* **does not** include:

- Ponds or lagoons designed and constructed specifically for waste treatment systems.
- Wetlands that were converted to cropland before December 23, 1985.

Point source dischargers are regulated by NPDES permits. An NPDES permit:

- Identifies wastewater discharges to surface waters from the point source facility.
- Sets requirements designed to protect water quality (such as discharge limits, management practices, and record-keeping requirements) that the discharger must meet.
- Allows an operation to discharge pollutants as long as the operation meets the requirements in the permit.
- If a facility discharges pollutants without having a permit, or has a permit but does not meet the requirements, it is violating the CWA. Its owner or operator could be subject to enforcement.
- We repeat, under the CWA, CAFOs are defined as point source dischargers. The revised NPDES CAFO regulation requires all CAFOs to apply for a permit. So if you own or operate a CAFO, you must apply for and comply with the conditions in an NPDES permit. If the owner and operator are different people, only one needs to apply for a permit.

The NPDES regulation describes which operations quality as CAFOs and sets the basic requirements included in CAFOs permits.

- *Important point:* Every CAFO has a duty to apply for a permit. Owners or operators of CAFOs that do not discharge must still contact their permitting authority and provide certain information to avoid permitting requirements. *Note:* We discuss CAFOs that do not need a permit later. Different kinds of CAFOs have different deadlines for when their operators must apply for NPDES permits. *Note:* We discuss, later in the text, when an NPDES permit must be obtained.

Key term: A *permitting authority* is the agency responsible for issuing NPDES permits in a state.

2.2.2 EFFLUENT LIMITATION GUIDELINES FOR CAFOs

The USEPA has preset some of the minimum requirements that go into each permit in regulations called "effluent limitations guidelines" (ELGs) for CAFOs and certain other industries. When the permitting authority issues a permit for a CAFO, it does not set permit requirements on its own. Instead, it places the requirements of the ELGs directly into the permit. These requirements may consist of both limits on the amount of a pollutant that can be discharged (numerical limits called "discharge limits") and other ELG requirements (management practices and record-keeping requirements). The state permitting authority may also set additional requirements needed to protect water quality or other requirements that apply under state or local law.

The ELGs for CAFOs include both discharge limits and certain management practice requirements. For most animal types, however, the ELGs for CAFOs apply only to large CAFOs. [Note, for example, that for duck CAFOs, the ELGs apply to all operations with 5,000 or more ducks, whether they are large, medium, or small CAFOs.] Permitting authorities set effluent limitations for medium and small CAFOs on a case-by-case basis, depending on the specific situation at the CAFO and based on the best professional judgment of the permitting authority. In many cases, those requirements are similar to the requirements for large CAFOs.

2.2.3 IMPORTANCE OF THE REGULATIONS

To reflect changes in the animal production industry since the original regulations were passed in the 1970s, the USEPA has revised these regulations. Out of 257,000 animal feeding operations (AFOs) in the United States today, about 15,500 are CAFOs. These operations generate manure, litter, and process wastewater that can contain such pollutants as nitrogen, phosphorus, metals, and bacteria. If CAFO operators don't properly manage these materials, they could release pollutants into the environment through spills, overflows, or runoff. These releases, in turn, might pollute surface waters and threaten the health of people and animals. On the other hand, when operators properly manage manure, litter, and process wastewater, they help to prevent water pollution and its negative impacts. The CAFO regulations were revised to reflect current practices in the industry and to set basic standards for CAFO operators to properly manage the manure, litter, and process wastewater generated at their operations.

Key term: *Process wastewater* is water used directly or indirectly in the operation of an AFO, including:

- Spillage or overflow from animal or poultry watering systems.
- Washing, cleaning, or flushing pens, barns, manure pits, or other facilities.
- Direct contact swimming, washing, or spray cooling of animals.
- Dust control.

Process wastewater also includes any water that comes into contact with any raw materials, products, or by-products, including manure, litter, feed, milk, eggs, and bedding.

The revised regulations focus on CAFOs that pose the greatest risk to water quality. By regulating mainly large CAFOs and some smaller CAFOs that pose a high risk to water quality, the EPA regulates close to 60% of all manure generated by operations that confine animals.

2.2.4 OTHER LAWS THAT REGULATE CAFOs

As previously mentioned, states, counties, and towns might have more requirements or more specific requirements designed to address particular circumstances. Each individual permitting authority can set additional requirements in any permit if it finds them necessary. State regulations must include the federal requirements, but they can also be broader, stricter, or more specific. State, county, and local requirements are discussed in greater detail in Sections 2.3 and 2.4.

An NPDES permit might include other federal requirements that apply to point source dischargers (for example, requirements under the Endangered Species Act, the National Historic Preservation Act, or the Total Maximum Daily Load [TMDL] program). CAFOs might also be subject to other federal requirements under, for example, the Federal Insecticide, Fungicide, and Rodenticide Act (FIFRA) or the Spill Prevention, Containment, and Countermeasure regulations. CAFO operators and owners must work with their permitting authority to ensure of compliance with all requirements that apply to their operation.

2.2.5 APPLICABILITY OF THE REGULATION

Federal CAFO regulations apply to owners and operators of AFOs that are CAFOs because they meet certain conditions. If an AFO meets those conditions, it is regulated and the owner or operator must apply for an NPDES permit. The following sections describe animal operations that are regulated.

2.2.5.1 Animal Feeding Operations Covered by the Regulations

All CAFOs are covered by these regulations. A CAFO is a specific kind of AFO. The regulations describe which AFOs are considered CAFOs. To be regulated as a CAFO, an operation must first meet the regulatory definition of an AFO.

2.2.5.1.1 AFO
40 CFR 122.23 (b)(1) defines an AFO as an operation that meets both of these conditions:

1. The animals are confined for at least 45 days during any 12-month period.

The 45 days of confinement do not have to be 45 days in a row, and the 12-month period can be any consecutive 12 months.

2. Crops, forage growth, and other vegetation are not grown in the area where the animals are confined.

This does not mean that any vegetation at all in a confinement area would keep an operation from being defined as an AFO. For example, a confinement area such as a pen or feedlot that has only "incidental vegetation" (as defined by the permitting authority) would still be an AFO as long as the animals are confined for at least 45 days in any 12-month period.

Pasture and rangeland operations are not AFOs because the animals in there operations are not confined or concentrated in an area where manure builds up. However, a pasture or grazing-based operation might also have additional areas, such as feedlots, barns, or pens, that meet the conditions necessary to be defined as an AFO.

Winter feedlots can be AFOs, even if the feedlot area is used to grow crops or forage when animals are not confined there. In the case of winter feedlots, the "no vegetation" condition applies to the time when the animals are confined there.

The AFO definition is not limited to the animal types discussed in the regulations. An operation that confines any type of animal and meets both of the conditions in the definition is an AFO. In addition to confinement areas at animal production facilities, confinement areas at auction houses, sale barns, livestock marketing areas, horse show arenas, and stable areas of racetracks can be considered AFOs if they meet both of the conditions in the definition.

- *Important point:* An animal confined for any portion of a day should be counted as being confined for that day. For example, a facility maintains a herd of beef cattle on pasture. This facility also includes a hospital area where cattle are confined for medication. Cattle are confined in the hospital area for 5 days each month for medication. The cattle are confined for a total of 2 hours each time they are medicated. These cattle are counted as being confined for 60 days each year (5 days/month × 12 months) even though they are not confined for a full day.

2.2.5.1.2   CAFO

For a facility to be a CAFO, it must first meet the regulatory definition of an AFO. A CAFO is an AFO that has certain characteristics. The two ways for an AFO to be considered a CAFO are:

- An AFO may be **defined** as a CAFO, **or**
- An AFO may be **designated** as a CAFO.

2.2.5.1.2.1   AFOs Defined as CAFOs

An AFO can be defined as a CAFO if it has a certain number of animals and it meets the other criteria contained in the regulations. The regulations set thresholds for size categories based on the number of animals confined at the operation for a total of 45 days or more in any 12-month period. Tables provided later in this chapter show the thresholds for large, medium, and small CAFOs for different kinds of animals.

Large CAFOs

An operation is defined as a large CAFO if it:

- Meets the regulatory definition of an AFO **and**
- Meets the large CAFO threshold for that animal type

Medium CAFOs

An operation is defined as a medium CAFO if it:

- Meets the regulatory definition of an AFO
- Meets the medium CAFO thresholds for that animal type **and**
- Meets at least one of the following two criteria (called "discharge criteria"):
  - A man-made ditch, pipe, or similar device carries manure or process wastewater from the operation to surface water, **or**
  - The animals come into contact with surface water that runs through the area where they're confined.

The discharge criteria apply to only the parts of the operation where animals are confined, manure or raw materials are stored, and waste is contained. For example, if you dig a ditch or install a pipe that drains water from your confinement area into a stream or lake, your operation would meet the first discharge criterion. Open tile drains in the areas where animals are confined, wastes are collected and stored, or raw materials are kept also meet the first criterion if the tile drains carry pollutants from these areas to surface water. Your operation meets the second discharge criterion if a stream runs through the confinement area and the animals have direct access to the stream.

- *Important point:* If you own two or more AFOs that are next to each other or use a common waste disposal area or system, you should count all the animals at all the operations together to determine whether your operations fall within the thresholds for the CAFO size categories. If both of your operations use a common waste disposal area or system, they are counted as one even if they're not next to each other. (Two operations under common ownership are considered to have a common waste disposal system if the manure, litter, or process wastewater from the two operations is mixed before disposal or land application or if the manure, litter, or process wastewater from the two operations is applied to the same land application area. Common waste disposal systems also include any other type of system where the wastes from two operations are commingled for handling or disposal.)

  Also, if an operation is entirely located on one site but ownership of the operation is split between two or more people, you should count all the animals at that operation to know if it falls within the thresholds for the CAFO size categories.

2.2.5.1.2.2   AFOs Designated as CAFOs
The second way for an AFO to be a CAFO is to be designated as a CAFO. If an AFO doesn't meet the definition of a large or medium CAFO but the permitting authority finds it to be a significant contributor of pollutants to surface waters, the permitting authority may designate that operation as a CAFO. To designate an AFO as a CAFO, the permitting authority must inspect the AFO and must find that the operation is a significant contributor of pollutants to surface waters.

Medium CAFOs
AFOs that fall within the size thresholds for medium CAFOs but don't meet either of the two discharge criteria may be designated as CAFOs by the permitting authority.

Small CAFOs
AFOs that don't confine enough animals to meet the medium CAFO size threshold may be CAFOs only by designation. The permitting authority may designate a small AFO as a small CAFO only if the AFO is a significant contributor of pollutants to surface waters **and** it meets at least one of two discharge criteria:

- A man-made ditch, pipe, or similar device carries manure or process wastewater from the operation to surface water or
- The animals come into contact with surface water that runs through the area where they're confined.

- *Important point:* An AFO might not meet the definition of a CAFO if:
  - It doesn't confine enough animals.
  - It doesn't meet the discharge criteria (for medium CAFOs).
  - It confines a type of animal not included in the large or medium CAFO definition.

The EPA and USDA promote efforts by states to use approaches other than NPDES permitting to help medium and small AFOs avoid having conditions that would result in those facilities being defined or designated as CAFOs. For example, the voluntary development and implementation of a CNMP prepared in accordance with the CNMP Technical Guidance issued by the USDA's NRCS should, in most instances, meet the minimum standard requirements of an NPDES permit.

2.2.6 CAFO Thresholds for Specific Animal Sectors

The USEPA has set thresholds for operations that confine different kinds of animals. The thresholds are used with discharge criteria to determine which AFOs are defined as large or medium CAFOs and which should be designated as medium or small CAFOs. Tables 2.1 through 2.12 show these thresholds.

- *Important point:* The thresholds in the regulations are for the actual number of animals confined, not the number of animals that could be confined. For example, if you raise cattle at a feedlot and you have the capacity to raise as many as 1,500 head at one time, but you never have more than 1,100 head any one time, your operation confines 1,100 head. If you have 3 chicken houses, confine 25,000 chickens in each house, and produce 6 flocks of chickens each year, your operation still confines only 75,000 chickens at one time, even through you might produce half a million chickens each year.

- *Important point:* If you confine more than one kind of animal at your operation, you should count each kind of animal separately. If you confine enough of any one kind of animal to meet the threshold for the animal sector (and your operation meets any other qualifying conditions), your operation is covered by the CAFO regulations. In this case, your permit will apply to the manure, litter, and process wastewater generated from all the animals confined at your operation, not just the sector that meets the size threshold. For example, if an AFO confines 800 beef cattle, 1,000 sows, and 150,000 broilers, the AFO is a large CAFO because it meets the large CAFO threshold for chicken operations. In this case, the permit applies to all manure, litter, and process wastewater produced by the confined broilers, sows, and beef cattle. The permit, however, would not apply to any animals pastured at this operation.

Cattle (other than mature dairy cows)

The thresholds in Table 2.1 apply to operations that confine any kind of cattle other than mature dairy cows, including heifers, steers, bulls, and cow/calf pairs. For example, these thresholds apply to beef cattle operations such as feedlots and backgrounding yards, veal calf operations, and contract dairy operations. Except for cow/calf pairs, each animal is counted as one animal, regardless of its age or weight. In the case of cow/calf pairs, the pair is counted as one animal until the calf is weaned. After the calf is weaned, the cow and calf count as individual animals.

- *Example:* An 850-head beef feedlot that also confines 100 cow/calf pairs where the calves have not been weaned has 950 cattle other than mature dairy cows. This is not a large CAFO. However, an 850-head beef feedlot that also confines an additional 100 cows and 100 weaned calves has 1,050 animals. This is a large CAFO.

TABLE 2.1
Cattle (other than mature dairy cows): Size Category Thresholds

An AFO that has ...	is a ...	by ...
At least 1,000 cattle, dairy heifers, cow/calf pairs, or veal calves	Large CAFO	Regulatory definition
From 300 to 999 cattle, dairy heifers, cow/calf pairs, or veal calves **and** meets one of the medium category discharge criteria	Medium CAFO	Regulatory definition
From 300 to 999 cattle, dairy heifers, cow/calf pairs, or veal calves **and** has been designated by the permitting authority	Medium CAFO	Designation
Fewer than 300 cattle, dairy heifers, cow/calf pairs, or veal calves **and** has been designated by the permitting authority	Small CAFO	Designation

Source: USEPA (2004).

TABLE 2.2

Mature dairy cows: Size Category Thresholds

An AFO that has ...	is a ...	by ...
At least 700 mature dairy cows	Large CAFO	Regulatory definition
From 200 to 699 mature dairy cows **and** meets one of the medium category discharge criteria	Medium CAFO	Regulatory definition
From 200 to 699 mature dairy cows **and** has been designated by the permitting authority	Medium CAFO	Designation
Fewer than 200 mature dairy cows **and** has been designated by the permitting authority	Small CAFO	Designation

Source: USEPA (2004).

TABLE 2.3

Swine (55 pounds or more): Size Category Thresholds

An AFO that has ...	is a ...	by ...
At least 2,500 swine weighing 55 pounds or more	Large CAFO	Regulatory definition
From 750 to 2,499 swine weighing 55 pounds or more **and** meets one of the medium category discharge criteria	Medium CAFO	Regulatory definition
From 750 to 2,499 swine weighing 55 pounds or more **and** has been designated by the permitting authority	Medium CAFO	Designation
Fewer than 750 swine weighing 55 pounds or more **and** has been designated by the permitting authority	Small CAFO	Designation

Source: USEPA (2004).

Mature dairy cows

The thresholds in Table 2.2 apply to operations that confine mature dairy cows. Mature daily cows include both milked and "dry" cows. Thresholds for AFOs that house any other kind of cattle, including heifers and veal calves, are shown in Table 2.1.

Swine (55 pounds or more)

The thresholds in Table 2.3 apply to operations that confine swine that weigh at least 55 pounds. These operations include farrow-finish operations, wean-finish operations, farrowing operations, breeding operations, grow-finish operations, and other specialized AFOs that confine mature swine. AFOs that house immature swine (less than 55 pounds) might also be subject to the thresholds shown in Table 2.4.

Swine (less than 55 pounds)

The thresholds in Table 2.4 apply to operations that confine swine that weigh less than 55 pounds. These thresholds typically apply to swine nurseries but may also apply to other facilities that confine swine of all sizes but primarily confine large numbers of immature swine. For example, an operation with 1,000 sows, 50 boars, and 14,000 newborn pigs is a large CAFO.

Remember that AFOs that house "mature" swine (55 pounds or more) are already subject to the thresholds in the sector "Swine (55 pounds or more)" (Table 2.3). So a swine operation could

TABLE 2.4
Swine (less than 55 pounds): Size Category Thresholds

An AFO that has ...	is a ...	by ...
At least 10,000 swine weighing less than 55 pounds	Large CAFO	Regulatory definition
From 3,000 to 9,999 swine weighing less than 55 pounds **and** meets one of the medium category discharge crime	Medium CAFO	Regulatory definition
From 3,000 to 9,999 swine weighing less than 55 pounds **and** has been designated by the permitting authority	Medium CAFO	Designation
Fewer than 3,000 swine weighing less than 55 pounds **and** has been designated by the permitting authority	Small CAFO	Designation

Source: USEPA (2004).

TABLE 2.5
Horses: Size Category Thresholds

An AFO that has ...	is a ...	by ...
At least 500 horses	Large CAFO	Regulatory definition
From 150 to 499 horses **and** meets one of the medium category discharge criteria	Medium CAFO	Regulatory definition
From 150 to 499 horses **and** has been designated by the permitting authority	Medium CAFO	Designation
Fewer than 150 horses and has been designated by the permitting authority	Small CAFO	Designation

Source: USEPA (2004).

be defined as a CAFO because of the number of swine weighing 55 pounds or more, the number of swine weighing less than 55 pounds, or both.

- *Important point:* If an operation confines some swine that weigh more than 55 pounds and some that weigh less than 55 pounds, what is the operaton's status?

Assuming that the operation is already an AFO, the next step is to count the number of each type of animal on the operation. Does the operation confine more than 2,500 swine, each weighing 55 pounds or more? Does the operation confine more than 10,000 swine, each weighing less than 55 pounds? If the answer to either or both questions is "yes," the AFO is defined as a large CAFO.

Horses

The thresholds in Table 2.5 apply to operations that confine horses. The confinement area does not include areas like pastures. Most horse operations confine their animals only for short-term stabling or visits to stalls for shoeing, veterinary care, or similar activities, although pregnant mares stalled for urine collection are an exception. A horse might not be confined for enough days for the operation to meet the criteria for being an AFO. Data from the USDA National Animal Health Monitoring System suggest that practically all large horse CAFOs (those with more than 500 horses in confinement) are racetracks.

Sheep or lambs

The thresholds in Table 2.6 apply to operations that confine sheep and lambs. Count all confined sheep and lambs to determine whether the operation meets these thresholds. Confinement areas do

TABLE 2.6
Sheep or Lambs: Size Category Thresholds

An AFO that has ...	is a ...	by ...
At least 10,000 sheep or lambs	Large CAFO	Regulatory definition
From 3,000 to 9,999 sheep or lambs **and** meets one of the medium category discharge criteria	Medium CAFO	Regulatory definition
From 3,000 to 9,999 sheep or lambs **and** has been designated by the permitting authority	Medium CAFO	Designation
Fewer than 3,000 sheep or lambs **and** has been designated by the permitting authority	Small CAFO	Designation

Source: USEPA (2004).

TABLE 2.7
Turkeys: Size Category Thresholds

An AFO that has ...	is a ...	by ...
At least 55,000 turkeys	Large CAFO	Regulatory definition
From 16,500 to 54,999 turkeys **and** meets one of the medium category discharge criteria	Medium CAFO	Regulatory definition
From 16,500 to 54,999 turkeys **and** has been designated by the permitting authority	Medium CAFO	Designation
Fewer than 16,500 turkeys **and** has been designated by the permitting authority	Small CAFO	Designation

Source: USEPA (2004).

not include grazing areas. Many operations confine animals only for shearing, veterinary care and lambing, and before sale or processing. The animals might not be confined for enough days for the operation to be considered an AFO. Animals must be confined for 45 days or more in a 12-month period for an operation to be considered an AFO.

Turkeys

The thresholds in Table 2.7 apply to operations that confine turkeys. Most turkey operations today confine their birds in confinement houses, but turkeys are also raised on lots. Count all birds, including poults and breeders, to determine whether the operation meets the thresholds.

Chickens (operations with a liquid manure handling system)

The thresholds in Table 2.8 apply to operations that confine laying hens or broiler chickens and use a liquid manure handling system (for example, caged housing where manure is flushed to a lagoon). Liquid manure handling systems are relatively common among layer operations and are rarely used in other chicken operations. Operations that do not use liquid manure handling systems are subject to thresholds for the sector "Laying hens (operations with other than a liquid manure handling system)" (Table 2.9) or "Chickens other than laying hens (operations with other than a liquid manure handling system)" (Table 2.10). For pullets, see "Chickens other than laying hens (operations with other than a liquid manure handling system)" (Table 2.10).

TABLE 2.8

Chickens (Operations with a Liquid Manure Handling System): Size Category Thresholds

An AFO that has ...	is a ...	by ...
At least 30,000 chickens and uses a liquid manure handling system	Large CAFO	Regulatory definition
From 9,000 to 29,999 chickens, uses a liquid manure handling system, **and** meets one of the medium category discharge criteria	Medium CAFO	Regulatory definition
From 9,000 to 29,999 chickens, uses a liquid manure handling system, **and** has been designated by the permitting authority	Medium CAFO	Designation
Fewer than 9,000 chickens, uses a liquid manure handling system, **and** has been designated by the permitting authority	Small CAFO	Designation

Source: USEPA (2004).

TABLE 2.9

Laying Hens (Operations with Other than a Liquid Manure Handling System): Size Category Thresholds

An AFO that has ...	is a ...	by ...
At least 82,000 laying hens and **does not** use a liquid manure handling system	Large CAFO	Regulatory definition
From 25,000 to 81,999 laying hens, **does not** use a liquid manure handling system, **and** meets one of the medium category discharge criteria	Medium CAFO	Regulatory definition
From 25,000 to 81,999 laying hens, **does not** use a liquid manure handling system, **and** has been designated by the permitting authority	Medium CAFO	Designation
Fewer than 25,000 laying hens, **does not** use a liquid manure handling system, **and** has been designated by the permitting authority	Small CAFO	Designation

Source: USEPA (2004).

TABLE 2.10

Chickens Other than Laying Hens (Operations with Other than a Liquid Manure Handling System): Size Category Thresholds

An AFO that has ...	is a ...	by ...
At least 125,000 chickens other than laying hens and **does not** use a liquid manure handling system	Large CAFO	Regulatory definition
From 37,500 to 124,999 chickens other than laying hens, **does not** use a liquid manure handling system, **and** meets one of the medium category discharge criteria	Medium CAFO	Regulatory definition
From 37,500 to 124,999 chickens other than laying hens, **does not** use a liquid manure handling system, **and** has been designated by the permitting authority	Medium CAFO	Designation
Fewer than 37,500 chickens other than laying hens, **does not** use a liquid manure handling system, **and** has been designated by the permitting authority	Small CAFO	Designation

Source: USEPA (2004).

Key term: The term *manure handling system* refers to the manure collection and storage practices used at a chicken or duck operation. Examples of a liquid manure handling system include:

- An operation where ducks are raised outside with swimming areas or ponds.
- An operation with a stream running through an open lot
- An operation with confinement buildings where water is used to flush the manure to a lagoon, pond, or some other liquid storage structure.

In the CAFO regulations, the terms *wet lots, wet systems,* and *liquid manure handling systems* refer to the same set of management practices and are used interchangeably.

AFOs with liquid manure handling systems are large CAFOs if they have 30,000 laying hens or broilers or 5,000 ducks.

Key term: The term *manure handling system* refers to the manure collection and storage practices used at a chicken or duck operation. Operations using the following practices are considered to have *other than a liquid manure handling system:*

- Confinement buildings with a mesh or slatted floor over a concrete pit where the manure is scraped.
- Dry bedding on a solid floor where manure and bedding are not combined with water for flushing to a storage structure.

When chicken or duck operations use such practices and do not use any liquid manure handling systems, such as flushing to lagoons or storage ponds, these operations are considered to have *other than liquid manure handling systems.* They might also be called *dry manure systems* or *dry operations.*

AFOs with other than liquid manure handling systems are large CAFOs if they have 30,000 or more ducks, 82,0000 or more laying hens, or 125,000 or more chickens other than laying hens.

Laying hens (operations with other than a liquid manure handling system)

The thresholds in Table 2.9 apply to layer operations that do not use a liquid manure handling system. These operations include scrape-out and belt manure handling systems, high-rise cage housing, and litter-based housing. A chicken operation that uses a liquid manure handling system is subject to thresholds for the sector "Chickens (operations with a liquid manure handling system)" (Table 2.8). Nonlayer operations, including broiler operations that do not use a liquid manure handling system, are subject to thresholds in the sector "Chickens other than laying hens (operations with other than a liquid manure handling system)" (Table 2.10).

Chickens other than laying hens
(operations with other than a liquid manure handling system)

The thresholds in Table 2.10 apply to operations that confine broilers, roasters, pullets, or breeders and do not use a liquid manure handling system. These chicken operations typically use enclosed housing and dry litter systems. A chicken operation that uses a liquid manure handling system is subject to thresholds for the sector "Chickens (operations with a liquid manure handling system)" (Table 2.8). A layer operation that does not use liquid manure handling is subject to thresholds for the sector "Laying hens (operations with other than a liquid manure handling system)" (Table 2.9).

Example:

- A chicken operation produces 6 flocks of 100,000 broilers each year. The operation does not use a liquid manure handling system. Because the operation confines 100,000 broilers at a time, the operation is a medium CAFO if it meets one of the two discharge criteria.

TABLE 2.11

Ducks (Operations with a Liquid Manure Handling System): Size Category Thresholds

An AFO that has ...	is a ...	by ...
At least 5,000 ducks **and** uses a liquid manure handling system	Large CAFO	Regulatory definition
From 1,500 to 4,999 ducks, uses a liquid manure handling system, **and** meets one of the medium category discharge criteria	Medium CAFO	Regulatory definition
From 1,500 to 4,999 ducks, uses a liquid manure handling system, **and** has been designated by the permitting authority	Medium CAFO	Designation
Fewer than 1,500 ducks, uses a liquid manure handling system, **and** has been designated by the permitting authority	Small CAFO	Designation

Source: USEPA (2004).

TABLE 2.12

Ducks (Operations with Other than a Liquid Manure Handling System): Size Category Thresholds

An AFO that has ...	is a ...	by ...
At least 30,000 ducks, **does not** use a liquid manure handling system, **and** meets one of the medium category discharge criteria	Large CAFO	Regulatory definition
From 10,000 to 29,999 ducks, **does not** use a liquid manure handling system, **and** has been designated by the permitting authority	Medium CAFO	Designation
Fewer than 10,000 ducks, **does not** use a liquid manure handling system, **and** has been designated by the permitting authority	Small CAFO	Designation

Source: USEPA (2004).

- Another chicken operation has 60,000 laying hens and an additional 60,000 pullets and does not use a liquid manure handling system. This operation is also a medium CAFO if it meets one of the discharge criteria.
- A third operation also has 60,000 laying hens and an additional 60,000 pullets. This operation uses a lagoon for manure storage, and thus it has a liquid manure handling system. This operation is a large CAFO.

Chicken operations with uncovered litter stockpiles are treated as having liquid manure handling systems and are subject to the large CAFO threshold of 30,000 chickens for operations with a liquid manure handling system. By covering such stockpiles, a chicken operation becomes eligible for the higher thresholds for operations with other than a liquid manure handling system.

Ducks (operations with a liquid manure handling system)

The thresholds in Table 2.11 apply to duck operations that use a liquid manure handling system. These include operations with "wet" lots, lots with storage ponds, lots with swimming areas, and operations that flush manure from confinement buildings to lagoons. Count all birds to determine whether the operation meets the thresholds. A duck operation that does not use a liquid manure handling system is subject to thresholds for the sector "Ducks (operations with other than a liquid manure handling system)" (Table 2.12).

Ducks (operations with other than a liquid manure handling system)

The thresholds in Table 2.12 apply to any duck operation that does not use a liquid manure handling system. Count all birds to determine whether the operation meets the thresholds. A duck operation that uses a liquid manure handling system is subject to thresholds for the sector "Ducks (operations with a liquid manure handling system)" (Table 2.11).

2.2.7   OTHER KINDS OF OPERATIONS CONSIDERED TO BE CAFOS

An AFO with a kind of animal not identified in the regulations might still be considered a CAFO. Animals not identified in the regulations include, for example, ostriches, llamas, and bison. The only way for such an AFO to be a CAFO is for the permitting authority to designate it as a CAFO.

2.2.8   CAFOS THAT DO NOT NEED PERMITS

Large CAFOs that do not have the potential to discharge don't need NPDES permits. A large CAFO doesn't need an NPDES permit if (1) it provides evidence to the permitting authority that it has no potential to discharge manure, litter, or process wastewater to surface waters; (2) the permitting authority agrees; and (3) the permitting authority gives notice that the CAFO has "no potential to discharge" manure, litter, or process wastewater. "No potential to discharge" means that the CAFO must not discharge manure, litter, or process wastewater from either the production areas or any land application areas to surface waters, even by accident or because of human error.

A large CAFO can qualify for a "no potential to discharge" determination if:

- The owner or operator can show that no possibility exists for any manure, litter, or wastewater to be added to surface waters under any circumstances or conditions.
- The operation has not had a discharge for at least the past 5 years.

- *Important point:* Medium and small CAFOs cannot qualify for a "no potential to discharge" determination because those operations must have discharge to be defined or designated as CAFOs in the first place. The "no potential to discharge" status is intended to provide relief where truly no potential exists for a CAFO's manure or wastewater to reach surface water under any circumstances or conditions. For example, if a CAFO meets the following conditions, the owner might be able to demonstrate to the permitting authority that the CAFO has no potential to discharge:
  - Is located in an arid or semiarid environment
  - Stores all its manure or litter in a permanent, covered containment structure that precludes wind dispersal and prevents precipitation from contacting the manure or litter
  - Has sufficient containment to hold all process wastewater and contaminated storm water
  - Does not land apply CAFO manure or litter because, for example, the CAFO sends all its manure or litter to a regulated, off-site fertilizer plant or composting facility

2.2.9   EXCLUSION FROM REGULATIONS

Large CAFOs

If you own or operate a large CAFO, the only way to avoid CAFO requirements is to request and be granted a "no potential to discharge" determination.

Medium AFOs

If you own or operate a medium-sized AFO, you can avoid having your operation defined or designated as a CAFO by:

- Eliminating any condition that meets the discharge criteria and
- Reducing or eliminating your operation's discharges of pollutants to surface waters to minimize the chance that the permitting authority will find that your operation is a "significant contributor of pollutants to waters of the United States"

Small AFOs

If you own or operate a small-sized AFO, you can avoid having your operation designated as a CAFO by:

- Eliminating any condition that meets the discharge criteria or
- Reducing or eliminating your operation's discharges of pollutants to surface waters to minimize the chance that the permitting authority will find that your operation is a "significant contributor of pollutants to waters of the United States"

The EPA's policy promotes state efforts to use non-NPDES programs to help medium and small AFOs protect water quality and encourages owners and operators to take part in voluntary programs that promote sustainable agriculture and reduce environmental harm. These programs can help owners or operators of medium- and small-sized AFOs reduce risks to water quality and avoid NPDES permitting requirements. For example, if you voluntarily develop and implement a CNMP using the USDA's guidance, your CNMP might help you avoid the conditions that would cause your AFO to be regulated under the CAFO regulations. Funding is available for CNMP development through the USDA's Environmental Quality Incentives Program.

2.2.10 Parts of a CAFO That Are Regulated

CAFO regulations apply to both the production areas and land application areas CAFO. The production areas include all areas where you confine animals, store manure and raw materials, and contain wastes.

- Examples of areas where you might confine animals are open lots, housed lots, feedlots, confinement houses, stall barns, free stall barns, milkrooms, milking centers, cowyards, barnyards, exercise yards, medication pens, walkers, animal walkways, and stables.
- Examples of areas where you might store manure are lagoons, runoff ponds, storage sheds, stockpiles, manure pits, liquid impoundments, static piles, and composting piles.
- Examples of areas where you might store raw materials are feed silos, silage bunkers, and storage areas for bedding materials.
- Examples of areas where you might contain wastes are lagoons, holding ponds, and evaporation ponds that you use to control runoff of rainwater from your animal confinement and manure storage areas.

An egg-washing or egg-processing facility is part of the production area. Any area where you store, handle, treat, or dispose of dead animals is also part of the production area.

Land application areas that are covered by CAFO regulations include any land under your control where you apply or might apply manure, litter, or process wastewater. Land is under your control if you own, rent, or lease it, regardless of whether it is adjacent to the production area or at a different site.

2.2.11 Federal NPDES Permit

Note: In Section 2.2.1, we introduced and discussed general requirements covered under NPDES. In the following sections, we discuss the actual procedures and practices involved with CAFO NPDES compliance.

The forms needed to apply for an NPDES permit are obtained from the permitting authority. Under the federal NPDES regulations, two kinds of permits are possible: general permits and individual permits. Each permitting authority adopts its own rules about what types of permits operations need, so contacting the permitting authority is necessary.

2.2.11.1   NPDES General Permit

An NPDES general permit has one set of requirements for a group of facilities. For example, all CAFOs or all poultry CAFOs in a particular area, such as an entire state or a watershed within the state, might be covered under one general permit. The permitting authority sets the permit conditions, issues a draft permit, and requests comments from the public. The permitting authority makes changes to the draft permit based on the public comments and then issues the final permit. The general permit specifies what kinds of operations can be covered. Owners and operators of eligible operations can then apply for coverage under the permit.

If an NPDES general permit is available in the state and the operation meets the eligibility requirements, a notice of intent (NOI) must be filled out and submitted to the permitting authority to apply for coverage under the general permit. The general permit explains how to apply for coverage and when coverage will become effective.

2.2.11.2   NPDES Individual Permit

An NPDES individual permit contains requirements designed specifically for one CAFO. An NPDES individual permit must be applied for if:

- A general NPDES permit is not available.
- The CAFO is not eligible to be covered under the general NPDES permit.
- The CAFO owner or operator wants an individual NPDES permit.
- The permitting authority requires an individual permit.

To apply for an individual permit, either NPDES Forms 1 and 2B or similar forms required by the state must be filled out. The forms must be completed and submitted to the permitting authority. When the permitting authority receives the permit application, it uses the information submitted to draft a permit for the operation. The permitting authority bases the permit requirements on the unique conditions at the particular operation. After a public comment period on the draft permit, the permitting authority modifies the draft, if necessary, and then issues a final NPDES individual permit.

2.2.11.3   Information to Include in the NOI or Permit Application

When applying for a general or individual NPDES permit, the following information must be given to the permitting authority:

- The name of the CAFO's owner or operator
- The CAFO's location and mailing address
- The latitude and longitude of the entrance to the CAFO's production area
- A topographic map of the area where the CAFO is located, with the location of the production area specifically marked

- *Important point:* Check the USEPA's web site at http://cfpub.epa.gov/npdes/stormwater/latlong.cfm to find out how to determine the latitude and longitude and where to get a topographic map for your location.

- The number of each kind of animal in confinement
- The kinds of structures used to contain or store manure, litter, and process wastewater and the total amount that each structure can store
- The total number of acres under the control of the CAFO that are available for land application of manure, litter, and process wastewater
- An estimate of the amount (tons or gallons) of manure, litter, and process wastewater the operation generates each year
- An estimate of the amount (tons or gallons) of manure, litter, and process wastewater transferred to other persons each year

Permit applications submitted after December 31, 2006, must also contain a statement certifying that a nutrient management plan has been developed and will be implemented. A current nutrient management plan must be in place for as long as the operation is covered by an NPDES permit.

The items listed above are the minimum that must be submitted. The permitting authority may require submission of additional information.

2.2.11.4 Permit Application Deadline

The permit application deadline depends on whether the operation is an existing CAFO, a newly defined CAFO, a new discharger, or a new source or has been designated as a CAFO by the permitting authority. Each of these categories has a different deadline for applying for an NPDES permit. Read the descriptions to determine when application for an NPDES permit is necessary.

You are responsible for applying for NPDES permit coverage for your CAFO. The federal regulations do not require your permitting authority to notify you that you must apply. For an individual permit, the permitting authority issues a permit after it receives a permit application from the facility seeking coverage. For a general permit, the permitting authority issues the general permit, and then operators submit NOIs to be covered under the permit. In both instances, the permitting authority is required to provide public notification that a permit has been drafted. In addition, although permitting authorities are not required to do so, many are likely to conduct outreach to communicate who must obtain a permit and how to do so. Ultimately, however, the responsibility to seek permit coverage lies with the CAFO owner or operator. Failure to seek coverage by the permitting deadlines described below could result in liability under the CWA, and penalties may apply.

2.2.11.4.1 Existing CAFOs

Existing CAFOs are operations that were defined as CAFOs under the 1976 NPDES CAFO regulations. If you operate an existing CAFO, you should already have an NPDES permit. You must reapply for a new permit 180 days before your existing permit expires, unless your permit indicates otherwise. Existing operations that appropriately claimed the 25-year, 24-hour storm permit exemption under the 1976 regulations would have until no later than February 13, 2006, to apply for a permit. See Table 2.13 to figure out whether your operation was previously regulated. If your CAFO was covered under the 1976 NPDES CAFO regulations but you don't have an NPDES permit, you must immediately apply for an NPDES permit.

2.2.11.4.2 Newly Defined CAFOs

Newly defined CAFOs are operations that are defined as CAFOs as of April 14, 2003 (the effective date of the revised regulations) but were not defined as CAFOs under the old NPDES regulation. Your operation might be a newly defined CAFO if it is a dry-waste chicken operation, a stand-alone dairy heifer operation, or a swine nursery that existed before April 14, 2003. Your operation might also be a newly defined CAFO if you were entitled to the 25-year, 24-hour storm permitting exemption under the old regulation. That exemption has been eliminated. Table 2.14 shows which operations are newly defined CAFOs.

TABLE 2.13
Size Category Thresholds for Existing CAFOs

| Sector | Existing CAFOs (covered under the 1976 NPDES CAFO regulations) | |
	Large	Medium
Slaughter and feeder cattle	1,000 or more	300–999
Mature dairy cows	700 or more	200–699
Swine (55 pounds or more)	2,500 or more	750–2,499
Horses	500	150–499
Sheep or lambs	10,000 or more	3,000–9,999
Turkeys	55,000 or more	16,500–54,999
Laying hens or broilers (continuous overflow watering)	100,000 or more	30,000–99,999
Laying hens or broilers (liquid manure handling system)	30,000 or more	9,000–29,999
Ducks	5,000 or more	1,500–4,999

Source: USEPA (2004).

2.2.11.4.3   New Dischargers

New dischargers are operations that met the CAFO definition after the revised regulations went into effect (after April 14, 2003) but are not new sources. Your operation might be a new discharger, for example, if it is a newly constructed medium CAFO, because medium and small CAFOs in most animal sectors are never defined as new sources (see Section 2.2.11.4.4, "New Sources"). It might also be a new discharger if it is an existing AFO and you increase the number of animals or otherwise change the operation so that it meets the CAFO definition. Three different permit application deadlines have been set for new dischargers:

1. If you build a new operation that is not subject to the ELGs (for example, it meets the definition of a medium CAFO or it confines animals other than the types covered by the ELGs), you must apply for an NPDES permit at least 180 days before you begin to operate your new CAFO.
2. If you increase the number of animals or make other changes at your operation so that it meets the definition of a CAFO, and the CAFO is not in a newly defined sector (see Table 2.14), you have 90 days after you make the change to your operation to apply for an NPDES permit.
3. If you increase the number of animals or make other changes at your operation so that it meets the definition of a CAFO, but the changes you make would not have made your operation a CAFO under the old regulations, you have until April 13, 2006, or 90 days after you make the changes at your operation, whichever is later, to apply for an NPDES permit. For example, your operation would fit this description if you're increasing the number of animals so that it will become a CAFO and the CAFO is in a newly defined sector (see Table 2.14).

2.2.11.4.4   New Sources

A large CAFO is a new source if construction began after April 14, 2003, on a site where no other source is located. An operation may also be a new source if it expands its operations, specifically if

TABLE 2.14
Size Category Thresholds for Newly Defined CAFOs

Sector	Newly defined	
	Large	Medium
Swine (less than 55 pounds)	10,000 or more	300–9,999
Laying hens (operations that do not have liquid manure handling systems)	82,000 or more	25,000–81,999
Chickens other than laying hens (operations that do not have liquid manure handling systems)	125,000 or more	37,500–124,999
Dairy heifers	1,000	300–999

Source: USEPA (2004).

the process or production equipment is totally replaced or if it adds new processes that are substantially independent of an existing source at the same site.

In most cases, only large CAFOs can be considered new sources. The term *new source* is used only in connection with facilities that are subject to New Source Performance Standards (NSPS), and in most cases, only large CAFOs are subject to the CAFO NSPS (see 40 CFR Part 412). For most animal sectors, a newly constructed operation that is either a medium or small CAFO is a new discharger rather than a new source.

- *Important point:* For duck CAFOs, operations with 5,000 or more birds are subject to NSPS if they meet the new source definition. This threshold corresponds to large duck CAFOs with liquid manure handling systems and large, medium, and some small duck CAFOs with other than liquid manure handling systems.

Example of new sources:

- A brand-new large swine CAFO that is constructed where no CAFO previously existed.
- A 500-head dairy AFO that expands to add 3,000 mature dairy cattle and includes new construction that will replace the existing milking and manure handling equipment.
- An existing 75,000-bird turkey CAFO that expands to add a 7,000-bird, wet lot duck CAFO with a separate waste handling system. In this case, the permit would continue to apply to the turkey facilities and would add new source requirements for the duck lot.

If you plan to own or operate a new source CAFO, you must apply for a permit at least 180 days before you begin to operate the CAFO.

2.2.11.4.5   Designated CAFOs
Designated CAFOs are small and medium AFOs that the permitting authority has designated as CAFOs. If your permitting authority has notified you that it has designated your operation as a CAFO, you must apply for a permit within 90 days after receiving the notice.

2.2.11.4.6   NPDES Permit Expiration Dates
Individual NPDES permits are usually written for 5-year terms and are reissued every 5 years.

General NPDES permits also are usually written for 5-year terms. Because a general NPDES permit is created for multiple permittees, however, it could have been issued several years before the NOI was submitted. In this case, the general NPDES permit might expire less than 5 years after submission of the NOI.

To reapply for a permit when it is due to expire, a new application form (for an individual permit) or a new NOI (to be covered under a general permit) must be submitted 180 days before the permit's expiration date. If this deadline has been met and the permitting authority fails to reissue the NPDES permit before the expiration date, the current NPDES permit remains in effect until the permitting authority acts on the new application.

Some permitting authorities might have other deadlines or procedures for reissuing CAFO NPDES permits. For example, some general permits are automatically continued without submitting a new NOI. Check the reapplication procedures specified in the permit, and contact the permitting authority to find out exactly what must be done to get a new permit before the current permit is due to expire.

An NPDES permit must be in effect for an operation as long as it is an operating CAFO. Under only a few situations can NPDES permit coverage be discontinued:

- The operation is closed.
- The operation is permanently changed so that it no longer meets the definition of a CAFO.
- The operated is changed so that it cannot and will not discharge. In this case, a "no potential to discharge" determination from the permitting authority is needed before NPDES permit coverage can be discontinued.

Under all circumstances, an NPDES permit must be in effect until all manure, litter, and process wastewater generated at the CAFO is properly disposed of. If an operation still has the potential to discharge when its permit is due to expire, the permit must be reapplied for. Once the manure, litter, and process waters have been properly disposed of, the CAFO can petition the permitting authority to terminate the permit.

2.3   STATE REGULATIONS

States play a key role in the regulation of CAFOs through their delegated NPDES permitting authority under the CWA. Simply, state regulatory agencies with authorized NPDES programs are principally responsible for implementing and enforcing the 2003 Final CAFO Rule. The requirements of the CAFO Rule are implemented by issuing NPDES permits. The Rule is implemented by states with authorized NPDES permit programs for CAFOs. Currently, 45 states and one territory have authorized CAFOs. In states without an authorized NPDES program for CAFOs and in Indian country, the USEPA implements the rule.

State planning and zoning enabling legislation plays the predominant role in defining what police powers local governments may or may not use in regulating land use. In many states, that legislation exempts agriculture from regulation through county zoning ordinances. To the consternation of many family farm activists, communities, and environmental groups, such exemptions have allowed CAFO owners a level of freedom in siting their facilities.

These exemptions for agriculture, for the most part, "originated in county zoning enabling legislation dating back to the 1950s as a result of a long standing history of trust and good will toward family farms from society in general" (Schwab, 1998, p. 27). Note, however, that most of these exemptions were enacted during a period "when modern confinement operations did not factor into legislator's perceptions of the beneficiaries of this generosity" (Schwab, 1998, p. 28). Their assumption was, instead, that most family farmers could be expected to be go about their business in a responsible manner—their actions would not have a major impact on their rural neighbors or the environment. While this may be a pie-in-the-sky viewpoint, a realistic desire existed to facilitate the survival of small farms for the good of rural society.

Although a complete inventory of various state approaches to all the issues posed by CAFOs is beyond the scope of this text, planners can check their own state regulations and applicable state environmental regulations. What follows are brief highlights of right-to-farm laws and their impact on protecting farmers from nuisance suits, as long as those farmers are engaged in widely accepted, conventional farming practices.

2.3.1 RIGHT-TO-FARM LAWS

In addition to providing farm families with a psychological sense of security that farming is a valued and accepted activity in their communities, right-to-farm laws are intended to discourage nonfarming neighbors from suing farmers. More specifically, right-to-farm laws were originally designed to protect agricultural operations existing within a state or within a given area of state by allowing owners or operators of those operations who meet the legal requirements of the right-to-farm law a defense to nuisance suits that might be brought against the operation. Most laws include a number of additional protections. Right-to-farm provisions may also be included in state zoning enabling laws, and farmers with land enrolled in an agricultural district may have stronger right-to-farm protection than other farmers. A growing number of counties and municipalities are passing their own right-to-farm legislation to supplement the protection provided by state law.

Right-to-farm statutes were originally developed in the 1970s as state lawmakers were becoming more aware of and concerned about the loss of agricultural land. During this period in history, many different conflicts occurred in potential uses of agricultural land, and the rising tide of urban encroachment into traditional agricultural areas was recognized as a critical issue. The common law of nuisance forbids individuals from using their property in a way that causes harm to others. A private nuisance refers to an activity that interferes with an individual's reasonable use or enjoyment of his or her property. A public nuisance is an activity that threatens the public health, safety, or welfare or damages community resources, such as public roads, parks, and water supplies. Nonagricultural persons were beginning to move into traditional agricultural areas and with them came new complaints concerning unavoidable aspects of agriculture, such as complaints concerning odor, flies, dust, noise from field work, spraying of farm chemicals, slow-moving farm machinery, and other necessary by-products of farming operations.

A successful nuisance lawsuit results in an injunction, which stops the activity causing the nuisance, provides monetary compensation, or both. In a private nuisance lawsuit involving complaints against a farming operation, the court must decide whether the farm practices at issue are unreasonable. To make this decision, courts generally weigh the importance of the activity to the farmer against the extent of harm to the neighbor or community. If the farm operation is found to be a nuisance, as mentioned, courts have the option of closing the operation, altering the way it conducts its business, or assessing penalties to compensate the neighboring landowner for the nuisance. Sometimes, even if a lawsuit fails, the cost of defending against the suit could threaten or even close the farming operation.

From state to state, the state statutes are strikingly similar. They attempt to limit the circumstances under which agricultural operations can be deemed nuisances. They do this by mentioning the need to conserve and protect agricultural land and encouraging, developing, and improving agricultural land for food production. Most states mention the fact that, as nonagricultural land uses have extended into agricultural area, increases in nuisance suits have occurred. In addition to citing the potential loss of agricultural operations, some states also mention the potential for problems in investments being made in farm improvements with exposure to nuisance litigation.

- *Interesting point:* Preliminary results from a state-by-state survey indicate that laws or right-to-farm acts that protect farms from nuisance suits over their operations exist in 43 states. Twelve states have specific laws protecting farmers from "takings" by government or neighbors (Edelman & Warner, 1999).

2.3.1.1 Types of Right-to-Farm Laws

Several types of right-to-farm laws exist: traditional, laws requiring generally accepted agricultural management practices, laws protecting specific types of agricultural activities, laws protecting feedlots, and laws protecting operations located within agricultural districts.

Traditional right-to-farm laws protect an agricultural operation if it has been in existence for 1 year prior to a change in the surrounding area that has given rise to the nuisance claim. Agricultural activities classified as a nuisance when the activities began and activities that are negligently or improperly conducted are not protected under traditional right-to-farm laws.

Some right-to-farm laws require the use of generally accepted agricultural management practices (GAAMPs) in order to be protected from nuisance litigation. These laws usually create a presumption of reasonableness on the part of an operation if standard practices are followed. GAAMPs are similar to best management practices. The outstanding issues involved when a state chooses to use the GAAMPs approach is the question of who establishes the GAAMPs. Some state laws require the state department of agriculture to set those standards. Other laws are silent on who establishes the standards. Silence on this issue leaves the farmer to, in litigation on the nature of the operation, place into evidence information concerning what the standard or acceptable practice might be and information that will support that he or she followed those practices.

- *Important point:* Some laws reflect that if an operation conforms with federal, state, and local laws and regulations concerning agricultural practices of permit requirements, the agricultural practice is a good agricultural practice when no adverse effects on public health or safety occur.

In some states, the Agricultural Commissioner establishes the GAAMPs for the state, presumably by rule or regulation. Some state statutes require the Commissioner to take into consideration information from the extension service, colleges of agriculture, and other relevant entities. In addition, some states require that the farmer cooperate with the NRCS and the state department of natural resources or other industry organizations with roles in establishing acceptable standards for the agricultural industry.

In still other states, right-to-farm laws list specific agricultural activities that are protected from nuisance litigation. Examples or agricultural by-product creations might include odor from livestock, manure, fertilizer, feed, noise from livestock or farm equipment used in the normal fashion, dust created during plowing or cultivation operation, use of chemicals if in conformity with established practices, and water pollution from livestock or crop production.

Animal feedlots are specifically protected in some states, particularly if the problems complained about are odor- or waste-related. Most nuisance suits brought against agricultural operations involve odors from animal feeding or some question concerning the handling of waste. For example, Iowa's law defines "feedlots" and offers protection to activities occurring in relation to those feedlots. Other states offering specific protection to animal feedlots are Oklahoma, Wyoming, Tennessee, and Kansas.

Finally, some right-to-farm laws require that, for an agricultural operation to have protection, the operation must be located within an acknowledged and approved agricultural district. These laws are usually part of a broader farmland preservation statutory program. For example, in Iowa, to form an agricultural district, farmers within that district must agree to restrictions on converting their land to nonagricultural uses for a period of time. The districts are created by a local county board after being petitioned by a group of farmers for the creation of the agricultural district. Some state laws grant absolute protection from nuisance suits for operations conducted within the confines of a properly created agricultural district. These types of laws exist in Delaware, Illinois, Iowa, Maryland, Minnesota, Ohio, Oregon, Virginia, and Wisconsin.

- *Important point:* Although we usually think of right-to-farm laws as having been created at the state level, some localities have passed specific right-to-farm ordinances. Some states allow local protections, but other states do not give local governments the power to regulate agricultural operations at any level.

2.3.1.2 Right-to-Farm Laws: Attributes

Most right-to-farm laws require that the farming operation must have been in existence before any change in the surrounding area occurred. "Changes in the surrounding area" usually refers to development in the area, someone moving in, a private business being opened, or another activity. Some laws require that an "established date of operation" be set. This date is the date upon which agricultural activities began on the site. If the operation should expand or change its operations in significant ways, a new established date of operation may be set. States usually require the agricultural operation to have been in existence at least 1 year before the change in the surrounding neighborhood. Some laws also require unchanged operation for more than 1 year, whereas other laws require only a prior existence with no specific time requirements.

Another pivotal feature required to obtain and keep protection is that the farm and farming practices must not have significantly changed. The change must have occurred in the surrounding neighborhood. Most right-to-farm protection has been given to operations that can point to change that occurred in the surrounding neighborhood while the farming operation remained unaffected. If the farming operation is changing, either in size or farming methods used, the protection from right-to-farm statures may be lost. Similarly, if an operation expands or adopts changes in technology, it is likely to lose its protected status. Questions predictably arise when an operation expands or uses a changed technology on the farm without necessarily incorporating any expansion. States have begun passing laws addressing these issues. Those laws may require:

- A new time period to run after each expansion
- That a "reasonable" expansion will not affect the original established date of operation so long as significant differences in environmental pressures on neighbors and livestock have not occurred
- That the operation ensure that its waste handling capabilities will not exceed the minimum recommendation of the extension service
- That complete relocation of the operation has not occurred

These new provisions:

- Allow expansions but give each expansion a separate established date of operation
- Provide no protection for expanded operations
- Provide no protection if the operation substantially increases in size
- Provide no change in established date, even if expansions or adoption of new technology has occurred

In other words, the states are all over the map on whether and to what extent a change in established date of operation will occur with expansion or adoption of technology on the farming site.

Most laws require that the farming operation be run in a reasonable manner. The operation cannot be handled in a negligent or improper manner. The problem then becomes answering the age-old question of what is reasonable and proper. What is reasonable and proper to one farmer may not be reasonable and proper to another farmer, to the extension service or other agricultural professional, or to the nonfarming community. Water pollution and erosion are usually not protected by right-to-farm laws. Most laws do not allow farmers to hide behind right-to-farm laws if they are conducting operations that are causing or may cause water pollution and soil erosion. In addition, most right-to-farm laws require operations to be in compliance with all relevant local laws and regulations, which can include zoning ordinances and waste disposal rules.

- *Important point:* While right-to-farm laws offer the farmer a defense in nuisance suits, the laws do not protect the farmer from a suit being filed. Some states are enacting statues that shift the costs and attorney fees onto the person who brings the nuisance suit if they are unsuccessful in proving their case. These statues are called *fee-shifting statutes.* These types of statutes can offer an additional deterrent to the bringing of nuisance suits against agricultural operations.

2.3.2.3 Right-to-Farm Laws: Criticisms

Most right-to-farm laws could benefit from improvement in definition of terminology and in clarity of purpose and language. For example, do current large CAFOs qualify as agricultural operations according to the framers' intentions? The agricultural community is still not well versed in the mechanism for usage of a right-to-farm statute, preferring to think of the statutes as general blanket protection for all agricultural activities; however, the statutes were never intended to be applied in that manner.

2.3.2.4 Right-to-Farm Laws: Case Law

As of September 1998, only a few dozen reported cases concerning interpretation of right-to-farm laws had appeared in the casebooks. While the number of reported cases has increased over time, still relatively few cases are on the books. Whether this phenomenon indicates that the protections offered agricultural operations under right-to-farm laws serve as a deterrent against unsubstantiated nuisance claims, or whether a rising number of ongoing nuisance claims against agricultural operations are in progress, the claims are either not going on to appellate courts for eventual reporting or are being settled out of courts. What will happen is still in question.

Critics of CAFOs, however, have continued to question how much legal protection these facilities ought to enjoy if their impacts far exceed those typically associated with a family farm. "As agriculture has become more industrialized, it was probably inevitable that legal challenges to right-to-farm laws would follow" (Schwab, 1998, p. 31).

- *Important point:* Among the reported cases, the courts have found that the right-to-farm protection does not apply if the activity in question was simply not covered specifically by the right-to-farm statute, if the nuisance resulted from changes in the farm, if the neighbors were already present during and before the complained-of activity, if the activity in question was not an agricultural activity, if the GAAMPs were not being followed, or if the operation was being conducted in an improper manner.

2.3.2.4.1 The Bormann Case

The shortage of reported cases in the right-to-farm area came to a complete halt with the Iowa Supreme Court's decision in *Bormann v. Board of Supervisors in and for Kossuth County, Iowa.* On September 23, 1998, the Iowa Supreme Court handed down a decision in *Bormann* that held unconstitutional a provision of the Iowa right-to-farm statutes. The provision allowed right-to-farm protections in properly designated "agricultural areas."

Because this case directly affects the subject matter we are addressing in this book, and thus could affect the user of this book, we have included the text of the case findings here.

Bormann v. Board of Supervisors in & for Kossuth County, Iowa (1998)

Opinion

In this appeal we are asked to decide whether a statutory immunity from nuisance suits results in a taking of private property for public use without just compensation in violation of federal and Iowa constitutional provisions. We think it does. We therefore reverse a district court ruling holding otherwise and remand. In doing so, we need not reach a second constitutional challenge.

I. Facts and Proceedings

The facts are not in dispute. In September 1994, Gerald and Joan Girres applied to the Kossuth County Board of Supervisors for establishment of an "agricultural area" that would include land they owned as well as property owned by Mike Girres, Norma Jean Thul, Gerald Thilges, Shirley Thilges, Thelma Thilges, Edwin Thilges, Ralph Reding, Loretta Reding, Bernard Thilges, Jacob Thilges, John Goecke, and Patricia Goecke (applicants). See Iowa Code § 352.6 (1993). The real property involved consisted of 960 acres. On November 10, 1994, the Board denied the application, making the following findings and conclusions:

a. The Board finds that the policy in favor of agricultural land preservation is not furthered by an Agricultural Area designation in this case as there are no present or foreseeable nonagricultural development pressures in the area for which the designation is requested.

b. The Board also finds that the Agricultural Area designation and the nuisance protections provided therein will have a direct and permanent impact on the existing and long-held private property rights of the adjacent property owners.

c. Thus, the Board concludes that the policy in favor of agricultural land preservation as set forth in Iowa Code chapter 352 is outweighed by the policy in favor of the preservation of private property rights.

d. Accordingly, the Board finds that the adoption of the Agricultural Area designation in this case in inconsistent with the purposes of Iowa Code chapter 352.

Two months later, in January 1995, the applicants tried again with more success. The Board approved the agricultural area designation by a 3-2 vote—one of which was based on the "flip [of] a nickel." In granting the designation, the Board this time found that the "application to create the agricultural area is consistent with the purposes of Chapter 352."

In April 1995, several neighbors of the new agricultural area filed a writ of certiorari and declaratory judgement action in district court. The defendants were the Board and individual board members Joe Rahm, Al Dudding, Laurel Fatz, James Black, and Donald McGregor (Board).

The plaintiffs, Clarence and Caroline Bormann and Leonard and Cecilia McGuire (neighbors), challenged the board's action in a number of respects. The neighbors alleged the Board's action violated their constitutionally inalienable right to protect property under the Iowa Constitution, deprived them of property without due process or just compensation under both the federal and Iowa Constitutions, denied them due process under the federal and Iowa Constitutions, ran afoul of res judicata principles, and was "arbitrary and capricious." The applicants intervened.

Based on stipulated facts, memoranda and oral argument, the district court determined that the Board's action was "arbitrary and capricious." The neighbors then sought, and received, a certification of appeal from this court.

II. Scope of Review

The neighbors sued at law and titled their petition as one for writ of certiorari and one for declaratory judgment. In the petition for writ of certiorari, the neighbors asked that a writ of certiorari issue because the Board's decision was "in access of" the Board's "jurisdiction" and was "contrary to law" and "illegal" because the decision "violates the Fifth Amendment to the United States Constitution, and article I, section 18 of the Iowa Constitution" in that the decision "effects a taking of the [neighbors'] private property for a use that is not public." The petition asked that the decision be annulled and decreed to be void.

In the petition for declaratory relief, the neighbors sought a declaration that the Board's decision violates the "Fifth Amendment to the United States Constitution, the Fourteenth Amendment to the United States Constitution, and article I, section 18 of the Iowa Constitution."

Iowa Rule of Civil Procedure 306 authorizes the district court to issue a writ of certiorari "where an inferior tribunal, *board or officer* exercising judicial functions, is alleged to have exceeded its, or his proper jurisdiction or otherwise acted illegally." Our scope of review is limited to sustaining a board's decision or annulling it in whole or in part. *Grant v. Fritz* (Emphasis added.) (Iowa 1972). In addition, the fact that the plaintiff has another remedy does not preclude granting the writ.

Thus, here, a petition for a writ of certiorari is appropriate to test the legality of the Board's decision. Our scope of review is limited to sustaining the Board's decision or annulling it in whole or in part. In addition, the fact that the neighbors may have another adequate remedy, like declaratory judgment, does not preclude our granting relief under Rule 306.

Iowa Rule of Civil Procedure 261 (declaratory judgement) authorizes "[c]ourts of record within their respective jurisdiction [to] declare rights, status, and other legal relations whether or not further relief is or could be claimed."

The purpose of a declaratory judgment is to determine rights in advance. *Miehls v. City of Independence* (1958). The essential difference between such an action and the usual action is that not actual wrong need have been committed or loss incurred to sustain declaratory judgment relief. But there must be no uncertainty that the loss will occur or that the right asserted will be invaded. As with a writ of certiorari, the fact that the plaintiff has another adequate remedy does not preclude declaratory judgment relief where it is appropriate.

We think the facts here are sufficient for us to proceed under either remedy. In addition, because the facts are not in dispute, we need not concern ourselves with whether we employ a correction-of-errors-at-law review or a de novo review. Our only question is a legal one.

III. The Takings Challenge

A. The Parties' Contentions

The Board's approval of the agricultural area here triggered the provisions of Iowa Code section 352.11(1)(a). More specifically, the approval gave the applicants immunity from nuisance suits. The neighbors contend that the approval with the attendant nuisance immunity results in a taking of private property without the payment of just compensation in violation of federal and state constitutional provisions.

The neighbors concede, as they must, that their challenge to section 352.11(1)(a) is a facial one because the neighbors have presented neither allegations nor proof of nuisance. However, the neighbors strenuously argue that in a facial challenge context courts have developed certain bright-line tests that spare them from this heavy burden. Specifically, the neighbors say, these bright-line tests provide that a governmental action resulting in the condemnation or the imposition of certain specific property interests constitutes automatic or per se takings.

Here, the neighbors argue further, that the section 352.11(1)(a) immunity provision gives the applicants the right to create or maintain a nuisance over the neighbor's property, in effect creating an easement in favor of the applicants. The creation of the easement, the neighbors conclude, results in an automatic or per se taking under a claim of regulatory taking.

The Board and applicants respond that a per se taking occurs only when there has been a permanent physical invasion of the property or the owner has been denied all economically beneficial or productive use of the property. They insist the record reflects neither has occurred. Thus, they contend, the court must apply a balancing test enunciated in *Penn Cent. Transp. Co. v. City of New York* (1978). They argue that under that balancing test the neighbors lose.

B. The Relevant Constitutional and Statutory Provisions

1. The Constitutional Provisions

The Fifth Amendment to the Federal Constitution pertinently provides that "[n]o person shall be ... deprived of life, liberty, or property without due process of law; nor shall private property be taken for public use, without just compensation." The Fourteenth Amendment to the Federal Constitution prohibits a state from "depriving any person of life, liberty, or property without due process of law." The Fourteenth Amendment makes the Fifth Amendment applicable to the states and their political subdivisions. *Chicago B. & Q.R.R. v. City of Chicago* (1897).

Article I, section 9 of the Iowa Constitution pertinently provides that "no person shall be deprived of life, liberty, or property, without due process of law." Article I, section 18 of the Iowa Constitution provides:

Eminent domain-drainage ditches and levees. Private property shall not be taken for public use without just compensation first being made, or secured to be made to the owner thereof, as soon as the damages shall be assessed by a jury.

2. The Statutory Provisions

Iowa Code section 352.6 sets forth the procedure for obtaining an agricultural area designation. The application is to the county board of supervisors. This provision also prescribes the conditions under which a county board of supervisors may designate farmland as an agricultural area. *Id.* An agricultural area includes, among other activities, raising and storing crops, the care and feeding of livestock, the treatment or disposal of wastes resulting from livestock, and the creation of noise, odor, dust, or fumes. Iowa Code § 352.2(6).

Iowa Code section 352.11(1)(a) provides the immunity from nuisance suits:

A farm or farm operation located in an agricultural area shall not be found to be a nuisance regardless of the established date of operation or expansion of the agricultural activities of the farm or farm operation. This paragraph shall apply to a farm operation conducted within an agricultural area for six years following the exclusion of land within an agricultural area other than by withdrawal as provided in section 351.9.

The immunity does not apply to a nuisance resulting from a violation of a federal statute, regulation, state statute, or rule. Iowa Code § 352.11(1)(b). Nor does the immunity apply to a nuisance resulting from the negligent operation of the farm or farm operation. *Id.* Additionally, there is no immunity from suits because of an injury or damage to a person or property caused by the farm or farm operation before the creation of the agricultural area. *Id.* Finally, there is no immunity from suit "for an injury or damage sustained by the person [bringing suit] because of the pollution or change in condition of the waters of a stream, the overflowing of the person's land, or excessive soil erosion into another person's land, unless the injury or damage is caused by an act of God." *Id.*

Iowa Code section 657.1 defines nuisance and provides for civil remedies:

Whatever is injurious to health, indecent, or unreasonably offensive to the senses, or an obstruction to the free use of property, so as essentially to unreasonably interfere with the comfortable enjoyment or life of property, is a nuisance, and a civil action by ordinary proceedings may be brought to enjoin and abate the same and to recover damages sustained on account thereof.

Iowa Code section 657.2 is a laundry list of the conduct or conditions that are deemed to be a nuisance. Those that are relevant to nuisances resulting from farming and farm operations include:

1. The erecting, continuing, or using any building or other place for the exercise of any trade, employment, or manufacture, which, by occasioning noxious exhalations, unreasonably offensive smells, or other annoyances, becomes injurious and dangerous to the health, comfort, or property of individuals or the public.

2. The causing or suffering any offal, filth, or noisome substance to be collected or to remain in any place to the prejudice of others.

...

4. The corrupting or rendering unwholesome or impure the water of any river, stream, or pond, or unlawfully diverting the same from its natural course or state, to the injury or prejudice of others.

Iowa Code § 657.2.

Our cases recognize that the statutory definition of nuisance does not "modify the common-law's application to nuisances." *Weinhold v. Wolff* (Iowa 1996). Rather, the statutory provisions "are skeletal in form, and [we] look to the common law to fill in the gaps." *Id.*

There are two kinds of nuisances: public and private. We cited the differences between the two in *Guzman v. Des Moines Hotel Partners*:

A public or common nuisance is a species of catchall criminal offenses, consisting of an interference with the rights of a community at large. This may include anything from the obstruction of a highway to a public gaming house or indecent exposures. A private nuisance, on the other hand, is a civil wrong based on a disturbance of rights in land. ... The essence of a private nuisance is an interference with the use and enjoyment of land. Examples include vibrations, blasting, destruction of crops, flooding, pollution, and disturbance of the comfort of the plaintiff, as by unpleasant odors, smoke, or dust.

(citations omitted). We are dealing here with private nuisances.

To fully understand the issues we about to discuss, we think it would aid our analysis to distinguish between the concepts of "private nuisance" and "trespass." We made this distinction in *Ryan v. City of Emmetsburg*:

As distinguished from trespass, which is an actionable invasion of interests in the exclusive possession of land, a private nuisance is an actionable invasion of interests in the use and enjoyment of land. Trespass comprehends an actual physical invasion by tangible matter. An invasion which constitutes a nuisance is usually by intangible substances, such as noises or odors.

232 Iowa (1942).

In *Ryan*, we also distinguished between the concepts of "nuisance" and "negligence." Negligence is a type of liability-forming conduct, for example, a failure to act reasonably to prevent harm. *Id.* In contrast, nuisance is a liability-producing condition. *Id.* Negligence may or may not accompany a nuisance; negligence, however, is not an essential element of nuisance. *Id.* If the condition constituting the nuisance exists, the person responsible for it is liable for resulting damages to others even though the person acted reasonably to prevent or minimize the deleterious effect of the nuisance. *Id.*

C. The Framework of Analysis

As the neighbors point out, the federal and state constitutional provisions we set out earlier provide the following framework for a "takings" analysis: (1) Is there a constitutionally protected private property interest at stake? (2) Has this private property interest been "taken" by the government for public use? and (3) If the protected property interest has been taken, has just compensation been paid to the owner? The neighbors contend there is a constitutionally protected private right which the Board has taken from them without paying just compensation. That taking, the neighbors contend, results from the Board's approval of the agricultural area triggering the nuisance immunity in section 352.11(1)(a). The Board and the applicants concede the neighbors have received no compensation so we need not concern ourselves with the third step of the analysis: Has just compensation been paid to the owner?

1. Is There a Constitutionally Protected Private Property Interest at Stake?

a. Does the Immunity Provision in Section 352.11(1)(a) Against Nuisance Suits Create a Property Right?

Textually, the federal and Iowa Constitutions prohibit the government from taking property for public use without just compensation. Property for just compensation purposes means "the group of rights inhering in the citizens' relation to the physical thing, as the right to possess, use and dispose of it." *United States v. General Motors Corp.* (1945). In short, property for just compensation purposes includes "every sort of interest the citizen may possess." *Id.; see also Liddick v. Council Bluffs* (1942) ("[P]roperty is not alone the corporeal thing, but consists also in certain rights therein created and sanctioned by law, of which, with respect to land, the principal ones are the rights of use and enjoyment. ...").

State law determines what constitutes a property right. *Webb's Fabulous Pharmacies, Inc. v. Beckwith* (US, 1980). Thus, in this case, Iowa law defines what is property.

The property interest at stake here is that of an easement, which is an interest in land. Over one hundred years ago, this court held that the right to maintain a nuisance is an easement. *Churchill v. Burlington Water Co.* (Iowa, 1895). *Churchill* defines an easement as a privilege without profit, which the owner of one neighboring tenement [has] of another, existing in respect of their several tenements, by which the servient owner is obliged to suffer, or not do something on his own land, for the advantage of the dominant owner. *Id.*

Churchill's holding that the right to maintain a nuisance is an easement and its definition of an easement are consistent with the *Restatement of Property*.

An easement is an interest in land which entitles the owner of the easement to *use* or enjoy land in the possession of another. ... It may entitle him to do acts which he would otherwise not be privileged to do, or it may merely entitle him to prevent the owner of the land subject to the easement from doing acts which he would otherwise be privileged to do. An easement which entitles the owner to do acts which, were it not for the easement, he would not be privileged to do, is an affirmative easement. ... [The easement] may entitle [its] owner to do acts on his own land which, were it not for the easement, would constitute a nuisance.

Restatement of Property § 451 cmt. a, at 2911-12 (1944) (emphasis added).

Another feature of easements is that easements run with the land:

The land which is entitled to the easement or service is called a dominant tenement, and the land which is burdened with the servitude is called the servient tenement. Neither easements [n]or servitudes are personal, but they are accessory to, and run with, the land. The first with the dominant tenement, and the second with the servient tenement.

Dawson v. McKinnon, 226 Iowa 756, 767, 285 N.W. 258, 263 (1939).

Thus, the nuisance immunity provision in section 352.11.(1)(a) creates an easement in the property affected by the nuisance (the servient tenement) in favor of the applicants' land (the dominant tenement). This is because the immunity allows the applicants to do acts on their own land which, were it not for the easement, would constitute a nuisance. For example, in their farming operations the applicants would be allowed to generate "offensive smells" on their property which without the easement would permit affected property owners to sue the applicants for nuisances. *See* Iowa Code § 352.2(6); *see also Buchanan v. Simplot Feeders Ltd. Partnership* (Wash. 1998) (holding that Washington's Right-to-Farm Act gives farm quasi easement, against urban developments that subsequently locate next to farm, to continue nuisance activities) (dictum).

b. Is an Easement a Protected Property Right Subject to the Requirements of the Just Compensation Clauses of the Federal and Iowa Constitutions?

Easements are property interests subject to the just compensation requirements of the Fifth Amendment to the Federal Constitution. *United States v. Welch* (U.S. 1910). Easements are also property interests subject to the just compensation requirements of our own Constitution. *Simkins v. City of Davenport* (Iowa 1975).

c. Has the Easement Resulted in a Taking?

(1) Takings Jurisprudence, Generally

There are two categories of state action that must be compensated without any further inquiry into additional factors, such as the economic impact of the governmental conduct on the landowner or whether the regulation substantially advances a legitimate state interest. The two categories include regulations that (1) involve a permanent physical invasion of the property or (2) deny the owner all economically beneficial or productive use of the land. *Lucas v. South Caroline Coastal Council* (U.S. 1992). These two categories are what the neighbors term "per se" takings. The per se rule regarding the first category—physical invasion—was firmly established in *Loretto v. Teleprompter Manhattan CATV Corp.* (U.S. 1982)

Presumably, in all other cases involving "regulatory takings" challenges, the United States Supreme Court engages in a case-by-case examination in determining at which point the exercise of the police power becomes a taking. *Id.* This ad hoc approach calls for a balancing test that is essentially one of reasonableness. The test focuses on three factors: (1) the economic impact of the regulation on the claimant's property; (2) the regulation's interference with investment-backed expectations; and (3) the character of the governmental action. *Penn cent. Transp. Co. v. [City of New York]* (U.S. 1978). According to some commentators, a court must first find that the regulation substantially advances legitimate state interests before the court may test the regulation against the three factors in *Penn Central. See, e.g.,* Craig A. Peterson, *Land Use Regulatory "Takings" Revisited: The New Supreme Court Approaches,* 39 Hastings L.J. 335, 351 (1988).

(2) Physical Invasion

The Board and applicants contend the neighbors' argument fails under both categories of per se takings: physical invasion and denial of all economically beneficial or produce use of the property. The neighbors do not contend the record supports a finding that the challenged statute denies them all economically beneficial or productive use of their property. Accordingly, we restrict our discussion to the physical invasion category.

According to one commentator,

[t]he term "regulatory taking" refers to situations in which the government exercises its "police powers" to restrict the use of land or other forms of property. This is often accomplished through implementation of land use planning, zoning and building codes. In contrast, a governmental entity exercises its eminent domain power or acts in an "enter-

prise capacity, where it takes unto itself private resources and uses them for the common good." Where the private landowner will not sell the land, the government entity seeks condemnation of the property and pays a fair purchase price to be determined in court. On the other hand, an inverse condemnation claim is sought by a landowner when the government fails to seek a condemnation action in court.

John W. Shonkwiler & Terry Morgan, *Land Use Litigation* § 1.02, at 6 (1986) [hereinafter Shonkwiler]. The neighbors' challenge here is one of inverse condemnation.

We think it would aid our analysis of the neighbors' takings argument to discuss those cases where a government entity acting in its enterprise capacity has appropriate private property without first exercising its eminent domain power.

(a) Trespassory Invasions of Private Property by Government Enterprise

Generally, when the government has physically invaded property in carrying out a public project and has not compensated the landowner, the United States Supreme Court will find that a per se taking has occurred. *See* Shonkwiler § 10.01(1), at 369. For example, in *Pumpelly v. Green Bay & Mississippi Canal Co.,* the Court held there was a taking where the defendant's construction of a dam, pursuant to state authority, permanently flooded the plaintiff's property (U.S. 1871). In so holding, the Court enunciated the following rule:

[W]here real estate is actually invaded by superinduced additions of water, earth, sand, or other material, or by having any artificial structure placed on it, so as to effectually destroy or impair its usefulness, it is a taking, within the meaning of the constitution.

Id.

In a more recent case, the Court applied the same rule to a state law that authorized third parties to physically intrude upon private property. *Loretto,* 458 U.S. at 432 n.9, 102 S. Ct. at 3174 n.9, 73 L. Ed. 2d at 880 n.9 (holding that a New York statute requiring the owners of apartment buildings to permit cable television operators to install transmission facilities on their property was in violation of the Just Compensation Clause).

(b) Nontrespassory Invasions of Private Property by Government Enterprise

To constitute a per se taking, the government need not physically invade the surface of the land. *See* Shonkwiler § 10.02(2), at 370. For example, in *United States v. Causby,* the Court held that the frequent and regular flights of government planes over the plaintiffs' land had created an easement in the lands for the benefit of the government (U.S. 1946). The plaintiffs owned a small chicken farm near an airport leased by the government for use by army and navy aircraft.

The glide path of one of the runways passed right over the plaintiffs' land at a height of only eighty-three feet. As a result of the aircraft's noise, the plaintiffs had to abandon their commercial chicken operation. *Id.*

The Court held that the flights' interference with the use of the plaintiffs' land constituted a taking of a flight easement that had to be compensated on the basis of diminution in the land's value resulting from the easement. *Id.* at 261-62, 66 S. Ct. at 1066, 90 L. Ed. at 1210. In the course of its opinion, the Court stated:

[T]the flight of airplanes, which skim the surface but do not touch it, is as much an appropriation of the use of the land as a more conventional entry upon it. … The reason is that there [is] an intrusion so immediate and direct as to subtract from the owner's full enjoyment of the property and to limit his exploitation of it. … The superadjacent airspace at this low altitude is so close to that land that continuous invasions of it affect the use of the surface of the land itself. We think the landowner, as an incident to his ownership, has a claim to it and invasions of it are in the same category as invasions of the surface.... Flights over private land are not a taking, unless they are so low and so frequent as to be a direct and immediate interference with the enjoyment and use of the land. We need not speculate on that phase of the present case. For the findings of the Court of Claims plainly establish that there was a diminution in value of the property and that the frequent, low-level flights were the direct and immediate cause. We agree with the Court of Claims that a servitude has been imposed upon the land.

Id. at 265-67, 66 S. Ct. at 1067-68, 90 L. Ed. At 1212-13; *accord Griggs v. Allegheny County*, 369 U.S. 84, 89, 82 S. Ct. 531, 533-34, 7 L. Ed. 2d 585, __(1962); *see also Portsmouth Harbor Land & Hotel Co. v. United States*, 260 U.S. 327, 43 Ct. 13, 67 L. Ed. 287 (1922) (holding that firing, and imminent threat of firing, of navy coastal guns over plaintiff's property imposed a "servitude" upon the plaintiff's land and thus amounted to a taking of some interest for public use); *Dolezal v. City of Cedar Rapids* (Iowa 1973) (recognizing a navigation easement as one that permits free flights over land including those so low and so frequent as to amount to a taking of property); 2A Philip Nichols, *Eminent Domain* § 6.06, at 6-92 (3d rev. ed. 1998) ("Physical invasions of property are not limited to human or even vehicular entry. To the contrary, the majority of cases involve the transmission of smoke, dust, earth, water, sewage or some other agent onto the impacted property. Regardless of the agent, the result of the invasion may be diminution in values of the property, partial or complete (and permanent and temporary) appropriation, or complete destruction.") [hereinafter Nichols].

In *Fitzgerrald v. City of Iowa City* (Iowa 1992), we had occasion to consider a physical invasion claim involving overflying aircraft. As in *Causby*, the plaintiffs in *Fitzgerrald* claimed the overflying aircraft so adversely affected the use and enjoyment of their property that a taking had resulted. We rejected the claim because the plaintiffs had failed to prove a "measurable decrease in market value" due to the overflying aircraft. *Id.* at 665. Nevertheless, we cited *Causby* for the proposition that "[i]n some circumstances, overflying aircraft may amount to a physical invasion." *Id.* We recognized that when interferences with property from overflying aircraft result in a measurable decrease in property market value, a taking has occurred. *Id.* at 663. In such cases, we said "the right to recovery is not for the nuisance that must be endured but for the loss of value that has resulted." *Id.* The loss-in-value measure of damages is what we would ordinarily use in eminent domain cases. *Id.* As mentioned, *Causby* used this same measure of damages.

The United States Supreme Court has allowed compensation for other kinds of interferences short of physical taking or touching of land. *See* William B. Stoebuck, *Condemnation by Nuisance: The Airport Cases in Retrospect and Prospect*, 71 Dick. L. Rev. 207, 220-21 (1967) [hereinafter Stoebuck]. For example, in *United States v. Welch*, the plaintiff had a passage easement over a neighbor's property (U.S. 1910). The passage was the plaintiff's only access to a county road. The government flooded the neighbor's property thereby cutting off the plaintiff's only access to the road. The Court held the plaintiff was entitled to compensation for the easement. *Id.* at 339, 30 S. Ct. at 527, 54 L. Ed. at 789-90. Because the benefited land—plaintiff's property—was not physically touched, this case is "a clear example of condemnation without any physical taking." Stoebuck, at 221; *see Nollan v. California Coastal Comm 'n*, (U.S. 1987) (holding that requiring property owner to give easement of access across his property to obtain a building permit was a physical taking of private property that required compensation).

In *Pennsylvania Coal Co. v. Mahon*, a state statute prohibited coal mining if it were done in a manner to cause subsidence of any dwelling (U.S. 1922). The plaintiff had a contract to mine coal under a dwelling but the statute prevented the plaintiff from doing so. *Id.* The Court held the statute was an attempt to condemn property—the right to mine coal—without compensation. *Id.* at 414, 43 S. Ct. at 159-60, 67 L. Ed. at 326. *Mahon* "is a situation in which, by denying an owner the occupancy and use of his property interest, the government takes the interest without any semblance of physical intrusion." Stoebuck, at 221.

Richards v. Washington Terminal Co. presents a factual scenario closer to the facts in this case (U.S. 1914). In *Richards*, the plaintiff owned residential property along the tracks of a railroad that had the power of eminent domain. The property lay near the mouth of a tunnel. The Court recognized that two kinds of the railroad's activities had partially destroyed the plaintiff's interest in the enjoyment of his property. The first kind involved smoke, dust, cinders, and vibrations invading the plaintiff's property at all points at which the property abutted the tracks. The second kind involved gases and smoke emitted from engines in the tunnel that contaminated the air and invaded the plaintiff's property. A fanning system inside the tunnel forced the emission of the gases and smoke from the tunnel. As to the first activity, the Court denied compensation because it was the kind of harm normally incident to railroading operations. *Id.* at 54-55, 34 S. Ct. at 657-58, 58 L. Ed. at _. As to the second activity—gases and smoke from the tunnel—the Court concluded the plaintiff was entitled to compensation for the "special and peculiar damage" resulting in diminution of the value of the plaintiff's property. *Id.* at 557, 34 S. Ct at 658, 58 L. Ed. at _.

Richards is viewed as recognizing the taking of a property interest or right "to be free from 'special and peculiar' governmental interference with enjoyment." Stoebuck, at 220. The taking involved "no kind of physical taking or touching—none whatever." *Id.* Viewed in this light, *Richards* "entirely does away with the requirement of a physical taking or touching." *Id.*; *see* Nichols § 6.01, at 6-9 n. 11 ("It is not necessary, in order to render a statute obnoxious to the restraint of the Constitution, that it must in terms or in effect authorize an actual physical taking of the property or thing itself, so long as it affects its free use and enjoyment. ...")

(c) Liability of Government for a Taking by the Operation of a Nuisance-Producing Governmental Enterprise

With regard to private nuisances, [t]he power of the legislature to control and regulate nuisances is not without restriction, and it must be exercised within constitutional limitations. The power cannot be exercised arbitrarily, or oppressively, or unreasonably.... It has been broadly stated, as an additional limitation to the power of the legislature, that ... the legislature may not authorize the use of property in such a manner as unreasonably and arbitrarily to infringe on the rights of others, as by the creation of a nuisance. So it has been held that the legislature has no power to authorize the maintenance of a nuisance injurious to private property without due compensation.

66 C.J.S. *Nuisances* § 7, at 738 (1950).

Thus, the state cannot regulate property so as to insulate the users from potential private nuisance claims without providing just compensation to persons injured by the nuisance. The Supreme Court firmly established this principle in *Richards,* holding that "while the legislature may legalize what otherwise would be a public nuisance, it may not confer immunity from action for a private nuisance of such a character as to amount in effect to a taking." *Richards*, 233 U.S. at 553, 34 S. Ct. at 657, 58 L. Ed, at __; *see also Pennsylvania R.R. v. Angel,* (N.J. Eq. 1886) ("[A]n act of the legislature cannot confer upon individuals or private corporations, acting primarily for their own profit, although for public benefit as well, any right to deprive persons of the ordinary enjoyment of their property, except upon condition that just compensation be first made to the owners.").

A number of state courts have decided takings cases on the basis that the government entity operated a nuisance-producing enterprise. *See, e.g., Thornburg v. Port of Portland* (Or. 1962) ("[A] taking occurs whenever government acts in such a way as substantially to deprive an owner of the useful possession of that which he owns, either by repeated trespasses or by repeated nontrespassory invasions called 'nuisance.'"). Significantly, a large number of these cases deal with smoke and odors from sewage disposal plants and city dumps. One commentator describes the case this way:

Typically, a city sewage plant or dump in the vicinity of, but not necessarily directly adjacent to, the plaintiff's land has wafted its noxious smoke, odors, dust, or ashes, usually combinations of these, over the plaintiff's land, with the obvious result of lessening its enjoyment. No physical touching is present, nor do the courts try to equate the municipal acts with touchings. [Several states] have allowed eminent domain compensation in cases of this kind.... More significant than a court's language is the result it announces, and in this respect all the decisions stand for the proposition that nuisance-type activities are a taking....

Stoebuck, at 226-27; *see also* Nichols § 6.07, at 6-112 to 6-113 ("[G]eneration of offensive odors, gases, smoke ... may constitute a taking.").

The commentator ascribes a name to the theory of these cases: condemnation by nuisance. Stoebuck, at 226. And the commentator has formulated the theory this way: "governmental activity by an entity having the power of eminent domain, which activity constitutes a nuisance according to the law of torts, is a taking of property for public use, even though such activity may be authorized by legislation." *Id.* at 208-09; *see also City of Georgetown v. Ammerman* (Ky. 1911) (holding that odors from city dump adjacent to plaintiff's property created a nuisance that was a taking of the property); *Ivester v. City of Winston-Salem* (N.C. 1939) (holding as part of fundamental law of North Caroline that odors from disposal plant next to plaintiff's property constituted a nuisance and were a taking; North Carolina has no constitutional provision for a "taking"); *Brewster v. City of Forney* (Tex. Ct. App. 1920) (holding under Texas Constitution that odors from a nearby sewage disposal plant resulting in a taking of plaintiff's property); Nichols § 6.07, at 6-112 (stating under broad view of property—right to use, exclude, and dispose—there need not be a physical taking of the property or even dispossession; any substantial interference with the elemental rights growing out of the property ownership is considered a taking).

One court long ago anticipated the so-called condemnation by nuisance theory this way:

Whether you flood the farmer's fields so that they cannot be cultivated, or pollute the bleacher's steam so that his fabrics are stained, or fill one's dwelling with smells and noise so that it cannot be occupied in comfort, you equally take away the owner's property. In neither instance has the owner any less of material things than he had before, but in each case the utility of his property has been impaired by a direct invasion of the bounds of his private dominion. This is the taking of his property in a constitutional sense.

Pennsylvania R.R. v. Angel, 7A. at 433-34.

Our own definition of a taking is in accord with this concept:

[A] "taking" does not necessarily mean the appropriation of the fee. It may be anything which substantially deprives one of the use and enjoyment of his property or a portion thereof.

Phelps v. Board of Supervisors of County of Muscatine (Iowa 1973) (holding that construction of a bridge and cause-way over river in such a manner as to allegedly cause greater flooding on adjacent property than previously was a "taking" within the meaning of the Iowa Constitution).

As mentioned, the Board's approval of the applicants' application for an agricultural area triggered the provisions of section 352.11(1)(a). The approval gave the applicants immunity from nuisance suits. (Significantly, section 352.2(6) allows an agricultural area to include activities such as the creation of noise, odor, dust, or fumes.) This immunity resulted in the Board's taking of easements in the neighbors' properties for the benefit of the applicants. The easements entitle the applicants to do acts on their property, which, were it not for the easement, would constitute a nuisance. This amounts to a taking of private property for public use without the payment of just compensation in violation of the Fifth Amendment to the Federal Constitution. This also amounts to a taking of private property for public use in violation of article I, section 18 of the Iowa Constitution.

In enacting section 352.11(1)(a), the legislature has exceeded its authority. It has exceeded its authority by authorizing the use of property in such a way as to infringe on the rights of others by allowing the creation of a nuisance without the payment of just compensation. The authorization is in violation of the Fifth Amendment to the Federal Constitution and article I, section 18 of the Iowa Constitution.

The district court erred in concluding otherwise.

D. The Remedy

In *Agins v. Tiburon*, the California Supreme Court held that when legislation results in a taking, the landowner's remedy is to seek a declaratory judgment action that the legislation is invalid because it makes no provision for payment of just compensation (Cal. 1979); *see* 1 Nichols, *Eminent Domain* § 1.42(1), at 1-157 (3d rev. ed. 1997). The court, however, refused for policy reasons to allow the landowner to sue in inverse condemnation for temporary takings damages. Temporary takings damages represent the damages the landowner suffers up to the time the court declares a statute invalid because it violates constitutional provisions for payments of just compensation. This was the holding in *Agins* under both the federal and state just compensation clauses. *Id.; see* 26Am. Jur. 2d *Eminent Domain* § 137 (1996) ("The constitutional requirement of just compensation may not be evaded or impaired by any form of legislation, and statutes which conflict with the right to just compensation will generally be declared invalid.").

Later, the United States Supreme Court had occasion to review the California rule in *First English Evangelical Lutheran Church of Glendale v. County of Los Angeles, California* (U.S. 1987). The Court held that invalidation of the offending legislation without compensation for the taking is a constitutional insufficient remedy for a taking under the Federal Just Compensation Clause. In addition to invalidation, the landowner is entitled to takings damages (temporary taking) that occurred before the ultimate invalidation of the challenged legislation. *Id.* at 319-21, 107 S. Ct. at 2388-89, 96 L. Ed. 2d at 266-68.

Here the neighbors seek no compensation. Rather, they seek only invalidation of that portion of section 352.11(1)(a) that provides immunity against nuisance suits. We therefore need not concern ourselves with damages for any temporary taking. Accordingly, we hold unconstitutional and invalidate that portion of section 352.11(1)(a) that provides for immunity against nuisance suits. We reach this result under the Fifth Amendment to the Federal Constitution and also under article I, section 18 of the Iowa Constitution.

We reverse and remand for an order declaring that potion of Iowa Code section 352.11(1)(a) that provides for immunity against nuisances unconstitutional and without any force or effect.

We reach this holding with a full recognition of the deference we owe to the General Assembly. That branch of government—with some participation by the executive branch—holds the responsibility to sort through the practical realities and, through the political process, reach consensus in highly controversial public decisions. Those decisions demand our sincere respect. The rule is therefore that "[a] challenger must show beyond a reasonable doubt that the statute violates the constitution and must negate every reasonable basis that might support the statute." *Johnson v. Veterans' Plaza Authority* (Iowa 1995). The rule finding constitutionality in close cases cannot control the present one, however, because, with all respect, this is not a close case. When all the varnish is removed, the challenged statutory scheme amounts to a commandeering of valuable property rights without compensating the owners, and sacrificing those rights for the economic advantage of a few. In short, it appropriates valuable private property interests and awards them to strangers.

The same public that constituted the other branches of state government to make political decisions with an eye on economic consequences expects the court to resolve constitutional challenges on a purely legal basis. We recognize that political and economic fallout from our holding will be substantial. But we are convinced our responsibility is clear because the challenged scheme is plainly—we think flagrantly—unconstitutional.

REVERSED AND REMANDED.

All justices concur except Larson and Andreasen, JJ., who take no part.

2.4 LOCAL REGULATIONS AND THE PUBLIC'S ROLE

Notwithstanding that the range of governmental activity available at the local level is extremely limited, three areas exist in which local governments (i.e., "the Public") can play a role and get involved, including (1) comprehensive plan considerations with regard to CAFOs, (2) zoning criteria associated with CAFOs, and (3) issues associated with the use of local health and environmental codes to regulate CAFOs (Schwab, 1998). Because each locality's CAFO regulations are site-specific, or location-specific, and are beyond the scope of this text, we do not delve into the voluminous minutiae involved with each location's regulatory requirements. Instead, we focus on the public's role and involvement in regard to CAFOs.

Beyond the previously mentioned areas of community regulatory involvement in the CAFO local planning process, the USEPA's CAFO Final Rule (FR/Vol 68. No. 29/Rules and Regulations, February 12, 2003) stresses a public role and involvement in the entire implementation of the NPDES Program, including the implementation of NPDES permitting of CAFOs. The NPDES regulations in 40 CFR parts 122, 123, and 124 establish public participation in USEPA and state permit issuance, in enforcement, and in the approval and modification of state NPDES programs. These opportunities for public involvement are long-standing elements of the NPDES program.

2.4.1 PUBLIC INVOLVEMENT

Sections 123.61-62 of the NPDES regulations specify procedures for review and approval of state NPDES programs. In the case of state authorization or a substantial program modification, the USEPA is required to issue a public notice, provide an opportunity for public comment, and provide for a public hearing if there is deemed to be significant public interest. To the extent that these final regulations require a substantial modification to a site's existing NPDES program authorization, the public has an opportunity to comment on the proposed modifications.

Section 123.64 of the NPDES regulation provides an avenue for direct public involvement. For example, any individual or organization having an interest may petition the USEPA to withdraw a state NPDES program for alleged failure of the State to implement the NPDES permit program, including failure to implement the CAFO permit program.

Section 124.10 establishes public notice requirements for NPDES permits, including those issued to CAFOs. Under these existing regulations, the public may submit comments on draft individual and general permits and may request a public hearing on such permits. Various sections of part 122 and § 124.52 allow the Director to determine on a case-by-case basis that certain operations may be required to obtain an individual permit rather than coverage under a general permit. Section 124.52 specifically lists CAFOs as an example point source where such a decision many be made. Furthermore, § 122.28(b)(3) authorizes any interested person to petition the Director to require an entity authorized by a general permit to apply for and obtain an individual permit. Section 11.28(b)(3) also provides an example case where an individual permit may be required, including a case where the discharge is a significant contributor of pollutants.

To help foster public confidence that environmental guidelines are being followed, the USEPA expects permitting authorities to make certain information available to the public upon request. For example, all CAFOs (large, medium, and small), whether covered by a general or an individual permit, report annually to the permitting authority the following information (which is available to the public):

- The number and type of animals, whether in open confinement or housed under roof
- The estimated amount of total manure, litter, and process wastewater generated by the CAFO in the previous 12 months
- The estimated amount of total manure, litter, and process wastewater transferred to other persons by the CAFO in the previous 12 months
- The total number of acres for land application covered by the nutrient management plan
- The total number of acres under control of the CAFO that were used for land application of manure, litter, and process wastewater in the previous 12 months
- A summary of all manure, litter, and process wastewater discharges from the production area that have occurred in the previous 12 months, including date, time, and approximate volume
- A statement indicating whether the current version of the CAFO's CNMP was developed or approved by a certified nutrient management planner (USDA and USEPA, 1999)

- *Important point:* Most information in CNMPs deals with issues that are of legitimate concern to residents of the surrounding area, such as the location and timing of manure application.

2.5 CNMPs

According to the NRCS (2004), a CNMP is a "Conservation (Farm) Plan" specific to CAFOs. A CNMP addresses the management and treatment necessary for a CAFO to meet its production goals and to protect soil and water resources on the farm and leaving the farm. The CNMP process is part of the national strategy by the USDA and the USEPA to better address the resource concerns associated with AFOs, preserve the livestock industry, and create a more uniform planning system nationwide. CNMPs may need to be approved by a "certified" technical service. Certification programs are currently under development at the national and state level.

The objective of a CNMP is to document CAFO owners' and operators' plans to manage manure and organic by-products by combining conservation practices and management activities into a conservation system that, when implemented, will achieve the goal of the producer and protect or improve water quality.

The NRCS (2004) points out that, in developing a CNMP with a CAFO owner or operator, alternatives are developed that address treatment of the resources of concern and are in accordance with the applicable NRCS technical standards. The CAFO owner or operator selects from these alternatives to create a CNMP that best meets his or her management objectives and environmental concerns.

CNMP implementation may require additional design, analysis, or evaluations. The certified conservation planner must maintain a relationship with the producer throughout CNMP implementation to address changes or new challenges. Evaluation of the effectiveness of the CNMP may begin during the implementation phase and may not end until several years after the last practice is applied. Follow-up and evaluation determine whether the implemented alternative meets the client needs and solves the conservation problems in a manner beneficial to the resources.

2.5.1 LIVESTOCK OPERATIONS EXPECTED TO NEED A CNMP

The *Unified National Strategy for Animal Feeding Operations* stipulates that all AFOs should have CNMPs to minimize the impacts of manure and manure nutrients on water quality. Obviously, the first step before implementing a CNMP is to determine if one is actually required. How do we do this? The NRCS (2005) points out that the best information source available on farms and on the characteristics of farms in the United States is the Census of Agriculture. The Census of Agriculture has information about the number and types of livestock on each farm. However, the census provides no information on how the animals are raised or to what extent or how long animals are held in confinement. Consequently, one cannot identify whether a farm in the census database is an AFO.

Eventually, farms that are expected to need a CNMP were identified on the basis of the number and types of livestock on the farm and an estimate of the amount of manure produced annually by those livestock. The 1997 Census of Agriculture, which is the most recent census available, was used to make the determination. Farms with significant numbers of fattened cattle, poultry, and swine would clearly need a CNMP, since these livestock types are almost always raised in confined settings. Dairies would also be expected to need CNMPs, since milk cows are confined for at least a portion of time each day for milking. Farms with an incidental number of these confined livestock types, however, would not be expected to implement a CNMP, even if the animals were confined. Similarly, most farms with pastured livestock types, such as beef cattle, horses, and sheep, would not meet the USEPA definition of an AFO, and so would not need CNMPs. However, some of the farms with pastured livestock types would be expected to need CNMPs if they produce a significant amount of recoverable manure.

The NRCS (2005) points out that three criteria were developed to identify farms that may need a CNMP, with each criterion addressing a separate segment of the livestock operations as represented in the census database.

The *first criterion* is used to identify farms with too few livestock to be considered as a farm that would need a CNMP. It is based on a profile of farms with livestock in the United States (shown in Table 2.15). The profile reveals that, of the 1,911,859 farms in the United States in 1997, two-thirds—1,315,051 farms (69%)—reported some kind of livestock on the farm or reported livestock sales. About 27% of those farms (361,031 farms) were "farms with few livestock." Farms with few livestock were farms with:

- Less than 4 animal units of any combination of fattened cattle, milk cows, swine, chickens, and turkeys
- Less than 8 animal units of cattle other than fattened cattle or milk cows
- Less than 10 horses, ponies, mules, burros, or donkeys
- Less than 25 sheep, lambs, or goats
- Less than $5,000 in gross sales of specialty livestock products

- *Important point:* Remember, an animal unit (AU) represents 1,000 pounds of live weight.

TABLE 2.15

Number of Farms with Livestock/Livestock Sales, 1997 Census of Agriculture

	Farms with few livestock	Farms specialty livestock types	Farms with pastured livestock types & few other livestock types	Farms with confined livestock	All farms with livestock
Alabama	8,142	236	21,415	4,038	33,831
Alaska	192	38	85	37	352
Arizona	1,603	67	2,338	233	4,241
Arkansas	7,209	314	21,391	6,491	35,405
California	10,881	817	12,964	3,478	28,140
Colorado	6,576	166	12,905	1,457	21,104
Connecticut	1,052	38	592	400	2,082
Delaware	314	8	186	981	1,489

TABLE 2.15 (continued)
Number of Farms with Livestock/Livestock Sales, 1997 Census of Agriculture

	Farms with few livestock	Farms specialty livestock types	Farms with pastured livestock types & few other livestock types	Farms with confined livestock	All farms with livestock
Florida	6,670	673	11,812	1,241	20,396
Georgia	7,100	177	15,950	4,984	28,211
Hawaii	752	50	498	147	1,447
Idaho	5,936	169	8,460	1,644	16,209
Illinois	10,403	135	13,128	11,197	34,863
Indiana	11,573	164	11,207	10,006	32,950
Iowa	9,697	156	19,354	26,081	55,288
Kansas	8,465	100	28,483	4,939	41,987
Kentucky	16,044	45	36,138	4,816	57,043
Louisiana	4,327	305	11,277	1,254	17,163
Maine	1,474	58	818	709	3,059
Maryland	2,732	73	2,554	2,440	7,799
Massachusetts	1,555	71	689	541	2,856
Michigan	10,466	326	6,958	6,565	10,554
Minnesota	10,554	330	12,930	19,171	42,985
Mississippi	5,025	411	15,089	2,578	23,103
Missouri	16,608	139	49,727	9,627	76,101
Montana	4,120	141	13,078	772	18,111
New Hampshire	997	32	460	315	1,804
Nebraska	5,011	101	19,929	9,893	34,934
Nevada	764	13	1,418	141	2,336
New Jersey	2,862	65	1,193	374	4,494
New Mexico	3,674	41	6,661	454	10,830
New York	6,709	211	5,626	9,076	21,622
North Carolina	9,447	187	15,309	6,435	31,378
North Dakota	2,184	195	12,114	2,269	16,762
Ohio	15,088	203	13,937	10,996	40,224
Oklahoma	15,166	91	46,256	3,440	64,953
Oregon	11,570	278	11,367	1,093	24,308
Pennsylvania	10,122	247	9,307	14,215	33,890
Rhode Island	218	10	107	65	400
South Carolina	4,561	71	7,410	1,415	13,457
South Dakota	2,782	147	15,293	5,789	24,011
Tennessee	18,530	107	38,217	3,566	60,420
Texas	42,210	495	114,373	6,516	163,594
Utah	4,117	193	5,907	1,197	11,414
Vermont	1,305	40	943	1,940	4,228
Virginia	8,599	91	20,178	3,359	32,227
Washington	8,262	249	7,577	1,497	17,585
West Virginia	5,304	34	8,368	959	14,665
Wisconsin	10,483	471	9,250	26,628	46,832
Wyoming	1,596	55	6,140	362	8,153
All states	**361,031**	**8,834**	**707,365**	**237,821**	**1,315,051**

Source: NRCS (2005).

About 75% of the farms with few livestock had only pastured livestock types; 23% had at least some fattened cattle, milk cows, swine, chickens, or turkeys; and about 2% primarily had specialty livestock with gross sales of specialty livestock products below $5,000. The average of gross livestock sales per farm was only $2,149, and no livestock sales were reported for 34% of the farms. These farms are expected to be too small to need a CNMP.

The *second criterion* for a farm that would need a CNMP was based on the amount of recoverable manure produced. Recoverable manure is the portion of manure that could be collected from the facility for land application or other use. Recoverable manure and manure nutrients were estimated for each farm in the census using the parameters shown in Table 2.16.

TABLE 2.16
Parameters Used to Calculate the Quantity of Manure and Nutrients As Excreted

Livestock type	Number of animals per AU	Tons of manure per AU per year		Pounds of nutrient per wet weight tons of manure	
		Wet weight	Oven dry weight	Nitrogen	Phosphorus
Fattened cattle	1.14	10.59	1.27	10.98	3.37
Beef calves	4	11.32	1.36	8.52	2.33
Beef heifers	1.14	12.05	1.45	6.06	1.30
Beef breeding cows and bulls	1	11.50	1.33	10.95	3.79
Beef stockers and grass-fed beef	1.73	11.32	1.36	8.52	2.33
Horses, ponies, mules, donkeys and burros	1.25	11.32	1.36	8.52	2.33
Sheep and goats	8	11.32	1.36	8.52	2.33
Milk cows	0.74	15.24	2.20	10.69	1.92
Dairy cows	4	12.05	1.45	6.06	1.30
Dairy heifers	0.94	12.05	1.45	6.06	1.30
Dairy stockers and grass-fed animals marketed as beef	1.73	12.05	1.45	6.06	1.30
Hogs for breeding	2.67	6.11	0.55	13.26	4.28
Hogs for slaughter	9.09	14.69	1.33	11.30	3.29
Chicken layers	250	11.45	2.86	26.93	9.98
Chicken pullets, less than 3 months old	455	8.32	2.08	27.20	10.53
Chicken pullets, more than 3 months old	250	8.32	2.08	27.20	10.53
Chicken broilers	455	14.97	3.74	26.83	7.80
Turkeys for breeding	50	9.12	2.28	22.41	13.21
Turkeys for slaughter	67	8.18	2.04	30.36	11.83

Source: NRCS (2005).
* Includes nitrogen and phosphorus in urine.

Included are estimates of recoverable manure for beef cattle and other pastured livestock types. The calculation is heavily influenced by recoverability factors, which range from 5% to 20% for pastured livestock types with more than 1 AU per acre of pastureland and rangeland. The criterion used to identify a farm expected to need a CNMP is production of more than 200 pounds of recoverable manure nitrogen annually. This criterion is equivalent to production of more than about 120 pounds of recoverable manure phosphorus annually. Farms at this threshold generate about 11 tons of manure (transport and handling weight) per year, which is less than a pickup truck load per month. (The actual amount varies by livestock type. The 11-ton estimate was empirically obtained by summarizing estimates from 3,218 farms with 190 to 200 pounds of recoverable manure nitrogen.)

Using this criterion, 255,070 farms were identified as farms that are expected to need a CNMP based on the amount of recoverable manure produced. However, this does not include farms with specialty livestock types because recoverable manure was not estimated for specialty livestock types.

The *third criterion* was developed to identify farms with specialty livestock types that may need a CNMP. Farms with specialty livestock types were defined to be farms with $5,000 or more in gross sales of livestock products from fish, bees, rabbits, mink, poultry other than chickens and turkeys, and exotic livestock that had few other livestock types on the farm (see Table 2.15). There were 8,834 of these farms in 1997. The dominant specialty livestock type—based on gross sales—was fish and other aquaculture species on 2,449 farms (28%), colonies of bees on 2,331 farms (26%), poultry other than chickens and turkeys (such as ducks and geese) on 1,490 farms (17%), mink and rabbits on 641 farms (7%), and other exotic livestock on 1,923 farms (22%). Obviously, farms specializing in aquaculture or honey production would not need a CNMP. The two remaining groups—farms with poultry other than chickens and turkeys and farms with mink and rabbits—are most likely to be raising animals in confined settings, and so were identified as farms that may need a CNMP.

Including these 2,131 farms with specialty livestock types, the total number of census farms that are expected to need a CNMP is 257,201. These farms are referred to as (CNMP farms) throughout this text. Table 2.17 provides a breakdown by livestock type.

TABLE 2.17

CNMP Farms by Dominant Livestock Type

Category of CNMP farm	Number
Farms with more than 35 AU of the dominant livestock type	
Fattened cattle	10,159
Milk cows	79,318
Swine	32,955
Turkeys	3,213
Broilers	16,251
Layers/pullets	5,326
Confined heifers/veal	4,011
Small farms with confined livestock types dominant	42,565
Farms with pastured livestock types dominant*	61,272
Farms with specialty livestock types	2,131
All CNMP farms	**257,201**

Source: NRCS (2005).

* Includes 24,697 farms with pastured livestock types and few other livestock and 36,575 farms with 4–35 AU of confined livestock types with beef cattle (other than fattened cattle) as the dominant livestock type.

2.5.2   CNMP Technical Guidance

The NRCS' *National Planning Procedures Handbook,* Subpart B, part 600.53, provides general criteria for CNMP development. At a minimum, CNMPs will meet the following criteria. They must:

- Provide documentation that addresses the items outlined in Section 600.6, Exhibit 15, Comprehensive Nutrient Management Pan-Format and Content.
- Document a CAFO owner's or operator's consideration of the six CNMP elements. While regulators recognize that a CNMP may not contain all six elements, the elements need to be considered by the CAFO owner or operator during development of the CNMP, and the owner's or operator's decisions regarding each must be documented. These elements are:
  - Manure and wastewater handling and storage
  - Land treatment practices
  - Nutrient management
  - Feed management
  - Other utilization activities
- Contain actions that address water quality criteria for feed production areas and for land on which manure and organic by-products will be applied. This includes addressing soil erosion to reduce the transport of nutrients within or off of a field to which manure is applied. For CAFO owners and operators who do not land-apply manure or organic by-products, the CNMP would address only the feedlot and production areas.
- Meet requirements of the NRCS Field Office Technical Guide conservation practice standards for all practices contained in the CNMP.
- Meet all applicable local, tribal, state, and federal regulations. When applicable, ensure that USEPA and NPDES or state permit requirements (i.e., minimum standards and special conditions) are addressed.

The USDA and USEPA agree that the following six elements should be included in a CNMP, as necessary (FR, 1998). The specific practices used to implement each component may vary to reflect the site-specific conditions and needs of the watershed.

1. **Feed Management**—Animal diets and feed may be modified to reduce the amounts of nutrients in manure. Feed management can include the use of low phosphorus corn and enzymes such as phytase, which can be added to nonruminant animal diets to increase phosphorus usage. Reduced inputs and greater use of phosphorus by animals reduce the amount of phosphorus excreted and produce manure with a nitrogen-phosphorus ratio closer to that required by crop and forage plants.
2. **Manure Handling and Storage**—Manure must be handled and stored properly to prevent water pollution from CAFOs. Manure and wastewater handling and storage practices should also consider odor and other environmental and public health problems. Handling and storage considerations should include:
   - *Divert clean water*—Siting and management practices should divert clean water from contact with feedlots and holding pens, animal manure, or manure storage systems. Clean water can include rainfall falling on roofs of facilities, runoff from adjacent lands, or other sources.
   - *Prevent leakage*—Construction and maintenance of buildings, collection systems, conveyance systems, and permanent and temporary storage facilities should prevent leakage of organic matter, nutrients, and pathogens to groundwater or surface water.

- *Provide adequate storage*—Liquid manure systems should safely store the quantity and contents of animal manure and wastewater produced, contaminated runoff from the facility, and rainfall. Dry manure, such as that produced in certain poultry and beef operations, should be stored in production buildings or storage facilities, or otherwise stored in such a way so as to prevent polluted runoff. Location of manure storage systems should consider proximity to water bodies, floodplains, and other environmentally sensitive areas.
- *Manure treatments*—Manure should be handled and treated to reduce the loss of nutrients to the atmosphere during storage, to make the material a more stable fertilizer when land-applied or to reduce pathogens, vector attraction, and odors, as appropriate.
- *Management of dead animals*—Dead animals should be disposed of in a way that does not adversely affect groundwater or surface water or create public health concerns. Composting, rendering, and other practices are common methods used for dead animal disposal.

3. **Land Application of Manure**—Land application is the most common, and usually most desirable, method of using manure because of the value of the nutrients and organic matter. Land application should be planned to ensure that the proper amounts of all nutrients are applied in a way that does not cause harm to the environment or to public health. Land application in accordance with the CNMP should minimize water quality and public health risk. Considerations for appropriate land application should include:

- *Nutrient balance*—The primary purpose of nutrient management is to achieve the level of nutrients (especially nitrogen and phosphorus) required to grow the planned crop by balancing the nutrients that are already in the soil and from other sources with those that will be applied in manure, biosolids, and commercial fertilizer. At a minimum, nutrient management should prevent the application of nutrients at rates that will exceed the capacity of the soil and planned crops to assimilate nutrients and prevent pollution. Soils and manure should be tested to determine nutrient content.
- *Timing and methods of application*—Care must be taken when land-applying manure to prevent it from entering streams and other water bodies or environmentally sensitive areas. The timing and methods of application should minimize the loss of nutrients to groundwater or surface water and the loss of nitrogen to the atmosphere. Manure application equipment should be calibrated to ensure that the quantity of material being applied is what is planned.

4. **Land Management**—Tillage, crop residue management, grazing management, and other conservation practices should be used to minimize movement to groundwater and surface water of soil, organic materials, nutrients, and pathogens from lands where manure is applied. Forest riparian buffers, filter strips, field borders, contour buffer strips, and other conservation buffer practices should be installed to intercept, store, and utilize nutrients or other pollutants that may migrate from fields on which manure is applied.

5. **Record Keeping**—CAFO operators should keep records that indicate the quantity of manure produced and how the manure was used, including where, when, and amount of nutrients applied. Soil and manure testing should be incorporated into the record keeping system. Records should be kept when manure leaves the CAFO.

6. **Other Utilization Options**—Where the potential for environmentally sound land application is limited, alternative uses of manure, such as the sale of manure to other farmers, the composting and sale of compost to home owners, and the use of manure for power generation, may also be appropriate. All manure-use options should be designed and implemented to reduce the risk to all environmental resources and must comply with federal, state, tribal, and local laws.

2.6   SUMMARY

CAFO regulations are utterly necessary for environmental and health protection. However, without adequate enforcement, regulations are meaningless. Increasing recognition of the problems that CAFOs can cause drive both regulation and enforcement. This chapter began to discuss one of the biggest causes for CAFO concern: Manure. This topic will be discussed in detail in the next chapter.

CHAPTER REVIEW QUESTIONS

1. How do federal, state, and local regulations work together to manage CAFO operations?
2. What's the difference between point-source and non-point-source pollution?
3. Why must CAFO owners and operators apply for an NPDES permit? What overarching regulation controls NPDES use?
4. What's an ELG? Why is it important?
5. What's process wastewater?
6. Why might a CAFO need to comply with Endangered Species Act or National Historic Preservation Act requirements? What other federal regulations might apply?
7. Define *confinement* as it applies to CAFOs. What constitutes a day of confinement?
8. What's the difference between definition and designation for AFO and CAFO classification?
9. How do the regulations apply to side-by-side CAFOs that share a waste system?
10. How is discharge criteria important to AFO or CAFO definition or designation?
11. Discuss thresholds, discharge criteria, AFOs, and CAFOs.
12. In counting actual animals confined, what's the important criteria: actual number or maximum number possible? How does turnover (as in number of broods per year) affect animal count?
13. What's the counting procedure for an AFO or a CAFO that houses more than one kind of animal?
14. What are the important determining factors between small, medium, and large CAFOs?
15. What's a manure handling system? A liquid manure handling system? An other than liquid manure handling system? A dry manure system or dry operation?
16. What's the connection between discharge criteria and liquid (or not liquid) manure handling systems?
17. How do exotic animals (such as llamas, ostriches, and bison) fit into AFO or CAFO classification?
18. What CAFOs don't need permits? Why? How can a CAFO meet that classification? Why can't small or medium CAFOs fit this classification?
19. What operation elements are subject to regulation? Why?
20. What different agencies issue NPDES permits? What constitutes a Permitting Authority? How long does a NPDES permit stay in effect? What's the renewal process? How can one be discontinued?
21. What's an NOI? Why is it important?
22. What information should be included on a NPDES application? Why is each important?
23. What's a nutrient management plan? When is it an issue? What should it accomplish?
24. What are the different permitting considerations for existing CAFOs? For new dischargers? For designated CAFOs? For newly defined CAFOs? For new sources?
25. What's the state regulatory role for CAFOs and NPDES permitting? Discuss planning and zoning legislation in terms of agriculture.
26. What do right-to-farm laws mean for those in rural areas?

27. What are the different types of right-to-farm laws? How are they applied?
28. Discuss the importance of change under right-to-farm laws.
29. Why are fee-shifting statutes important in right-to-farm cases?
30. What factors affect a farming operation's right-to-farm applicability?
31. What are the key issues of the Bormann case?
32. What are three areas of special concern for local legislation and the public?
33. Why are CNMPs important in public involvement?
34. What's a CNMP's objective? What issues do CNMPs address?
35. How can identifying what operations need CNMPs be accomplished? What criteria are involved?
36. What six elements are essential to correct CNMP application? Why is each important?

THOUGHT-PROVOKING QUESTIONS

1. What are the cracks in the network of federal, state, and local regulations as they relate to CAFOs?
2. For those who own or operate farm operations, would an AFO or CAFO designation or definition be best? Why? Under what circumstances would one designation be better than another?
3. Are the current regulations for CAFOs enough? How thoroughly are they enforced? What states exhibit the most serious CAFO-related problems?
4. How does under-funding for the USEPA and other watchdog and protection agencies affect environmental and health concerns?
5. Define, describe, and discuss exemption from zoning for agriculture.
6. What's the history and the impact of right-to-farm laws for rural communities? Why were they enacted?
7. What are the critical attributes for right-to-farm laws? Why?
8. How could unclear language and purpose undermine the purpose of right-to-farm legislation?
9. What are the long-term legal ramifications of the Bormann case? Why?
10. Does the venue of the CNMP provide enough in the way of public voice for the community? Why or why not?

REFERENCES

Bormann v. Board of Supervisors for Kossuth County, (1998), No. 192/96-2276, Supreme Court of Iowa.

Edelman, M., & Warner, M. 1999. *Preliminary State Policing Survey Results. National Survey of Animal Confinement Policies: Report of the Animal Confinement Policy National Task Force.* Washington, DC: National Public Policing Education Committee.

FR *(Federal Register)*/Volume 63, No. 182/September 21, 1998/Unified National Strategy for Animal Feeding Operations.

FR *(Federal Register)*/Volume 68, No. 29/Wednesday, February 12, 2003/Rules and Regulations.

NRCS. 2004. *National Planning Procedures Handbook.* Washington, DC: United States Department of Agriculture. www.nrcs.usda.gov/programs/ afo/cnmp_guide_600.50.html. Last accessed December 31, 2004.

NRCS. 2005. Costs associated with development and implementation of comprehensive nutrient management plans (CNMP): Livestock operations that are expected to need a CNMP. www. NRCS.USDA.gov/technical/land/pubs/cnmplb.pdf]. Last accessed January 1, 2005.

Proulx, A. 2002. *That Old Ace in the Hole.* New York: Scribner.

Schwab, J. 1998. Planning and zoning for concentrated animal feeding operations. PAS report No. 482. Chicago, Illinois: American Planning Association.

U.S. Department of Agriculture (USDA) and U.S. Environmental Protection Agency (USEPA). 1999. *Unified National Strategy for Animal Feeding Operations.* Washington, DC: USDA and USEPA.

3 Manure Characteristics

"Don't hogs on small farms stink?" ...

Sure, but they are spread out and they are in the open air. The smell is nothing compared to closing in a massive number of animals. You drive past a herd a cattle grazing in a pasture. There's no smell. You drive past a feedlot—it stinks. With the hog farms, we are talking a large number a confined animals.

(Proulx, 2002, p. 114)

3.1 INTRODUCTION

Livestock manure is a valuable source of nutrients for crops and can improve soil productivity. However, the benefits of manure are widely misunderstood. For example, as a source of primary nutrients, manure offers much less, pound for pound, than a container of inorganic fertilizer. Manure properties depend on several factors: animal species, diet, digestibility, protein and fiber content, and animal care, housing, environment, and stage of production. Characterized in several ways, manure has important properties for collection, storage, handling, and use, including the solids content (the percent of solids per unit of liquid) and the size and makeup of manure solids (fixed and volatile solids, suspended solids, and dissolved solids). Manure contains primary nutrients—nitrogen, phosphate, and potash—but in small amounts. For example, you would need eight times as much horse manure as 5-10-10 fertilizer to supply a given amount of nitrogen. It literally takes a pile of manure to provide an adequate supply of primary nutrients. Manure components can be characterized as organic and inorganic. To help control diseases and parasites, human waste should not be mixed with animal manures (especially pig, dog, and cat manure).

The livestock manure described in this chapter is of an organic nature and agricultural origin. The information and data presented here can be used for planning and designing waste management systems and system components and for selecting waste handling equipment.

3.2 LIVESTOCK WASTE CHARACTERIZATION VARIABLES

Livestock waste characteristics are subject to wide variation; both greater and lesser values than those presented can be expected, even though we usually assign a single value for a specific waste characteristic. This value is usually presented as a reasonable value for facility design and equipment selection for situations where site-specific data are not available. Because of data variations (and the multiple reasons for such variations), planners and designers must be provided with the information they need—when justified by the situation—for seeking and establishing more appropriate values. Therefore, in this chapter, much attention is given to describing data variation and causal factors.

- *Important point:* Onsite manure/waste sampling, testing, and data collection are valuable assets in manure/waste management system planning and design and should be used where possible. Such sampling can result in greater certainty and confidence in the system design and in economic benefits to the owner. However, caution must be exercised to assure that representative data and samples are collected. Characteristics of as-excreted manure are greatly influenced by the effects of weather, season, species, diet, degree of confinement, and stage of the production and reproduction cycle. Characteristics of stored and treated manure are strongly affected by such actions as sedimentation, flotation, and biological degradation in storage and treatment facilities.

3.2.1 Definitions of Manure/Waste Characterization Terms

Table 3.1 gives definitions and descriptions of manure/waste characterization terms. It includes abbreviations, definitions, units of measurement, methods of measurement, and other considerations for the physical and chemical properties of manure and residue. (*Note:* Much of the data presented in this section is adapted from the U.S. Department of Agriculture's (USDA's) Natural Resources Conservation Service's (NRCS's) *Agricultural Waste Management Field Handbook* [NRCS/USDA, 1992]).

TABLE 3.1
Definitions and Descriptions of Manure Characterization Terms

Term	Abbreviation	Units of measure*	Definition	Method of measurement
			Physical properties	
Weight	Wt	lb	Quantity or mass.	Scale or balance.
Volume	Vol	ft^3; gal	Space occupied in cubic units.	Place in or compare to container of known volume; calculate from dimensions of containment facility.
Moisture content	MC	%	Part of waste material that is removed by evaporation and oven drying at 217°F (103°C).	Evaporate free water on steam table and dry in oven at 217°F for 24 hours or until constant weight.
Total solids	TS	%	Residue remaining after water is removed from waste material by evaporation; dry matter.	Evaporate free water on steam table and dry in oven at 217°F for 24 hours or until constant weight.
Volatile solids	VS, TVS	%; % w.b.; % d.w.	Part of total solids that is driven off as volatile (combustible) gases when heated to 1112°F (600°C); organic matter.	Place total solids residue in furnace at 1112°F for at least 1 hr.
Fixed solids	FS, TFS	%; % w.b.; % d.w.	That part of total solids that remains after volatile gases are driven off at 1112°F (600°C); ash.	Determine weight (mass) of residue after volatile solids have been removed as combustible gases when heated at 1112°F for at least 1 hr.
Dissolved solids	DS, TDS	%; % w.b.; % d.w.	That part of total solids passing through the filter in a filtration procedure.	Pass a measured quantity of waste material through 0.45 micron filter using appropriate procedure; evaporate filtrate and dry residue to constant weight at 217°F.
Suspended solids	SS, TSS	%; % w.b.; % d.w.	That part of total solids that is removed in a filtration procedure.	Determined by calculating the difference between total solids and dissolved solids.

TABLE 3.1 (continued)
Definitions and Descriptions of Manure Characterization Terms

Term	Abbreviation	Units of measure*	Definition	Method of measurement
			Chemical properties	
Ammoniacal nitrogen (total ammonia); ammonia nitrogen	NH_3-N	mg/L µg/L mg/L µg/L	Both NH_3 and NH nitrogen compounds.	Common lab procedure uses digestion, oxidation, and reduction to convert all or selected nitrogen forms to ammonium that is released and measured as ammonia.
Ammonium nitrogen	NH_4-N	mg/L µg/L	The positively ionized (cation) form of ammoniacal nitrogen.	
Total kjeldahl nitrogen	TKN	mg/L µg/L	The sum of organic nitrogen and ammoniacal nitrogen.	
Nitrate nitrogen	NO3-N	mg/L µg/L	The negatively ionized (anion) form of nitrogen that is highly mobile.	
Total nitrogen	TN; N	%; lb	The summation of nitrogen from all the various nitrogen compounds listed above.	
Phosphorus	P	%; lb	Acid-forming element that combines readily with oxygen to form the oxide P_2O_5. As a plant nutrient, it promotes rapid growth, hastens maturity, and stimulates flower, seed, and fruit production.	Lab procedure uses digestion and/or reduction to convert phosphorus to a colored complex; result measured by spectrophotometer.
Potassium	K	%; lb	As a plant nutrient, available potassium stimulates the growth of strong stems, imparts resistance to disease, increases the yield of tubers and seed, and is necessary to form starch, sugar, and oil and transfer them through plants.	Lab digestion procedure followed by flame photometric analysis to determine elemental concentration.
5-day biochemical oxygen demand	BOD_5	lb of O_2	That quantity of oxygen needed to satisfy biochemical oxidation of organic matter in waste sample in 5 days at 68°C (20°C).	Extensive lab procedure of incubating waste sample in oxygenated water for 5 days and measuring amount of dissolved oxygen consumed.
Chemical oxygen demand	COD	lb of O_2	Measure of oxygen-consuming capacity of organic and some inorganic components of waste materials.	Relatively rapid lab procedure using chemical oxidants and heat to fully oxidize organic components of waste.

Source: Adapted from NRCS (1992).

* % w.b. is percent measured on a wet basis, and % d.b. is percent measured on a dry basis.

The first four physical properties—weight (Wt), volume (Vol), total solids (TS), and moisture content (MC)—are important to agricultural producers and facility planners and designers. They describe the amount and consistency of the material to be dealt with by equipment and in treatment and storage facilities. The first three of the chemical constituents—nitrogen (N), phosphorus (P), and potassium (K)—are also of great value to waste systems planners, producers, and designers. Land application of agricultural manure/waste is the primary waste use procedure, and N, P, and K are the principal components considered in development of an agricultural manure/waste management plan.

TS and the fractions of TS that are volatile solids (VS) and fixed solids (FS) are presented. VS and FS are sometimes referred to, respectively, as total volatile solids (TVS) and total fixed solids (TFS). Characterization of these solids gives evidence of the origin of the waste, its age and previous treatment, its compatibility with certain biological treatment procedures, and its possible adaptation to mechanical handling alternatives.

Manure with a very high water content may be characterized according to the amounts of solids that are dissolved or suspended. Dissolved solids (DS) or total dissolved solids (TDS) are in solution. Suspended solids (SS) or total suspended solids (TSS) float or are kept buoyant by the velocity or turbulence of the wastewater.

Table 3.1 also lists physical and chemical properties of livestock and other organic agricultural wastes. Because microorganisms are not commonly used as a design factor for no-discharge waste management systems that use manure or wastes on agricultural land, data on biological properties are not presented in this table.

Key terms: The terms *manure, waste,* and *residue* are sometimes used synonymously. In this text, *manure* refers to combinations of feces and urine only, and *manure/waste* includes manure plus other material, such as bedding, soil, wasted feed, and water that is wasted or used for sanitary and flushing purposes. Small amounts of wasted feed, water, dust, hair, and feathers are unavoidably added to manure and are undetectable in the production facility. These small additions must be considered to be a part of manure and a part of the "as-excreted" characteristics presented. *Litter* is a specific form of poultry waste that results from "floor" production of birds after an initial layer of a bedding material, such as wood shavings, is placed on the floor at the beginning of and perhaps during the production cycle.

Manure is often given descriptive names that reflect their moisture content, such as liquid, slurry, semisolid, and solid. Manures with a moisture content of 95% or more exhibit qualities very much like water and are called *liquid manure/wastes.* Properly designed and managed lagoon treatment systems should have less than 1% solids, typically from 0.1% to 0.5%. However, overloaded lagoons commonly reach as high as 2% solids (NWPS-18, 2004).

Manure/wastes with a moisture content of about 75% or less exhibit the properties of a solid and hold a definite angle of repose; these wastes can be stacked and picked up with a front-end loader. They are called *solid manure* or *solid manure/waste.*

Manure/wastes with a moisture content between about 75% and 95%—25% and 5% solids—are *semiliquid (slurry)* or *semisolid.*

- *Important point:* Because manure/wastes are heterogeneous and inconsistent in their physical properties, the moisture content and range indicated above must be considered generalizations subject to variation and interpretation.

- *Important point:* Because of the high moisture content of as-excreted manure and treated waste, their specific weight is very similar to that of water—62.4 pounds per cubic foot. Some manure/waste with considerable solids content can have a specific weight of as much as 105% that of water. Some dry wastes, such as litter, with significant void space can have a specific weight of much less than that of water. Assuming that wet and moist wastes weigh 60 to 65 pounds per cubic foot is a convenient and useful estimate for planning waste management systems.

- *Important point:* Odors are associated with all livestock production facilities. Animal manure/waste is common source of significant odors, but other sources, such as poor quality or spoiled feed and dead animals, can also be at fault. Freshly voided manure is seldom a cause of objectionable odor, but manure that accumulates or is sorted under anaerobic conditions does develop unpleasant odors. Such wastes can cause complaints at the production facility when the waste is removed from storage or when it is spread on the fields. Manure-covered animals and ventilation air exhausted from production facilities can also be significant sources of odor. The best insurance against undesirable odor emissions is waste management practices that quickly and thoroughly remove wastes from production facilities and place them in treatment or storage facilities or apply them directly to soil.

3.2.2   UNITS OF MEASURE AND CONVERSION FACTORS

Manure/waste production from livestock is expressed in pounds per day per 1,000 pounds of livestock live weight (lb/d/1000#). Volume of manure/waste materials is expressed in cubic feet per day per 1,000 pounds of live weight (ft^3/d/1000#). English units are used exclusively for weight, volume, and concentration data for manure, waste, and residue.

The concentration of various components in waste is commonly expressed as milligrams per liter (mg/L) or parts per million (ppm). One mg/L is 1 mg (weight) in 1 million parts (volume); for example, in 1 L. One ppm is 1 part by weight in 1 million parts by weight. Therefore, mg/L equals ppm if a solution has a specific gravity equal to that of water.

Generally, substances in solution up to concentrations of about 7,000 mg/L do not materially change the specific gravity of the liquid, and mg/L and ppm are numerically interchangeable. Concentrations are sometimes expressed as mg/kg or mg/1000g, which are the same as ppm.

Occasionally, the concentration is expressed as a percent. A 1% concentration equals 10,000 ppm. Very low concentrations are sometimes expressed as micrograms per liter (μg/L). A microgram is 1 millionth of a gram.

Various solid fractions of a manure/waste or residue, when expressed in units of pounds per day or as a concentration, generally are measured on a wet weight basis (% w.b.), a percentage of the "as is" or wet weight of the material. In some cases, however, data are recorded on a dry weight basis (% d.w.), a percentage of the dry weight of the material. The difference in these two values for a specific material is most likely very large. Nutrient and other chemical fractions of a manure/waste material, expressed as a concentration, may be on a wet weight or dry weight basis, or expressed as pounds per 1,000 gallons of waste.

Key term: *parts per million (ppm)*—1 ppm is 1 part by weight in 1 million parts by weight.

Key term: *milligrams per liter (mg/L)*—1 mg/L is 1 milligram (weight) in 1 million parts (volume), i.e., 1 L. Therefore, ppm = mg/L when a solution has the same specific gravity as water. Generally, substances in solution up to concentrations of about 7,000 mg/L do not materially change the specific gravity of water. To that limit, ppm and mg/L are numerically interchangeable. A 1% solution has a concentration of 10,000 ppm, which equals 1 g in 100 g of water.

Key term: *Electrical conductivity*—electrical conductance is expressed in reciprocal ohms (mhos); electrical conductivity (EC) is expressed in mhos/cm. 1 mhos/cm = 1,000 millimhos/cm (mmhos/cm) = 1,000,000 micromhos per centimeter (umhos/cm). 1 mmhos/cm equals a concentration of approximately 640 ppm dissolved salts.

Conversion Factors

Length

Unit of measure	Symbol	Conversion table mm	cm	m	km	in	ft	mi
millimeter	mm	1	0.1	0.001	—	0.0394	0.003	—
centimeter	cm	10	1	0.01	—	0.394	0.033	—
meter	m	1000	100	1	0.001	39.37	3.281	—
kilometer	km	—	—	1000	1	—	3,281	0.621
inch	in	25.4	2.54	0.0254	—	1	0.083	—
foot	ft	304.8	30.48	0.305	—	12	1	—
mile	mi	—	—	1609	1.609	—	5280	1

Area

Unit of measure	Symbol	Conversion table m²	ha	km²	ft²	acre	mi²
square meter	m²	1	—	—	10.76	—	—
hectare	ha	10,000	1	0.01	107,640	2.47	0.00386
square kilometer	km²	1×10^6	100	1	—	247	0.386
square foot	ft²	0.093	—	—	1	—	—
acre	acre	4,049	0.405	—	43,560	1	0.00156
square mile	mi²	—	259	2.59	—	640	1

Volume

Unit of measure	Symbol	Conversion table km³	m³	L	Mgal	acre-ft	ft³	gal
cubic kilometer	km³	1	1×10^9	—	—	811,000	—	—
cubic meter	m³	—	1	1000	—	—	35.3	264
liter	L	0.001	1	—	—	0.0353	0.264	
million U.S. gallons	Mgal	—	—	—	1	3.07	134,000	1×10^6
acre-foot	acre-ft	—	1,233	—	0.3259	1	43,560	325,848
cubic foot	ft³	—	0.0283	28.3	—	—	1	748
gallon	gal	—	—	3.785	—	—	0.134	1

Flow rate

Unit of measure	Symbol	Conversion table km³/yr	m³/s	L/s	mgd	gpm	cfs	acre-ft/day
cubic kilometers/yr	km³/yr	1	31.7	—	723	—	1,119	2,220
cubic meters/second	m³/s (m³/sec)	0.0316	1	1000	22.8	15,800	35.3	70.1
liters/second	L/s (L/sec)	—	0.001	1	0.0228	15.8	0.0353	(0.070)
million U.S. gal/day	mgd (Mgal/d)	—	0.044	43.8	1	694	1.547	3.07
U.S gal/min	gpm (gal/min)	—	—	0.063	—	1	0.0022	0.0044
cubic ft/second	cfs (ft³/s)	—	0.0283	28.3	0.647	449	1	1.985
acre-ft/day	acre-ft/day	—	—	14.26	0.326	226.3	0.504	1

Conversion Factors (continued)

Weights				Conversion table			
Unit of measure	Symbol	T	lb	kg	g	mg	µg
ton (short)	T	1	2000	907	—	—	—
pound	lb	—	1	0.454	453,592	—	—
kilogram	kg	—	—	2.205	1	1000	1×10^6
gram	g	—	—	0.001	1	1000	1×10^6
milligram	mg	—	—	—	0.001	1	1000
microgram	µg	—	—	—	—	0.001	1

Note: 1 short ton = 2,000 lb; 1 long ton = 2,240 lb; 1 metric ton = 1,000,000 g = 1,000 kg = 2,205 lb.

Miscellaneous

1 acre-inch	= 27,154 gallons
1 horsepower	= 0.746 kilowatts
1 horsepower	= 550 foot-pounds per second
Degrees C	= 5/9 (F° - 32°)
Degrees F	= 9/5 (C° + 32°)
1 gram	= 15.43 grains
1 ppm	= 8.345 pounds per million gallons of water = 0.2268 pounds per acre-inch
1 U.S. gallon	= 8.345 pounds

Amounts of the major nutrients (N, P, and K), are always presented in terms of the nutrient itself. Only the nitrogen quantity in the ammonium compound (NH_4) is considered when expressed as ammonium nitrogen (NH_4-N).

Commercial fertilizer formulations for nitrogen, phosphorus, and potassium and recommendations are expressed in terms of N, P_2O_5, and K_2O. When comparing the nutrient content of a manure, waste, or residue with commercial fertilizer, use the conversion factors listed in Table 3.2 and make comparisons on the basis of similar elements, ions, and compounds.

"In response to economic and environmental concerns, and as a result of increased educational, cost-share and regulatory programs, the number of livestock producers who test their manure has increased dramatically in the last five to ten years" (Minnesota Department of Agriculture, 2005). In fact, many states and localities require producers to have a manure nutrient management plan for their operation. Having accurate manure analysis helps the accuracy of the plan and the likelihood of plan approval by the state, can save producers money, and protects water quality.

- *Important point:* The accuracy of a chemical analysis of manure is only as good as the sample sent to the lab. The sample collected should closely represent the material used as a fertilizer. Manure collected at one point production can be completely different from manure collected at another point. Manure characteristics can also change with the seasons. For the most-accurate results, manure should be sampled and analyzed as close as possible to the time when it will be used (Oklahoma State University, 2005).

Agricultural manure/waste, like wastewater biosolids, must be viewed not only as merely a disposal problem but also as a product of some value—a valuable resource. Applied at proper rates to cropland, manure/wastes improve the physical condition of the soil and reduce the need for commercial fertilizers.

TABLE 3.2
Factors for Determining Nutrient Equivalency

Multiply	By	To get
NH_3	0.824	N
NH_4	0.778	N
NO_3	0.226	N
N	1.216	NH_3
N	1.285	NH_4
N	4.425	NO_3
PO_4	0.326	P
P_2O_5	0.437	P
P	3.067	PO_4
P	2.288	P_2O_5
K_2O	0.830	K
K	1.205	K_2O
ppm	0.0083	lb/1000 gal

Livestock produce valuable amounts of fertilizer. Agricultural wastes, such as manure, are rich in plant nutrients. A recent report by Cornell University showed that approximately 75% of the nitrogen, 60% of the phosphorus, and 80% of the potassium fed to dairy cattle is excreted in manure (poultry and swine have higher values for phosphorus and potassium). In addition, manure supplies calcium, manganese, magnesium, zinc, copper, sulfur, and other micronutrients. Actual nutrient content of manure varies with type of animal, feed, manure storage system, and method of manure application.

Assuming no nutrient loss during handling and a value of $0.22 per pound for nitrogen, $0.20 per pound for phosphoric acid (P_2O_5), and $0.10 per pound for potash (K_2O) (based on 1991 pricing data):

- A 100-head beef herd produces $4,410 worth of fertilizer per year.
- A 100-head dairy herd produces $4,810 worth of fertilizer per year.
- A 100,000-bird broiler operation produces $3,485 worth of fertilizer per year.

Without manure analysis, farmers may buy more commercial fertilizer than needed or spread too much manure on their fields. Either practice can result in over fertilization which, in turn, can depress crop yields and cut profits. Improper spreading of manure can pollute surface water and groundwater, and contamination of wells by nitrates and bacteria can increase health risks.

To get an analysis of manure, take the following steps:

1. Contact the county extension agent or your local testing lab for a nutrient management kit. The kit may contain a manure sampling jar, soil test bags, record sheets, and instructions. Note that a fee may be charged for each soil sample tested.
2. Collect a *representative* manure sample. For daily spreading, take many small samples over a representative period. In a manure pack, collect samples from a variety of locations in the pile. Be sure to collect both manure and bedding materials. Agitate liquid manure systems before collecting samples.

3. Follow the specific instructions included in the kit for collecting samples from your liquid, solid, or semisolid system with a minimum of mess and effort. The small samples collected should be mixed together in a clean bucket. Place a portion of the mixture in the sample jar.
4. Keep samples cool and deliver them to the county extension agent early in the week to avoid storage over weekends. Also avoid sending samples for testing before a holiday.

Collect samples well in advance of the date manure is planned to be spread, so the test results can be used to calibrate the manure spreader. With liquid waste systems, collecting samples may be easiest when the manure is pumped into the spreader. Use these test results to calibrate the spreader for future applications of manure or to determine if additional chemical fertilizer is needed.

3.2.2.1 Sampling Techniques

Remember, taking a representative sample is critical to obtaining a reliable manure analysis. Manure nutrient composition can vary significantly within the same storage. We describe several recommended sampling techniques:

3.2.2.1.1 Broiler or Pullet House Litter

Dry litter varies across the width of the house—material near the curtains is different from that under feeders and waterers. Differences also occur between brood and grow-out areas and even the north and south sides of a house. These differences must be considered to get a representative sample. The following techniques allow samples to be taken with birds in the house.

1. Trench Method
Using a shovel (a narrow spade works as well), dig a trench as wide as the shovel across half of the broiler house (Figure 3.1). Start at the centerline of the house and dig a trench in the litter to the sidewall. If cake is present, cut the caked litter to the width of the shovel and collect it too. Place the entire contents of the trench on a tarp or drop cloth. Thoroughly mix the litter using a hoe. Place a portion of this well-mixed litter into a zipper-closing plastic bag. Place it in a second bag. Use the litter remaining on the tarp to backfill the trench.

2. Zigzag Method
Walk the entire house in a zigzag pattern (Figure 3.1) and grab 15 to 20 subsamples with a shovel or coffee can. Collect the entire depth of the litter, but do not remove soil beneath the litter. Place subsamples in a plastic bucket, and mix thoroughly. Take a small sample from the bucket and place it in a zipper-closing plastic bag. Place this bag in a second plastic bag.

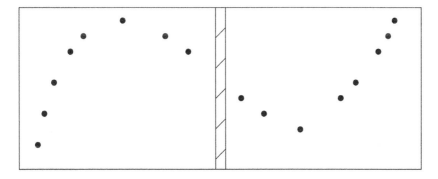

FIGURE 3.1 Taking poultry liter samples in the house using trench and zigzag methods. (Adaptation from Sampling Animal Manure, Oklahoma Cooperative Extension Service.)

3.2.2.1.2   Litter Inside a Breeder House (partially slatted)

A composite sample from a partially slatted breeder house can be sampled by collecting sub-samples from both slatted and litter areas. In all, collect at least 20 subsamples to get a representative sample of the building. Since two thirds of the house is under slats and one third is litter area, collect 14 cores from under the slats and 7 samples from the litter area. Sample through the slats using a soil probe or section of pipe. Collect litter samples similar to the zigzag method above. Place the slat and litter samples in a plastic bucket and mix thoroughly. Take a small sample from the bucket and place in a zipper-closing plastic bag. Place this bag in a second plastic bag.

3.2.2.1.3   Lagoon Effluent

If effluent is pumped from the top of a lagoon only, a sample need only be taken from the upper 2 ft. Samples taken from the upper layer of the lagoon should represent the contents of the layer for several weeks, although lagoons do change from month to month. The lagoon effluent should be sampled during the season when irrigation is intended. For instance, if the plan is to irrigate Bermuda grass in May and wheat in August, then take two effluent samples—April-May for the Bermuda grass, and July-August for the wheat.

1. Bucket-Toss Method

A simple effluent sampler is a rope attached to a small plastic bucket. Throw the bucket out into the lagoon, let it sink, and then slowly pull the bucket back to shore, being careful not to collect scum or solids with the sample. Then swirl the bucket and pour a subsample into a plastic container.

2. Dipper Method

Dipping is less accurate than the bucket-toss method. If you object to handling effluent covered ropes, use a plastic bottle securely taped to a long pole. Make sure that the pole is long enough to reach over any scum collected at the edge of the lagoon. Dip out a number of samples at different depths and locations, and then mix the samples together in a bucket. Swirl the bucket and pour a subsample into a plastic container.

3.2.2.1.4   Entire Lagoon Contents

Sometimes producers need to analyze the entire contents of a lagoon, or they need to measure chemicals deeper than 2 ft in the lagoon. Lagoons separate into layers. The bottom of the lagoon contains sludge. A scum or crust may form at the top of the lagoon. Between the sludge and scum is a large volume of liquid. To determine the total contents of a lagoon for diagnostic purposes, you must put together a sample from all the layers. Two choices are available—collect a complete column of the lagoon profile or collect material from each layer and mix it into a composite sample based on the mass of each layer. Either way means getting out on the lagoon in a boat.

1. Column Sampler

A number of column samplers are commercially available. All are basically a long hollow tube that is slowly lowered into the lagoon. Once the sampler reaches the bottom, the tube is closed off, so an entire column can be raised from the lagoon. Ensure that the sampler is long enough to reach the bottom of the lagoon and is large enough to collect an undisturbed sludge sample.

2. Grab Sampler

A discrete or grab sample is a small sample taken from one layer. The idea is to grab the sample without disturbing layers above or below it. Discrete samplers use water pressure to force sludge or liquid into the sampler. The "Sidewinder" sampler is an easy-to-build grab sampler for lagoons. Once collected, discrete samples can be analyzed separately or combined into a composite sample for the whole lagoon.

3.2.2.1.5   Waste Storage Pond/Settling Basin Slurry

Layers form in a waste storage pond just as they do in a lagoon. Sampling the entire contents of a pond requires the same techniques as a lagoon. Storage ponds are mixed before slurry is spread on the field as fertilizer. The buck-toss or dipper methods can be used to collect samples from ponds, but the pond must first be agitated. Solids contents change as the pond is pumped. Take small samples over the entire pumping period and mix them into a larger sample. Removal a small subsample from the well-mixed sample and place it in a plastic container.

3.2.2.1.6 Prefabricated Storage Structure Slurry
Aboveground storage structures are agitated before spreading. The return line on a pump agitator should have a valve to allow the taking of samples. Take a number of small samples while employing the storage structure. To collect samples from a propeller-agitated pit, use the bucket-throw or dip method. Remove a small subsample from the well-mixed sample and place in a plastic container.

3.2.2.1.7 Slurry/Semisolid from Pits beneath Slotted Floors
Column samplers used to sample lagoons work in storage pits as well. Homemade column samplers work just as effectively, though. Take a section of plastic pipe narrow enough to slip through the floor slots, but wide enough to collect undisturbed solids. Lower the pipe through the slots until it hits the bottom of the pit. Cap the upper end, trapping a column of manure. Empty the entire contents of the pipe into a plastic bucket. Take samples from a number of locations throughout the pit. Swirl or mix the contents of the bucket and pour a subsample into a plastic container.

3.2.2.1.8 Solid/Semisolid Manure off Feedlot Surfaces
Using a soil probe, take a minimum of 20 cores randomly from the pen surface. Walk the entire area of the pen in a zigzag pattern to make sure that cores are removed from all areas. Be careful to remove only manure, not the hardened soil beneath. Collect cores in a plastic bucket and mix them thoroughly. Take a small sample from the bucket and place it in a zipper-closing plastic bag. Place the bag in a second plastic bag. Manure characteristics change with the age of cattle and other management differences, so representative pens of the same age and similar management practices should be sampled.

3.2.2.1.9 Solid Manure from Stockpiles and Dry Stacks
Using a shovel, remove samples from several locations of at least 18 inches into the pile. Place the subsamples in a plastic bucket and mix. Do not allow the material to dry. Place a portion of the sample in a plastic bag. For added safety, place the bag in a second plastic bag.

3.2.2.1.10 Liquid and Slurry during Land Application
Sometimes getting a representative sample is easier when samples are collected during application. However, the total N concentration of samples collected in the field may be lower than samples taken from storage, because some ammonia is lost during application. Contact your local extension educator or crop consultant before using samples collected in the field for fertilizer recommendations.

1. Catch Cans in the Field
This technique is especially useful if slurry is spread from a honey wagon or tank truck. Randomly place a number of cans in the field. Collect waste from the cans and mix it in a large bucket immediately after spreading. Swirl the bucket to mix the contents, and pour a subsample into a plastic container.

2. Slurry or Liquid from a Big Gun Sprayer
Some big gun sprayers have a valve at the spray riser that is used to drain the hose. Place a bucket under the valve and open the valve while the gun is running. Open the valve slowly because big guns operate at high pressures. Collect a number of samples while pumping and mix together. Take a subsample from the well-mixed material and place in a plastic container.

3. Sample Liquids from a Sprinkler Nozzle
Impact sprinklers and low energy precision application spray nozzles work at lower pressures than big guns, so collecting a sample directly from the spray stream is safe. Place a bucket or cylinder directly in the stream. In large irrigation systems, collect a number of samples at different locations. Mix samples into a composite. Take a subsample of the well-mixed liquid and place it in a plastic container.

3.2.2.2   After Sampling

After collecting the samples, ship liquid and slurry samples in quart-sized plastic bottles with screw-top lids. Only fill bottles half full to allow for gas expansion, and squeeze flexible bottles slightly before screwing on the lid. Place solid and semisolid samples in zipper-closing plastic bags, and place a second plastic bag over both liquid and solid samples for extra safety. Use cardboard boxes to ship sample bottles and bags. Pack the box tightly with expanded Styrofoam peanuts or shredded paper and seal with strapping tape.

Preservatives are generally not needed for manure samples used for fertilizer recommendations. Other analyses may require special shipping and preservation. This is especially true when collecting samples for biological or bacteriological analysis.

3.2.2.3   Lab Analysis

The manure sample should be analyzed for nitrogen, phosphorus, potassium, moisture content, calcium, manganese, magnesium, sulfur, zinc, and copper. Nutrient content, primarily that of nitrogen, phosphorus, and potassium, is important when calculating land application rates and determining treatment techniques. A copy of the results are sent directly to the applicant and to the county extension agent. The agent will be able to answer questions and help plan fertilization and nutrient management programs.

3.3   MANURE CHARATERISTICS

- *Important point:* Whenever locally derived values for animal waste characteristics are available, this information should be given preference over the more general data used in this section.

- *Important point:* Carbon:Nitrogen ratios used in this section were established using the ash content in percent (dry weight basis) to determine the amount of carbon. The formula used, which estimates carbon in percent (dry weight basis), was:

$$C = \frac{100\% \text{ ash}}{1.8}$$

Total dissolved salts values are based on the work of Arrington and Pachek (1980).

3.3.1   As-Excreted Manure

"As-excreted" manure characteristics are the most reliable data available. Daily as-excreted manure production data are presented where possible in pounds per day per 1,000 pounds livestock live weight (lb/d/1000#) for typical commercial animals and birds. Units of cubic feet per day per 1,000 pounds live weight (ft³/d/1000#) allow waste production to be calculated on a volumetric basis. MC and TS are given as a percentage of the total wet weight (% w.b.) of the manure. TS are also given in units of lb/d/1000#. Other solids data and the nutrient content of the manure are presented in units of lb/d/1000# on a wet weight basis.

Manure and waste properties resulting from other situations, such as flushed manure, feedlot manure, and poultry litter, are the result of certain "foreign" materials being added or some manure components being lost from the as-excreted manure. Much of the variation in livestock waste characterization data in this section, and in other references, results largely from uncertain and unpredictable additions to and losses from the as-excreted manure.

Livestock manure and waste produced in confinement and semiconfinement facilities are of primary concern and are given the greatest consideration in this section. Manure from unconfined animals and poultry, such as those on pasture or range, are of lesser significance because handling and distribution problems are not commonly encountered.

3.3.2 Foreign Material in Manure

Foreign material commonly added to manure in the production facility are (1) bedding, or litter, (2) wasted and spilled feed and water, (3) flush water, (4) rainfall, and (5) soil. These materials are often added in sufficient quantities to change the basic physical and chemical characteristics of the manure. The resulting combination of manure and foreign material is called *waste*. Dust, hair, and feathers are also added to manure and waste in limited amounts. Hair and feathers, especially, can cause clogging problems in manure handling equipment and facilities though the quantities may be small. Other adulterants are various wood, glass, and plastic items and dead animals and birds.

3.3.2.1 Bedding

Livestock producers use a wide range of bedding materials, with the choice influenced by availability, cost, and performance properties. Both organic and inorganic materials are used successfully. Unit weights of materials commonly used for bedding dairy cattle are given in Table 3.3.

Quantities of bedding materials used for dairy cattle are shown in Table 3.4. The total weight of dairy manure and bedding is the sum of the weights of both parts. The total volume of dairy manure

TABLE 3.3
Unit Weights of Common Bedding Materials

| | lb/ft³ | |
Material	Loose	Chopped
Legume hay	4.25	6.5
Nonlegume hay	4.00	6.0
Straw	2.50	7.0
Wood shavings	9.00	
Sawdust	12.00	
Soil	75.00	
Sand	105.00	
Ground limestone	95.00	

Source: NRCS (1992).

TABLE 3.4
Daily Bedding Requirements for Dairy Cattle

| | Barn type (lb/d/1000#) | | |
Material	Stanchion stall	Free stall	Loose housing
Loose hay or straw	5.4	9.3	
Chipped hay or straw	5.7	2.7	11.0
Shavings or sawdust	3.1		
Sand, soil, or limestone	1.5		

Source: NRCS (1992).

TABLE 3.5
Dairy Waste Characterization—As Excreted*

| Component | Units | Cow | | |
		Lactating	Dry	Heifer
Weight	lb/d/1000#	80.00	82.00	85.00
Volume	ft³/d/1000#	1.30	1.30	1.30
Moisture	%	87.50	88.40	89.30
TS	% w.b.	12.50	11.60	10.70
	lb/d/1000#	10.00	9.50	9.14
VS	lb/d/1000#	8.50	8.10	7.77
FS	lb/d/1000#	1.50	1.40	1.37
COD	lb/d/1000#	8.90	8.50	8.30
BOD_5	lb/d/1000#	1.60	1.20	1.30
N	lb/d/1000#	0.45	0.36	0.31
P	lb/d/1000#	0.07	0.05	0.04
K	lb/d/1000#	0.26	0.23	0.24
TDS		0.85		
C:N ratio	10	13	14	

Source: NRCS (1992).

* Increase solids and nutrients by 4% for each 1% feed waste more than 5%.

and bedding is the sum of the manure volume plus half of the bedding volume. Only half of the bedding volume is used to compensate for the void space in bedding materials.

- *Important point:* Broiler producers replace the bedding material after three to six batches, or once or twice a year. The typical 20,000-bird house requires about 10 tons of wood shavings for a bedding depth of 3 to 4 inches.

3.3.2.2 Wasted Feed and Water

Wasted feed has a great influence on the organic content of manure. Feed consumed by animals is 50% to 90% digested, but spilled feed is undigested. A pound of spilled feed results in as much waste as 2 to 10 lb of feed consumed. Small quantities, about 3%, of wasted feed are common and very difficult to see. Wastage of 5% is common and can be observed. Obvious feed wastage is indicative of 10% or more waste. Anticipated feed waste of more than 5% should be compensated for as noted on the "as excreted" manure data summaries (Tables 3.5, 3.8, 3.11, 3.14, 3.17, 3.18, 3.19, 3.20).

Wasted water must be expected and controlled. Excess MC and increased waste volume can hamper equipment operation and limit the capacity of manure handling and storage facilities. Faulty waterers and leaky distribution lines cause severe limitations and problems in the manure management system. Excess water from foggers and misters used for cooling stock in hot weather may also be of concern in some instances.

3.3.2.3 Soil

Soil is another natural adulterant of livestock manure. Its presence is most common on dairies on which the cows have access to paddocks and pastures. Dry soil adheres to the cows' bodies in limited amounts. Wet soil or mud adheres even more, and either falls off or is washed off at the dairy barn. Soil and other inorganic materials used for freestall base and bedding are also added to the

TABLE 3.6
Dairy Waste Characterization—Milking Center

Component	Units MH	Milking center*			
		MH+MP	MH**	+ MP	+ HA***
Volume	ft³/d/1000#	0.22	0.60	1.40	1.60
Moisture	%	99.72	99.40	99.70	98.50
TS	% w.b.	0.28	0.60	0.30	1.50
VS	lb/1000 gal	12.90	35.00	18.30	99.96
FS	lb/1000 gal	10.60	15.00	6.70	24.99
COD	lb/1000 gal	25.30	41.70		
BOD_5	lb/1000 gal		8.37		
N	lb/1000 gal	0.72	1.67	1.00	7.50
P	lb/1000 gal	0.58	0.83	0.23	0.83
K	lb/1000 gal	1.50	2.50	0.57	3.33
C:N ratio	10	12	10	7	

Source: NCRS (1992).
* MH: Milk house; MP: Milking parlor; HA: Holding area.
** Holding area scraped and flushed—manure excreted.
*** Holding area scraped and flushed—manure included.

manure. Soil or other inorganic materials commonly added to manure can result in waste that has double the fixed solids content of "as excreted" dairy manure.

3.3.3 DAIRY

Manure characteristics for lactating and dry cows and for heifers are listed in Table 3.5. These data are appropriate for herds of moderate to high milk production. Quantities of dairy manure vary widely from small cows to large cows and between cows at low production levels and high production levels. Dairy feeding systems and equipment often allow considerable feed waste, which in most cases is added to the manure. Feed waste of 10% can result in an additional 40% of total solids in dairy waste. Dairy cow stalls are often covered with bedding materials that improve animal comfort and cleanliness. Virtually all of the organic and inorganic bedding materials used for this purpose will eventually be pushed, kicked, or carried from the stalls and added to the manure. The characteristics of these bedding materials are imparted to the manure. Quantities of bedding materials added to cow stalls and resting areas are shown in Table 3.4.

Milking centers—the milk house, milking parlor, and holding area—can produce about 50% of waste volume, but only about 15% of the total solids in a dairy enterprise (Table 3.6). Because this very dilute wastewater has different characteristics than the waste from the cow yard, it is sometimes managed by a different procedure. Values used to compute characteristics from mikhouses came from research by Cornell University completed in 1979 in New York.

About 5 to 10 gal of fresh water per day for each cow milked are used in a milking center where flushing of wastes is not practiced. However, where manure flush cleaning and automatic cow washing are used, water use can be 150 gal/d/cow or more. Dairies employing flush cleaning systems use water in approximately the following percentages for various cleaning operations:

- Parlor cleanup and sanitation   10%
- Cow washing   30%
- Manure flushing   50%
- Miscellaneous   10%

TABLE 3.7
Dairy Waste Characterization–Lagoon

Component	Units	Lagoon		
		Anaerobic		
		Supernatant	Sludge	Aerobic* supernatant
Moisture	%	99.75	90.00	99.95
TS	%w.b.	0.25	10.00	0.05
VS	lb/1000 gal	9.16	383.18	1.67
FS	lb/1000 gal	11.66	449.82	2.50
COD	lb/1000 gal	12.50	433.16	1.25
BOD_5	lb/1000 gal	2.92		0.29
N	lb/1000 gal	1.67	20.83	0.17
NH_4-N	lb/1000 gal	1.00	4.17	0.10
P	lb/1000 gal	0.48	9.16	0.08
K	lb/1000 gal	4.17	12.50	
C:N ratio		3	10	

Source: NRCS (1992).
* Milk house and milking parlor wastes only.

TABLE 3.8
Beef Waste Characterization—As Excreted*

Component	Units	Feeder, yearling 750 to 1,100 lb		450 to 750 lb	Cow
		High forage diet	High energy diet		
Weight	lb/d/1000#	59.10	51.20	58.20	63.00
Volume	ft³/d/1000#	0.95	0.82	0.93	1.00
Moisture	%	88.40	88.40	87.00	88.40
TS	% w.b.	11.60	11.60	13.00	11.60
	lb/d/1000#	6.78	5.91	7.54	7.30
VS	lb/d/1000#	6.04	5.44	6.41	6.20
FS	lb/d/1000#	0.74	0.47	1.13	1.10
COD	lb/d/1000#	6.11	5.61	6.00	6.00
BOD_5	lb/d/1000#	1.36	1.36	1.30	1.20
N	lb/d/1000#	1.31	0.30	0.30	0.33
P	lb/d/1000#	0.11	0.094	0.10	0.12
K	lb/d/1000#	0.24	0.21	0.20	0.26
C:N ratio		11	10	12	10

Source: NRCS (1992).
* Average daily production for weight range noted. Increase solids and nutrients by 4% for each 1% feed waste more than 5%.

Lagoons that receive a significant loading of manure, such as from the holding area or the cow feed yard, generally operate in an anaerobic mode (Table 3.7). Supernatant (upper liquid layer of the lagoon) concentration in an anaerobic lagoon is much greater than that in an aerobic lagoon. Anaerobic dairy lagoon sludge accumulates at a rate of about 0.073 cubic foot per pound of TS

TABLE 3.9
Beef Waste Characterization—Feedlot Manure

| Component | Units | Unsurfaced lot* | Surfaced lot** | |
			High forage diet	High energy diet
Weight	lb/d/1000#	17.50	11.70	5.30
Moisture	%	45.00	53.30	52.10
TS	% w.b.	55.00	46.70	47.90
	lb/d/1000#	9.60	5.50	2.50
VS	lb/d/1000#	4.80	3.85	1.75
FS	lb/d/1000#	4.80	1.65	0.75
N	lb/d/1000#	0.21		
P	lb/d/1000#	0.14		
K	lb/d/1000#	0.03		
C:N ratio		13		

Source: NRCS/USDA (1996).
* Dry climate (annual rainfall less than 15 inches); annual manure removal.
** Dry climate; semiannual manure removal.

added to the lagoon, equivalent to about 266 ft^3 per year for each 1,000 lb lactating cow equivalent (100% of waste placed in lagoon).

If a dairy waste lagoon receives wastewater only from the milk house or the milking parlor, the lagoon generally exhibits a very dilute supernatant and operates in an aerobic mode (Table 3.7). The rate of sludge accumulation in such lagoons is slow.

3.3.4 BEEF

Table 3.8 lists characteristics of as-excreted beef manure. Beef waste of primary concern are those from the feedlots (Table 3.9). The characteristics of these solid wastes vary widely because of factors that include climate, diet, feedlot surface, animal density, and cleaning frequency. The soil in unsurfaced beef feedlots is readily incorporated into the manure because of the animal movement and cleaning operations. Wasted feed is an important factor in the characterization of beef wastes.

Beef feedlot runoff water also exhibits wide variations in character (Tables 3.10a and 3.10b). The influencing factors that are responsible for feedlot waste variations are similar to those listed for solid wastes. Surfaced feedlots produce more runoff than unsurfaced lots.

3.3.5 SWINE

Swine waste and waste management systems have been widely studied, and much has been reported on swine manure properties. Table 3.11 lists characteristics of as-excreted swine manure from feeding and breeding stock. Breeding stock manure characteristics, also shown in Table 3.11, are subject to less variation than are the characteristics of manure for growing animals. Wasted feed also significantly changes manure characteristics. A 10% feed waste increases manure total solids by 40%.

Ration components can make a significant difference in manure characteristics. Corn, the principal grain in swine rations, is highly digestible (90%). Table 3.11 was developed for corn-based rations. If a grain of lower digestibility, such as barley (79%), is substituted for 50% of the corn in the ration, the TS of the manure increase 41% and the VS increase 43% above that of a ration based on corn. Wasted feed further increases the necessary size of storage units and lagoon facilities needed.

TABLE 3.10a
Beef Waste Characterization—Feedlot Runoff Pond

Component	Units	Runoff pond	
		Supernatant	Sludge
Moisture	%	99.70	82.80
TS	% w.b.	0.30	17.20
VS	lb/1000 gal	7.50	644.83
FS	lb/1000 gal	17.50	788.12
COD	lb/1000 gal	11.67	644.83
N	lb/1000 gal	1.67	51.66
NH_4-N	lb/1000 gal	1.50	
P	lb/1000 gal	X	17.50
K	lb/1000 gal	7.50	14.17

Source: NRCS/USDA (1996).

TABLE 3.10b
Nitrogen Content of Cattle Feedlot Runoff[1]

	lb N/acre-inch		
Annual rainfall	Below average conditions[2]	Average conditions[3]	Above-average conditions[4]
< 25 inches	360	110	60
25 to 35 inches	60	30	15
> 35 inches	15	10	5

Source: Alexander & Margheim (1974).

[1] Applies to waste storage ponds that trap rainfall runoff from uncovered, unpaved feedlots. Cattle feeding areas make up 90% or more of the drainage area. Similar estimates were not made for phosphorus and potassium. Phosphorus content of the runoff will vary inversely with the amount of solids retained on the lot or in settling facilities.

[2] No settling facilities are between the feedlot and pond, or the facilities are ineffective. Feedlot topography and other characteristics are conducive to high solids transport or cause a long contact time between runoff and feedlot surface. High cattle density—more that 250 head per acre.

[3] Sediment traps, low gradient channels, or natural conditions that remove appreciable amounts of solids from runoff. Average runoff and solids transport characteristics. Average cattle density—125 to 250 head per acre.

[4] Highly effective solids removal measures, such as vegetated filter strips or settling basins that drain liquid waste through a pipe to storage pond. Low cattle density—less than 120 head per acre.

A common procedure for collecting and storing swine waste under slatted floors is in deep or shallow tanks that may be allowed to overflow to lagoons or longer-term storage units. Daily accumulation of such waste cannot be accurately predicted. Table 3.12 presents concentration data on solids and nutrients in swine waste in tanks. Using these concentrations and the volume of waste on hand, plans for use of the waste can be made.

Swine waste storage structures and facilities must make allowances for wasted water. Small pigs, especially, play with automatic waterers and can waste up to 3 gal of water per day per head. Table 3.13 gives data on the nature of rainfall runoff and settling basin sludge from surfaced swine feedlots exposed to precipitation.

TABLE 3.11
Swine Waste Characterization—As Excreted*

Component	Units	Grower 40–220 lb	Replacement gilt	Sow Gestation	Sow Lactation	Boar	Nursing/ nursery pig 0–40 lb
Weight	lb/d/1000#	63.40	32.80	27.20	60.00	20.50	106.00
Volume	ft³/d/1000#	1.00	0.53	0.44	0.96	0.33	1.70
Moisture	%	90.00	90.00	90.00	90.00	90.00	90.00
TS	% w.b.	10.00	10.00	9.20	10.00	9.30	10.00
	lb/d/1000#	6.34	3.28	2.50	6.00	1.90	10.60
VS	lb/d/1000#	5.40	2.92	2.13	5.40	1.70	8.80
FS	lb/d/1000#	0.94	0.36	0.37	0.60	0.30	1.80
COD	lb/d/1000#	6.06	3.12	2.37	5.73	1.37	9.80
BOD_5	lb/d/1000#	2.08	1.08	0.83	2.00	0.65	3.40
N	lb/d/1000#	0.42	0.24	0.19	0.47	0.15	0.60
P	lb/d/1000#	0.16	0.08	0.06	0.15	0.05	0.25
K	lb/d/1000#	0.22	0.13	0.12	0.30	0.10	0.35
TDS	1.29						
C:N ratio	7	7	6	6	6	8	

Source: NRCS/USDA (1996).

* Average daily production for weight range noted. Increase solids and nutrients by 4% for each 1% feed waste more than 5%.

TABLE 3.12
Swine Waste Characterization—Storage Tanks under Slats

Component	Units	Farrow	Nursery	Grow/finish	Breeding/gestation
Moisture	%	96.50	96.00	91.00	97.00
TS	% w.b.	3.50	4.00	9.00	3.00
VS	lb/1000 gal	189.85	233.27	562.35	149.96
FS	lb/1000 gal	101.64	99.97	187.45	99.97
N	lb/1000 gal	29.16	40.00	52.48	25.00
NH_4-N	lb/1000 gal	23.32	33.32		
P	lb/1000 gal	15.00	13.32	22.50	10.00
K	lb/1000 gal	23.32	13.32	18.33	17.50
C: N ratio	x	4	3	6	3

Source: NRCS/USDA (1996).

Anaerobic lagoons are used extensively for swine waste in the United States. Supernatant, the upper liquid layer of properly operating swine lagoons, is often brown or purple. Its characteristics are listed in Table 3.13. Light yellowish-green lagoon supernatant is generally less concentrated, and black is generally more concentrated than indicated in the table.

Sludge accumulates in a good anaerobic swine lagoon at a rate of 0.0485 ft³ of TS placed in the lagoon, about 12 ft³ per grower or finisher equivalent annually.

3.3.6 POULTRY

Because of the high degree of industry integration, standardized rations, and complete confinement, layer and broiler manure characteristics vary less than those of other species. Turkey production

TABLE 3.13

Swine Waste Characterization—Anaerobic Lagoon; Feedlot Runoff

Components	Units	Anaerobic lagoon		Feedlot runoff*	
		Supernatant	Sludge	Runoff water	Settling basin sludge
Moisture	%	99.75	92.40	98.50	88.8
TS	% w.b.	0.25	7.60	1.50	11.2
VS	lb/1000 gal	10.00	379.89	90.7**	
FS	lb/1000 gal	10.83	253.27	21.3**	
COD	lb/1000 gal	10.00	538.18		
BOD$_5$	lb/1000 gal	3.33			
N	lb/1000 gal	2.91	25.00	2.00**	5.6**
NH$_4$-N	lb/1000 gal	1.83	6.33	1.20**	4.5**
P	lb/1000 gal	0.63	22.50	0.38**	2.2**
K	lb/1000 gal	3.16	63.31	1.10**	10.0**
C:N ratio		2	8		

Source: NRCS/USDA (1996).

* Semihumid climate (approx. 30 in annual rainfall); annual sludge removal.

** lb/yr/1000

TABLE 3.14

Poultry Waste Characterization—As Excreted*

Component	Units	Layer	Pullet	Broiler	Turkey	Duck
Weight	lb/d/1000#	60.50	45.60	80.00	43.60	
Volume	ft^3/d/1000#	0.93	0.73	1.26	0.69	
Moisture	%	75.00	75.00	75.00	75.00	
TS	% w.b.	25.00	25.00	25.00	25.00	
	lb/d/1000#	15.10	11.40	20.00	10.90	12.0
VS	lb/d/1000#	10.80	9.70	15.00	9.70	7.0
FS	lb/d/1000#	4.30	1.70	5.00	1.25	5.0
COD	lb/d/1000#	13.70	12.20	19.00	12.30	9.5
BOD$_5$	lb/d/1000#	3.70	3.30	5.10	3.30	2.5
N	lb/d/1000#	0.83	0.62	1.10	0.74	0.7
P	lb/d/1000#	0.31	0.24	0.34	0.28	0.3
K	lb/d/1000#	0.34	0.26	0.46	0.28	0.5
TDS		2.89				
C:N ratio		7	9	8	7	6

Source: NRCS/USDA (1996).

* Increase solids and nutrients by 4% for each 1% feed waste more than 5%.

is approaching the same status. Table 3.14 presents waste characteristics for as-excreted poultry manure.

Table 3.15 lists data for poultry flocks that use a litter (floor) system. Bedding materials, whether wood, crop, or other residue, are largely organic matter with little nutrient components. Litter moisture in a well-managed house generally is in the range of 25% to 35%. Higher moisture levels in the litter result in greater weight and reduced levels of nitrogen.

TABLE 3.15

Poultry Waste Characterization—Litter

Component	Units	Layer high-rise*	Broiler	Turkey	Broiler breeder**	Duck**
Weight	lb/d/1000#	24.00	35.00	24.30		
Moisture	%	50.00	24.00	34.00	34.00	11.20
TS	% w.b.	50.00	76.00	66.00	66.00	88.80
	lb/d/1000#	12.00	26.50	16.10		
VS	lb/d/1000#	21.40	58.60			
FS	lb/d/1000#	5.10	30.20			
N	lb/d/1000#	0.425	0.68	0.88	1.06	2.31
NH_4-N	lb/d/1000#			0.01		
P	lb/d/1000#	0.275	0.34	0.40	1.32	
K	lb/d/1000#	0.30	0.40	0.45	1.19	
C: N ratio			9	14		

Source: NRCS/USDA (1996).

* No bedding or litter material added to waste.

** All values % w.b.

Most broiler houses are now cleaned out one or two times a year. Growers generally have five or six flocks of broilers each year, and taking the "cake" out after each flock is a common practice. The cake is generally 1 to 2 in of material. About 2 or 3 in of new litter is placed on the floor before the next flock. Much of the waste characterization data for broiler litter are based on five or six cycles per year.

When a grower manages for a more-frequent, complete cleanout, the data in Table 3.15 need adjustment. The birds still produce the same amount of N, P, and K per day. However, with a more frequent cleanout the density and moisture content of the litter is different and the nutrients are less concentrated. The nutrient amount is less compared to the litter volume because less time is allowed for the nutrients to accumulate. A further complication is that nitrogen is lost to the atmosphere during storage while fresh manure is continually being deposited.

High-rise layer houses use no bedding and store manure for up to a year. Bird densities in high-rise houses have increased greatly in recent years, and the manure characteristics have been subject to great change. Use of current data for high-rise characterization is important.

As in other livestock operations, feed waste greatly increases the volume and organic content of the waste. A 10% wastage of feed, when added to the manure, increases TS by 42%.

Poultry lagoon supernatant and sludge characteristics are listed in Table 3.16. Anaerobic lagoon supernatant from good layer and pullet lagoons is brown, rosy, or burgundy. Yellowish-green supernatant is less concentrated. Blackish supernatant is more concentrated and generally has a higher value than those shown.

Layer lagoon sludge is much denser than pullet lagoon sludge because of its high grit or limestone content. Layer lagoon sludge accumulates at rate of about 0.0294 ft³/lb of waste TS added to the lagoon, and pullet lagoon sludge accumulates at a rate of 0.0454 ft³/lb of TS. This is equivalent to about 0.6 ft³ per layer and 0.3 ft³ per pullet annually.

3.3.7 Veal

Data on manure characteristics from veal production are shown in Table 3.17. Sanitation in veal production is an extremely important factor, and waste management facilities should be planned for handling as much as 3 gallons of wash water per day per calf.

TABLE 3.16
Poultry Waste Characterization—Anaerobic Lagoon

Component	Units	Layer		Pullet	
		Supernatant	Sludge	Supernatant	Sludge
Moisture	%	99.50	86.90	99.70	92.60
TS	% w.b.	0.50	13.10	0.30	7.40
VS	lb/1000 gal	18.33	404.06	10.83	314.09
FS	lb/1000 gal	23.32	687.32	14.17	302.42
N	lb/1000 gal	6.25	32.50	3.00	24.17
NH_4-N	lb/1000 gal	4.58	7.66	2.24	4.91
P	lb/1000 gal	0.83	45.82	0.75	27.49
K	lb/1000 gal	8.33	6.00	7.00	6.17
C:N ratio	2	7	2		7

Source: NRCS/USDA (1996).

TABLE 3.17
Veal Waste Characterization—As Excreted

Component	Units	Veal feeder
Weight	lb/d/1000#	60.00
Volume	ft³/d/1000#	0.96
Moisture	%	97.50
TS	lb/d/1000#	2.50
	lb/d/1000#	1.50
VS	lb/d/1000#	0.85
FS	lb/d/1000#	0.65
COD	lb/d/1000#	1.50
BOD_5	lb/d/1000#	0.37
N	lb/d/1000#	0.20
P	lb/d/1000#	0.03
K	lb/d/1000#	0.25
C: N ratio		2

Source: NRCS/USDA (1996).

3.3.8 SHEEP

As-excreted manure characteristics for sheep are limited to those for the feeder lamb (Table 3.18). In some cases, bedding may be a significant component of sheep waste.

3.3.9 HORSE

Table 3.19 lists characteristics of as-excreted horse manure. Because large amounts of bedding are used in the stabling of most horses, qualities and quantities of wastes from these stables generally are dominated by the kind and volume of bedding used.

TABLE 3.18
Lamb Waste Characterization—As Excreted*

Component	Units	Lamb
Weight	lb/d/1000#	40.00
Volume	ft³/d/1000#	0.63
Moisture	%	75.00
TS	% w.b.	25.00
	lb/d/1000#	10.00
VS	lb/d/1000#	8.30
FS	lb/d/1000#	1.76
COD	lb/d/1000#	11.00
BOD_5	lb/d/1000#	1.00
N	lb/d/1000#	0.45
P	lb/d/1000#	0.07
K	lb/d/1000#	0.30
C:N ratio		10

Source: NRCS/USDA (1996).

* Increase solids and nutrients by 4% for each 1% feed waste more than 5%.

TABLE 3.19
Horse Waste Characterization—As Excreted*

Component	Units	Horse
Weight	lb/d/1000#	50.00
Volume	ft³/d/1000#	0.80
Moisture	%	78.00
TS	% w.b.	22.00
	lb/d/1000#	11.00
VS	lb/d/1000#	9.35
FS	lb/d/1000#	1.65
N	lb/d/1000#	0.28
P	lb/d/1000#	0.05
K	lb/d/1000#	0.19
C:N ratio		19

Source: NRCS/USDA (1996).

* Increase solids and nutrients by 4% for each 1% feed waste more than 5%.

3.3.10 RABBIT

Some properties of rabbit manure are listed in Table 3.20. These properties refer only to feces; no urine has been included. Reliable information on daily production of rabbit manure, feces, and urine is not available.

TABLE 3.20
Rabbit Waste Characterization—As Excreted*

Component	Units	Rabbit
VS	% d.b.	0.86
FS	% d.b.	0.14
COD	% d.b.	1.00
N	% d.b.	0.03
P	% d.b.	0.02
K	% d.b.	0.03
C:N ratio		16

Source: NRCS/USDA (1996).

* Increase solids and nutrients by 4% for each 1% feed waste
   more than 5%.

3.3.11 FLUSH WATER

Hydraulic manure transport, or flush cleaning, is an effective method of manure collection and handling, but it uses relatively large quantities of water. Small quantities of manure can be diluted 5 to 10 times in the cleaning process; therefore, waste handling problems are multiplied.

Because the resulting quantity of waste or wastewater is large, lagoons and irrigation equipment are usually parts of waste management systems using flush cleaning. While fresh water is required for cleaning in many instances, recycled lagoon liquid (supernatant) can be used and can greatly reduce the volume of fresh water needed for waste management. Where necessary, the approval of appropriate state and local authorities should be requested before lagoon supernatant recycling is implemented.

Because quantities of flush water vary widely between operations, estimated values should be based on local calculations or measurement. Estimates of flush water requirements for various mechanisms and for various species may be made from the following equations and test results:

Swine—(siphon, gated tank, or tipping tank)

$$Q = 0.5L \times W$$

where:
Q = Flush water vol. gal/flush.
L = Gutter length, ft.
W = Gutter width, ft.

Dairy

	Gated tank	Pump flush
Gal/d/ft² alley surface	2.5	15.0
Gal/d/cow	80.0	550.0

Dairies with gated tank flush cleaning and automatic cow washing commonly use 100 to 150 gal/d/cow, but multiple flushing and alternative equipment may double this amount.

Poultry—(pump flush) 1.0 to 1.5 gal/bird/flush.

3.4 SUMMARY

Manure characteristics are essential to making full use of livestock wastes, a valuable resource.

CHAPTER REVIEW QUESTIONS

1. What manure characteristics are important for fertilizer?
2. What properties affect manure collection, handling, and storage?
3. What variables can affect CAFO animal wastes? What causes these variations?
4. What is the importance of on-site sampling, testing, and data collection?
5. What physical and chemical properties are important in determining animal waste characteristics?
6. How does water content affect manure sampling, collection, and storage?
7. What is the difference between manure, manure/waste, and residue? Why is this distinction important?
8. How does water content, as opposed to solids content, affect the weight of as-excreted manure?
9. When is manure/waste a significant source of problem odor?
10. What is the difference between wet weight and dry weight? Is it significant?
11. What conversion factors should be used when making nutrient content comparisons?
12. Why is manure sampling important?
13. What is a manure nutrient management plan?
14. What does sampling technique have to do with analysis accuracy?
15. Discuss manure as a disposal problem. How could it affect the bottom line?
16. Describe the steps to take to get a manure analysis.
17. What sampling techniques are used for broiler or pullet house bedding? For litter inside a breeder house? For lagoon effluent? For entire lagoon contents? For waste storage pond or settling basin slurry? For prefabricated storage structure slurry? For slurry and semisolid wastes from pits beneath slotted floors? For solid and semisolid manure off feedlot surfaces? For solid manure from stockpiles or dry stacks? For liquid and slurry during land application?
18. For what elements should a manure sample be analyzed?
19. What manure type provides the most reliable data?
20. What foreign materials are commonly added to or found in manure?
21. How do these materials affect manure collection, sampling, storage, and use?
22. What characteristics are common to dairy manures? Beef? Swine? Poultry? Veal? Horse? Rabbit?
23. What are the advantages and disadvantages of hydraulic manure transport?

THOUGHT-PROVOKING QUESTIONS

1. What are the advantages and disadvantages of manure as a soil additive? How does it compare to inorganic fertilizers? What can manure provide that inorganic fertilizers don't? What are advantages of inorganic fertilizer?
2. What is the long-term effect of using inorganic fertilizers instead of manure?

REFERENCES

Agricultural Waste Management Field Handbook, Part 651, Chapter 4, Agricultural Waste Characteristics, USDA, NRCS 1992 (w/revisions 1996). Washington, DC. ftp://ftp.wcc.nrcs.usda.gov/downloads/waste-mgmt/AWMFH/awmfh-chapr4.pdf. Accessed January 8, 2005.

Alexander, E.L., & Margheim, G.A. 1994. Personal communication with C.E. Fogg, in *Agricultural Waste Management Field Handbook* (1992), Part 651.0405. Washington, DC: National Resource Conservation Service.

Arrington, R.M. and Pachek, C.E., 1980. Soil nutrient content of manures in an arid climate. Conference on confined animal production and water quality. GPAC publication 151. Great Plains Agricultural Council, Denver, Co. pp. 259–266.

Minnesota Department of Agriculture. 2005. Manure testing laboratory certification program. www.mda.state.mn.us/appd/mncert.htm.

NRCS 1992. *Agricultural Waste Management Field Handbook*, Part 651, Chapter 4. Washington, DC: National Resource Conservation Service.

NRCS/USDA. 1996. *Agricultural Waste Management Field Handbook* (rev). Washington, DC: National Resource Conservation Service and United States Department of Agriculture.

NWPS-18. 2004. *Manure Characteristics*. Ames, IAQ: Midwest Plan Service.

OSU. 2005. Sampling animal manure. Oklahoma State University. www.osuextra.ocm. Last accessed January 5, 2005.

Proulx, A. 2002. *That Old Ace in the Hole*. New York: Scribner.

4 Manure Storage

Cattle produce nearly 1 billion tons of organic waste each year. The average feedlot steer produces more than 47 pounds of manure every twenty-four hours. Nearly 500,000 pounds of manure are produced daily on a standard 10,000-head feedlot. This is the rough equivalent of what a city of 110,000 would produce in human waste. There are 42,000 feedlots in 13 U.S. states.

(Ensminger, 1991, p. 187)

4.1 INTRODUCTION

Manure and wastewater storage and handling needs are highly specific to the condition and location of each facility, and they differ from farm to farm. The capability to store manure reduces or eliminates the need to collect, remove, and spread manure frequently and gives the producer control over when manure must be removed and applied to land—manure nutrients are best used when applied just before or during the growing season of the crop. Simply, according to NWPS 18-1, Section 2 (2001), "the primary reason to store manure is to allow the producer to land apply the manure at a time that is compatible with the climatic and cropping characteristics of the land receiving the manure and the producer's time availability...the type of crop and method of manure application are important considerations in planning manure storage facilities" (p. 1). As livestock operations have increased in size and manure management systems have evolved from solid and semisolid systems to liquid systems, the need for storage has become more pronounced.

- *Important point:* Land application during periods of saturated, wet, frozen, or snow-covered soil conditions is not recommended and is prohibited in some states.

Manure and wastewater storage and handling includes components and activities associated with the production facility, feedlot, manure and wastewater storage and treatment structures and areas, and any areas or mechanisms used to facilitate transfer of manure and wastewater. For most concentrated animal feeding operations (CAFOs), addressing this element requires a combination of conservation practices, management activities, and facility upgrades designed to meet the production needs of the livestock operation, while addressing environmental concerns specific to each operation.

- *Important point:* Manure storage provides for better use of farm labor; better use of equipment; increased ability to apply manure nutrients when they are most needed by the crop; increased ability to apply manure at times that avoid adverse climatic and soil conditions, such as during saturated or frozen soil; and minimizes the frequency of manure storage agitation and application events and the corresponding frequency of released odors.

Various issues are associated with manure storage duration. For example, *long-term storage* (6 to 12 months) offers the operator the greatest flexibility by accommodating long winter seasons and fits most cropping schedules. It also provides maximum flexibility for scheduling custom spreading operations and, for operators who irrigate, it provides storage until the growing season. *Mid-term storage* (3 to 6 months) accommodates short periods with frozen, snow-covered, or saturated soil. However, this storage period may not work with traditional crop rotations. In addition, some pastures, grassland, or hay land might be needed for spreading during the growing season. *Short-term storage* (3 months or less) is best for warm climates with no long periods of frozen or saturated soil. Warm climate regions also provide more easily available pastures, grassland, and hay land for spreading. Frequent spreading requires, as needed, the availability of equipment and labor.

The actual *size* of a manure storage system depends on the volume of manure produced per day. This depends on animal numbers or animal units (AU), species, and size. Manure volume is best determined by actually measuring the amount of manure that is hauled per year from a production facility, but it can be estimated from published values developed from experimental and field studies if actual values are not available.

- *Important point:* In determining a storage volume for a farm, at least five factors should be considered: (1) regulatory requirements, (2) animal type, size, and number, (3) storage time needed, (4) wastewater use and water wastage expected, and (5) rainwater runoff from manure-spread lots and evaporation.

Siting is another factor to be considered when constructing manure storage and handling systems, because most states have specific regulatory requirements that affect the siting and construction of manure storage facilities. The extent of the various restrictions is usually a function of operation size. Consider the construction of earthen manure storage, for example. Site selection and construction of this type of manure storage requires soil and site investigation and evaluation procedures beyond what might be needed for a facility constructed of concrete or steel. These procedures ensure that soil leaching rates of nutrients and manure liquids will be acceptable. Additionally, an environmental assessment should be made to help minimize risks. The environmental assessment takes into consideration various levels of risk, including proximity to surface water and drinking water wells. It also answers questions such as: Is the construction area a floodplain? Are homes, public use areas, or businesses nearby at high or low risk? Are neighbors downwind of prevailing weather winds? Is the site elevation lower or higher than the storage elevation? Is the storage facility site highly visible because of location close to a road, or are the topography, vegetation, or buildings configured to visually screen the storage facility? Is water runoff situated in a manner that diverts it around, instead of flowing into, the manure storage?

Manure storages often must take into consideration *separation distance requirements*. Specific requirements vary by site and can be determined by checking with local and state authorities. *Setback distances* are intended to reduce the environmental impact of a manure storage facility on off-farm dwellings, businesses, and public entities, such as roads, parks, and churches. These distances usually depend on the size of the livestock and poultry operation, type of manure stored, and whether the storage is covered. *Setback distance* is measured from the manure storage structure to the nearest nonowned residence, public building, or entity. The regulatory agency may require that public notice be given to neighbors within a certain distance (for example, within 1.5 times the required setback) before construction begins or before the operation permit can be obtained.

- *Important point:* The regulatory setback distance, with respect to locating a site for animal manure storage, is based on horizontal distance from manure storage to some feature, such as neighboring residences or water supplies.

- *Important point:* Separation distances may also be required from groundwater sources. A vertical separation (usually 3 to 5 ft) may be required between the bottom of an earthen storage facility and the seasonal high water table.

Locating a manure storage structure in a floodplain presents several potential hazards. The facility may be inundated and structurally compromised because of flooding. In most locations, regulations prevent or limit construction in floodplains, because floodwater can exert unbalanced, inward hydraulic pressure on earthen impoundments if the manure level inside the basin is lower than the water level outside the basin. Such pressure can impair the seal and compromise structural integrity.

Before constructing an earthen storage or lagoon, a detailed soil and site evaluation is needed. When a site is found to have unsuitable soil materials for construction, a clay liner, synthetic liner, or soil amendment that modifies the existing soil properties may provide a means to attain the required permeability.

Key term: *Permeability* is a measure of how readily liquids can pass through a soil. Silt loams and sandy loam soils are relatively permeable. An impermeable soil is one that does not allow liquids to pass through easily. Clay loams and loams are relatively impermeable. Manure storages should be located over impermeable soil so that seepage will not pass down through the soil into the groundwater.

In this chapter, we describe various types of manure/wastewater storage and handling systems. In addition, using data derived from the National Resources Conservation Service (NRCS/USDA) document *Manure and Wastewater Handling and Storage Costs* (2005), we focus on the costs involved in operational needs of the various manure storage/handling systems.

4.2 TYPES OF MANURE STORAGE AND HANDLING SYSTEMS

Livestock and poultry manure is handled either as liquid, slurry, or solid in U.S. production facilities. Storage structures for these handling mediums include earthen slurry basins or pits, lagoons, runoff holding pounds, unroofed slurry pits or tanks (not roofed), below-building pits (concrete), slurry pits, reception pits, roofed tanks (earthen or concrete), uncovered storage (steel or concrete) and covered storage (earthen or concrete pad) for solid manure.

- *Important point:* Handling liquid and slurry manure is considered to be easier to automate than handling solid manure.

4.2.1 RUNOFF HOLDING PONDS

Runoff holding ponds are typically located in areas where they can use natural surface-drainage conditions. These storage systems temporarily store runoff water from a feedlot until it can be applied to the land. Lot runoff must pass through a settling facility before going to the holding pond. The holding pond is not intended to receive roof water, cropland drainage, or other unpolluted waters. Holding ponds do not treat manure as lagoons do; the only store it until it can be spread.

Runoff holding ponds are typically of earthen or concrete construction. They must be sealed to prevent seepage into groundwater. Holding-pond bottoms tend to seal naturally. The storage volume should be adequate to hold the runoff expected from the lot for the length of storage. The required storage capacity depends on desired length of storage, source of liquids and runoff water, rainfall duration and frequency, and the balance between rainfall and evaporation. The runoff holding pond is usually emptied by pumping and land application using some type of irrigation. The holding pond should be emptied before it is full, when wastewater can infiltrate the soil. Soil should not be frozen, frosted, overly wet, or snow covered. The holding pond should be emptied completely to assure maximum capacity for runoff.

Advantages of runoff holding ponds include their applicability for storm events in arid regions, storage of feedlot storm runoff, and their ability to be managed with irrigation equipment. Disadvantages include the need to separate solids from liquids using a settling basin, soil evaluation requirements, the need for proper soil material, and seal construction. In addition, runoff holding ponds are not appropriate for regions with shallow water tables or high-risk geology.

4.2.2 LAGOONS

Lagoons are earthen structures that provide biological treatment, as well as storage, of manure liquid. Liquid waste is biodegraded to convert organic matter (wasted feed, feces, and urine) in animal manure to a more stable end product. As with earthen storages, a seal on the bottom and sides is needed in some soils to meet permeability requirements.

Advantages of lagoon storage include frequent crop irrigation in western states, feasibility for long-term storage, ability to be sized for lot runoff and fresh water inputs, biological treatment of manure, ability to be managed with irrigation equipment, and a relatively low storage cost per animal unit. Disadvantages include the large land area needed for construction; the high phosphorus levels in sludge if not regularly agitated and removed; the requirement for soil evaluation, proper soil material, and seal construction; the need for an appropriate site or soil type; lack of aesthetics, public concerns over odor; and relatively high emissions of nitrogen and greenhouse gases.

- *Important point:* Lagoons in colder climates must be larger than lagoons in warmer climates because bacterial action is greater and more efficient at warmer temperatures.

4.2.3 EARTHEN SLURRY BASINS OR PITS

Depending on soil and site conditions, earthen slurry basins and pits are low-cost storage options. However, they are designed to store manure only and are not treatment systems. The design is similar to but smaller than lagoons. A seal on the bottom and sides may be needed to meet permeability standards required by regulation. A larger land area is needed for construction of earthen structures because of the need for berms, and front and back berm slopes must be gentle enough to be properly maintained and to prevent erosion. Maintenance requirements are also greater than those for fabricated structures because of the need to maintain a well-trimmed vegetative cover on and around the berm to allow easy visual inspection and prevent tree roots from penetrating the berm. Planned access points for agitation and pumping should be part of the design to minimize soil erosion and damage to the storage line.

Advantages of storing manure in earthen slurry basins or pits include: the relatively high nutrient density compared to lagoons, low to moderate nutrient loss, manure may be injected or incorporated, they are less expensive than concrete or steel tanks, and they can be sized for lot runoff and minimal fresh water inputs. Disadvantages include: higher odor potential because of greater surface area; rainfall adds extra water; they may be difficult to agitate properly; they require soils evaluation, proper soil material, and seal construction; they require relatively expensive application equipment; the large number of loads that must be removed when the storage is emptied; and they are not appropriate for regions with a shallow water table or high-risk geology.

4.2.4 SLURRY PITS OR TANKS

Unroofed slurry pits and tanks are fabricated tanks constructed with concrete or coated metal (glass-lined steel). Manure is typically drained, scraped, or flushed from the production building to the storage structures. These storage structures are typically pre-engineered packaged units that are less subject to specific design and construction procedures by the livestock producer.

- *Important point:* Dangerous manure gases are likely to be released during agitation of slurry-stored manure.

The advantages of slurry pits and tanks include: a relatively high nutrient density. Low to moderate nutrient loss and manure may be injected or incorporated. Disadvantages include: higher costs than earthen structures, more odor than covered storage, rainfall adds extra water, may not be compatible with systems having significant lot runoff or high water use, and require relatively expensive application equipment.

- *Important point:* Slurry manure systems require less storage volume than lagoons.

4.2.5 BELOW BUILDING PITS

The below building pit, often called a "deep pit," is a common type of slurry storage. Here, manure is deposited directly through slotted floors into the pit below.

Advantages of below building pits include relatively high nutrient density, low to moderate nutrient loss, manure may be injected or incorporated, and no rainfall effects. Disadvantages include: more expensive than earthen storage, odor, animal and worker health problems may result with prolonged exposure to manure gases, may require pit ventilation, not appropriate for regions with a shallow water table, high-risk soil conditions or geology, relatively expensive application equipment, and manure solids are more difficult to remove.

4.2.6 SLURRY PITS, RECEPTION PITS, OR ROOFED TANKS

With slurry pits, reception pits, and roofed tank structures, manure is usually scraped from the production buildings and may flow into the tanks by gravity or be pumped into the tanks from a collection sump or reception pit. These manure storage structures are either earthen or concrete. Adequate agitation is necessary to suspend solids and facilitate complete removal of the contents of these manure tanks. Fabricated tanks are usually the least costly to cover if odor becomes an issue.

Advantages of slurry pits, reception pits, and roofed tanks include: relatively high nutrient density, low to moderate nutrient loss, manure may be injected or incorporated, and they are not subject to rainfall effects. Disadvantages include: they are more expensive than earthen storage, may have more odor, may require pit ventilation, may not be compatible with systems that allow significant lot runoff or high water use, and they require relatively expensive application equipment.

4.2.7 SOLID MANURE

Typical uncovered solid manure storage structures store litter from poultry (turkeys, layers, broilers, and ducks), separated or scraped solids from swine and dairy operations, manure collected from outside beef feedlots, and other solid floor/manure pack shelters involving large amounts of bedding. Uncovered solid manure structures are typically used in low-rainfall (arid) areas where they are well-drained; the material is stacked or stockpiled for subsequent spreading. Regulations may required contaminated runoff from these structures be collected and disposed of in an environmentally sound manner.

Advantages of uncovered solid manure storage structures include: less expensive than roofed storage, high nutrient density, owners do not have to haul water, low nutrient loss (but higher than a covered storage), and are most applicable in arid regions. Disadvantages include: rainfall/runoff contamination potential, runoff controls may be required, not applicable as sole storage for systems with lot runoff or high water use, bedding may be required, and are less applicable in humid regions.

4.2.8 ROOFED OR COVERED SOLID MANURE

In higher rainfall regions, roofed or covered solid manure storage structures usually have concrete bottoms and may have concrete walls to confine the solids and provide push walls for stacking and loading of solids. These storage structures are an option where adequate amounts of bedding are used to make the manure a stackable solid. However, bedding contributes to the volume of manure that must be stored.

Advantages of solid manure (roofed or covered) storage structures include high nutrient density, lack of need for workers to haul water, little or no seepage, low nutrient loss, and no runoff from stacked manure. Disadvantages include higher costs than open stacks, inapplicability as sole storage for systems with lot runoff or high water use, and possible requirements of bedding.

- *Important point:* Water used in cleaning increases the volume of manure storage facilities need.

- *Important point:* Solid manure systems with bedding usually produce fewer odors than manure slurries since bacterial action produces fewer odorous compounds as the result of moisture content.

4.3   MANURE HANDLING AND STORAGE COSTS

[*Note:* In this section, a group of "virtual" farms were established to provide a generalized approach, including various assumptions, to estimating needs and costs for manure storage and handling systems by identifying major cost items involved. This approach was guided by the NRCS *Agricultural Waste Management Field Handbook* (AWMFH) (NRCS, 1992). In particular, much of the data contained within our virtual assessment is based on the NRCS's *Costs Associated with Development and Implementation of Comprehensive Nutrient Management Plan* (CNMP; NRCS, 2005)].

This assessment does not address federal, state, and local regulatory requirements associated with animal feeding operations. Many states have, or are in the process of adopting, regulations that would require some livestock operations to implement systems that are equivalent to a CNMP or part of a CNMP. Some of these regulations impose stricter requirements than represented by the NRCS CNMP guidelines. Consideration of regulatory trends was given, however, to the determination of CNMP needs particularly for large operations.

- *Important point:* This assessment did not attempt to account for the implementation of CNMPs or elements of CNMPs since 1997. Consequently, part of the costs presented in this assessment may have already been borne by some livestock operations.

- *Important point:* Cost estimates may be overstated somewhat because they do not account for innovation and technological advances that are expected to occur as the CNMP initiative is implemented.

No attempt was made to account for payment by recipients for manure exported off the farm or charges to the livestock operation by recipients for accepting the manure. A variety of payment arrangements presently exist, depending on traditions and markets established in the production region, the type of manure, and existing state and local regulations. In some cases, livestock operators are responsible for applying the manure to recipients' land. For the purposes of this cost assessment, it is assumed that all manure exported off a farm would be given and accepted without payment, the livestock operation bears the cost of transporting the manure to the manure-receiving farm, and the off-farm land application cost is borne by the recipient.

- *Important point:* CNMP development and implementation costs are not estimates of the costs to producers of complying with USEPA regulations.

No account was made of the financial benefits that might be realized because of operational needs implementation, including any savings in commercial fertilizer costs for the additional acreage that receives manure applications. The nutrient value of manure is considered one of the many benefits of implementing operations needs.

No attempt was made to adjust costs for inflation, even though some cost increases will certainly occur over the 10-year implementation period. To make this adjustment, one would need to know the rate at which operational needs would be implemented, which depends on regulatory incentives, financial incentives, and the availability of technical assistance. Cost estimates reported here may therefore be understated to some extent, depending on the rate of inflation and implementation over the next 10 years. This cost assessment also does not account for cost savings that could be realized by improvements in feed management.

The virtual model used here shows that alternatives to land application of manure are needed in some regions of the country. Under the assumptions of the model simulation, 248 counties do not have adequate land to assimilate the manure produced in those counties when applied at rates that meet operational needs criteria. Most of these counties are co-located, reducing the opportunity to transport the manure to surrounding counties for land application. The amount of county-level excess manure represents about 16% of the total recoverable manure nutrients produced by all operations needs farms in the country. Included in the cost assessment are estimates of the cost of transporting this county-level excess manure off the farm.

In our virtual CNMP model, we define the basic set of production technologies in terms of *representative farms* for each livestock type. *Representative farms* define broad groups of livestock production facilities that, within a livestock sector, have similar characteristics for managing livestock and managing manure—in other words, a hypothetical farm with a typical animal waste handling system for a given livestock type. This set of representative farms was expanded to a larger set of model farms by adding the dimensions of size and location. Size categories for the dominant livestock type were selected to reflect differences in production technologies by farm size. Geographic regions generally reflected major production regions, with further delineation by climate in areas where climate would be expected to influence the kind of production system found. Not all representative farms are present in each size class and location. Each model farm is thus a representative farm of a certain size in a specified location.

Representative farms were derived from two sources of information: farmer surveys and expert judgment. Results from farmer surveys were available for dairies, swine, and layers. These surveys were not conducted for the specific purpose of inventorying manure-handling practices on farms but did include questions about the production technologies in use and a few questions about manure management. A team of USDA experts evaluated the survey results and identified the dominant manure management technologies, basing them on manure handling characteristics as much as possible. Only the most dominant technologies were included; technologies that occurred relatively infrequently in survey results were discarded. Farmer survey results were not available for fattened cattle, veal, confined heifers, broilers, pullets, or turkeys. For these livestock types, representative farms were derived by the team of USDA experts based on their knowledge of industry practices.

In addition to providing a structure for deriving operational needs for the manure and wastewater handling and storage element, this analytical framework was used to assign costs related to manure testing and record keeping. A slightly expanded version of the framework was used to estimate operational needs development costs.

Model Farms for Dairy

Five representative farms were derived for dairy based on a 1996 National Animal Health Monitoring System (NAHMS) survey of 2,542 dairies in 20 states (USDA/Animal and Plant Health Inspection Service [APHIS] 1996). The survey included questions about the manure storage facilities on the farm and the frequency of manure spreading. Production technologies for dairies were therefore defined in terms of manure storage. The five representative farms are:

1. Essentially no storage, frequent spreading
2. Solids storage (typically outside, separate from pens, but may include some manure pack and dry lot conditions); no appreciable liquid storage
3. Liquid to slurry storage in deep pit or aboveground tank; some solids storage; no earthen basins, ponds, or lagoons; typically less than monthly spreading
4. Primarily liquid manure stored in basin, pond, or lagoon; some solids storage for outside areas; typically less than monthly spreading
5. Liquid system (any combination of 3 and 4); primarily used in the West and Southeast; often associated with manure pack; and solids spreading in the West.

Model Farms for Layers

Three representative farms were derived for layers based on a 1999 NAHMS survey of 526 layer farms in 15 states (USDA/APHIS, 1999). The survey included a question about the type of facility used relative to manure collection and handling. Production technologies for layers were therefore defined in these terms. Five types of systems were identified in the survey, but they were combined into three groups of representative farms because of similar operational needs and cost assumptions. The three representative farm types are:

- High rise (pit at ground level with elevated house) or shallow pit (house not elevated)
- Flush system to lagoon
- Manure belt or scraper system.

Model Farms for Swine

Five representative farms were derived for swine based on two farmer surveys: a 1995 NAHMS survey of 1,477 swine farms in 16 states (USDA/APHIS, 1995) and a 1998 Agricultural Resource Management Survey (ARMS) on 1,600 swine farms in 21 states (USDA/ERS, 2000). The surveys included questions about the type of facility used to rear swine and the type of manure handling and storage system. Production technologies for swine were therefore defined in these terms. The initial breakdown was made using the NAHMS survey results. The ARMS results were used to update the representation of confinement facilities that had storage ponds or lagoons and were used to estimate representation in the West. The representative farms are:

1. Total confinement with liquid system, including lagoon
2. Total confinement with slurry system; no lagoon
3. Open building with outside access and liquid to slurry system (holding pit under slat or open flush gutter)
4. Open building with outside access and semisolid to solid wastes (mechanical scraper/tractor scrape/hand clean)
5. Pasture or lot with or without hut

Model Farms for Other Confined Livestock Types

Survey results for the remaining confined livestock types are not available. A team of USDA experts defined the predominant production technologies for each livestock type. Representative farms were defined as follows:

Fattened cattle

1. Dry lot (small) scraped on a frequent basis, manure stacked until application
2. Dry lot with manure pack and occasional complete clean out and removal; at least rudimentary runoff collection/storage.

Confined heifers

1. Confinement barns with bedded manure; solids handling
2. Small open lots with scraped solids and minimal runoff control

Veal

1. Confinement house with liquid and slurry components

Turkeys

1. Confinement house
2. Turkey ranching (building with open sides and lot)

Broilers

1. Standard broiler house; complete litter clean out or cake out

Pullets

1. High-rise or shallow-pit confinement house

Model Farms for Pastured Livestock Types

1. Pasture with heavy use area
2. Pasture with windbreak or shelterbelt
3. Pasture with lot and scrape-and-stack manure handling
4. Pasture with barn for shelter

Major cost items for manure storage and handling are broken down into the following components:

- Mortality management (poultry and swine)
- Lot upgrades
- Clean water diversions (including roof runoff management, earthen berms, and grassed waterways)
- Liquid treatment (small dairies)
- Collection and transfer (including solids, liquid, contaminated runoff, and pumping)
- Settling basins
- Solids storage
- Liquid storage
- Slurry storage
- Runoff storage ponds

Cost estimates for conservation practices for pastured livestock are included in the manure handling and storage element. Components for farms with pastured livestock types include:

- Fencing
- Water well
- Watering facility
- Heavy use area protection
- Windbreak or shelter break establishment
- Solids storage
- Filter strip

Manure handling and storage costs for the system are associated with the dominant livestock type on each farm. However, many of these farms have other confined livestock types on the farm. The costs associated with addressing needs for the secondary livestock types on the farm, for the most part, are incorporated into the system costs for the dominant livestock type. Any additional costs are assumed minor and are not estimated. For several compounds, however, costs are based on the amount of recoverable manure produced on the farm (handling and transport weight), which includes recoverable manure from all livestock types on the farm.

4.3.1   MORTALITY MANAGEMENT

For our "virtual" farms, the cost of mortality management is included for all poultry and swine farms. For dairy and fattened cattle, existing mortality management practices are assumed to be adequate in most cases. Various acceptable methods were used to manage poultry and swine mortality, such as composting, incineration, burial pits, and freezing. Composting was selected as the representative technology for assessing system operational costs.

- *Important point:* For illustrative and descriptive reasons and to broaden applicability, throughout this section, for our virtual farms we substitute the general term "operational costs" or "operational needs" for CNMP.

4.3.1.1   Poultry

The cost of mortality management for poultry was determined on a per-house basis. A concrete slab covered with a timber structure comprised the composting facility. Capital and operating costs of the structure were based on costs reported by the North Carolina Cooperative Extension (1999) for a 100,000-broiler flock. The cost of the timber structure and concrete floor was $3,600, and the cost of water service for the facility was $150, resulting in an annual capital cost of $559. Operating costs included labor (27.5 hours per flock at $10 per hour per year) and machinery rental ($20 per hour at 51 hours per year), for a total of $2,533 per year. For the 25,000-bird broiler house used as the standard house size in this study, annual costs were $140 for capital and $633 for operating costs.

Costs for the other poultry livestock types were estimated by prorating the cost for broilers based on capacity needed for the other poultry types. The capacity needed was estimated using a method published by the North Carolina Cooperative Extension (1996). Maximum capacity was estimated by multiplying the expected daily death rate by the market weight (maximum weight), and then multiplying by the number of birds per house. Although mortality takes place throughout the production cycle with birds at various weights, for most operations the majority of the mass that must be dealt with occurs near the end of production, when birds are closest to their market weight. To ensure adequate composter space, capacity is based on the greatest demand to handle bird mortality. Calculations are shown in Table 4.1.

TABLE 4.1
Capacity Need Calculations

Poultry type	Birds per house	Market weight (lb/bird)	Mortality rate (%)	Mortality rate (lb/d)	Annual capital cost per house ($)	Annual operating cost per house ($)
Broilers	25,000	4.5	0.1	113	140	633
Layers and pullets	50,000	4.0	0.033	66	82	371
Turkeys for slaughter	5,000	19.2	0.080	77	96	433
Turkeys for breeding	8,000	18.8	0.100	150	187	846

Needs for mortality management for poultry were judged to be lower for larger operations and higher for turkey operations. Needs were assigned as follows:

- 45% for broiler and pullet farms with less than 220 AU
- 15% for broiler and pullet farms with more than 220 AU
- 45% for layer farms with less than 400 AU
- 15% for layer farms with more than 400 AU
- 60% for turkey farms with less than 220 AU
- 30% for turkey farms with more than 220 AU

4.3.1.2 Swine

Estimates of mortality management costs for swine were based on a composting facility consisting of a concrete pad with walls constructed of large, round bales; a tarp cover; and a fence to keep animals out. Included in the system are a carcass cutter and grinder. (Costs for this system are described by Ken Foster in *Cost Analysis of Swine Mortality Composting*, Purdue University. Last accessed January 15, 2005 @ pasture.ecas.purdue.edu/~epalas/swine/ econ/ compostin.htm.)

The annual cost of the cutter and grinder is $1,248 and is incurred only once per operation, regardless of the size of the operation. Other capital costs (concrete slab, fence, tarp, bales) were reported by Foster for a farrow-to-finish operation with a maximum capacity of about 250 AU at $549 per year. Annual operating costs (labor, sawdust, fuel, and utilities) for this system were reported at $350 per year. On an animal unit basis, these costs convert to $2.20 per AU for the additional capital costs and $1.40 per AU for operating costs.

Because swine operations have only recently begun to address mortality management practices as an integral part of their operation, operational needs were set at 70% for all sizes and types of swine operations.

4.3.2 FEEDLOT UPGRADES

The cost of feedlot upgrades was applied only to cattle on feed (fattened cattle and confined heifers) and consists of improving the open lot area where cattle are held to ensure the proper functioning of collection systems. It includes grading to enhance drainage and a concrete pad to protect drainage collection and diversion areas during manure collection activities. These lot upgrades exclude the costs of berm construction for diverting contaminated water into the storage pond, which are costed separately.

A 750-head fattened cattle operation was used as a basis for deriving representative costs for this component. Costs were estimated assuming installation of 111 cubic yards of concrete (6,000 ft^2) at $200 per cubic yard and 1,700 cubic yards of earthmoving and shaping at $2.00 per cubic yard. (These costs were based on data taken from Iowa State University's *Beef Feedlot Systems Manual*. Last accessed January 16, 2005 @ www.extension.iastate.edu/publications/pm1867.pdf.) The total capital cost is thus $35,600 per 750-head operation, or $34 per head. The amortized annual cost is $5 per head.

Most operations typically have addressed this component as a part of their existing management systems, so needs were judged to be comparatively low:

- 15% for fattened cattle farms with a scrape-and-stack operation
- 30% for confined heifer farms with a scrape-and-stack operation
- 30% for the smaller, fattened cattle farms with manure pack
- 5% for the larger, fattened cattle farms with manure pack

4.3.3 CLEAN WATER DIVERSIONS

Clean water diversions are used to minimize the amount of rainfall runoff that can come in contact with areas of the animal production operation where manure and wastewater are present, primarily, the open lot areas. The types of clean water diversions used in this study were roof runoff management, earth berms with a surface outlet, earth berms with underground pipe outlets, and grassed waterways. Because diversions were only essential for operations with an open lot, clean water diversions were not applied to operations that only confined animals in buildings.

4.3.3.1 Roof Runoff Management

Gutters and downspouts were used to capture rainfall on the roofs of buildings and route it away from the production area. This kind of clean water diversion was applied to dairy, turkey, and swine operations that provided outside access to animals. Fattened cattle operations were not included because these operations typically raise animals in a feedlot without any buildings or structures within the confinement area.

The per-unit costs used were based on data taken from the NRCS *Field Office Technical Guide,* Section 1, Annual Cost List (NRCS, 2005c). The installation cost for a standard gutter and downspout used in most areas of the United States is $2.25 per foot. In areas of higher rainfall, such as the Southeast, a larger gutter is needed, at a cost of $4.50 per foot. (Since downspouts are often damaged by animals and machinery, repairs and maintenance were assumed to be 7%, to the maintenance costs estimated as 3% of all capital costs, which when added brought the total percentage for maintenance cost for this component to 10%.) The estimated quantities of gutters and downspouts used per type and location of facility were based on average building size and typical building capacities. Dairy costs were based on 200 ft of gutters and 40 ft of downspouts for a 100-cow dairy and were converted to a per-head basis. The annual capital cost for dairies, including maintenance and repair, was $2.37 per head in the Southeast and $1.18 per head in other regions. For turkey ranches, the annual capital cost was $473 per house, assuming 800 ft of gutter and 160 ft of downspouts per house. For swine farms with buildings and outside access, the annual capital cost was $0.85 per animal unit, based on 200 ft of gutter and 40 ft of downspouts for a 140-AU operation.

Roof runoff management has been a neglected component on some systems but is commonly present on other systems. Larger operations are expected to have fewer needs than smaller operations. Operational needs were assigned as follows:

- 30% for swine farms with buildings and outside access
- 90% for turkey ranches
- 80% for Dairy Belt dairies #1 and #2 (solids systems) with up to 270 AU
- 45% for Dairy Belt dairies #1 and #2 with more than 270 AU
- 40% for all other dairies

4.3.3.2 Earthen Berms with Underground Pipe Outlets

This type of clean water diversion was used for fattened cattle operations with a manure pack method of managing waste, as well as for all dairy operations. These operations generally take advantage of the profile of the land to provide drainage within the lot. Often, these operations have dry or intermittent streams (swales) that run through the feedlot areas. To control clean water upgradient of the lot, a small earthen berm is installed across the swale above the feedlot or lot to catch the clean runoff and then let the water out through an underground pipe to some point downstream of the feedlot area.

The cost of installing the earthen berm associated with this system includes the cost of hauling and shaping activities. The berm used for this type of system is considerably shorter than those for other diversion practices because its only function is to create a temporary pool that drains out through the underground pipe. Although the berm length is considerably shorter than other berms described in this section, it is usually higher, to create sufficient hydraulic pressure to discharge through a long pipeline. The assumed dimensions of the berm were based on a trapezoidal shape with an 8-ft top width, 3 horizontal to 1 vertical side slopes, and 3 ft of average height (1.9 cubic yards per foot of length) for a length of 30 ft per berm. The cost per cubic yard was $2 installed, or $115 per berm. The estimate for the underground outlet pipe was based on a 12-in diameter corrugated metal pipe, and unit costs reflect the cost of pipe and installation activities, such as excavation, pipe laying, and backfill. Lengths were estimated based on professional judgement of a typical distance through a feedlot based on the particular size of an operation. Larger operations could require more than one berm and pipe outlet per feedlot. Per-unit costs were taken from the NRCS *Field Office Technical Guide,* Section 1, Annual Cost List (NRCS, 2005c). Cost estimates were developed for three different size operations.

Using three cost estimates, the following rules were established for assigning costs to farms on a per-head basis:

- If the number of head is less than 100, the cost per head is $5.07.
- If the number of head is between 100 and 300, the cost per head is $4.47.
- If the number of head is more than 300, the cost per head is $3.58.

Number of animals	Linear feet of pipe	Pipe cost per foot ($)	Number of 30-foot berms	Berm cost ($)	Total cost installed ($)	Cost per animal ($)	Annual cost per animal ($)
75	200	12	1	115	2,515	34	5.07
150	360	12	1	115	4,435	30	4.47
600	1,200	12	3	345	14,745	25	3.58

Most of these operations already have this practice in place or do not need it because of the terrain characteristics near the facility. Some systems in some regions of the country, however, were judged to have relatively high needs. Operational needs were assigned as follows:

- 20% for smaller fattened cattle farms
- 10% for larger fattened cattle farms
- 50% for dairy representative farm #1 in Dairy Belt
- 50% for dairy representative farm #2 in Dairy Belt with < 270 AU
- 30% for dairy representative farm #2 in Dairy Belt with > 270 AU
- 20% for dairy representative farm #2 in West and Southeast
- 30% for dairy representative farm #3
- 40% for dairy representative farm #4
- 20% for dairy representative farm #5 in Southeast and in West with < 270 AU
- 10% for dairy representative farm #5 in West with > 270 AU

4.3.3.3   Grassed Waterways

Grassed waterways are shaped channels that are seeded to establish vegetation. They are used for clean water diversion in areas that receive sufficient annual rainfall such that vegetation can be maintained naturally and where the runoff-contributing watershed is relatively small. These waterways are more efficient than earthen berms because they can handle larger flows without concern of erosion. This typical practice is used east of the Mississippi River and represents the clean water diversion treatment needs for fattened cattle operations and confined heifer operations that use a stack-and-scrape manure management system. Only 15% of these operations were assumed to need to install this practice because of its common use.

All grassed waterways were assumed to be 30 ft wide. The length varies by the size of the operation. Per-unit costs were taken from the NRCS *Field Office Technical Guide,* Section 1, Annual Cost List (NRCS, 2005c). The cost of installing a grassed waterway involves grading and shaping the channel, which costs $115 per acre, and seeding, which costs $125 per acre. The total cost is $240 an acre. Lengths were estimated based on professional judgment of a typical distance to bypass a feedlot for two sizes of farms and then converted to a per-head cost:

- The $0.20 cost per head was assigned to all operations with less than 500 head
- The $0.08 cost per head was assigned to operations with more than 500 head.

Number of animals	Linear feet of waterway	Acres	Total cost installed ($)	Cost per animal ($)	Annual cost per animal ($)
150	1,200	0.83	199	30	0.20
600	1,800	1.24	298	44	0.08

4.3.3.4   Earthen Berms with Surface Outlet

Earthen berms with a surface outlet are shaped mounds of uniform cross section made of soil to serve as an intercept upslope of an open lot to divert clean water around the lot to a stable natural outlet. This clean water diversion practice was used only on turkey and swine operations that have open lots as part of their production areas. Per-unit costs were taken from the NRCS *Field Office Technical Guide,* Section 1, Annual Cost List (NRCS, 2005c).

All open lots were assumed to have a diversion along two sides. Installation involved primarily earth-hauling and shaping activities. The assumed dimensions of the berm were based on a trapezoidal shape with an 8-ft top width, 3 horizontal to 1 vertical side slopes, and 2 ft of height for a running volume of 1 cubic yard of diversion per foot of length. The cost per linear foot was $2.00 installed.

For a swine operation with open lot access and 900 animals (100 AU), 460 ft^2 of loafing area is typically provided per AU, or 46,000 ft^2. Assuming a square lot, the dimension of a side would be 214 ft. Assuming the diversion would be wrapped around two sides, the total length would be 428 ft, for a total cost of $856. The amortized annual cost would be $128 per year, or $1.28 per AU per year. Operational needs for these operations were judged to be 20% for swine representative farm #4 (building with outside access) and 50% for swine farm #5 (pasture or lot).

A typical turkey operation would raise approximately 5,000 birds per house. One house is equivalent to 75 AU. Assuming the lot area provided 460 ft^2 per AU (the same as the proportional area per AU provided for swine), the area of a turkey lot would be 34,500 ft^2, or a lot with sides measuring 185 ft. The total length of the berm would be 370 ft and would cost $740. The amortized annual cost would be $111 per year per house. Operational needs were judged to be 40% for turkey ranches.

4.3.4   LIQUID TREATMENT

Small dairy operations that remove solids daily or weekly would continue to handle their manure as a solid and use a liquid treatment approach to handle the liquid component. Generally, cows on these operations are kept on pasture most of the day. However, they are brought in to be milked and, as a result, spend some time in an open lot. During storms, runoff from the open lots contains manure and related wastes, but this is normally a small volume. Milk-house washings would also generate small amounts of wastewater. For these operations, it was assumed that the runoff and milk-house washings could be handled with a biofilter, a small, vegetated area that functions like a wetland by capturing the runoff and bioprocessing it through infiltration of nutrients into the soil for use by the vegetation. Use of a biofilter for liquid treatment precludes the need for collection, transfer, or storage of liquid wastes on these farms.

For the purposes of this simulation, the biofilter was assumed to be a vegetated filter strip of 12,000 ft^2, at $0.20 per square foot, for a cost of $3,000. The construction of the filter would be accomplished by land grading equipment. Based on an average size operation of 75 milk cows, the capital cost is $6.00 per cow annually.

A liquid treatment component was included for dairy representative farms #1 and #2 with less than 135 AU per farm. Operational needs were judged to be high for this component: 65% for farm #1 and 75% for farm #2.

4.3.5   COLLECTION AND TRANSFER

The collection and transfer component addresses the installation and operation of practices associated with handling manure and wastewater within production areas. The type of collection used depends on the type of animal feeding operation, consistency of the manure handled, and type of management system used. Management systems for animals raised in buildings address a single manure consistency, either a liquid/slurry or a solid. Operations that use open lots generally need to address both solids and liquids because manure and contaminated runoff are generally handled separately.

Operational costs were determined for three types of collection systems: solids collection, liquid collection with flush systems, and contaminated runoff collection. For the last two types of collection systems, a liquid pumping system is needed to transfer the wastewater to a storage structure and from the storage structure to land application equipment. For solids, manure is transferred to a solids storage facility during collection.

Almost all model farms include a collection component, a transfer component, or both. Representative farms that predominantly handle manure as slurry, however, have storage pits either under the building or adjacent to the housing facility, requiring only rinsing to collect the manure. For these representative farms, adequate collection structures were assumed, with only a transfer component needed. These farms include veal, swine representative farms #2 and #3, dairy representative farms #1 and #2 with more than 135 AU, and dairy representative farm #3. Dairy representative farms #1 and #2 with fewer than 135 AU have liquid treatment components (filter strips for milkhouse washings) and would not need collection or transfer component therefore.

4.3.5.1   Solids Collection

Solids collection is a component for all operations *except* swine and dairy farms with complete liquid or slurry systems, layer farms with liquid systems, and veal farms. Generally, most operations have adequate collection systems already in place, so operational needs are expected to be low. Operational needs were judged to be 10% for all but these cases:

- 2% for broiler farms
- 15% for turkey farms (representative farms #1 and #2).

Solids collection for dairy, fattened cattle, confined heifers, and swine raised in buildings with outside access or in pastures or lots was assumed to consist of a tractor scraper used to collect and pile the manure on a concrete slab. Costs are based on the amount of manure to be handled. The scrape operation costs are based on a 37-hp tractor with scraper at a purchase price of $22,000. Assuming this equipment is dedicated 80% to this function, the annual cost is $3,591. Conventional guidelines for estimating annual operating costs—fuel, oil, and labor—for equipment used on an intermittent basis, as in this case, is 15% of the purchase price (Tilmon and German, 1997). Thus, the annual operating costs were estimated at $3,300 per year. The cost per ton was determined for a 10-head dairy operation, which was then used for all dairy, fattened cattle, confined heifer, and swine farms with a solids collection component. A 150-head dairy oration has about 200 AU and produces about 580 tons of manure at transport and handling weight (assuming about 2.2 tons of manure as excreted at oven-dry weight, converting to a handling weight by multiplying by 2, and adjusting for recoverability with a 0.65 recovery factor). Thus, capital costs are $6.20 per ton of solids and operating costs are $5.70 per ton.

The cost of solids collection for broiler, pullet, turkey, and layer operations with a high-rise or shallow pit production system that raise poultry in confinement buildings was based on the assumption that the buildings are partly cleaned out after each flock and are completely cleaned out once per year. A custom rate was used and, because most of the cost is labor, it was categorized as an operating cost, even though a portion of the cost covers the cost of the equipment. The custom rate used was based on several sources of information obtained from the University Extension Service and private industry sources. The rates varied from $0.20 to $0.07 per square foot, depending on the size of the house and regional location. However, the predominant price range was from $0.04 to $0.065 (including both annual cleanout and four to five cake-outs per year.) Averaging the costs from the sources considered provided a custom rate of $0.053 per square foot of house. An average size broiler and turkey house is about 20,000 ft^2, producing an annual cleanout cost estimate of $1,060 per house. The average size of a layer or pullet house with a 50,000-bird capacity is about 24,000 ft^2, producing an annual cleanout cost of $1,272 per house.

For layer operations that use mechanical belt systems installed beneath the layer cages, manure falls directly onto the belt and the belt periodically empties itself onto a stacking area. For layer operations that use scraper-type systems, the litter produced is removed from the building by mechanical scrapers and deposited in a stacking area. Solids collection for these two types of operations was viewed as the activity to move the litter deposited in the stacking areas at the ends of buildings to a central storage area or directly into trucks for transport off the farm. Costs were based on equipment rental rates for a 150-hp front-end loader (3 yard bucket) at $15.08 per hour and an operator cost of $10.00 per hour. Based on a weekly manure production of about 42 tons of litter per house (50,000 birds), the time needed to move the litter is approximately 1.5 hours per week per house for 78 hours per year, or $1,956 per house annually.

4.3.5.2 Liquid Collection with Flush Systems

Dairy, swine, and layer operations that handle their wastes as liquids commonly use the flush system. Waste is collected by flushing floor gutters within the barn to move waste and water to a collection tank, where it is transferred to a holding pond or lagoon by gravity or transfer pump. Existing flush operations are assumed to have most of the system in place. Therefore, systems would only need to be upgraded to be consistent with any modifications in the storage and handling systems. Components assumed to be needed were a flush tank, collection tank, transfer pipe, and pit agitation pump. Operational needs were judged to be comparatively low for the following representative farms with flush systems:

- 10% for swine representative farm #1 (liquid system with lagoon or storage pond)
- 10% for layer representative farm #2 (flush to lagoon)
- 30% for dairy representative farm #4 (liquid system with lagoon or storage pond) with less than 270 AU
- 40% for dairy representative farm #5 (liquid system with lagoon or storage pond) with less than 270 AU
- 20% for dairy representative farm #4 or #5 with more than 270 AU.

Costs for three sizes of dairy farms were used as the basis for flush cost systems. The base system for the smallest operations included two collection tanks (10 ft wide by 20 ft long and 8 ft deep);

TABLE 4.2
Cost Estimates for Liquid Collection
with Flush Systems for Dairy Farms

	Operation		
Cost component	100-head ($)	200-head ($)	300-head ($)
Flush tank	7,801	15,602	23,403
Collection tanks	5,721	11,442	17,163
Collection pipe	562	562	562
PTO impeller	5,367	5,367	5,367
Total capital cost	19,451	32,973	46,495
Annual capital cost	2,899	4,914	6,929
Annual operating cost	1,185	2,369	3,554
Annual capital cost/head	28.99	24.57	23.10

Source: NRCS/USDA (2004).

a transfer pipe (50 ft of 100-lb/in^2 polyvinyl chloride [PVC]); and an agitation pump (power take-off [PTO] driven impeller). Costs for larger systems would account for the increased size needed to handle more animals. Operating costs cover fuel, oil, electricity, and pump maintenance. For these systems, the cost of the pipe used to transfer the waste to the field for application was treated as a hauling cost, and the cost of pumping to the field for irrigation is covered under the pumping transfer system costs. The dairy liquid collection costs are summarized in Table 4.2.

The costs shown in Table 4.2 were applied to dairy representative farms #4 and #5. Dairies with less than 150 head were assigned a capital cost of $28.99 per head. Dairies with 150 to 250 head were assigned a capital cost of $24.57 per head. Dairies with more than 250 head were assigned a capital cost of $23.10 per head. Operating costs for all size farms were $11.84 per head.

The same components are also needed for swine operations with liquid wastes (swine representative farm #1) and layer farms with liquid wastes (layer representative farm #2). The costs above were converted to an animal unit basis for these swine farms and to a per-house basis for the layer farms. The annual capital cost was $20.70 per AU for swine farms with less than 200 AU, $17.55 per AU for farms with 200 to 400 AU, and $16.50 per AU for farms with more than 400 AU; annual operating costs were $8.46 per AU for all size groups. For layers, the annual capital cost was $3,157 per house, and the annual operating cost was $1,291 per house.

4.3.5.3 Contaminated Runoff Collection

Earthen berms are used to divert rainfall runoff that has come in contact with manure in the production area to a storage pond. The contaminated water diversion would be located on the down-gradient end of the production area. The types of contaminated water diversions typically used are earthen berms with a surface outlet and earthen berms with pipe outlets.

Contaminated water diversions are necessary components for all fattened cattle and confined heifer representative farms, as well as for turkey ranches and swine farms with pastures or lots (swine farm #5). Lots on dairy farms and swine farms with a building and open access are assumed to be small enough that contaminated water diversions would not be needed or would be incorporated into the structure of the runoff storage pond.

Typically, turkey operations and swine operations with pastures or lots would use an earthen berm with a surface outlet that diverts the runoff to a small storage pond. Construction is similar to earthen berms with surface outlets used for clean water diversion. Based on the costs previously presented for clean water diversion berms, the annual capital cost would be $111 per house for

turkey ranches and $1.28 per animal unit for swine. Operational needs were judged to be compara-
tively high for these farms, as follows:

- 50% for swine representative farm #5
- 90% for turkey ranches.

Fattened cattle and confined heifer operations use similar systems; however, they would generally outlet the captured contaminated runoff through a pipe into a holding pond. These operations
generally take advantage of the relief profile of the land to provide drainage within the lot. On the
downslope end of the lot, an earthen berm is constructed that channels all lot rainfall runoff to a
pipe outlet that conveys the contaminated runoff water to a holding pond or lagoon.

Earthen berm costs were calculated using the following assumptions: the shape was trapezoidal
with an 8-ft top width, the side slopes were 3 horizontal to 1 vertical, and the height was 2 ft. The
unit cost of the berm is $2.00 per linear foot, according to the NRCS *Field Office Technical Guide,*
Section 1, Annual Cost List (NRCS, 2005c). The berm length was equal to the downslope width of
the lot. The following approach was used to determine the length of berm: first it was assumed that
each animal unit was provided 460 ft^2 of lot space; total lot size was computed by multiplying the
number of AU by 460, and then the square root of the area was taken to represent the berm length.
The outlet pipe was assumed to be a 12-in diameter corrugated metal pipe (CMP). The unit cost for
pipe, $12 per foot, reflects the cost of the pipe and installation activities, such as excavation, pipe
laying, and backfill. The length of pipe needed on any particular site varies depending on the distance from the berm to the storage pond. To simulate this variation, the length of pipe is assumed
to be 20% of the length of diversion.

Three size categories were used for assigning costs to the fattened cattle and heifer farms (see
chart below). Using these three cost estimates, the following rules were established for assigning
capital costs to farms on a per-head basis:

- If the number of heads is less than 200, the cost per head is $1.31.
- If the number of heads is between 200 and 450, the cost per head is $0.80.
- If the number of heads is more than 450, the cost per head is $0.56.

The majority of fattened cattle and confined heifer operations were judged to need contaminated water diversions. Operational needs followed:

- 55% for coined heifer and fattened cattle farms with scrape-and-stack manure handling
  system in the South and West
- 40% for confined heifer and fattened cattle farms with scrape-and-stack manure handling
  system in the Midwest and Northeast

	Size 1	Size 2	Size 3
Animal number (head)	116	308	616
Area of lot (ft^2)	53,130	141,080	283,360
Length of berm (ft)	230	376	532
Cost of berm ($)	460	752	1,064
Cost of berm per head ($)	3.96	2.44	1.72
Linear feet of pipe	46	75	106
CMP cost per foot ($)	12	12	12
Cost of pipe installed per head ($)	4.76	2.93	2.07
Annual cost per head ($)	1.31	0.80	0.56

- 60% of the smaller fattened cattle operations with manure pack
- 50% of the larger fattened cattle operations with manure pack

4.3.5.4 Pumping Transfer System

All model farms that must handle waste or wastewater in liquid or slurry form should facilitate the transfer of that liquid or slurry from the storage structure (storage pit, holding pond, lagoon, or runoff storage pond) to the appropriate conveyance for land application. Some operations own a pump for this purpose, but smaller operations would likely rent the equipment. Costs were therefore estimated on a per-ton basis using a standard rental rate. Several rental rates were obtained from the literature, and rates varied depending on the geographic location; but the rates were all within about 15% of each other. The average rate was $240 per 8-hour day, or $17.50 per hour. The pumping rate used in the land application section was 500 gal/min, which converts to about 1.5 tons per minute (267 gal/ton), or 90 tons per hour, after allowing for about 20% downtime for setup or moving the pump. Thus, the capital cost of the pump would be about $0.20 per ton. Operating costs would be minimal, consisting primarily of fuel costs. An operating cost of $0.06 per ton was based on the cost of 3 gal of fuel ($1.65 per gallon) per hour.

These costs would be appropriate for operations that use irrigation systems to land-apply wastewater. However, for smaller operations that use a tank truck and sprayer to land apply wastes, additional downtime should be factored into the costs to account for multiple trips to the field to empty the liquid storage facility. During these trips, the operator would still pay a rental charge, but the pump would be idle. In the section on nutrient management costs, we assumed that operations with less than 1,000 tons of liquid waste per year would use a tank truck and sprayer for land application. Assuming the pump would only be operated 40% of the time for these smaller operations, the pumping rate would be about 45 tons per hour and thus capital costs would be $0.40 per ton. Operating costs would remain the same, at $0.06 per ton.

Operation needs for pumping transfer systems were assumed to be the same as the needs for storage (runoff storage pond, slurry storage, or liquid storage ponds or lagoons).

4.3.6 Storage of Solid Wastes

The part of the manure that can be handled as a solid, including bedding material, is collected from production areas and stored until it can be applied to land. To efficiently use manure nutrients to fertilize crops, the window of opportunity to land apply manure is limited. Therefore, an essential part of an operational system is manure storage facilities that have enough capacity to hold manure until the proper time for land application.

Solids storage is included as an operational systems component for dairy representative farms #1 and #2, fattened cattle and confined heifer farms with scrape-and-stack systems for manure handling, swine representative farm #4 (building with outside access), and for all poultry operations except layer farms with flush-to-lagoon systems. Fattened cattle farms and dairy farms in the West with manure pack systems do not need a separate solids storage component, since the manure pack is the method of storage. Similarly, swine farm #5 does not need a storage component because the solids can be collected from the lot or pasture at the time of application.

Conservation practice standards used in operational needs development do not require a minimum period of storage because the storage requirements would vary depending on the crop-growing season, crops being grown, climate, and type of management system in place. These factors determine what the storage capacity should be on a particular farm. For purposes of this assessment, however, general minimum storage capacities were established so that cost estimates could be made.

Consistent with typical management practices used in the poultry industry, the storage capacity is assumed to be 1 year of litter production for all poultry types. For other animal sectors, the storage period is generally less than 1 year, because the solid can be handled more frequently and the limiting period of storage is dictated by availability of cropland to receive the manure. For most of the country, 180 days (50% of the storage period for poultry) was assumed to represent the typical length of storage, because that period allows storage of manure through the winter and wet months of the year. Model farms in the Southeast, in most cases, can produce some type of crop year round and therefore would not need a 180-day storage capacity, so storage time was set at 90 days. (For this purpose, the Southeast states are Texas, Louisiana, Mississippi, Alabama, Georgia, Florida, and South Carolina.)

Storage costs were determined as the cost per ton of solids using the hauling weight to approximate the tons to be stored. The cost per ton was determined using a typical storage facility for a broiler operation. This cost per ton was then applied to all livestock types after adjusting for storage time needed. For example, the cost per ton, which was based on a 365-day storage capacity, was multiplied by 0.5 to estimate the cost per ton for operations that only needed a 180-day storage capacity.

The solid storage structure for a typical broiler house was used as the basis for calculating the costs of storage needs for all model farms. The storage cost for broilers was based on a 1,600-ft^2 timber shed with end bays, push walls, and a concrete floor. The shed cost $12,403, or $1,863 per year per house. Using selected information on tons of manure at transport weight, the average amount of manure per poultry house was determined at about 267 tons per year, including bedding. Thus, the cost per ton is about $7 for all poultry farms. For other livestock operations, except those in the Southeast, the cost per ton is $3.50 after adjusting for the needed storage capacity. Similarly, the cost per ton in the Southeast is $1.75 per ton. The total storage cost for each operation was determined by multiplying these costs-per-ton values by the total tons of recoverable solid manure (at hauling weight) produced in a year.

Generally, the majority of operations are expected to have an adequate solids storage system already in place. The major exception is dairy farms in the Dairy Belt that reported no solids storage in a formal farmer survey. Operational needs for solids storage were judged at:

- 100% for dairy farm #1 in the Dairy Belt
- 20% for dairy farm #2 with 35 to 135 AU and all sizes in the West
- 40% for dairy farm #2 in the Dairy Belt with 135 to 270 AU
- 10% for dairy farm #2 in the Southeast with more than 135 AU
- 25% for fattened cattle and confined heifer farms with scrape-and-stack systems
- 40% for confined heifers in confinement barns
- 60% for swine representative farm #4
- 55% for layer farms in the Southeast, West, and South Central regions with less than 400 AU
- 30% for layer farms in the Southeast, West, and South Central regions with more than 400 AU
- 40% for layer farms in the North Central and Northeast regions with less than 400 AU
- 20% for layer farms in the North Central and Northeast regions with more than 400 AU
- 40% for broiler farms in the East and pullet farms in the North with less than 440 AU
- 50% for broiler farms in the West and turkey farms with less than 440 AU
- 60% for pullet farms in the South and West with less than 440 AU
- 25% for all broiler farms, pullet farms, and turkey farms with more than 440 AU

4.3.7 Storage of Slurry Wastes, Liquid Wastes, and Contaminated Runoff

Slurry wastes, liquid wastes, and contaminated runoff are normally stored in earthen or fabricated structures. Earthen structures are also used to treat manure in anaerobic, aerobic, or aerated lagoons. While lagoons and earthen storages appear similar, the design process for each is different.

In this virtual study, the nonsolid storage facilities were designated as liquid storage, slurry storage, and runoff storage ponds. Liquid and slurry systems are differentiated by the consistency of the material being stored, as determined by the livestock type, and the total solids and slurry manure varies by livestock type. Liquid storages and runoff storage ponds are identical in appearance. Liquid storage ponds as described here generally store more wash water than runoff water, whereas runoff storage ponds generally store more runoff water than wash water. A runoff storage pond for a small dairy captures wash water as well.

4.3.7.1 Liquid Storage

The category of liquid storage includes both liquid storage and treatment lagoons. Most treatment lagoons provide a storage function as well as a treatment function. The design concept for anaerobic lagoons is to size the structure based on the treatment volume needed to degrade the organic material. Additional volume is added for long-term storage of sludge (decay residuals) and storage volumes.

Liquid storage in ponds or lagoons is a component of manure management systems for some swine, dairy, and layer model farms. These systems are typically flush systems in which wastewater is gravity fed or pumped to storage ponds or lagoons. Most of these operations are assumed to have adequate liquid storage or treatment systems in place. However, some may be in disrepair, undercapacity, or need to be replaced entirely. Operational needs for liquid storage, with the exception noted below, were judged at:

- 20% for dairy farm #4 in the Dairy Belt with 35 to 135 AU
- 30% for dairy farm #4 in the Dairy Belt with 135 to 270 AU
- 40% for dairy farm #4 in the Dairy Belt with more than 270 AU
- 30% for dairy farm #5 in the Southeast
- 30% for dairy farm #5 in the West with less than 270 AU
- 20% for dairy farm #5 in the West with more than 270 AU
- 40% for layer farm #3 (flush to lagoon)
- 20% for swine farm #1 for all sizes and regions

An assumption was made that managements of a portion of the operations would choose to convert from one method of handling manure to another method as long as improvements are being made to the operation. The changes that take place cannot be predicted, so the general assumption was that the method of handling manure would remain the same after operational needs implementation. In the case of representative farm #2 for the largest dairies in the Dairy Belt, however, labor costs associated with properly handling the manure as a solid would be too high, and the operator would most likely convert to a liquid system. Thus, operational needs are 100% for the liquid storage component on these farms.

The cost of constructing a pond or lagoon was estimated for each model farm using a representative number of animals per farm. For dairy farms, the representative number of animals was estimated as 137% of the number of milk cows, which accounts for the dairy herd plus dry cows (17%) and calves and heifers (20%). Storage capacity was assumed to be 180 days for all systems. The calculated annual cost was then converted to a per head basis (dairy), a per-animal-unit basis (fattened cattle), or a per-house basis (layers).

Pond or lagoon sizes were developed using the NRCS Animal Waste Management (AWM) engineering design program (NRCS, 2005a). AWM integrates all aspects of the sizing process to meet current NRCS conservation practice standard criteria for waste storage facilities and waste treatment lagoons. Where appropriate, a treatment component was included in the design. Categories were further defined to reflect regional differences. A typical set of climate data (monthly precipitation and evaporation) was selected for each region, representative of the model farm. AWM then calculated manure volume for 180-day storage, 180-day normal rainfall on the pond surface, the rainfall on the pond surface from a 25-year, 24-hour storm event, and as appropriate, the 180-day runoff volume for the most critical 6-month period of the year based on location. Where the

TABLE 4.3
Per-Unit Cost Estimates for Liquid Storage

Livestock type	Region	Numbers animals per farm used to design pond	Storage unit size (gal)	Total installation cost ($)	Annual installation cost ($)	Cost per unit ($)
Dairy	Dairy Belt	300	4,342,477	65,137	9,707	32.36 per head
Dairy	Dairy Belt	200	2,893,414	52,081	7,762	38.81 per head
Dairy	Dairy Belt	100	1,321,828	23,793	3,546	35.46 per head
Dairy	SE	100	1,580,733	28,453	4,240	42.40 per head
Dairy	SE	300	4,573,781	68,607	10,224	34.08 per head
Dairy	West	100	1,607,863	28,942	4,313	43.13 per head
Dairy	West	200	3,130,253	46,954	6,997	34.99 per head
Dairy	West	300	5,216,732	78,251	11,662	38.87 per head
Layers	SE	50,000	7,054,470	105,817	15,770	15,770 per house
Layers	SE	200,000	26,515,403	397,731	59,274	14,818 per house
Layers	SC	200,000	25,387,588	380,814	56,752	14,188 per house
Swine	SE	83 AU	1,165,377	17,481	2,605	31.39 per AU
Swine	SE	248 AU	3,222,244	48,334	7,203	29.04 per AU
Swine	NC-NE	415 AU	5,384,140	80,762	12,036	29.00 per AU
Swine	NC-NE	2,075 AU	26,408,062	396,121	59,034	28.45 per AU
Swine	West	415 AU	6,577,275	98,659	14,703	35.43 per AU
Swine	West	2,075 AU	32,348,499	485,227	72,313	34.85 per AU

Source: Adapted from NRCS/USDA (2004).

liquid is recycled for flushing, AWM allows the designer to reduce inputs. The AWM program also adjusted volumes for evaporation. The results from AWM gave pond and lagoon dimensions and final volume in gallons.

The installation costs were based on actual cost data for equivalent systems. The costs per gallon were calculated from the total cost of an installed pond or lagoon by the design storage volume. Costs were obtained from various locations across the country from NRCS engineers who had firsthand knowledge of an actual system. The costs used in this assessment reflect averages of the information received from sources across the country. Various systems were included in the development of costs, including partially excavated ponds, complete earthen-filled ponds, and flexible membrane-lined ponds. Installation costs per gallon were: 2.2 cents per gallon for pond and lagoons with a capacity of less than 1 million gallons, 1.8 cents per gallon for capacities from 1 million to 3 million gallons, and 1.5 cents per gallon for capacities greater than 3 million gallons. Costs associated with liquid storage are shown in Table 4.3.

4.3.7.2 Slurry Storage

Slurry storage in earthen pits, concrete tanks, or small storage ponds is a component of manure management systems for some swine, dairy, and veal model farms. These storage facilities are commonly beneath a slatted floor. Storage facilities were designed for 120 days of storage to reflect common practice in the industry. Most of the dairy operations for representative farm 3 and veal farms originally were slurry systems, so most are assumed to already have adequate storage systems. For swine farms with slurry systems, the majority were assumed to need extensive upgrades to meet the 120-day storage requirement. Operational needs for slurry storage were judged at:

TABLE 4.4

Per-Unit Cost Estimate for Slurry Storage

Livestock type	Region	Numbers AU per farm used to design storage unit	Storage unit size (gal)	Total installation cost ($)	Annual installation cost ($)	Cost per unit ($)
Dairy	Dairy Belt	200 head	1,122,000	20,196	3,010	15.05 per head
Dairy	Dairy Belt	300 head	1,683,000	30,294	4,515	15.05 per head
Dairy	Dairy Belt	100 head	561,000	12,342	1,839	18.39 per head
Swine	SE	83	287,363	6,322	942	11.35 per AU
Swine	SE	248	708,225	15,581	2,322	9.36 per AU
Swine	NC-NE	415	1,101,176	19,821	2,954	7.12 per AU
Swine	NC-NE	2,075	5,245,933	78,689	11,727	5.65 per AU
Swine	West	415	1,068,808	19,239	2,867	6.91 per AU
Swine	West	2,075	5,037,143	75,557	11,260	5.43 per AU
Swine	NC-NE	450	2,148,585	32,229	4,803	10.67 per AU
Veal	All	415	1,101,176	19,821	2,954	7.12 per AU

Source: Adapted From NRCS/USDA (2004).

- 20% for dairy farm #3 in the Dairy Belt with 35 to 135 AU
- 30% for dairy farm #3 in the Dairy belt with 135 to 270 AU
- 40% for dairy farm #3 in the Dairy Belt with more than 270 AU
- 30% for veal farms
- 50% for swine farm #3
- 60% for swine farm #2

Slurry storage facility costs were estimated in the same manner as liquid storage ponds and lagoons, using the same approach and the same costs per gallon. Costs associated with slurry storage are shown in Table 4.4 for each model farm.

4.3.7.3 Runoff Storage Ponds

Open lots where animals are held produce contaminated water during rainfall events in the form of runoff. Runoff storage ponds are constructed to capture and store this contaminated water. They are needed for pasture-based swine operations (swine farm #5) and swine operations with lots (swine farm #4), turkey ranches, dairy farms #1 and #2, fattened cattle and confined heifer farms with scrape-and-stack manure management systems, fattened cattle manure management system, and fattened cattle feedlots with manure pack. These ponds also collect the wash water used around dairies.

A majority of these farms do not have runoff storage ponds or have inadequate existing ponds. Operation needs for these farms were judged to be high, at:

- 80% for dairy farms #1 and #2
- 90% for turkey ranches
- 70% for fattened cattle farm #2
- 40% for fattened cattle farm #1 and confined heifer farm #2 (scrape and stack) in the Northeast and Midwest
- 50% for fattened cattle farm #1 and confined heifer farm #2 (scrape and stack) in the Southeast and West
- 50% for swine farms #4 and #5

TABLE 4.5
Per-Unit Cost Estimate for Runoff Storage Ponds

Livestock type	Region	Numbers AU per farm used to design pond	Pond size (gal)	Total installation cost ($)	Annual installation cost ($)	Cost per unit ($)
Dairy	Dairy Belt	200 head	1,355,750	24,404	3,637	18.18 per head
Dairy	Southeast	200 head	1,337,331	24,072	3,587	17.94 per head
Dairy	West	200 head	731,983	16,104	2,400	12.00 per head
Swine	Southeast	83	241,281	5,308	791	9.53 per AU
Swine	West	450	632,799	13,922	2,075	4.61 per AU
Swine	Midwest	450	1,398,349	25,170	3,751	8.34 per AU
Confined heifers	Northeast	50	395,232	8,695	1,296	25.92 per AU
Confined heifers	Southeast	50	400,076	8,802	1,312	26.23 per AU
Confined heifers	Midwest	50	308,505	6,787	1,011	20.23 per AU
Fattened cattle	Northeast	50	395,232	8,695	1,296	25.92 per AU
Fattened cattle	Southeast	50	400,076	8,802	1,312	26.23 per AU
Fattened cattle	Midwest	50	308,505	6,787	1,011	20.23 per AU
Fattened cattle	Southeast	100	535,736	11,786	1,756	17.56 per AU
Fattened cattle	Midwest	50	234,919	5,168	770	15.40 per AU
Fattened cattle	Midwest	100	399,713	8,794	1,311	13.11 per AU
Fattened cattle	Northern Plains	350	791,552	17,414	2,595	7.41 per AU
Fattened cattle	Northern Plains	750	1,608,964	28,961	4,316	5.75 per AU
Fattened cattle	Central Plains	750	1,673,838	30,129	4,490	5.99 per AU
Fattened cattle	Central Plains	1,500	3,321,639	49,825	7,425	4.95 per AU
Fattened cattle	West	250	317,391	6,983	1,041	4.16 per AU
Fattened cattle	West	750	1,136,631	20,459	3,049	4.07 per AU
Turkeys	East	500	1,350,897	24,316	3,624	540.87 per AU
Turkeys	Midwest	500	1,167,101	21,008	3,131	467.28 per AU
Turkeys	California	1100	2,285,140	41,133	6,130	415.87 per AU
Turkeys	West other than California	600	1,374,213	24,736	3,686	458.50 per AU

Source: Adapted from NRCS/USDA (2004).

Costs for runoff storage ponds for dairy, fattened cattle, swine farms, and confined heifer farms were estimated in the same manner as liquid storage ponds and lagoons, using the same approach and the same costs per gallon. Costs associated with runoff storage ponds are shown in Table 4.5.

4.3.8 Settling Basins

Settling basins are expected to be a component for all farms with runoff storage ponds. Runoff from open lots generally carries manure solids and sometimes soil particles with it. If these solids are allowed to reach the runoff storage ponds, the operator of the system is faced with the problem of handling a primarily liquid wastewater that contains some solids, making land application of the liquid more difficult because of plugging of irrigation or spray nozzles. The operator must also address the removal of residual solids from the liquid holding pond periodically to ensure design capacity is maintained, which incurs another cost for the operator. Because animal operations that use open lots must already handle both solids and liquids, most operations would prefer to separate solids from the lot runoff before it can enter the runoff storage pond. By separating the solids from the runoff, the solids can be managed more effectively, and the storage pond can be sized and operated more efficiently. Although some operations would continue to handle the runoff as a composite mixture, the added costs of dealing with the solids in the runoff storage pond would easily offset the cost of installing a settling basin. Operational needs for settling basins were the same as those for runoff storage ponds.

Settling basin area need was based on a typical open lot area size for a given animal operation size and the expected routed rainfall runoff volume associated with a 10-year, 24-hour rainfall event on the open lot. Four size classes of operations—100 AU, 200 AU, 500 AU, and 1,000 AU—were used to calculate costs on a per AU basis. The cost of the basin construction (land grading, excavation, and placement of earthen fill) would be about $0.04 per gallon of temporary storage volume. The concrete bottom was assumed to be 6 in thick, with wire mesh reinforcement, at a cost of $200 per cubic yard ($3.70 per square foot) installed. The outlet structure was costed at $780. The costs per AU follow:

These costs were assigned to operational needs farms based on the size of operation, at:

- $5.49 per AU for farms with less than 135 AU
- $4.28 per AU for farms with 135 to 300 AU
- $2.63 per AU for farms with 300 to 1,000 AU
- $2.01 per AU for farms with more than 1,000 AU

4.3.9   CONSERVATION PRACTICES FOR PASTURED LIVESTOCK

Pastured livestock operations differ from conventional feeding operations in that the animals are raised primarily on pastures or ranges, rather than in a confined environment. However, pastured and range animals sometimes are confined in a more conventional sense to provide for ease of management. For example, in areas of the country where winter is severe, common practice keeps pastured or range animals in a confined area with a dependable water supply and access by the farmer to provide supplemental feed. As a result, concentrations of manure accumulate in these confined areas, generally near feed bunks and watering sources. Sometimes these confinement areas are located adjacent to streams and watercourses. The focus of an operational plan for these operations ensures a dependable source of water away from the streams to eliminate direct contact with watercourses and provide for collection and handling of recoverable manure generated in these concentrated areas.

AU's used for sizing	Storage volume	Size of concrete bottom ($)	Total cost ($)	Annual cost ($)
100	17,000	600	3,682	5.49
200	50,000	800	5,743	4.28
500	108,000	1,000	8,828	2.63
1,000	206,700	1,200	13,492	2.01

Costs associated with conservation practices for pastured livestock are grouped under the manure and wastewater storage and handling element, although they include some costs associated with pasture management that would be expected to be included in an operational plan for these farms. As many as 24,697 farms with pastured livestock and few other livestock qualified as farms that may need an operational plan because of the amount of recoverable manure that would potentially be produced. An additional 36,575 farms had less than 35 AU of confined livestock types but had beef cattle as the dominant livestock type on the farm. These two groups of farms comprise the set of farms for which operational needs components for pastured livestock are applied.

Operational needs and costs associated with conservation practices for pastured livestock were derived using the same approach as used for the manure and wastewater storage and handling element. The methods used to estimate operational needs-related costs are presented in the following sections for each component. All costs, except where noted otherwise, were based on the NRCS *Field Office Technical Guide's* (NRCS, 2005c) average cost lists for individual components or practices. All capital costs were amortized over 10 years at 8% interest.

4.3.9.1   Fencing

To properly control the access of animals to water, feed, and loafing areas, a planned system of fencing is needed that is consistent with each individual animal feeding operation's management strategy. Often the need is primarily focused to exclude animals from direct access to a stream. However, with exclusion from the stream, alternative water sources must be provided, and generally, additional fencing is needed to control the movement of animals relative to the new water source. About one-third of the pastured livestock operations were judged to need additional fencing.

The amount of fencing needed depends on the particular operation. For a typical 150-AU cattle operation, about a mile of fence is assumed necessary to supplement existing fencing and replace fencing in disrepair, or 35.2 ft per AU. Based on NCRS Conservation Practice Standard *Fence* (Code 382) (2004) (available at nrcs.usda.gov/ references.), the cost of fencing was $0.80 per foot of fence, for a total cost of $28.16 per AU, or $4.20 annually per AU.

4.3.9.2   Water Wells

An alternative water source must be provided if livestock are excluded from direct access to streams and watercourses. Numerous methods are used to provide this alternative water source, with no consistency of method demonstrated in any particular region of the country. Methods include installing water wells dedicated to providing water for the pasture confinement area, using in-stream pumps to transfer water from the immediate stream corridor, developing natural spring areas that are located away from the stream corridor, and pumping and piping water from an existing water system.

Before discussing our assumption for a new rural well installation for livestock watering needs, we point out that well water may not be the pristine water source many think it is. [*Note:* The following is adapted from Spellman, 2001]

Some common groundwater quality concerns in rural areas are excessive hardness, a high concentration of salt or ion, or the presence of hydrogen sulfide, methane gas, petroleum organic compounds, or bacteria. Some are naturally occurring; others are introduced by human activities. In many rural areas, homeowners have little recourse other than to chemically treat well water to remove or reduce the level of these constituents or to abandon the well supply. Hardness, iron, and sulfur are common constituents that can be treated.

Salt contamination is difficult and expensive to remedy unless the well drawing saline water from a deep aquifer also penetrates one or more freshwater aquifers at lesser depths. In such cases, the deep saline aquifer can be sealed off and the well can be drilled in the freshwater aquifer instead. In many parts of the country, however, when a well is drilled deeper into bedrock to obtain larger supplies, saline water is more likely encountered than is additional freshwater.

Road-salt contamination of groundwater has increased in the last 30 years and is of major concern in northern areas. Highway departments mix salt with sand to spread on rural roads for deicing. Salt is readily soluble in water, runs off highways into lakes and streams, and percolates to the water table. Probably more serious than the spreading of salt is the stockpiling of uncovered salt and sand mixtures. This practice produces concentrated saltwater runoff that percolates to underlying aquifers and nearby wells. Many stockpiles are within small towns or near rural housing areas, where nearby domestic wells can become contaminated.

A chronic problem in many rural homes and farms is *leaking* or *spilled fuel oil* that eventually contaminates the owner's well. Many homes have a fuel tank, buried or aboveground, adjacent to the house and within a few feet of the well. Spill or accumulated leakage eventually can migrate to the aquifer and can be drawn into the well, making it unusable for years. Usually, the only solution is to obtain a new water source. In some instances, however, reducing the pumping rate to reduce drawdown allows the oil to float on the water surface safely above the well's intake area.

Perhaps the problem that poses the greatest hazard to a well owner is *flammable gas* in the well. Small volumes of natural gas, usually methane, can be carried along with the water into wells tapping carbonate or shale rock. In some areas, the gas dissipates soon after installation of the well but, in other areas, a large continual source of natural gas remains. Because methane is flammable and cannot be detected by smell, precautions are needed to prevent explosions and fire. Venting of the wellhead to the open air is the simplest precaution, but because gas can also accumulate in pump enclosures, pressure tanks, and basements, additional venting may be needed. For this reason, a home should never be built over a well.

The most common water-quality problem in rural well water supplies is *bacterial contamination* from septic-tank effluent. According to the USEPA (2000), contamination of drinking water by septic effluent may be one of the foremost water-quality problems in the nation.

Probably the second most serious water-contamination problem in rural farm homes and locations is *barnyard waste*. If a barnyard is upslope from a well, barnyard waste that infiltrates to the aquifer may reach the well. Pumping, too, can cause migration of contaminants to the well. On many farmsteads built more than 100 years ago, the builders were careful to place the supply well upslope from the barnyard. Unfortunately, many present-day owners have not remembered this basic principle and have constructed new houses and wells downslope of their barnyards.

The last three decades have seen a significant increase in small part-time farms and rural dwellings, as large farms have been sold and divided into smaller units. Many modern rural homes are constructed on former cropland on which heavy applications of *herbicides* and *fertilizers* have been made. How these chemicals move through the soil and groundwater, how quickly they decompose, and how their harmful effects are neutralized are not well understood.

Also common is the farming practice of applying fertilizers and pesticides to croplands immediately adjacent to barnyards or farmyards. Residue from these applications can infiltrate to an aquifer and can be drawn into a supply well for a barn or house. Decreasing the use of fertilizers and pesticides in the vicinity of wells can help minimize this problem.

Farm owners should also be careful to properly dispose of wastewater from used containers of *toxic chemicals*. Many farms have their own disposal sites, commonly pits or a wooded area, for garbage and the boxes, sacks, bottles, cans, and drums that contained chemicals. Unfortunately, these owner disposal sites can contaminate farm water supplies.

- *Important point:* Based on the possibility of contaminants entering rural water wells, NRCS Conservation Practice Standard Water Well (Code 642) (2003) (available at www. nm.nrcs.usda.gov/technical/fotg/section-4/standards.) makes the following point regarding sanitary protection.

4.3.9.3 Sanitary Protection

Wells shall be located at safe distances from any potential sources of contamination or pollution, including unsealed abandoned wells. The allowable distance shall be based on consideration of site-specific hydro-geologic factors and shall comply with requirements of all applicable state and local regulations and construction codes. Wells should be located a safe distance from sources of contamination. The table below shows the <u>minimum</u> setback requirement for installation of wells.

Notwithstanding the possible well contaminant sources listed above, for this virtual assessment, a new well is assumed to be installed. The use of a dedicated well is generally the method of choice because of its reliability in providing a consistent quantity and quality of water (springs go dry, stream flows and quality fluctuate). Costs were based on criteria for well development in NRCS Conservation Practice Standard *Water Well* (Code 642) (NRCS, 2003). Well depth was assumed at 250 ft. (Actual depths vary from 100 ft to over 1,000 ft around the country; however, most wells used for livestock watering are installed near riparian areas where the depth to a reliable, potable water table is relatively shallow.) Using $22 per foot as the cost of installing a well, the average cost of a well 250 ft deep is $5,500, or $820 annually per farm. Representative farms #3 and #4 were judged not to need to construct a well, as an alternative water source is most likely readily available. For representative farms #1 and #2, it was judged that about 40% of the operations would need to implement this practice.

Minimum Horizontal Distance between Wellhead and Source of Contamination

Source of contamination	Minimum distance (ft)
Waste disposal lagoon	300
Cesspool	150
Silo pit, seepage pit	150
Livestock and poultry yards	100
Manure pile, privy	100
Septic tank and disposal field	100
Gravity sewer or drain	50
Standing water	10

4.3.9.4   Water Facility

Along with the need to provide an additional source of water is the need to provide temporary water storage and a watering facility for animals. The amount of water storage needed depends on the source and reliability of water and the size of the herd. Watering facility design is based on the criteria established in NRCS Conservation Practice Standard *Watering Facility* (Code 614) (NRCS, 2000). *Water facility* is defined as a device (tank, trough, or other watertight container) for providing animal access to water.

In most situations, a watering facility consists of a corrugated metal trough with a concrete bottom and pad that stores the equivalent of 1 day of water needs. Storage needs were based on 30 gal per AU. For this assessment, costs per AU were based on storage requirements for a 150-AU herd, which would be 4,500 gal. The watering facility would consist of a circular corrugated metal tank 1.5 ft deep and 23 ft in diameter. The cost is $0.75 per gallon for a total cost per AU of $22.50, or an annual cost of $3.35 per AU.

- *Important point:* According to NRCS (Code 614), a trough or tank shall have adequate capacity to meet the water requirements of the livestock and wildlife. This will include the storage volume necessary to carry over between periods of replenishment.

In the Northern Plains and Mountain states, where winter confinement areas tend to be located a considerable distance from the operations' headquarters and where winter temperatures can drop and remain below freezing, special "frost free" watering facilities are needed. This type of facility is an enclosed fiberglass, insulated tank with a small drinking area for cow access. The need for more than 1 day of storage depends on how remote and accessible the confinement site is. For the purposes of this assessment, 1 day of storage was used to calculate the cost. Based on a per-unit cost of $3 per gallon, the total cost per AU is $90, or $13.41 per AU annually.

In some areas of the upper Midwest and New England, winter temperatures also drop to below freezing; however, because of the close proximity of the headquarters area to the confinement areas, more cost-effective alternative methods are available to ensure the water does not freeze (such as manual clearing of ice and electric heaters).

- *Important point:* Operational needs for watering facilities are the same as those for water wells.

4.3.9.5 Heavy Use Area Protection

The purpose of heavy use area protection is to stabilize areas of high traffic or use by equipment and animals. Associated with operational needs for pastured livestock, this practice generally would address the area surrounding the watering facility. The practice would not only protect the integrity of the watering facility but also provide an area for easier recoverability of manure. For the purposes of this assessment, heavy use area protection consists of a concrete pad surrounding the watering facility. Costs per AU were based on a 150-AU herd. The heavy use area would be a square pad, 43 ft on a side or 1,815 ft^2, 6 in thick. Subtracting out the area of the tank, the required installation is 1,414 ft^2, or 26.2 cubic yards of concrete. Based on an installation cost of $120 per cubic yard (which includes the minor grading and shaping required, forming, cost of concrete, and labor), the cost of the pad for the 150-AU herd would be $3,141, or $3.12 per head–AU costs would be $6.35 for a 50-AU herd and $2.32 for a 250-AU herd. The following function was derived for use in estimating the cost per AU:

x = herd size
a = annual cost per AU

If $x \leq 50$, then $a = \$6.35$
If $x \geq 250$, then $a = \$2.32$

$$\text{If} \quad 0 < x < 150, \text{ then } a = 6.35 - \left[\frac{(x-50)}{(150-50)} \times (6.35 - 3.12) \right]$$

$$\text{If } 150 < x < 250, \text{ then } a = 3.12 - \left[\frac{(x-50)}{(250-50)} \times (3.12 - 2.32) \right]$$

- *Important point:* Heavy use area protection is needed only for representative farm #1. Operational needs were judged at 50% for these operations.

4.3.9.6 Windbreak or Shelterbelt Establishments

One of the primary reasons that pastured livestock have been wintered in riparian areas is to provide shelter from the wind and weather. In moving pastured livestock directly out of the immediate stream corridor, certain regions of the country will be moving their animals away from natural cover and protection from the elements. Replacement of the needed protection is essential in implementing operational needs. The windbreaks or shelterbelts are installed along the edge of the confinement area on the side of the prevailing winds expected in the winter. The windbreaks or shelterbelts generally consist of linear plantings of single or multiple rows of trees or shrubs or sets of linear plantings. In practice, from three to seven parallel rows of trees of varying species are usually planted. This is primarily a concern in the West, Northern Plains, and Mountain states.

- *Important point:* According to NRCS Conservation Practice Standard (Code 380) Windbreak/Shelterbelt Establishment (2000), the purposes of planting windbreaks and shelterbelts are multifaceted:

  - to reduce soil erosion from wind
  - to protect plants from wind-related damage

- to alter the microenvironment for enhancing plant growth
- to manage snow deposition
- to provide shelter for structures, livestock, and recreational areas
- to enhance wildlife habitat by providing travel corridors
- to provide living noise screens
- to provide living visual screens
- to provide living barriers against airborne chemical drift
- to delineate property and field boundaries
- to improve irrigation efficiency
- to enhance aesthetics
- to increase carbon storage

The criteria used to determine the size and type of protection needed were based on NRCS Conservation Practice Standard *Windbreak/Shelterbelt Establishment* (Code 380) (NRCS, 2000). Cost estimates were calculated for three herd size categories: 50, 150, and 250 AU. For these herd sizes, the length of the windbreak or shelterbreak would be 600, 1,200, and 1,800 ft, respectively. Installation cost is $4.20 per foot. Thus, the annual cost per AU is $7.51 per AU for a 50-AU herd, $5.01 per AU for a 150-AU heard, and $4.51 per AU for a 250-AU head. The following function was derived for use in estimating the cost per AU:

x = herd size
a = annual cost per head

If $x < = 50$, then $a = \$7.51$
If $x > = 250$, then $a = \$4.51$

$$\text{If} \quad 0 < x < 150, \text{ then } a = 7.51 - \left[\frac{(x - 50)}{(150 - 50)} \times (7.51 - 5.01) \right]$$

$$\text{If } 150 < x < 250, \text{ then } a = 5.01 - \left[\frac{(x - 50)}{(250 - 50)} \times (5.01 - 4.51) \right]$$

Windbreak or shelterbreak establishment is only needed for representative farm #2. Operational needs were judged to be 50% for these operations.

4.3.9.7   Solids Storage

Most pasture operations would allow manure to accumulate through the period of temporary confinement, periodically removing the manure as it accumulates. A designated storage area is generally not needed to manage the manure produced. However, in regions that include the Midwest, Lake states, and the Northeast, manure cannot be periodically spread because of frozen and snow-covered ground. In these regions, temporary storage is needed for about 2 to 3 months while the animals are temporarily confined. Because the period of storage is during the winter, when the only precipitation expected is in the form of snow, a cover for the storage area is not considered essential. Therefore, a concrete slab 6 in thick was used for estimation. For a 150-AU herd, the relative size of a solid storage pad would be 1,600 ft². A 1,600 ft² pad 6 in thick would require 29.6 cubic yards of concrete. Based on a per-unit cots of $120 per cubic yard (which includes the minor grading and shaping required, forming, cost of concrete, and labor), the total cost of the storage pad would be

$3,556, or about $1.85 per ton of recoverable solids. Operational needs were judged to be 50% for operations in the Midwest, Northeast, and Lake states.

4.3.9.8 Filter Strip

For pasture operations in the Midwest, Lake states, and Northeast states, filter strips on the downslope edge of a temporary confinement area are needed to prevent removal of solids and dissolved nutrients from the lot with the runoff from snowmelt and spring rains. Costs per AU were based on a 50-AU head size. The filter strip was assumed to require an area 30 ft wide by 400 ft long, resulting in a treatment area of 12,000 ft^2, or 0.28 acres. The average cost of shaping and seeding is $1,500 per acre; thus, the total cost of the filter strip is $413, equivalent to an annual cost of $1.23 per AU. Because the typical location of these pasture operations is near stream corridors, vegetated areas are often already in place, assuming the lot areas have been set back from the stream. Operational needs were therefore judged to be only 30% for representative farms #3 and #4.

4.3.10 SUMMARY OF OPERATIONAL NEEDS COSTS FOR MANURE HANDLING AND STORAGE

Estimates of operational needs for each model farm were used to calculate estimates for each virtual farm in the Census of Agriculture. For farms with more than one representative farm type assigned to it, the probabilities associated with each representative farm were used as weights to obtain a weighted total.

The average annual per-farm cost estimates for each of the manure and wastewater handling and storage components are present in Table 4.6 according to livestock type. Manure storage components (solids, liquid, slurry, and runoff ponds) had the highest cost per farm for all but pastured livestock and swine farms. Liquid transfer costs were slightly higher than storage costs for swine farms. For dairies, liquid transfer costs were nearly as high as storage costs. Collections were a significant portion of the total costs for fattened cattle and turkey farms, and mortality management costs were a significant portion for swine, broiler, and turkey farms.

The annual average cost for the manure handling and storage element was estimated to be $2,409 per farm (Table 4.7). Capital costs were nearly 75% of the total cost, overall. The highest cost was for fattened cattle farms, at $9,112 per farm, and for turkey farms, at $7,940 per farm, reflecting the larger number of AU per farm for these two types of farms. Dairy farms had the highest cost per AU at $22 per milk cow animal unit. Swine farms had the next highest cost per AU at $18 per swine AU.

Costs differed most by farm size (Table 4.6). Large farms (producing more than 10 tons of phosphorus annually) had an average annual cost of $15,167 per farm, compared to an average annual cost of $3,397 per farm for medium-size farms and $1,070 per farm for small farms. The cost per AU on large farms, however, was lower than for medium-size and small farms, because of the economies of scale embodied in the assignment of per unit costs and the lower operational needs expected for the largest farms.

Per-farm costs were highest in the Pacific, Mountain, and Southern Plains regions (Table 4.7) and lowest in the Lake states and Corn Belt region. Total costs were highest in the Corn Belt region, the Lake states, and the Northern Plains, which together represented about 45% of the total costs for manure and wastewater handling and storage.

Overall, annual manure handling and storage costs totaled $645 million.

4.4 SUMMARY

Manure storage is an essential part of AFO and CAFO operation. In Chapter 5, we discuss AFO and CAFO siting concerns, another essential factor in AFO and CAFO operation.

TABLE 4.6

Annual Manure Handling and Storage Cost per Farm, by Livestock Type and Farm Size

Dominant livestock type or farm size class	Number of farms	AU for dominant livestock type	AU for other livestock type	Capital costs ($)**	Operating costs ($)**	Maintenance costs ($)**	Total costs ($)	Cost per AU of dominant livestock type ($)
Fattened cattle	10,159	858	440	7,629	1,254	229	9,112	11
Milk cows	79,318	149	46	2,620	551	79	3,249	22
Swine	32,955	236	40	3,451	585	104	4,139	18
Turkeys	3,213	638	49	5,305	2,476	159	7,940	12
Broilers	16,351	150	33	1,666	635	50	2,351	16
Layers/pullets	5,326	258	39	3,519	390	106	4,015	16
Confined heifers/veal	4,011	237	64	2,710	401	81	3,192	13
Small farms with confined livestock types	42,565	18	7	149	46	4	199	11
Pastured livestock types	61,272	107	10	NA	NA	NA	823	8
Specialty livestock types	2,131	NA	17	563	263	17	843	NA
Large farms	19,746	1,129	290	11,627	2,721	340	15,167	13
Medium farms	39,437	191	61	2,477	543	74	3,397	18
Small farms	198,018	63	17	773	126	23	1,070	17
All types	257,201	165	45	1,867	389	56	2,509	15

Source: NRCS/USDA (2004).

NA: Not available.

* Includes pastured livestock types.

** Costs for farms with pastured livestock types dominant were not broken down into capital and operating costs. Costs for these farms are presented in the total cost column.

TABLE 4.7

Annual Manure and Wastewater Handling and Storage Cost per Farm, by Farm Production Region

Farm production region	Number of farms	Capital costs ($)	Operating costs ($)	Maintenance costs ($)	Total costs ($)
Appalachian	22,899	2,155	545	65	2,987
Corn Belt	71,540	1,312	214	39	1,647
Delta states	12,352	1,468	436	44	2,181
Lake states	52,817	1,363	250	41	1,669
Mountain	7,964	4,184	980	126	6,177
Northeast	31,598	1,595	303	48	1,976
Northern Plains	26,309	2,012	345	60	3,088
Pacific	7,974	5,684	1,479	171	7,731
Southeast	12,807	2,074	549	62	2,901
Southern Plains	10,941	3,508	775	105	4,776
All types	**297,201**	**1,807**	**389**	**56**	**2,509**

Source: NRCS/USDA (2004).

CHAPTER REVIEW QUESTIONS

1. What are the advantages of manure storage for AFOs and CAFOs?
2. What agribusiness changes have increased the need for manure storage?
3. Why can't manures be land-applied year-round, rather than stored?
4. What components and activities are associated with manure storage?
5. How do AFOs and CAFOs commonly address these activities and components?
6. What issues are associated with long-term storage?
7. What issues are associated with mid-term storage?
8. What issues are associated with short-term storage?
9. How are manure storage needs determined?
10. What factors are used to determine farm storage volume?
11. What importance is siting in manure storage facility construction?
12. What considerations are needed for earthen manure storage that aren't important for concrete? Why?
13. What are distance separation requirements, and why are they important for storage siting?
14. What functions do runoff holding ponds serve? What are their characteristics?
15. What are runoff storage pond advantages and disadvantages?
16. What are lagoons? What functions do they serve? How do they differ from runoff storage ponds?
17. What are advantages and disadvantages of lagoons? What are the differences for lagoon size requirements determined for warm and cold climates?
18. What are earthen slurry basins? What functions do they serve? What specialized maintenance do they require? What are their advantages and disadvantages?
19. What are below-building pits? What are their advantages and disadvantages?
20. What are roofed tanks? What are their advantages and disadvantages? What special considerations are needed for operation?
21. What are the advantages and disadvantages of uncovered stored manures? Where are they best used?

22. What elements related to costs do the virtual farms used to estimate costs include? Explain why.
23. What are the sources for the representative farms?
24. What representative farms are included for dairy farms? For swine? For layers? For beef production? For other confined livestock types?
25. What place does "mortality management" have in virtual farms?
26. What are the capital costs associated with swine mortality management?
27. What are the benefits incurred with feedlot upgrades?
28. What are clean water diversions? Why and how are they used?
29. Why is roof runoff management important? What types of operations does it apply to?
30. How do earthen berms with underground pipe outlets provide clean water diversion? What are the advantages and disadvantages?
31. How do grass waterways provide clean water diversion? What are the advantages and disadvantages?
32. How do earthen berms with surface outlets provide clean water diversion? What are the advantages and disadvantages?
33. How does liquid treatment using a biofilter for open lots work to provide clean water diversion? What are the advantages and disadvantages?
34. What are the common collections systems in common use? Why?
35. What are the common transfer systems in common use? Why?
36. How are solid and liquid wastes handled differently? What costs are unique to each type of system?
37. How do flush systems work, and where are they used?
38. Where are contaminated runoff collection systems used, and how do they work?
39. What are the advantages and disadvantages of renting pumping equipment?
40. What are the advantages and disadvantages of purchasing pumping equipment?
41. How do you calculate the costs for each?
42. What is the purpose of manure storage?
43. What factors are critical to determining storage facility needs?
44. What minimums and maximums for storage times are suggested by conservation practice standards? What differences are suggested for various parts of the country? How does climate affect storage?
45. How are storage costs figured?
46. What are the differences between solid and liquid slurry storage?
47. What elements do solids contain?
48. What elements does slurry contain?
49. What elements do liquid wastes contain?
50. What effect does rainfall have on the various types of storage?
51. What are common industry practices for slurry storage? For solids storage? For liquid storage?
52. What are settling basins used for and why are they necessary?
53. How are settling basin area costs determined?
54. How are settling basin needs determined?
55. What are the differences between pastured livestock operations and feeding operations?
56. What are "landscape" concerns for pastured livestock?
57. What costs are associated with pastured livestock?
58. When do such operations need a management plan?
59. What groups of farms comprise the set of farms for which operational needs are applied?
60. Describe the need for fencing plans for pastured livestock operations.
61. When are alternative water sources needed?
62. Describe some of the methods used to provide alternate water sources.

63. What are the cost concerns with these individual methods?
64. What water quality concerns must be considered for well sources?
65. What pollution source concerns must be considered for well sources?
66. What contamination source concerns must be considered for well sources?
67. What bacterial, pesticide, herbicide, and fertilizer source concerns must be considered for well sources?
68. What waste source concerns must be considered for well sources?
69. What wastewater concerns are important for pastured livestock operations?
70. How do you determine the minimum horizontal distance between wellhead and contamination source?
71. How are costs determined for well installation?
72. Define "water facility."
73. What water facility considerations are of concern for pastured livestock operations?
74. How are water facility installation costs determined?
75. When are frost-free facilities necessary?
76. Describe the features essential for a frost-free facility.
77. Define "heavy use area protection." Why must some areas be provided with this protection?
78. How are use area costs determined?
79. Why are pastured livestock often wintered in riparian areas?
80. When do windbreaks or shelterbelts need to be provided? Why?
81. Discuss eight purposes of shelterbelts.
82. How are windbreak costs determined?
83. What are manure storage concerns for livestock housed in temporary confinements?
84. What regions are most affected by these concerns?
85. Why are filter strips important, and what can they accomplish?
86. How are filter strip costs determined?
87. How are annual manure handling and storage cost estimates determined?
88. What effect does economy of scale have on the average AU costs for large, medium, and small operations?
89. What regions average the highest and lowest average AU costs? What factors are involved for each region?

THOUGHT-PROVOKING QUESTIONS

1. What are your state's siting requirements? What restrictions does your state place on manure storage siting?
2. Grazing rights issues are not uncommon in some areas of the country. How do these issues affect pastured livestock operations?
3. What effects do different livestock types have on pasture? How does this affect the land?
4. What are the differences between how cattle affect the land and how buffalo affect the land?
5. What has happened to shortgrass and tallgrass prairie ecosystems over the last 150 years?

REFERENCES

Ensminger, M.E. 1991. Animal Science. Danville, IL: Interstate Publishers.

North Carolina Cooperative Extension. 1996. Worksheet to determine size of poultry mortality composter. Pub. EBAE 177-93.

North Carolina Cooperative Extension. 1999. A cost comparison of composting and incinerating methods for mortality disposal. Poultry Science Facts.

NRCS/USDA. 1992. Agricultural waste management field handbook. http://wwww.ftw.nrcs.usda.gov/awmfh. html.

NRCS. 2005. Animal Waste Management (AWM) engineering design program. www.nrcs. usda.gov.

NRCS. 2005. Cost associated with development and implementation of comprehensive nutrient management plant (CNMP). http://www.nrcs.usda.gov/technical/ECS/nutrient/gm-190.html. Accessed January 5, 2005.

NRCS. 2005. Field Office Technical Guide, Section 1, Annual Cost List. www.nrcs.usda. technical.

NRCS/USDA. 2004. Manure and wastewater handling and storage costs. www.nrcs.usda.gov/technical/land/ pubs/cnmp1h.pdf. Accessed November 5, 2004.

NRCS/USDA. 2005. Manure and wastewater handling and storage costs, Document 600.E.54a. www.nrcs. usda.gov/programs/afo/cnmp_guide_index.html. Accessed January 8, 2005.

NWPS 18-1. 2001. *Manure Storages.* Ames, IA: MidWestPlanService, Iowa State University.

Spellman, F.R. 2001. The Handbook for Waterworks Operator Certification, Vol. 1. Boca Raton, FL: CRC Press.

Tilmon, H.D., & German, C. 1997. Considerations in using custom services and machinery rental: custom rates and guidelines for computing machinery ownership costs. Univ. Delaware, Col. Agric. and Nat. Resource.

USDA/APHIS. 1995. Part I, reference of 1995 swine management practices. *Natl. Animal Health Monitor Sys.* http://www.aphis.usda.gov.

USDA/APHIS. 1996. Part I, reference of 1996 dairy management practices. *Natl. Animal Health Monitor. Sys.* http://www.aphis.usda.gov.

USDA/APHIS. 1999. Part I, reference of 1999 table egg layer management in the U.S. *Natl. Animal Health Monitor. Sys.* http://www.aphis.usda.gov.

USDA/ERS. 2000. Data from the 1998 agricultural resource management survey, hogs production practices and costs and returns report. Unpublished.

USEPA. 2000. State compendium: Programs and regulatory activities related to animal feeding operations. Office Waste Management.

5 AFO/CAFO Siting: Physical Factors

And he passes scores of anonymous, low, grey buildings with enormous fans at their ends set back from the road and surrounded by chain-link fence. From the air these guarded hog farms resembled strange grand pianos with six or ten white keys, the trapezoid shape of the body the effluent lagoon in the rear.

(Proulx, 2002, p. 2)

5.1 INTRODUCTION

Keeping in mind that animal farm operation (AFO) and concentrated AFO (CAFO) siting in any community involves a complex interplay of regulations, scientific considerations, economic questions and people's emotions, the actual siting of AFO and CAFO planning, design, implementation, and function are dependent on various physical factors. These factors include (1) soil physical and chemical properties and landscape features, (2) plant growth's role in management of nutrients in an agricultural waste management system, (3) the engineering suitability of the soil and foundation characteristics of the site and the potential for contamination of groundwater, and (4) the planning and design options for arranging and integrating waste management systems into an existing or proposed farmstead.

This chapter first describes soil agricultural waste interactions and those soil properties and characteristics that affect soil suitability and limitations for a farmstead. Second, it discusses the function and availability of plant nutrients as they occur in agricultural wastes and introduces the effects of trace elements and metals on plants. General guidance is given so components of waste can be converted to a plant-available form and nutrients harvested in the crop can be estimated. The impact of excess nutrients, dissolved solids, and trace elements on plants is given in relationship to agricultural waste application. Thirdly, the chapter covers geological and groundwater considerations that can affect the planning, design, and construction of an AFO/CAFO. Finally, we discuss the planning and design options for arranging and integrating components of agricultural waste management systems (AWMSs) into an existing or proposed farmstead.

5.2 ROLE OF SOILS IN MANURE MANAGEMENT

In establishing an animal manure management system, soil data should be collected early in the planning process. Essential soil data include soil maps and the physical and chemical properties that affect soil suitability and limitations.

- *Important point:* Soil maps are available in published soil surveys, or if not published, are available at the local Natural Resources Conservation Service (NRCS) field office. Soil suitability and limitation information can be obtained from published soil surveys, Section II of the *Field Office Technical Guide* (FOTG), and Field Office Communication System (FOCS) tables and soil data sets, soil interpretation records (SIRs), and the *National Soils Handbook* interpretation guides, part 603.

Soil information and maps may be inadequate for planning manure management system components. Manure management systems should not be implemented without adequate and complete soil maps or soil interpretive information. If soil data or maps are inadequate or unavailable, soil survey information must be obtained before completing a manure management system plan. This information will include a soil map of the area, a description of soil properties and their variability, and soil interpretive data.

5.2.1   SOIL BASICS

[*Note:* Much of the information in this section is adapted from *The Science of Environmental Pollution*, Spellman (1999).]

Any fundamental discussion about soil begins with a definition of what soil is. The word soil is derived through Old French from the Latin *solum*, which means floor or ground. A more concise definition is made difficult by the great diversity of soils throughout the globe. However, here is a generalized definition from the Soil Science Society of America:

> Soil is unconsolidated mineral matter on the surface of the earth that has been subjected to and influenced by genetic and environmental factors of parent material, climate, macro- and microorganisms, and topography, all acting over a period of time and producing a product—soil—that differs from the material from which it is derived in many physical, chemical, and biological properties and characteristics.

5.2.1.1   Soil Phases

Soil is heterogeneous material made up of three major components: a solid phase, a liquid phase, and a gaseous phase [see Figure 5.1(a)]. [*Note:* This phase relationship is important in dealing with soil pollution, because each of the three phases of soil are in equilibrium with the atmosphere and with rivers, lakes, and the oceans. Thus, the fate and transport of pollutants are influenced by each of these components.] All three phases influence the supply of plant nutrients to the plant root.

Soil is also commonly described as a mixture of air, water, mineral matter, and organic matter [see Figure 5.1(b)]; the relative proportions of these four compounds greatly influence the productivity of soils. The interface (where the regolith meets the atmosphere) of these materials that make up soil is what concerns us here.

Key term: *Regolith* is commonly used to describe the Earth's surface.

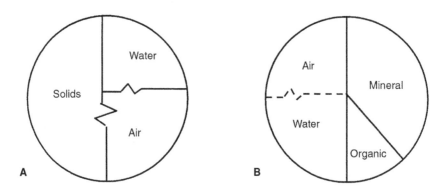

FIGURE 5.1   (A) Three phases of soil: solids, water, and air. Broken lines indicate that these phases are not constant but change with conditions; (B) Another view of soil (a loam surface soil). Makeup: air, water, and solids in mineral and organic content.

Keep in mind that the four major ingredients that make up soil are not mixed or blended like cake batter. Instead, pore spaces (vital to air and water circulation, providing space for roots to grow and microscopic organisms to live) are a major and critically important constituent of soil. Without significant pore space, soil would be too compacted to be productive. Ideally, pore space will be divided roughly equally between water and air, with about one-quarter of the soil volume consisting of air and one-quarter consisting of water. The relative proportions of air and water in a soil typically fluctuate significantly as water is added and lost. Compared to surface soils, subsoils tend to contain less total pore space, less organic matter, and a larger proportion of micropores, which tend to be filled with water.

Let's take a closer look at the four major components that make up soil.

The *mineral and organic phases* (solids) are the main nutrient reservoirs. They hold nutrients in the cation form (positively charged ions), such as potassium (K), nitrogen (N, as ammonium), sodium, calcium, magnesium, iron, manganese, zinc, and cobalt on negatively charged clay and organic colloidal particles. Anionic (negatively charged ions) nutrients, such as nitrogen (as nitrate), phosphorus (P), sulfur, boron, and molybdenum, are largely held by the organic fraction or mineral complexes. Mineral matter varies in size and is a major constituent of nonorganic soils. *Mineral matter* consists of large particles (rock fragments) including stones, gravel, and coarse sand. Many of the smaller mineral matter components are made of a single mineral. Minerals in the soil (for plant life) are the primary source of most of the chemical elements essential for plant growth.

Soil *organic matter* consists primarily of living organisms and the remains of plants, animals, and microorganisms that are continuously broken down (biodegraded) in the soil into new substances that are synthesized by other microorganisms. These other microorganisms continually use this organic matter and reduce it to carbon dioxide (via respiration) until it is depleted, making repeated additions of new plant and animal residues necessary to maintain soil organic matter (Brady & Weil, 1996).

The amount of plant-available nutrient held by a soil depends on its unique chemical and physical makeup. This makeup can be ascertained by a soil's cation-exchange capacity, pH, organic matter content, clay mineralogy, and water-holding capacity.

The presence of water in soil, the *liquid or solution phase*, is reflective of climatic factors and is essential for the survival and growth of plants and other soil organisms. Soil moisture is a major determinant of the productivity of terrestrial ecosystems and agricultural systems. Nutrients transported in the liquid phase are present in the solute form of the nutrient element. Oxygen and carbon dioxide can be dissolved in the soil solution and transported to and from the system. A large percentage of animal waste material is composed of water. Water moving through soil materials is a major force behind soils formation. Along with air, water, and dissolved nutrients, soil moisture is critical to the quality and quantity of local and regional water resources.

In the *gaseous phase,* soil air circulates through soil pores in the same way air circulates through a ventilation system. Only when the pores (ventilation ducts) become blocked by water or other substances does the air fail to circulate. Though soil pores normally connect to interface with the atmosphere, soil air is not the same as atmospheric air. It differs in composition from place to place. Soil air also normally has higher moisture content than the atmosphere. The content of carbon dioxide is usually higher and that of oxygen lower than accumulations of these gases found in the atmosphere. Gas exchange affects denitrification, mineralization of organic material, and soil microorganism growth rate.

5.2.2 Soil–Animal Waste Interaction

Soil–animal waste interactions are a complex set of relationships dependent on the soil environment, microbial populations, and the chemical and physical properties of the soil and waste material. The following discussion describes some of these relationships.

5.2.2.1   Filtration

Soil filtering systems are used to deplete biological oxygen demand (BOD), consume or remove such biostimulants as phosphates and nitrates, provide long-term storage of heavy metals, and deactivate pathogens and pesticides. Soils suitable for use as filtering systems have permeability slow enough to allow adequate time for purification of water percolating through the soil system.

A balance of air, water, and nutritive substances at a favorable temperature is important to a healthy microbial population and an effective filtration system. For example, overloading the filtration system with wastewater containing high amounts of suspended solids causes clogging of soil pores and a reduction of soil hydraulic conductivity. Management and timing of wastewater applications are essential to maintaining soil filter systems. Climate, suspended solids in the wastewater, and cropping systems must be considered to maintain soil porosity and hydraulic conductivity.

The wastewater application rate should not exceed the waste decomposition rate, which is dependent on soil temperature and moisture content. Periods of wetting and drying increase microbial decomposition and by-product uptake by the crop and decrease potential soil pore clogging. In areas where the temperature is warm for long periods, the application rates may be higher if crops or other means of using the by-products of waste decomposition are available.

Tillage practices that maintain or improve soil tilth and reduce soil compaction and crusting should be included in the land application part of animal waste management systems. These practices help to maintain soil permeability, infiltration, and aeration, which enhance the biological decomposition processes.

5.2.2.2   Biological Degradation

When animal waste is applied to soil, several factors affect biological degradation of animal waste organics. These factors interact during the biological degradation process and can be partitioned into soil and organic factors.

Soil factors that affect biological degradation are temperature, moisture, oxygen supply, pH, available nutrients (N, P, K, and micronutrients), porosity, permeability, microbial population, and bulk density. Organic factors are carbon-to-nitrogen ratio (C:N), lignin content, and BOD.

Key term: The carbon-to-nitrogen ratio (C:N) is the amount of carbon in a residue in relation to the amount of nitrogen. The rate of organic matter decomposition and timing of nutrient availability are influenced by the C:N ratio. Everything has a ratio of carbon to nitrogen in its tissues. The C:N ratio of soils, including everything organic in the soil, is 8-12:1.

The soil and organic factors interact and determine the environment for microbial growth and metabolism. The physical and chemical nature of this environment determines the specific types and numbers of soil microorganisms available to decompose organic material.

The decomposition rate of organic material is primarily controlled by the chemical and biological composition of the animal waste material, soil moisture and temperature, and available oxygen supply. Rapid decomposition of organic wastes and mineralization of organic nitrogen and phosphorus by soil microorganisms are dependent on an adequate supply of oxygen and soil moisture.

High loading rates or high BOD waste may consume most of the available oxygen and create an anaerobic environment. This process can cause significant shifts in microbial populations, microbial metabolisms, and mineralization by-products. Under anaerobic conditions, by-products may be toxic and can occur in sufficient concentrations to inhibit seed germination and retard plant growth, even after aerobic conditions have been restored.

- *Important point:* Toxins may accumulate if soil microbial life is degraded. The soil's ability to dechlorinate organic compounds can be impaired, especially in sulfate-rich anaerobic environments. Heavy application of animal wastes to low-pH soils can lead to buildup of ammonium and a concomitant reduced functioning of Nitrobacter. This effect can result in nitrite accumulation (Spellman, 1996).

5.2.2.3 Chemical Reactions

In organic animal waste material management, the chemical reactions that occur between the soil and the animal waste components must be taken into account. These reactions are broadly grouped as ion exchange, adsorption, precipitation, and complexation. The mechanisms and rates of these reactions are dependent on the physical, chemical, and biological properties of the soil and organic animal waste material.

Organic waste mineralization by-products consist of macro- and micro-plant nutrients, soluble salts, gases, and heavy metals. These by-products dissolve and enter soil water solutions as precipitation or irrigation water infiltrates the soil surface and percolates through the soil profile. The dissolved by-products are subject to the interactions of ionic exchange, adsorption, precipitation, and complexation. These processes store and exchange the macro- and micro-plant nutrient by-products of organic waste mineralization. They also intercept and attenuate heavy metals, salts, and other detrimental mineralization by-products that can adversely affect plant growth and crop production.

Ion exchange reactions involve both cations and anions (Table 5.1). Ionic exchange and adsorption is the replacement or interchange of ions bonded electrostatically to exchange sites on soil particles and soil organic materials with similarly charged ions in the soil solution. This ionic interchange occurs with little or no alteration to exchanging ions.

Cation exchange is the adsorption and exchange of nonmetal and metal cations to negatively charged site on soil particles and soil organic materials. Cation-exchange capacity (CEC) is the measure of a soil's potential to exchange cations and is related to soil mineralogy, pH, and organic matter content.

Anion exchange is the exchange and replacement of negatively charged ions to positively charged sites on soil particles. Anion exchange capacity is lower than cation exchange in most soils; anion exchange is important, however, because the anion exchange potential of a soil is related to its ability to retain and exchange nitrate nitrogen, sulfate, chloride, boron, molybdenum, and phosphorus.

Adsorption and *precipitation* are processes that remove an ion from a soil solution. Sorption occurs as ions attach to the solid soil surface through weak chemical and molecular bonds or as strong

TABLE 5.1
Common Exchangeable Soil Cations and Anions

Elements	Cations	Anions
Aluminum	Al^{+3}	
Boron	BO_3^{-3}	
Calcium	Ca^{+2}	
Carbon	CO_3^{-2}, HCO	
Chlorine	Cl^-	
Copper	Cu^+, CU^{+2}	
Hydrogen	H^+	OH^-
Iron	Fe^{+2}, Fe^{+3}	
Magnesium	Mg^{+2}	
Manganese	Mn^{+2}, Mn^{+3}	
Molybdenum	MoO_4^{-2}	
Nitrogen	NH_4^+	NO_2^-, NO_3^-
Phosphorus	HPO_4^{-2}, $H_2PO_4^-$	
Potassium	K^-	
Sulfur	SO_3^{-2}, SO_4^{-2}	
Zinc	Zn^{+2}	

chemical bonds. Precipitation is the deposition of soluble compounds in soil voids. It occurs when the amount of the dissolved compounds in the soil solution exceeds the solubility of those compounds.

Complexation is the combination of different atoms to form a new compound. In soils, it is the interaction of metals with soil organic matter and some oxides and carbonates, resulting in the formation of large, stable molecules. This process extracts phosphorus and heavy metals from the soil solution. These stable complexes act as sinks for phosphorus, heavy metals, and some soil micronutrients.

5.2.3 Soil–Animal Waste Mineralization Relationship

The mineralization of animal waste material is governed by the biological, chemical, and physical properties of soil and organic waste; soil moisture; and soil temperature. Organic waste mineralization is a process in which microbes digest organic waste, reduce the waste material to inorganic constituents, and convert it to more stable organic materials. Inorganic materials released during this process are the essential plant nutrients (N, P, K), macronutrients and micronutrients, salts, and heavy metals.

5.2.3.1 Microbial Activity

Soil–animal waste material microbial composition and microbial activity greatly influence the rate of organic waste mineralization. Soil moisture, temperature, and aeration regulate soil microbial activity and thus are factors that influence the rate of waste mineralization.

The highest potential microbial activity and the highest potential rate of organic waste mineralization occur in soils that are warm, moist, and well aerated. Lower potential rates should be expected when soils are dry, cold, or saturated with water.

Average annual soil surface temperature and seasonal temperature variations have a significant impact on the duration and rate of soil microbial activity. Average annual soil temperatures in the contiguous United States range from less than 32°F (0°C) to more than 72°F (22°C). Microbial activity is highest in soils with high average annual soil temperature and lowest in soils with low temperature.

In many areas, the mean winter soil temperature is 9°F (5°C) or more below the mean summer soil temperature. Microbial activity and organic waste mineralization in the soils in these areas are greatest during the summer months and least during the winter months. Thus, microbial activity decreases or increases as mean monthly soil temperature changes throughout the year.

- *Important point:* Agricultural wastes applied to cold or frozen soils mineralize very slowly, are difficult or impossible to incorporate, and are vulnerable to surface runoff and erosion. Potential agricultural waste contamination of surface water is highest when agricultural wastes are applied under these conditions.

Microbial activity is also highly dependent on the soil moisture content. Soils that are dry throughout most of the growing season have a low organic matter mineralization rate. Microbial activity in these soils is greatest immediately after rainfall or irrigation events and decreases as soil moisture decreases. Conversely, soils that are moist throughout most of the growing season have higher microbial activity and more capacity to mineralize organic waste. Wet soils or soils that are saturated with water during the growing season have potentially lower microbial activity than moist soils. This difference is not caused by a lack of soil moisture but is the result of low soil aeration that occurs when soil is saturated.

5.2.3.2 Nitrogen Mineralization

Organic nitrogen is converted to inorganic nitrogen and made available for plant growth during the waste mineralization process. This conversion process is a two-way reaction that both releases and consumes nitrogen.

Animal waste materials, especially livestock manure, increase the energy or food supplies available to the soil microbial population. This energy stimulates soil microbial activity, which consumes more available nitrogen than the mineralization processes release. Thus, high microbial activity during initial waste mineralization can cause a reduction of available nitrogen below that needed for plant growth. Nitrogen deficiency also occurs if waste mineralization cannot supply sufficient quantities of nitrogen to plants during periods of rapid growth. This is most apparent in spring as the soil warms and crops exhibit a short period of nitrogen deficiency.

- *Important point:* Ammonium nitrogen (NH4+) is the initial by-product of organic nitrogen mineralization. Ammonium is adsorbed to soil particles through the cation exchange and can be used by plants or microorganisms. Ammonium nitrogen is further oxidized by nitrifying bacteria to nitrate (NO_3^-). This form of nitrogen is not strongly adsorbed to soil particles or easily exchanged by anion exchange.

Nitrate forms of soil nitrogen are susceptible to leaching and can leach out of the plant root zone before they can be used for plant growth. Nitrate can contaminate if leached below the soil root zone or transported off the field by runoff to surface water. Soils with high permeability and intake rates, coarse texture, or shallow depth to a water table are the most susceptible to nitrate contamination of groundwater. Those with low permeability and intake rates, fine texture, or steep slopes have a high runoff potential and are the most susceptible to nitrogen runoff and erosional losses.

5.2.3.3   Phosphate Mineralization

Organic phosphorus in animal wastes is made available for plant growth through the mineralization process. Phosphorus is removed from the soil solution by adsorption to the surface of clay particles or complexation with carbonates, iron, aluminum, or more stable organic compounds.

Key term: *Mineralization* is the conversion of an element from an organic form to an inorganic state as a result of microbial decomposition.

Phosphorus mobility is dependent on the phosphorus adsorption and complexation capacity of a soil. Soils with slow permeability and high pH, lime, iron or aluminum oxides, amorphous materials, and organic matter content have the highest phosphorus adsorption capacity. Adsorbed phosphorus is considered unavailable for plant growth. Soil erosion and runoff can transport the sorbed and complexed phosphorus offsite and contaminate surface water. Adsorbed phosphorus in surface water may become available by changes in the water pH or redox potential. Conversely, soils with rapid permeability, low pH, and low organic matter have low phosphorus adsorption capacity, allowing phosphorus to leach below the root zone. However, this seldom occurs.

5.2.3.4   Potassium, Calcium, and Magnesium Mineralization

Potassium, calcium, and magnesium converted from organic to inorganic compounds during mineralization have similar reactions in the soil. Upon dissolution, they become cations that are attracted to negatively charged soil particles and soil organic matter. These minerals are made available for plant growth through the cation exchange process. Potassium is less mobile than nitrogen and more mobile than phosphorus. Leaching losses of potassium are not significant and have little potential to contaminate groundwater. Calcium and magnesium can leach into groundwater or aquifers, but they do not constitute a hazard to water quality.

5.2.3.5   Heavy Metal and Trace Element Mineralization

Heavy metals and trace elements are by-products of the organic mineralization process. Municipal sludge applied to the land is often a source of heavy metals. They are strongly adsorbed to clay particles or complexed (chelated) with soil organic matter and have very little potential to contaminate groundwater supplies and aquifers. This immobilization is strongest in soils with a high content of organic matter, pH greater than 6.0, and CEC of more than 5. However, application of organic waste containing high amounts of heavy metals can exceed the adsorptive capability of the soil and increase the potential for groundwater or aquifer contamination.

Sandy soils with low organic matter content and low pH have a low potential for retention of heavy metals. These soils have the highest potential for heavy metals and trace element contamination of aquifers and groundwater. Surface water contamination from heavy metals and trace elements is a potential hazard if agricultural wastes are applied to areas subject to a high rate of runoff or erosion.

5.2.4   Soil Characteristics

Note that no clear delineation or line of demarcation can be drawn between the properties of one soil and those of another. Instead, a gradation (sometimes quite subtle—like from one shade of white to another) occurs in soils as one moves from one soil to another (Spellman, 1999). Brady and Weil (1996) point out that "the gradation in soil properties can be compared to the gradation in the wavelengths of light as you move from one color to another. The changing is gradual, and yet we identify a boundary that differentiates what we call 'green' from what we call 'blue'" (p. 58).

Soil suitabilities and limitations for animal waste application are based on the most severely rated soil property or properties. A severe suitability rating does not necessarily imply that animal wastes cannot be used. It does, however, imply a need for careful planning and designing to overcome the severe limitation or hazard associated with one or more soil properties. Care must be taken in planning and designing animal waste management systems that are developed for soils with a moderate limitation or hazard suitability rating.

- *Important point:* In general, moderate limitations or suitability ratings require less management or capital cost to mitigate than do severe ratings.

- *Important point:* Slight is the rating given soils with properties favorable for the use of animal wastes. The degree of limitation is minor and can be overcome easily. Good performance and low maintenance can be expected.

Soil suitability for site-specific animal waste storage or treatment practices, such as a waste storage pond, waste treatment lagoon, or waste storage structure, are not discussed in this section. Soil variability within soil map delineations and mapping scales generally prevent using soil maps for evaluation of these site-specific animal waste management system components. Soil investigations conducted by a soil scientist or other qualified person are needed to determine and document site-specific soil information, such as soil type, observed and inferred soil properties, and the soil limitations or hazards for the site-specific components.

Non-site-specific animal waste utilization practices are those that apply animal wastes to fields or other land areas by spreading, injection, or irrigation. The suitability, limitations, or hazards associated with these practices are dependent upon and influenced by the geographical variability of the soil and soil properties within the area of application.

Soil suitability ratings for non-site-specific animal waste management system components and practices are determined from soil survey maps, SIRs, or *National Soils Handbook* interpretive guides. Soil variability within fields or geographic areas may require the collective assessment of soil suitability and limitation ratings for the application of animal wastes in the area under consideration. Soil features and their combined effect on the animal waste management system are important considerations when evaluating soil–animal waste suitability ratings for soils. A soil scientist should be consulted when assessing the effects of soil variability on design and function of an animal waste management system.

[*Note:* Following a brief description of specific soil characteristics, Table 5.3 lists soil characteristics and recommendations and limitations for land application of animal wastes.]

5.2.4.1   Available Water Capacity

Available water capacity is the amount of water that a soil can store and is available for use by plants, the water held between field capacity and the wilting point adjusted downward for rock fragments and for salts in solution. *Field capacity* is the water retained in a freely drained soil about 2 days after thorough wetting. The *wilting point* is the water content at which sunflower seedlings wilt irreversibly.

Available water is expressed as a volume fraction (0.20), as a percentage (20%), or as an amount (in inches). An example of a volume fraction is water in inches per inch of soil. If a soil has an available water fraction of 0.20, a 10-in. zone then contains 2 in. of available water.

Available water capacity is often stated for a common depth of rooting (where 80% of the roots occur). This depth is at 60 in. or more in areas of the western United States that are irrigated and at 40 in. in the higher rainfall areas of the eastern United States. Some publications use classes of available water capacity. These classes are specific to the areas in which they are used. Classes use such terms as *very high, high, medium,* and *low*.

* *Important point:* Available water capacity infers the capacity of a soil to store or retain soil water, liquid animal wastes, or mineralized animal waste solids in the soil solution. Applying animal wastes increases soil organic matter content, helps to stabilize soil structure, and enhances available water capacity.

Limitations for animal waste applications are slight if the available water capacity is more than 6.0 in. per 5 ft of soil depth, moderate if it is 3 to 6 in., and severe if it is less than 3 in. Soils for which the limitations are moderate have reduced plant growth potential, limited microbial activity, and low potential for retaining liquid and mineralized animal waste solids. Lower waste application rates diminish the potential for groundwater contamination and help to alleviate animal waste overloading.

Soils with severe limitations because of the available water capacity have low plant growth potential, very low potential for retaining liquid or mineralized animal waste solids, low microbial activity, and high potential for animal waste contamination of surface water and groundwater. Reducing waste application rates, splitting applications, and applying waste only during the growing season diminish the potential for groundwater and surface water contamination and help prevent animal waste overloading.

The volume of liquid animal waste application should not exceed the available water capacity of the root zone or the soil moisture deficit at the time of application. Low rates and frequent applications of liquid animal wastes on soil that has low available water capacity during periods of high soil moisture deficit can reduce potential for groundwater contamination.

5.2.4.2   Bulk Density

Soil bulk density is defined as the ratio of the mass of dry solids to the bulk volume of the soil occupied by those dry solids. Bulk density of the soil is an important site characterization parameter because it changes for a given soil and varies with the structural condition of the soil, particularly for conditions related to packing. Bulk density is expressed in grams per cubic centimeter (g/cm³). It affects infiltration, permeability, and available water capacity. Coarse textured soils have only a slight limitation because of bulk density. Medium to fine textured soils in which the bulk density in the surface layer and subsoil is less than 1.7 g/cm^3 have slight limitations for application of animal waste. Medium to fine textured soils in which the bulk density in these layers is more than 1.7 g/cm^3 have slight limitations for application of agricultural waste. Medium to fine textured soils in which the bulk density in these layers is more than 1.7 g/cm^3 have moderate limitations.

Animal waste application equipment may compact the soil if the waste is applied to soil by spreading or injection when soil moisture content is at or near field capacity. Animal wastes should be applied when soil moisture content is significantly less than field capacity to prevent compaction.

Animal wastes can be surface-applied to medium to fine textured soils with a bulk density of less than 1.7 g/cm^3. Liquid waste should be injected and application rates reduced when the bulk density of medium to fine textured soil is equal to or greater than 17 g/cm^3. Injection application and reduced application rates on these soils help to prevent liquid waste runoff and compensate for slow infiltration.

Incorporating wastes with high solids content with high levels of organic carbon reduces the soil surface bulk density and improves soil infiltration and surface permeability. The high bulk density associated with coarse textured soils does not impede or affect the application of agricultural wastes. The high permeability rate of coarse textured soils may affect the application rate because of the potential for groundwater contamination.

5.2.4.3   Cation-Exchange Capacity

CEC is a value given on a soil analysis report to indicate the soil's capacity to hold cation nutrients. CEC, however, is not something that is easily adjusted but is a value that indicates a condition or possibly a restriction that must be considered when working with that particular soil. Unfortunately, CEC is not a packaged product. CEC is determined by the amount of clay and humus present in the soil (the two main colloidal particles), and neither are practical to apply in large quantities.

Clay and humus are essentially the cation warehouse or reservoir of the soil and are very important because they improve the nutrient- and water-holding capacity of the soil. Sandy soils with very little organic matter have a low CEC, but heavy clay soils with high levels of organism matter have a much greater capacity to hold cations.

Soils with high CEC and organic soils can exchange and retain large amounts of cations released by agricultural waste mineralization processes. Conversely, soils in which CEC is low have low potential for exchanging and retaining these agricultural waste materials. The potential for agricultural waste contamination of underlying groundwater and aquifers is highest for soils with low CEC and lowest for those with high CEC.

The limitations for solid and liquid waste applications are slight for soils with a cation-exchange capacity of more than 15, moderate for those with a capacity of 5 to 15, and severe for those with CECs less than 5. Underlying groundwater supplies and aquifers can become contaminated when agricultural wastes are applied at high rates to soils with moderate or severe limitations because of their CEC. Reducing animal waste application rates can reduce the hazard for groundwater contamination.

5.2.4.4   Depth to Bedrock or Cemented Pan

The *depth to bedrock or a cemented pan* is the depth from the soil surface to soft or hard consolidated rock or a continuous indurated or strongly cemented pan. A shallow depth to bedrock or cemented pan often does not allow for sufficient filtration or retention of animal wastes or agricultural waste mineralization by-products. Bedrock or a cemented pan at a shallow depth, less than 40 in., limits plant growth and root penetration and reduces soil animal waste adsorptive capacity. Limitations for application of animal wastes are slight if bedrock or a cemented pan is at a depth of more than 40 in., moderate if it is at a depth of 20 to 40 in., and severe at a depth of less than 20 in.

Animal wastes continually applied to soils with moderate or severe limitations because of bedrock or a cemented pan can overload the soil retention capacity, allowing waste and mineralization by-products to accumulate at the bedrock or cemented pan soil interface. When this accumulation occurs over fractured bedrock or a fractured cemented pan, the potential for groundwater and aquifer contamination is high. Reducing waste application rates on soils with moderate limitations diminishes groundwater contamination and helps to alleviate the potential for animal waste over-loading. If the limitations are severe, reducing waste application rates and split applications can help to decrease overloading and reduce the potential for contamination.

5.2.4.5   Depth to High Water Table

Depth to high water table is the highest average depth from the soil surface to the zone of saturation during the wettest period of the year. This saturated zone must be more than 6 in. thick and must persist for more than a few weeks. A shallow depth to high water table may not allow for sufficient filtration or retention of animal wastes or animal waste mineralization by-products. A higher water table at a depth of less than 4 ft can limit plant and root growth and reduce soil's animal waste adsorptive capacity.

Limitations for application of animal wastes are slight if the water table is at a depth of more than 4 ft, moderate at a depth of 2 to 4 ft, and severe if at a depth of less than 2 ft. Depth and type of water table, time of year, and duration data should be collected if animal wastes are to be applied to soils suspected of having a water table within 4 ft of the soil surface.

Animal wastes applied to soils with moderate limitations because of the water table can over-load the soil's retention capacity and percolate through the soil profile, contaminating the water table. Reducing waste application rates on these soils helps to alleviate animal waste overloading and lessens the potential for groundwater contamination.

- *Important point:* The potential for contamination of shallow groundwater is very high if animal wastes are applied to soils with severe limitations. Careful application and management of animal wastes is recommended with these soils. Management should include frequent applications at very low rates.

5.2.4.6   Flooding

Flooding is the temporary covering of the soil surface by flowing water. Ponded and standing water or flowing water during and shortly after rain or snowmelt are not considered flooding. Flooding events transport surface-applied animal wastes off the application site or field and deposit these materials in streams, rivers, lakes, and other surface water bodies.

Soils with no or a low rate of flooding potential (5 times or less in 100 years) have slight limitations for the application of animal waste. Occasional flooding (5 to 50 times in 100 years) is a moderate limitation for the application of animals waste, and frequent flooding (50 to 100 times in 100 years) is a severe limitation.

Animal wastes should be applied during periods of the year when the probability of flooding is low. Liquid animal waste should be injected, and solid animal waste should be incorporated immediately after application. Incorporating animal wastes and applying wastes when the probability of flooding is low reduces the hazard to surface water.

5.2.4.7   Fraction Greater than 3 in. in Diameter—Rock Fragments, Stones, and Boulders

Rock fragments, stones, and boulders are soil fractions greater than 3 in. and are measured as a weight percent or estimated as a volume percentage of the whole soil. The upper size limit is unde-fined but, for practical purposes, is about 40 in. Stoniness is a soil-surface feature defined as the percent of stones and boulders (rock fragments greater than 10 in. in diameter) that cover the soil surface, and is represented as classes 1 through 6.

Limitations for animal waste application are slight if stoniness is class 1 (less than 0.1% of the surface covered with stones and boulders), moderate if it is class 2 (0.1 to 3.0% of the surface covered with stones and boulders), and severe if class 3, 4, 5, or 6 (more than 3% of the soil surface is covered with stones and boulders).

Rock fragments, stones, and boulders can restrict application equipment operations and trafficability and affect animal waste incorporation. Incorporating animal waste with high solids content can be difficult or impractical where:

- Rock fragments between 3 and 10 in. in diameter make up more than 15% by weight (10%, by volume) of the soil.
- Stones and boulders more than 10 in. in diameter make up more than 5% by weight (3%, by volume) of the soil.
- The soil is in stoniness class 2 or higher.

Because of this, animal wastes applied to these areas may be transported offsite by runoff and have the potential to contaminate adjacent surface water. Local evaluation of the site is required to determine if the size, shape, or distribution of the rock fragments, stones, and boulders will impede application or incorporation of animal wastes.

5.2.4.8   Intake Rate

Intake rate is the rate at which water enters the soil surface. Initial water intake is influenced by soil porosity, bulk density, moisture content, texture, structure, and permeability of the surface layer. Confined water intake rate is controlled by the permeability of underlying layers. Water intake potential is inferred from hydrologic soil groups and inversely related to the hydrologic group runoff potential. If animal wastes with large quantities of suspended solids are applied at high rates on soils with high or moderate intake potential, soil macropore space can clog and the soil intake rate is reduced. Conversely, application and incorporation of animal wastes to soils with slow water intake potential can increase soil structure and porosity, thus improving the potential water intake rate. The short-term effect may be pore clogging and resulting runoff if application rates are high on soil with a slow intake rate.

Soils in hydrologic groups B and C have moderate intake potential and slight limitations for application of animal wastes. Soils in hydrologic group D have slow intake potential and high runoff potential and generally have moderate limitations for the applications of animal wastes. Incorporating animal wastes applied to hydrologic group D soils helps to prevent the removal and transport of wastes by runoff and water erosion and can reduce the potential for surface water contamination. Liquid waste application rates should not exceed irrigation intake rates for soils in hydrologic groups B, C, or D. Application rates that exceed the irrigation intake rate may result in runoff of animal wastes, which have the potential to contaminate adjacent surface water.

Soils in hydrologic group A generally have moderate limitations for the application of animal wastes with high solids content and severe limitations for liquid wastes. Rapid intake of liquid and mineralized waste solids has the potential to contaminate underlying aquifers and groundwater supplies. Aquifer contamination potential can be reduced by reducing application rates, using split applications, and applying the waste only during periods of the year when evapotranspiration exceeds precipitation.

Soils in dual hydrologic groups (A/D, B/D, or C/D) have severe limitations for the application of animal wastes. Rapid and moderate infiltration of liquid and mineralized waste solids has the potential to contaminate underlying high water table and groundwater supplies. Water table depth, type, time of year, and duration data should be collected if animal wastes are to be applied to soils in dual hydrologic groups. Aquifer and water table contamination can be lessened by reducing application rates, using split applications, and applying only during periods of the year when evapotranspiration exceeds precipitation.

TABLE 5.2
Animal Waste-Soil Permeability Rate Limitations

Waste	Limitations (in/hr)		
	Slight	**Moderate**	**Severe**
Solids	< 2.0	2.0–6.0	> 6.0
Liquids	0.2–2.0	0.06–0.2 or 2.0–6.0	< 0.06 or > 6.0

Source: Adapted from NRCS/USDA (1992, 1996).

5.2.4.9 Permeability Rate

Permeability (hydraulic conductivity) is the quality of soil that enables water to move downward through the soil profile. Generally inferred from the permeability of the most slowly permeable horizons in the profile, permeability is estimated from soil physical properties and is expressed in inches per hour (in./hr). Permeability rates affect runoff, leaching, and decomposition rates of agricultural wastes applied to or incorporated in the surface layer. Application and incorporation of animal wastes improve soil surface intake and permeability; however, frequent applications at high rates can clog soil pores and reduce soil surface permeability and intake.

Animal wastes can be applied to soils with only slight limitations because of permeability. Animal wastes applied to soils with permeability of less than 0.2 in./hr should be incorporated (solids) or injected (liquids) into the soil to reduce potential surface water contamination from erosion and runoff. Split rate applications of liquid wastes applied to soils with permeability of more than 2 in./hr reduce the potential for contamination of shallow aquifers. Reducing the rate of application and using split applications of waste solids on soils with severe limitations for this use can reduce the potential for contamination of shallow aquifers. Table 5.2 shows the limitation ratings for solid and liquid wastes.

5.2.4.10 Soil pH

Soil pH affects plant nutrient availability, animal waste decomposition rates, and adsorption of heavy metals. Soils in which the surface pH is less than 6.5 have lower potential for plant growth and low heavy metal adsorption.

Limitations and recommendations are based on the lowest pH value of the surface layer. Limitations for the application of animal wastes are slight if the pH in the surface layer is more than 6.5, moderate if it is 3.5 to 6.5, and severe if less than 3.5. Continuous high application rates of animal wastes reduce soil pH.

- *Important point:* If large amounts of animal wastes are applied to small fields or land tracts, the soil pH should be monitored to prevent its reduction to levels that affect soil ratings and limitations for plant growth.

5.2.4.11 Ponding

Ponding is standing water in a closed depression that can be removed only by percolation, transpiration, or evaporation. Animal wastes applied to soils that are ponded have a very high potential for contaminating the ponded surface water. Application on these soils should be avoided if possible.

5.2.4.12   Salinity

Salinity is the concentration of dissolved salts in the soil solution and is related to electric conductivity, the standard measure of soil salinity, recorded as (Mhos/cm). High soil salinity interferes with the ability of plants to absorb water from the soil and to exchange plant nutrients. This interference reduces plant growth and seed germination and limits the crops that can be successfully grown. If soil salinity is a potential hazard or limitation, crops with a high tolerance to salinity should be used in the animal waste management system.

Salinity ratings are for the electric conductivity of the soil surface. Limitations for the application of animal wastes are very slight if salinity is measured as less than 4 mmhos/cm, slight if it is 4 to 8 mmhos/cm, moderate if 8 to 16 mmhos/cm, and severe if more than 16 mmhos/cm.

- *Important point:* Soils with moderate limitations affect the choice of crops that can be grown and cause reduced germination. Animal wastes with a high content of salt can be applied to moderately rated soils, but applications should be rotated among fields and rates should be reduced to prevent an increase in soil salinity and further degradation of plant growth.

Applying animal wastes with high salt content to soils with a severe rating should be avoided to prevent increasing soil salinity and further inhibiting plant growth and organic matter decomposition. However, limited amounts of animal wastes can be applied if applications are rotated among fields and soil salinity is monitored.

Animal wastes with low salt content and a high C:N ratio can be applied and have a beneficial impact on soils with a moderate or severe salinity rating. Application of low salt, high C:N ratio animal waste to these soils improves intake, permeability, available water capacity, and structure. It also reduces salt toxicity to plants.

5.2.4.13   Slope

Slope is the inclination of the soil surface from the horizontal expressed as a percentage. Slope influences runoff velocity, erosion, and the ease with which machinery can be used. Steep slopes limit application methods, rates, and machinery choices. Runoff velocity, soil-carrying capacity of runoff, and potential water erosion all increase as slopes become steeper.

Limitations for the application of animal wastes are slight if the slope is less than 8%, moderate if 8% to 15%, and severe if more than 15%. Animal wastes applied to soils with moderate limitations should be incorporated. This minimizes erosion and transport of waste materials by runoff, thus reducing the potential for surface water contamination.

Soils with severe slope limitations have limited cropping potential and are subject to excessive runoff and erosion. Animal wastes should be incorporated into these soils as soon as possible to reduce the potential for surface water contamination. Conservation practices that reduce potential water erosion and runoff help prevent the erosion and transport of animal wastes and should be incorporated in the animal waste management system.

5.2.4.14   Sodium Adsorption

Sodium adsorption is represented by the sodium adsorption ratio (SAR), the measured amount of sodium relative to calcium and magnesium in a water extract from a saturated soil paste. A high or moderate SAR, more than 4, interferes with the ability of the plant to absorb water from the soil and to exchange plant nutrients. This interference reduces plant growth and seed germination and limits the choice of crops that can be successfully grown. A SAR of more than 13 has a detrimental effect on soil intake, permeability, and structure.

Limitations for the application of animal wastes are slight if the SAR is less than 4, moderate if it is 4 to 13, and severe if it is greater than 13. Soils with moderate limitations affect the choice

TABLE 5.3

Soil Characteristics and Recommendations and
Limitations for Land Application of Animal Waste

Restricting feature (soil characteristics)	Soil condition	Degree of limitation	(Limitation or hazard) Recommendation	Impact
Droughty (available water capacity)	(in.) > 6	Slight	Apply waste.	Improves available water capacity.
	3.0 – 6.0	Moderate	(Low available water capacity and very low retention). Reduce application rates and use split applications.	Improves available water capacity. Contaminants can flow into groundwater and enter surface water.
	< 3.0	Severe	(Very low available water capacity and very low retention). Reduce application rates and use split applications.	Improves available water capacity. Contaminants can flow into groundwater and enter surface water.
Dense layer (bulk density)	(g/cc)			
Soil texture: Medium and fine	< 1.7	Slight	Apply when soil moisture content is such that the field is in tillable condition.	Reduces bulk density and minimizes compaction.
Coarse	All		(Compaction and runoff.) Apply when soil moisture content is such that the field is in tillable condition. Incorporate high solids content waste. Reduce application rate and inject liquid waste.	Reduces bulk density and minimizes compaction.
Low adsorption (cation-exchange [meq/100 g of soil] capacity)				
	> 15	Slight	Apply waste.	Increases cation-exchange capacity and organic matter content.
	5–15	Moderate	(Low adsorption and exchange of cations and heavy metals.) Reduce application rates.	Contaminants can flow into groundwater.
	< 5	Severe	(Very low adsorption and exchange of cations and heavy metals.) Reduce application rates.	Contaminants can flow into groundwater.
Thin layer/ cemented pan (depth to bedrock or cemented pan)	(in.)			
	> 40	Slight	Apply waste.	None.
	20–40	Moderate	(Moderate soil depth and limited root zone.) Reduce application rates.	Contaminants can flow into groundwater. Potential waste over-loading of the soil if applied at high rates.

TABLE 5.3 (continued)
Soil Characteristics and Recommendations and
Limitations for Land Application of Animal Waste

Restricting feature (soil characteristics)	Soil condition	Degree of limitation	(Limitation or hazard) Recommendation	Impact
	< 20	Severe	(Shallow soil depth and root zone.) Reduce application rates and use split applications.	Contaminants can flow into groundwater. Potential waste over-loading of the soil if applied in a single application at high rates.
Wetness (depth to high water table)	(ft)			
	> 4	Slight	Apply wastes.	None.
	2–4	Moderate	(Moderate soil depth and limited root zone.) Reduce application rates.	Contaminants can flow into groundwater.
	< 2	Severe	(Shallow soil depth and root zone.) Application of animal wastes not recommended.	Contaminants can flow into groundwater.
Flooding (flooding frequency)				
	None, rare (5 times or less in 100 years)	Slight	Apply waste.	None.
	Occasional (5 to 50 times in 100 years)	Moderate	(Flooding and transport of waste offsite.) Apply and incorporate waste during periods when flooding is unlikely.	Contaminants can enter surface water.
	Frequent (50 to 100 times in 100 years)	Severe	(Flooding and transport of waste offsite.) Apply and incorporate waste during periods when flooding is unlikely.	Contaminants are likely to enter surface water.
Too stoney or too cobbly (fraction, > 3 in in diameter; Rock fragments, 3–10 in. in diameter; Stones and boulders, > 10 in. in diameter)	% by weight (volume)			
(Rock fragments)	< 15 (< 10)	Slight	Apply waste.	None.
(Stones and boulders)	< 5 (< 3)			
(Rock fragments)	15–35 (10–25)	Moderate	(Restricted equipment operation.) Apply waste at reduced rates.	Contaminants can enter surface water.
(Stones and boulders)	5–15 (3–10)			
(Rock fragments)	> 25 (> 25)	Severe	(Restricted equipment trafficability and operation.) Apply waste at reduced rates.	Contaminants can enter surface water.
(Stones and boulders)	> 15 (> 10)			

TABLE 5.3 (continued)
Soil Characteristics and Recommendations and
Limitations for Land Application of Animal Waste

Restricting feature (soil characteristics)	Soil condition	Degree of limitation	(Limitation or hazard) Recommendation	Impact
(Stoniness)	Stoniness class			
	1	Slight	Apply waste.	None.
	2	Moderate	(Restricted equipment operation.) Apply waste at reduced rates.	Contaminants can enter surface water.
	3, 4, 5	Severe	(Restricted equipment trafficability and operation.) Apply waste at reduced rates.	Contaminants can enter surface water.
Intake (hydrologic soil group)				
Liquid and solid wastes	B and C	Slight	Apply solid waste. Do not exceed irrigation intake rates of liquid waste.	High application may cause clogged surface pores and reduced infiltration.
Solid wastes	A	Moderate	(Leaching of mineralized waste.) Reduce rate of application.	Application may cause clogged surface pores and reduced infiltration.
Liquid wastes		Severe	(Rapid infiltration and leaching vulnerability.) Split applications and reduce application rates.	Contaminants can flow into groundwater.
Liquid and high solid wastes	D	Moderate	(Slow infiltration and potential runoff.) Inject or incorporate agricultural wastes.	Improves infiltration and surface soil permeability. Contaminants can enter surface water.
Liquid & high solids Wastes	A/D, B/D, C/D	Severe	(Water table near the soil surface.) Reduce application rates.	Contaminants can flow into groundwater.
Poor filter or percs slowly				
(Permeability)	(in./hr)			
High solids waste	< 2.0	Slight	Apply waste.	Improves infiltration and surface soil permeability.
Liquid waste	0.6–2.0	Moderate	Apply waste.	Improves infiltration and surface soil permeability.
Liquid waste	0.2–0.6	Moderate	(Slow permeability and potential runoff vulnerability.)	Contaminants can enter surface water.
Liquid and high solid wastes	2.0–6.0	Moderate	(Leaching vulnerability.) Inject liquid waste and incorporate high solids content waste.	Contaminants can flow into groundwater.
Liquid waste	< 2.0	Severe	(Slow to very slow permeability and potential contamination of surface water.) Inject liquid waste and incorporate high solids content waste.	Contaminants can enter surface water.
Liquid and high solids wastes	> 6.0	Severe	(Rapid permeability and leaching vulnerability.) Split applications of liquid waste and reduce application rates of liquid and high solids content waste.	Contaminants can flow into groundwater. Reduced permeability from organic matter accumulation in pores.

TABLE 5.3 (continued)
Soil Characteristics and Recommendations and
Limitations for Land Application of Animal Waste

Restricting feature (soil characteristics)	Soil condition	Degree of limitation	(Limitation or hazard) Recommendation	Impact
Too acidic (pH)	> 6.0	Slight	Apply waste.	Very high application rates of wastes may lower soil pH.
	4.5 – 6.0	Moderate	(Increased availability of heavy metals & reduced plant growth potential.) Reduce application rates, apply lime, and incorporate.	Heavy metal contaminants can flow into groundwater.
	< 4.5	Severe	(Increased availability of heavy metals, reduced plant growth, and limited crop selection.) Reduce application rates, apply lime, and incorporate.	Heavy metal contaminants can flow into groundwater.
Ponding (ponding)	All	Severe	(Ponded water.) Application of agricultural wastes not recommended.	Contaminants can enter surface water.
Excess salt (salinity)	(mmhos/cm)			
	< 4	Slight	Apply waste.	None.
	4 – 8	Moderate	(Slight salinity—choice of crops and germination restricted.) Apply high C:N low salt wastes. **Saline wastes:** Rotate application fields and reduce rates.	**High C:N and low salt wastes:** Improve soil infiltration, permeability, and structure; reduce plant toxicity. **Saline wastes:** May increase soil salinity if applied at continuous high rates.
	> 8	Severe	(Salinity, crops limited to salt-tolerant grasses.) Apply high C:N, low salt wastes. **Saline wastes:** Rotate application fields and reduce rates.	**High C:N and low salt wastes:** Improve soil infiltration, permeability, and structure; reduce plant toxicity. **Saline wastes:** May increase soil salinity if applied at continuous high rates.
Slope (slope)	(%)			
	< 8	Slight	Apply waste.	None.
	8–15	Moderate	(Moderately steep slopes, potential water erosion.) Incorporate liquid and high solids waste and control runoff.	Contaminants can enter surface water.
	> 15	Severe	(Steep slopes, water erosion, and limited cropping potential.) Incorporate liquid and high solids waste and control runoff.	Contaminants can enter surface water.

TABLE 5.3 (continued)
Soil Characteristics and Recommendations and
Limitations for Land Application of Animal Waste

Restricting feature (soil characteristics)	Soil condition	Degree of limitation	(Limitation or hazard) Recommendation	Impact
Excessive sodium (sodium adsorption)	(SAR)			
	< 4	Slight	Apply waste.	None.
	4–13	Moderate	(Slight sodicity, choice of crops and germination restricted.) Apply high C:N low sodium wastes. Rotate application fields & reduce rates for sodic wastes.	**High C: N and low salt wastes:** Improve soil infiltration, permeability, and structure; reduce plant toxicity. **Saline wastes:** May increase soil salinity if applied at continuous high rates.
		Severe	(Sodicity, limited to sodium-tolerant grasses.) Apply high C:N, low sodium wastes. Rotate application fields & reduce rates for sodic wastes.	**High C: N & low salt wastes:** Improve soil infiltration, permeability, and structure; reduce plant toxicity. **Saline wastes:** may increase soil salinity if applied at continuous high rates.

Source: Adapted from NRCS/USDA (1992, 1996).

of crops that can be grown and reduce germination. To prevent increasing soil SAR and further degradation of soil properties, animal wastes that are high in sodium should not be applied to soils with moderate or severe ratings. Animal wastes with low sodium content and high C:N ratio can be applied and have a beneficial impact on soils with a moderate or severe SAR rating.

- *Important point:* Application of animal wastes with low salt content and high C:N ratio to these soils improves soil intake, permeability, and structure. It also reduces the plant toxicity effect of soil sodium.

5.3 ROLE OF PLANTS IN ANIMAL WASTE MANAGEMENT

Animal manure is a waste by-product that can be used as a plant nutrient. Properly managed and used animal wastes are a natural resource that can produce economic returns. Waste management systems, properly planned, designed, installed, and maintained, prevent or minimize degradation of soil, water, and air resources while providing chemical elements essential for plant growth.

The objectives of a complete system approach to animal waste management are to design a system that:

- Recycles nutrients in quantities that benefit plants
- Builds levels of soil organic matter
- Limits nutrient of harmful contaminant movement to surface water and groundwater
- Does not contaminate food crops with pathogens or toxic concentrations of metals or organics
- Provides a method in the soil environment to fix or transform nonessential elements and compounds into harmless forms

5.3.1 Animal Waste as a Resource for Plant Growth

Notwithstanding that, after centuries during which animal manure was highly prized for its fertilizing value, manure has become a liability on many American farms manure still has beneficial value when applied properly. The primary objective of applying animal waste to land is to recycle part of the plant nutrients contained in the waste material into harvestable plant forage, fruit, or dry matter. An important consideration is the relationship between the plant's nutrient requirement and the quantity of nutrients applied in the animal wastes. A plant does not use all the nutrients available to it in the root zone. The fraction of the total that is assimilated by the roots varies depending on the species of plant, growth stage, depth and distribution of its roots, moisture conditions, soil temperature, and many other factors. The uptake efficiency of plants generally is not high, often less than 50%. Perennial grasses tend to be more efficient in nutrient uptake than row crops. They grow during the period of waste application, which maximizes the nutrient removal from the applied waste product.

Another major objective in returning wastes to the land is enhancing the receiving soil's organic matter content. As soils are cultivated, the organic matter in the soil decreases. Throughout several years of continuous cultivation in which crop residue returns are low, the organic matter content of most soils decreases dramatically until a new equilibrium is reached. This greatly decreases the soil's ability to hold the key plant nutrients of nitrogen, phosphorus, and sulfur. These nutrients may move out of the root zone, and crop growth will suffer. The amount of crop residue that is produced and returned to the soil is reduced.

5.3.2 The Plant–Soil System

For centuries wastes have been spread on soil to recycle nutrients because of their positive effects on plant growth. Soils have the ability to retain plant nutrients contained in waste. Soil retention is an important storage mechanism, and the soil is enhanced by the organic matter supplied by waste. Plants absorb the nutrients in the waste, for the most part through the roots, and transform the soluble chemical elements, some of which are water contaminants, into plant tissue. This is the basis for addressing some of today's water quality concerns. Cropping systems and precisely calculated nutrient budgets can be tailored to meet planned waste application levels and crop nutrient needs and to reduce or eliminate losses from the plant–soil system.

5.3.2.1 Nutrient Transformation

Plant uptake is not the only form of nutrient transformation that takes place in the soil-plant system. The following processes can transform the chemical compounds derived from waste material:

1. Absorption by the roots and assimilation by the plant
2. Degradation by soil microorganisms and becoming a part of the soil organic component or broken down further into a gas, ion, or water
3. Fixation to soil minerals or attachment to soil exchange sites
4. Solubilization and movement with runoff water
5. Movement with eroded mineral or organic material
6. Downward leaching through the soil toward the groundwater
7. Escape from plant tissue into the atmosphere

Plants can play a role in all of these processes. Processes 4, 5, 6, and 7 are nutrient escape mechanisms. Plant species and cultivars can be selected to interrupt many of these mechanisms. An example of process 4 is that cultivated crops that are conservation tilled and planted on the contour with grass sod improve removal of soluble nutrients by soil infiltration.

Other mechanisms might be active in the removal of some solid constituents. Many soil conservation actions reduce erosion, which interrupts process 5. Deep, fibrous-rooted plants or plants that can actively take up nutrients beyond the normal growing season of most agricultural crops interrupt process 6 by preventing escape of leaching soluble nutrients.

- *Important point:* Plants can also be selected for their propensity to uptake a certain nutrient. Several crops are heavy users of nitrogen and accumulate nitrate; grass species vary significantly in their ability to remove and transform nitrogen within the soil. Alfalfa removes potassium and nitrogen in larger quantities and at a deeper rooting depth than do most agricultural crops.

Plants can also act as catalysts or provide a better environment to promote the transformation processes. Plant growth moderates soil temperature, reduces evaporation from soil surface, provides an energy source of carbohydrates, and aggregates soil particles, which promotes high soil aeration. All this provides a better climate for a wide variety of soil microorganisms, which aids process 2.

Process 3 is aided by plant growth as well, but generally this comes very slowly. The classic example is the difference in the cation-exchange capacity between a prairie soil and a forest soil derived from the same parent material. Jenny (1941) points out that the surface layer of prairie soil has a much higher organic matter content and cation-exchange capacity, at least double to sometimes nearly quadruple that of forest soil, yet what takes centuries to build up can be destroyed in less than two decades by erosion and excessive tillage. High-residue crops in crop rotations help to prevent large decreases in soil organic matter content and have beneficial effects on nutrient retention (Wild, 1988).

A classic example of nutrient transformation where microbial degradation and eventual escape of nitrogen gas occurs is exhibited in denitrification, an important process by which nitrogen in excess of crop requirements can be removed from the plant–soil system. This process requires the presence of nitrate-nitrogen, an organic carbon source, and anaerobic soil conditions. About one unit of organic carbon is required for each unit of nitrate-nitrogen to be denitrified (Firestone, 1982).

- *Important point:* Regarding soil microbes, Sir E. John Russell, in his 2002 book *Soil Conditions and Plant Growth,* says that in one tiny gram of soil treated with farmyard manure there are some 29 million bacteria; however, where chemical fertilizers were used, the number was cut almost in half. In an acre of rich earth, bacteria are estimated to weigh more than a quarter of a ton; as they die, their bodies become converted to humus, enriching the soil naturally.

Denitrification in land treatment systems is best accomplished if the nitrogen is in the nitrate form and the waste contains sufficient organic carbon to supply energy to the denitrifying microorganism. Where the nitrogen in the waste material is in the organic or ammonium form, an aerobic condition must be present to convert the nitrogen to the nitrate form. During the aerobic process, aerobic bacteria in the soil oxidize the organic carbon, leaving less carbon available for anaerobic microbial use when the system goes anaerobic.

Plant residue and roots are major sources of organic carbon for these microbial processes. The presence of living plants stimulates denitrification. This is attributed to two effects. First, low oxygen levels in the soil area immediately surrounding respiring plant roots create a condition in which denitrifying anaerobes can exist. Second, root excretions can serve as a food source of decomposable organic carbon for the denitrifying bacteria.

5.3.2.2 Soil Supports Plant Growth

One of the most obvious functions of soil is to provide support for plants. Roots anchored in soil enable growth in plants to remain upright. Optimum plant growth depends on the soil having the biological, chemical, and physical conditions necessary for the plant root system to readily absorb nutrients and water. For instance, plants require soil pore space for root extension. Plant root metabolism also depends on sufficient pore space to diffuse gases, such as oxygen and carbon dioxide. This allows for efficient root respiration, which keeps roots in a healthy condition for nutrient uptake. A decrease in soil pore space, such as that experienced with soil compaction, retards the diffusion of gases through the soil matrix, which greatly affects root growth.

Such inhibitory factors as toxic elements (aluminum or high concentrations of soluble salts) can limit or stop plant growth. Therefore, a plant's rate of absorption of nutrients involves many processes going on in the soil and plant roots.

5.3.3   Plant Nutrient Cycling

Nutrient cycling is the exchange of nutrient elements between living and nonliving parts of the ecosystem. Plants and microbes absorb nutrients and incorporate them into organic matter, and the microbes (with aid of animals) digest the organic matter and release the nutrients in mineral form. Nutrient cycling conserves the nutrient supply and results in repeated use of the nutrients in an ecosystem.

The process of element cycling by plants is complex and not totally understood. Some generally known points are:

- The process is not the same for all plants or for all elements.
- The complete process occurs within a healthy root system adequately supplied with carbohydrates and oxygen.
- The essential elements must be in an available form in the root zone in balanced amounts.
- Cycling varies from element to element and from crop to crop.
- Soil conditions, such as temperature, moisture supply, soil reaction, soil air composition, and soil structure, affect the rate at which elements are cycled.

5.3.3.1   Essential Plant Macronutrients

Plant growth can require up to 20 chemical elements. Plants get carbon, hydrogen, and oxygen from carbon dioxide and water. Macronutrients (elements that include nitrogen, phosphorus, potassium, sulfur, calcium, and magnesium) are needed in relatively large quantities. Micronutrients (or trace elements, including boron, chlorine, cobalt, copper, iron, manganese, molybdenum, silicon, sodium, vanadium, and zinc) are needed in small amounts, or not at all, depending on the plant (Tisdale et al., 1985).

Macronutrients and micronutrients are taken from the soil–water solution. Nitrogen is partly taken from the air by nitrogen-fixing plants associated with soil bacteria. As a whole, the 20 elements listed are termed essential elements; however, cobalt, silicon, sodium, and vanadium are essential elements for the growth of only particular plant species. A list of macronutrients and their major roles in plant growth are given in Table 5.1. The quantities of the macronutrients contained in harvested crops are given in Table 5.4.

5.3.3.2   Nonessential Elements

Besides the 20 essential elements, other elements nonessential for plant growth must be monitored where municipal sludge is used as a soil amendment. These too are referred to as *trace elements*. Because these elements occur as impurities, they are often inadvertently applied to soils through additions of various soil amendments. Animal waste contains certain elements that can be considered nonessential. Nickel, arsenic, and copper have been found in poultry litter. Dairy manure has elevated levels of aluminum.

TABLE 5.4

Major Macronutrient Elements and Role in Plants

Element	Role in Plants
Nitrogen (N)	Constituent of all proteins, chlorophyll, and in coenzymes and nucleic acids.
Phosphorus (P)	Important in energy transfer as part of adenosine triphosphate. Constituent of many proteins, coenzymes, nucleic acids, and metabolic substrates.
Potassium (K)	Little if any role as constituent of plant compounds. Functions in regulatory mechanisms as photosynthesis, carbohydrate translocation, protein synthesis, etc.
Calcium (Ca)	Cell wall component. Plays role in the structure and permeability of membranes.
Magnesium (Mg)	Constituent of chlorophyll and enzyme activator.
Sulfur (S)	Important constituent of plant proteins.

Source: Adapted from Foth (1990).

5.3.3.1.1 Nitrogen

The atmosphere is made up of 79% nitrogen, by volume, as inert N_2 gas. Although a large quantity of nitrogen exists in the atmosphere, it is also the nutrient that is absorbed from soil in the greatest quantity and is the most limiting nutrient for food production.

Even though nitrogen is abundant, it is still the nutrient most frequently limiting crop production because the plant-available forms of nitrogen in soil are constantly undergoing transformation. Crops remove more nitrogen from the soil than any other nutrient. The limitation is not related to the total amount of nitrogen available, but to the form the crop can use. Most of the nitrogen in plants is in the organic form and incorporated into amino acids, the building blocks of proteins. By weight, nitrogen makes up from 1% to 4% of a plant's harvested material.

Essentially all of the nitrogen absorbed from soil by plant roots is in the inorganic form of either nitrate (NO_3) or ammonium (NH_4). Generally, young plants absorb ammonium more readily than nitrate; however, as plants age, the reverse is true. Under favorable conditions for plant growth, soil microorganisms generally convert ammonium to nitrate, so nitrates generally are more abundant when growing conditions are most favorable. Once inside the root, ammonium and nitrate are converted to other compounds or transported to other parts of the plant.

5.3.3.1.2 Phosphorus

The earth's crust contains about 0.1% phosphorus. On this basis, the phosphorus in a plow layer is equivalent to the phosphorus in 20,000 bushels of corn for an acre. This does not include phosphorus that could be absorbed by roots at depths below the plow layer.

Phosphorus concentration in plant leaves ranges between 0.2% and 0.4% (Walsh & Beaton 1973). Phosphorus is important for plant growth because of its role in ribonucleic acid (RNA), the plant cells' genetic material, and its function in energy transfer with adenosine triphosphate (ATP).

Phosphorus is available for absorption by plants from the soil as the orthophosphate ions (H_2PO_4 and HPO_4). These ions react quickly with other compounds in soil to become much less available for plant uptake. The presence of aluminum, iron, calcium, and organic matter links phosphorus in highly insoluble compounds. The concentration of orthophosphate ions in the soil solution is very low, less than 0.05 mg/L, so equilibrium is established between the soluble ion and the adsorbed form in the soil.

Phosphorus immobility in soils is caused by several factors, including the presence of hydrous oxides of aluminum and iron; soils with a high clay content, especially ones high in kaolin; soils high in volcanic ash or allophone; low or high soil pH; and high exchangeable aluminum. Of these factors, the one most easily manipulated is soil pH. Maintaining a soil pH between 6.0 and 6.5

achieves the most plant-available phosphorus in a majority of soils. Knowing the extent each of the factors are at work in a particular soil gives the upper limit at which phosphorus loading can occur in the soil before soluble phosphorus leaching from the soil becomes a serious water quality concern.

The relative immobility of phosphorus in the soil profile allows some animal waste to be applied in excess of the crop's nutrient needs, resulting in soil phosphorus residual. Building soil phosphorus residual can be beneficial in soils that readily fix phosphorus into an insoluble, unavailable form for plant uptake. This phosphorus reservoir, if allowed to rise, gives a corresponding rise in the soluble phosphorus content in the soil. This addition of total phosphorus has to be tempered with some restraint.

* *Important point:* Manure applications can actually increase phosphorus leaching because organic phosphorus is more mobile through the soil profile than its inorganic counterparts. This would be particularly true on coarse-textured soils with low cation-exchange capacities and low iron, aluminum, and calcium content.

High phosphorus application rates appreciably increase the phosphorus concentration in the soil solution and availability for plant uptake into plant tissue, but this phosphorus rarely becomes toxic to the plant. Phosphorus toxicity depends on the plant species, phosphorus status of the plant, concentration of micronutrients, and soil salinity. Poor growth in plants with high phosphorus levels can cause reduced nodulation in legumes, inhibition of the growth of root hairs, and a decrease in the shoot to root ratio (Kirkham, 1985).

5.3.3.1.3 Potassium, Calcium, and Magnesium

Potassium, calcium, and magnesium have similar reactions in the soil. The similar size and uptake characteristics can cause plant fertility problems. An excess of any one of these elements in the soil impacts the uptake of the others. It is, therefore, extremely important not to create nutrient imbalances by overapplying one of these elements to the exclusion of the others. Upon mineralization from the organic material, each element produces cations that are attracted to negatively charged particles of clay and organic matter.

There is a wide range in the potassium content of soils and availability of potassium for plant growth. Some soils are very deficient in available potassium, whereas others are very sufficient. Potassium is much less mobile than nitrogen, but more so than phosphorus. Leaching losses of potassium generally are insignificant except in sandy and organic soils, because sandy soils have a low cation-exchange capacity and generally do not have clayey subsoil that can reabsorb the leaching potassium. Potassium can leach from organic soils because the bonding strength of the potassium cation to organic matter is weaker than that to clay (Tisdale et al., 1985).

Some potassium is leached from all soils, even in the humid regions in soils with strong fixing clays, but the losses do not appear to have any environmental consequences. Potassium leached from the surface soil is held in the lower horizons of the soil and returned to the surface via plant root uptake and translocation to aboveground plant parts.

The behaviors of calcium and magnesium in soil is similar to that of potassium. They are all released from minerals by weathering and occur as exchangeable cations. Calcium and magnesium can occur in drainage water, but this has not been reported to cause an environmental problem. In fact, it can be beneficial in some aquatic systems. Total dissolved salts may increase.

5.3.3.1.4 Sulfur

Sulfur exists in some soil minerals, including gypsum. Mineral weathering releases the sulfur as sulfate, which is absorbed by roots and microorganisms; that is, part of the sulfur applied to well-drained soils ends up in sulfate form. Soil bacteria and fungi oxidize sulfur. Plants absorb the oxidized sulfate ion. Sulfate concentrations between 3 and 5 mg/L in the soil are adequate for plant growth. Sulfates are moderately mobile and may be adsorbed on clay minerals, particularly the kaolinitic type, and on hydrous oxides of aluminum as well as iron, to a lesser extent. If the soils in the waste management system are irrigated, sulfates can leach into the subsoil and even into groundwater. Under poor drainage conditions, sulfates are converted mainly to hydrogen sulfide and lost to the atmosphere. In some instances, they are converted to elemental sulfur in waterlogged soils.

5.3.3.1.5 Trace Elements

Trace elements (micronutrients) are relatively immobile once they are incorporated into the soil. The one nonmetal, boron, is moderately mobile and moves out of the rooting depth of coarse-textured acidic soils and soils with low organic matter content. The levels of plant-available forms of all these elements are generally very low in relation to the total quantity present in soils. Some of these elements are not available for most plants to take up.

- *Important point:* The most abundant boron mineral in soils is tourmaline, a borosilicate. Boron is released by weathering and occurs in the soils solution mostly as undissociated boric acid. Boron in solution tends to equilibrate with adsorbed boron.

Soil reaction has the greatest influence on availability of trace elements taken up by plants. Except for molybdenum (Mo), the availability of trace elements for plant uptake increases as soil pH decreases. The opposite occurs for molybdenum. For most agricultural crops, a pH range between 6.0 and 7.0 is best. As soil acidity increases, macronutrients deficiencies and micronutrient toxicity can occur depending on the nutrient, its total quantity available in the soil, and the plant in question. In alkaline soils, crops can suffer from phosphorus and micronutrient deficiencies.

- *Important point:* Molybdenum is needed for nitrogen fixation in legumes.

Two nonessential elements of primary concern in municipal sludge (biosolids) are lead and cadmium. At the levels commonly found in soils or biosolids, these elements have no detrimental effect on plant growth, but they can cause serious health problems to the people or animals eating plants that are sufficiently contaminated with them. Lead can be harmful to livestock that inadvertently ingest contaminated soil or recently applied biosolids while grazing. Some plants take up cadmium quite readily (Table 5.5). If the plants are eaten, this element accumulates in the kidneys and can cause a chronic disease called *proteinuria*. This disease is marked by an increase of protein content in the urine.

Key term: *Biosolids–*from *Merriam-Webster's Collegiate Dictionary, Tenth Ed.* (1998): *n* (1977) solid organic matter recovered from a sewage treatment process and used especially as fertilizer [or soil amendment]—usually used in plural.

Note: In this text, *biosolids* is used in many places (activated sludge being the exception) to replace the standard term *sludge*. The authors view the term *sludge* as an inappropriate to use to describe biosolids. Biosolids can be reused; they have some value. Because biosolids have value, they certainly should not be classified as a "waste" product—and when biosolids for beneficial reuse is addressed, we make clear that they are not.

Another nonessential element of concern is nickel. In high enough concentrations in soil, it can become toxic to plants. Hydroxylic acid reacts with nickel to inhibit the activity of the urease molecule. This can interfere with plant metabolism of urea.

Two essential elements, zinc and copper, can also become toxic to plant growth in excessive soil concentrations. These elements become toxic because they are mutually competitive as well as competitive to other micronutrients at the carrier sites for plant root uptake. Excessive concentrations of either element in the available form induce a plant nutrient deficiency for the other. High soil

TABLE 5.5

**Relative Accumulation of Cadmium
into Edible Plant Parts by Different Crops***

High uptake	Moderate uptake	Low uptake	Very low uptake
Lettuce	Kale	Cabbage	Snapbean family
Spinach	Collards	Sweet corn	Pea
Chard	Beet roots	Broccoli	Melon family
Escarole	Turnip roots	Cauliflower	Tomato
Endive	Radish globes	Brussels sprouts	Pepper
Cress	Mustard	Celery	Eggplant
Turnip greens	Potato	Berry fruits	Tree fruits
Beet greens	Onion		
Carrots			

Source: USEPA (1983).

* The classification is based on the response of crops grown in acidic soils
with received cumulative cadmium (Cd) application of 4.5 lb/ac. We do not
imply that these higher uptake crops cannot be grown on soils of higher Cd
concentrations. Such crops can be safely grown if the soil is maintained at
pH of 6.5 or greater at the time of planting, because the tendency of the crop
to assimilate heavy metals is significantly reduced as the soil pH increases
above 6.5.

concentrations of copper or zinc, or both, can also induce iron and manganese deficiency symptoms (Tisdale et al., 1985).

- *Important point:* Copper and zinc are released from mineral weathering to the soil solution as micronutrient cations that can be adsorbed onto cation exchange sites.

In all, five elements of major concern have been targeted by the U.S. Environmental Protection Agency (USEPA) with biosolids application to agricultural land: cadmium, copper, nickel, lead, and zinc. Table 5.6 shows their recommended cumulative soil limits in kilograms per hectare and in pounds per acre. Note that these loading limits depend on the soil's cation-exchange capacity and a plow-layer pH maintained at 6.5 or above. Application of wastes with these elements should cease if any one of the elements' soil limit is reached (USEPA, 1983). Some states have adopted more conservative limits than those shown in Table 5.6. Consult state regulations before designing a waste utilization plan.

1. Table 5.6 values should not be used as definitive guidelines for fruit and vegetable production.
2. Interpolation should be used to obtain values in cation-exchange capacity (CEC) range 5–15.
3. The soil plow layer must be maintained at a pH of 6.5 or above at the time of each biosolids application.

Other trace elements have been identified as harmful to plant growth or potentially capable of occurring in high enough concentrations in plant tissue to harm plant consumers; these include aluminum, antimony, arsenic, boron, chromium, iron, mercury, manganese, and selenium. Generally,

TABLE 5.6

Recommended Cumulative Soil Test Limits for Metals of Major Concern Applied to Agricultural Cropland[1]

Metal	Soil cation-exchange capacity (meq/100 g)[2,3] lb/ac (kg/ha)		
	< 5	5 to 15	> 15
Pb	500 (560)	1,000 (1,120)	2,000 (2,240)
Zn	250 (280)	500 (560)	1,000 (1,120)
Cu	125 (140)	250 (280)	500 (560)
Ni	125 (140)	250 (280)	500 (560)
Cd	4.4 (5)	8.9 (10)	17.8 (20)

Source: USEPA (1983).

[1] Table 5.6 values should not be used as definitive guidelines for fruit and vegetable production.

[2] Interpolation should be used to obtain values in CEC range 5–15.

[3] Soil plow layer must be maintained at pH 6.5 or above at time of each biosolids application.

they do not occur in wastes, such as biosolids, in high enough concentrations to pose a problem or they are only minimally taken up by crops (USEPA, 1983).

As seen in Table 5.5 for cadmium uptake, plants differ in their capacity to absorb elements from the soil. They also differ greatly in their tolerance to trace element phytotoxic effects. Tables giving specific tolerance levels for plant uptake are needed for individual plant species. Almost any element in the soil solution is taken into the plant to some extent, whether needed or not. An ion in the soil goes from the soil particle to the soil solution, moves through the solution to the plant root, enters the root, and moves from the root through the plant to the location where it is used or retained.

5.3.3.1.6 Synthetic Organic Compounds

When dealing with municipal biosolids, one other constraint to application rates should be addressed. Most biosolids have synthetic organic compounds, such as chlorinated hydrocarbon pesticides, that can be slow to decompose and may be of concern from a human or animal health standpoint.

Polychlorinated biphenyls (PCBs) are also present in many biosolids. Federal regulations require soil incorporation of any biosolids with more than 10 ppm of PCBs, wherever animal feed crops are grown. PCBs are not taken up by plants but can adhere to plant surfaces and be ingested by animals and humans when the contaminated plant parts are eaten. Pesticide uptake by crops is minimal, and concentrations in wastes are much lower than those typically and intentionally applied to control pests on most croplands (USEPA, 1983).

5.3.4 BALANCING PLANT NUTRIENT NEEDS WITH ANIMAL WASTE APPLICATION

Proper animal waste land application practice balances the capacity of the soil and plants to transform chemical elements in waste produce with the amount that is applied or is residual in the system. A lack of plant nutrients in an available form for uptake can cause a deficiency in toxicity. Both situations decrease plant growth. An excess can also find its way through the food chain and be hazardous to the consumer or the environment. Those elements that are not transformed or retained in the soil can leave the system and become a contaminant to surface water and groundwater.

5.3.4.1 Deficiencies of Plant Nutrients

The deficiency of nutrients to the plants from agricultural waste application can occur by either the shortage of supplied elements contained in the material or the interference in the uptake of essential nutrients caused by the excessive supply of another. In the first case, an analysis of the waste material is needed to determine the amount of plant nutrients being supplied, and this amount is balanced with the quantity required by the crop. Using the Nutrient Management Standard (590) with a nutrient budget worksheet assures that all essential nutrients are being supplied to the crop. For the second case, an example in the section "Excesses of Plant Nutrients, Total Dissolved Solids, and Trace Elements" shows the antagonism that excessive uptake of ammonium ions from manure has on the calcium ion. High levels of copper, iron, and manganese in waste material can cause a plant deficiency of zinc because zinc uptake sites on the root are blocked by the other ions.

- *Important point:* NRCS Conservation Practice Standard Code 590, *Nutrient Management*, describes the management of the amount, source, placement, form, and timing of the application of plant nutrients and soil amendments. The purposes of Code 590 are to minimize the transport of applied nutrients into surface water or groundwater, to budget and supply adequate plant nutrients for optimum crop yield and quality, to properly use manure or organic by-products as plant nutrient sources, and to promote management practices that sustain the physical, biological, and chemical properties of the soil.

5.3.4.2 Excesses of Plant Nutrients, Total Dissolved Solids, and Trace Elements

The tolerance of plants to high levels of elements in plant tissue must also be accounted for in animal waste application. Heavy applications of animal waste can cause elevated levels of nitrates in plant tissue, which can lead to nitrate poisoning of livestock consuming that foliage.

The ability to accumulate nitrates differs from plant to plant or even within cultivars (a specially developed agricultural plant variety) for a species. Concentrations of nitrate nitrogen of less than 0.1% in plant dry matter are considered safe to feed livestock. Large applications of waste material on tall fescue, orchard grass, and sudangrass (for example, grass sorghum used for pasture or hay) can cause nitrate buildup. Cattle grazing these plants can, thus, be poisoned. When the concentration of nitrate nitrogen in the dry harvested material exceeds 0.4%, the forage is toxic.

Animal manure releases ammonia gas upon drying. Urea content in manure is unstable. As manure dries, the urea breaks down into ammonium. The release of gaseous NH_3 from manure can result in ammonia toxicity. Exposure of corn seeds to ammonia during the initial stages of germination can cause significant injury to the development of seedlings. High levels of NH_3 and NH_4 in the soil interfere with the uptake of the calcium ion, causing plants to exhibit calcium deficiency (Hensler et al., 1970). Part of the ammonium released is adsorbed on the cation exchange sites of the soil, releasing calcium, potassium, and magnesium ions into solution. High levels of these ions in the soil solution contribute to an increase in the soluble salt level as well as pH.

- *Important point:* Proper handling of manure is necessary to prevent toxicity from occurring. Manure may contain high levels of ammonium nitrogen; up to 50% is in the NH4 form. To prevent toxicity from occurring on young plant seedlings, the manure should be field-spread and either immediately incorporated into the soil to adsorb the NH4 on the cation exchange sites of the soil or allowed to air-dry on the soil surface. Surface drying greatly reduces the level of ammonia by volatilization. Direct planting into a manure-covered soil surface, such as with no-till planting, can lead to germination problems and seedling injury unless rainfall or surface drying has lessened the amount of ammonia in the manure.

Applying manure at rates based on nitrogen requirements of the crop helps to avoid excess NH_4 buildup in the seed zone. A 0.25-inch rain or irrigation application generally is sufficient to dissipate the high concentrations of NH_4 in the seed zone.

Sidedressing of manure on corn, either by injection or surface application, has been shown to be an effective way to apply the inorganic portion (NO_3 and NH_4) of nitrogen that is quickly made available for plant growth (Klausner & Guest, 1981). Injecting manure into soil conserves more of the ammonium nitrogen during periods of warm, dry weather and prevents ammonia toxicity to the growth of plants (Sutton et al., 1982).

Key term: Giving crops an extra boost of fertilizer is called *sidedressing*. Good sidedressing helps crops grow evenly and smoothly and helps deliver better harvests.

The soluble salt content of manure and biosolids is high and must be considered when these wastes are applied to cropland. The percent salt in waste may be estimated by multiplying the combined percentages of potassium, calcium, sodium, and magnesium as determined by laboratory analysis by a factor of two (USEPA, 1979).

Key term: The term *soluble salts* refers to the salts (ions) dissolved in the soil's water and a soil factor limiting crop growth in some regions.

Under conditions where only limited rainfall and irrigation are applied, salts are not adequately leached out of the root zone and can build up high enough quantities to cause plant injury. Plants that are salt sensitive, or only moderately tolerant, show progressive decline in growth and yields as levels of salinity increase (Tables 5.7, 5.8, and 5.9).

Some plant species are tolerant to salinity but are salt-sensitive during germination. If manure or biosolids are applied to land in areas that receive moderate rainfall or irrigation water during the growing season, soluble salts in the waste can disperse through the profile or leach below the root zone. If manure or biosolids are applied under a moisture-deficit condition, salt concentrations can build up.

TABLE 5.7
Salt Tolerance of Field Crops*

Field crop	ECe in millimhos per CM at 25°C (approximate)
Barley	21.2
Sugarbeets	18.1
Cotton	17.5
Safflower	15.7
Wheat	15.7
Sorghum	13.7
Soybean	10.4
Sesbania	10.4
Rice	9.3
Corn	8.2
Broadbean	7.8
Flax	7.8
Beans	5.0

Source: Adapted from NRCS/USDA (1992, 1996).

* The indicated salt tolerances apply to the period of rapid plant growth and maturation, from the late seeding stage onward. Crops in each category are ranked in order of decreasing salt tolerance.

TABLE 5.8
Salt Tolerance of Forage Crops*

Forage crop	ECe in millimhos per CM at 25°C (approximate)
Bermuda grass	19.0
Tall wheatgrass	19.0
Crested wheatgrass	19.0
Tall fescue	15.8
Barley hay	14.5
Perennial rye	14.2
Hardinggrass	14.0
Birdsfoot trefoil	12.0
Beardless wildrye	12.0
Alfalfa	9.2
Orchardgrass	9.0
Meadow foxtail	7.0
Clovers, alsike and red	4.2

Source: Adapted from NRCS/USDA (1992, 1996).

* The indicated salt tolerances apply to the period of rapid plant growth and maturation, from the late seeding stage onward. Crops in each category are ranked in order of decreasing salt tolerance.

TABLE 5.9
Salt Tolerance of Vegetable Crops*

Field crop	ECe in millimhos per CM at 25°C (approximate)
Beets	15.0
Spinach	9.0
Tomato	8.3
Broccoli	8.3
Cabbage	8.0
Potato	8.0
Sweet corn	8.0
Sweet potato	8.0
Lettuce	7.2
Bell pepper	6.4
Onions	6.0
Carrot	6.0
Green beans	5.2

Source: Adapted from NRCS/USDA (1992, 1996).

* The indicated salt tolerances apply to the period of rapid plant growth and maturation, from the late seeding stage onward. Crops in each category are ranked in order of decreasing salt tolerance.

A soil test, the electrical conductivity of saturated paste extract, is used to measure the total salt concentration in the soil. After prolonged application of manure, the soil electrical conductivity should be tested. Conductivity values of 2 mmhos/cm or less are considered low in salts and suitable for all crops. Above values of 4 mmhos/cm, plant growth is affected except for all but the most tolerant crops (Table 5.7 to Table 5.9). At these high conductivity values, irrigation amounts need to be increased to leach salts. Added water percolating through the profile may then cause concern with leaching of nitrates. Manure application rates may have to be adjusted (Stewart, 1974).

Key term: The *saturated paste extract test* involves the saturation of soil with water and subsequent vacuum extraction of the liquid phase for the determination of dissolved solids.

Trace element toxicity is of concern with waste application on agricultural land. Animal manure can have elevated amounts of aluminum, copper, and zinc. Biosolids can have elevated concentrations of several elements, most notably aluminum, cadmium, chromium, copper, iron, mercury, nickel, lead, and zinc. The element and concentration in biosolids depend on the predominant industry in the service area. If wastes with elevated levels of trace elements are applied over a long period at significant rates, trace element toxicity can occur to plants. Micronutrient and trace element toxicity to animals and humans can also occur where cadmium, copper, molybdenum, and selenium levels in plant tissue become elevated.

Table 5.10 lists some general crop-growth symptoms and crops most sensitive to the given trace elements. If such symptoms should occur, a plant tissue test should be done to confirm which

TABLE 5.10
General Effects of Trace Elements Toxicity on Common Crops

Element	Symptoms
Al	Overall stunting, dark-green leaves, purpling of stems, death of leaf tips, and coralloid and damaged root system.
As	Red-brown necrotic spots on old leaves, yellowing and browning of roots, depressed tillering.
B	Margin or leaf tip chlorosis, browning of leaf points, decaying growing points, and wilting and dying-off of older leaves.
Cd	Brown margin of leaves, chlorosis, reddish veins and petioles, curled leaves, and brown stunted roots.
Co	Interveinal chlorosis in new leaves followed by induced Fe chlorosis and white leaf margins and tips, and damaged root tips.
Cr	Chlorosis of new leaves, injured root growth.
Cu	Dark green leaves followed by induced Fe chlorosis; thick, short, or barbed-wire roots; depressed tillering.
F	Margin and leaf tip necrosis, chlorotic and red-brown points of leaves.
Fe	Dark green foliage, stunted growth of tops and roots, dark brown to purple leaves of some plants ("bronzing" disease of rice).
Hg	Severe stunting of seedlings and roots, leaf chlorosis and browning of leaf points.
Mn	Chlorosis and necrotic lesions on old leaves, blackish-brown or red necrotic spots, accumulation of MnO_2 particles in epidermal cells, drying tips of leaves, and stunted roots.
Mo	Yellowing or browning of leaves, depressed root growth, depressed tillering.
Ni	Interveinal chlorosis in new leaves, gray-green leaves, and brown and stunted roots.
Pb	Dark-green leaves, wilting of older leaves, stunted foliage, and brown short roots.
Rb	Dark-green leaves, stunted foliage, and increasing amount of shoots.
Se	Interveinal chlorosis or black spots at Se content at about 4 mg/L, complete bleaching or yellowing of younger leaves at higher Se content, pinkish spots on roots.
Zn	Chlorotic and necrotic leaf tips, interveinal chlorosis in new leaves, retarded growth of entire plant, injured roots resemble barbed wire.

Source: Adapted from NRCS/USDA (1992, 1996); Kabata & Pendias (1984).

TABLE 5.11

Interaction among Elements within Plants and Adjacent to Plant Roots

	Antagonistic elements	Synergistic elements
	Major Elements	
Ca	Al, B, Ba, Be, Cd, Co, Cr, Cs, Cu, F, Fe, Li, Mn, Ni	Cu, Mn, Zn
Mg	Al, Be, Ba, Cr, Mn, F, Zn, Ni, Co, Cu, Fe	Al, Zn
P	Al, As, B, Be, Cd, Cr, Cu, F, Fe, Hg, Mo, Mn, Ni Pb, Rb, Se, Si, Sr, Zn	Al, B, Cu, F, Fe, Mn, Mo, Zn
K	Al, B, Hg, Cd, Cr, F, Mo, Mn, Rb	(No evidence)
S	As, Ba, Fe, Mo, Pb, Se	F, Fe
N	B, F, Cu	B, Cu, Fe, Mo
Cl	Cr, I	(No evidence)
	Trace Elements	
Cu	Cd, Al, Zn, Se, Mo, Fe, Ni, Mn	Ni, Mn, Cd
Zn	Cd, Se, Mn, Fe, Ni, Cu	Ni, Cd
Cd	Zn, Cu, Al, Se, Mn, Fe, Ni	Cu, Zn, Pb, Mo, Fe
B	Si, Mo, Fe	Mn, Fe, N
Al	Cu, Cd	(No evidence)
Pb	—	Cd
Mn	Cu, Zn, Mo, Fe, Ar, Cr, Fe, Co, Cd, Al, Ni, Ar, Se	Cu, Cd, Al, Mo
Fe	Zn, Cr, Mo, Mn, Co, Cu, Cd, B, Si	Cd, B
Mo	Cu, Mn, Fe, B	Mn, B, Si
Co	Mn, Fe	(No evidence)
Ni	Mn, Zn, Cu, Cd	Cu, Zn, Cd

Source: Adapted from NRCS/USDA (1992/1996).

element is at fault. Many of the symptomatic signs are similar for two or more elements; knowing with certainty which element is in excess from observation of outward symptoms is extremely difficult. Much of the toxicity of such trace elements can be because of their antagonistic action against nutrient uptake and use by plants. Table 5.11 shows the interaction among elements within plants and adjacent to the plant roots.

5.3.5 APPLICATION OF ANIMAL WASTE

As mentioned, animal manure can be an economical source of crop nutrients. Three key steps must be taken to use manure in an environmentally and economically sound manner:

- Know the nutrient content of the manure
- Apply a uniform rate based on crop nutrient needs
- Adjust the rate of supplemental fertilizer to compensate for the nutrients applied in the manure

5.3.5.1 Field and Forage Crops

Manure and sewage have been used for centuries as fertilizers and soil amendments to produce food for human and animal consumption. Generally, manure and biosolids are applied to crops that are most responsive to nitrogen inputs. Responsive field crops include corn, sorghum, cotton, tobacco, sugar beets, and cane.

- *Important point:* Biosolids should not be used on tobacco. The liming effect of the biosolids can enhance the incidence of root diseases of tobacco and can also elevate cadmium levels in tobacco leaves, rendering it unfit for marketing (USDA, 1986).

Cereal grains generally do not receive fertilizer application through manure because spreading to deliver low rates of nitrogen is difficult. Small grains are prone to lodging (tipping over en masse under wet, windy conditions) because of the soft, weak cell walls derived from rapid tissue growth.

Legumes (including alfalfa, peanuts, soybeans, and clover) benefit less by manure and biosolids additions because they fix their own nitrogen. The legumes, however, use the nitrogen in waste products and produce less symbiotically-fixed nitrogen. Alfalfa, a heavy user of nitrogen, can cycle large amounts of soil nitrogen from a depth of up to 6 ft. Over 500 lb per acre of nitrogen uptake by alfalfa has been reported (Schertz & Miller, 1972; Schuman & Elliott, 1978).

The greatest danger of using manure and biosolids on legume forages is that the added nitrogen may promote the growth of the less desirable grasses in the stand. This is caused primarily by introducing another source of nitrogen but can also be a result of the physical smothering of legume plants by heavy application cover of manure.

Cattle grazing on magnesium deficient forage develop health problems. Grass tetany, a serious and often fatal disorder in lactating ruminants, is caused by low magnesium content in rapidly growing cool season grasses. High concentrations of nitrogen and potassium in manure applications to the forages aggravate the situation. Because of the high levels of available nitrogen and potassium in manure, early season applications on mixed grass-legume forages should be avoided until the later-growing legume is flourishing, because legumes contain higher concentrations of magnesium than grasses.

Key term: *Grass tetany* is a nutritional or metabolic disorder characterized by low blood magnesium, yet it is not just a simple magnesium deficiency. Also called *grass staggers, wheat pasture poisoning,* and *hypomagnesemia,* it primarily affects older cows and nursing calves under 8 weeks of age but can also occur in young or dry cows and growing calves. It happens most frequently when cattle are grazing lush, immature grass but occasionally occurs when cattle are fed dry forages (winter tetany) (Guyer et al., 1984).

Perennial grasses benefit greatly by the addition of manure and biosolids. Many are selected as vegetative filters because of their efficient interception and uptake of nutrients and generally longer active growing season. Others produce large quantities of biomass and thus can remove large amounts of nutrients, especially nitrogen, from the soil-plant system.

Bermudagrass pastures in the South have received annual rates of manure that supply over 400 lb of nitrogen per acre without experiencing excessive nitrate levels in the forage. However, runoff and leaching potentials are high with these application rates and must be considered in the utilization plan.

Grass sods also accumulate nitrogen. An experiment in England carried out for 300 years at Rothamsted showed a steady increase in soil nitrogen for about 125 years before leveling off when an old plowed field was retired to grass (Wild, 1988). However, where waste is spread on the soil surface, any ammonia nitrogen in the waste generally is lost to the air as a gas unless immediately incorporated.

Grass fields used for pasture or hay must have waste spread when the leaves of the plants are least likely to be contaminated with manure. If this is done, the grass quality is not lessened when harvested mechanically or grazed by animals (Simpson, 1986).

Spreading waste immediately after harvest and before regrowth is generally the best time for hay fields and pastures in a rotation system. This is especially important where composted biosolids are applied on pasture at rates of more than 30 tons per acre. Cattle and sheep ingesting the compost inadvertently can undergo copper deficiency symptoms (USDA, 1986).

TABLE 5.12
Summary of Joint EPA/FDA/USDA Guidelines for Biosolids Application for Fruits and Vegetables

Annual and cumulative Cd rates	Annual rate should not exceed 0.5 kg/ha (0.446 lb/ac). Cumulative Cd loadings should not exceed 5, 10, or 20 kg/ha, depending on CEC values of < 5, 5 to 15, and > 15 meq/100 g, respectively, and soil pH.
Soil pH	Soil pH (plow zone – top 6 in.) should be 6.5 or greater at time of each biosolids application.
PCBs	Biosolids with PCB concentrations of more than 10 ppm should be incorporated into the soil.
Pathogen reduction	Biosolids should be treated by pathogen reduction process before soil application. A waiting period of 12 to 18 months before a crop is grown may be required, depending on prior biosolids processing and disinfection.
Use of high-quality biosolids	High-quality biosolids should not contain more than 25 ppm Cd, 1,000 ppm Pb, and 10 ppm PCB (dry weight basis).
Cumulative lead (Pb) application rate	Cumulative Pb loading should not exceed 800 kg/ha (714 lb/ac).
Pathogenic organisms	A minimum requirement is that crops to be eaten raw should not be planted in biosolids-amended fields within 12 to 18 months after the last biosolids application. Further assurance of safe and wholesome food products can be achieved by increasing the time interval to 36 months. This is especially warranted in warm, humid climates.
Physical contamination and filth	Biosolids should be applied directly to soil and not directly to any human food crop. Crops grown for human consumption on biosolids-amended fields should be processed using good food industry practices, especially for root crops and low-growing fresh fruits and vegetables.
Soil monitoring	Soil monitoring should be performed on a regular basis, at least annually for pH. Every few years, soil tests should be run for Cd and Pb.
Choice of crop type	Plants that do not accumulate heavy metals are recommended.

Source: USEPA (1983).

Some reports show that manure applied to the soil surface has caused ammonium toxicity to growing crops (Klausner & Guest, 1981). Young corn plants 8 in. high showed ammonia burn after topdressing with dairy manure during a period of warm, dry weather. The symptom disappeared after a few days, with no apparent damage to the crop. This is very similar to corn burn affected during sidedressing by anhydrous ammonia. Liquid manure injected between corn rows is toxic to plant roots and causes temporary reduction in crop growth. Warming soil conditions dissipate the high ammonium levels, convert the ammonium to nitrates, and alleviate the temporary toxic conditions (Sawyer & Hoeft, 1990).

5.3.5.2 Horticultural Crops

Vegetables and fruits benefit from applications of animal wastes, though care must be taken because produce can be fouled or disease spread. Surface application of wastes to the soil around fruit trees will not cause either problem, but spray applications of liquid waste could.

Manure or biosolids applied and plowed under before planting will not cause most vegetables to be unduly contaminated with disease organisms as long as they are washed and prepared according to good food industry standards. However, scab disease may be promoted on the skin of potatoes with the addition of organic wastes. Well-rotted or composted manure can be used to avoid excessive scabbing if it is plowed under before the potatoes are planted (Martin & Leonard, 1949). Additional guidelines for the use of municipal biosolids are in Table 5.12.

5.3.5.3 Vegetated Filter Strips for Animal Waste Treatment

A *vegetated filter strip* is a strip of herbaceous vegetation situated between cropland, grazing land, or disturbed land (including forestland) and environmentally sensitive areas that is composed of grasses or other dense vegetation and offers resistance to shallow overland flow. Downgradient of an animal production facility or cropland where animal waste has been applied, these strips can filter nutrients, sediment, organics, agrichemicals, and pathogens from runoff received from the contributing areas.

Four processes are involved in the removal of the elements in the run-on water. The first process is deposition of sediment (solid material) in the strip. The decrease in flow velocity at the upslope edge of the vegetated filter strip greatly reduces the sediment transport capacity, and suspended solids are deposited. In the second process, the vegetation provides conditions that allow surface run-on water to enter the soil profile. Once infiltrated into the soil, the elements are entrapped by chemical, physical, and biological processes and are transformed into plant nutrients or organic components of the soil. In the third process, some soluble nutrients moving with the run-on water can be directly absorbed through the plant leaves and stems; in the fourth, the thick, upright vegetation adheres to solid particles carried in the runoff, physically filtering them out. In all of the processes, the nutrients taken from the run-on water by the plants transform a potential pollutant into vegetative biomass that can be used for forage, fiber, or mulch material.

Research results show that vegetated filter strips have a wide range of effectiveness (Adam et al., 1986; Dillaha et at., 1988; Doyle et al., 1977; Schwertz & Clausen, 1989; Young et al., 1980). Variations in effectiveness are associated with individual site conditions for both the vegetated filter strip site and the contributing area.

Land slope, soils, land use and management, climate, vegetation type and density, application rates for sites periodically loaded, and concentration and characteristics of constituents in incoming water are all important site characteristics that influence effectiveness. Operation and management of the contributing area, along with maintenance of the vegetated filter strip, influence the ability of the total system to reduce the concentration and amount of contaminants contained in the runoff from the site. Knowledge of site variables is essential before making planning decisions about how well vegetated filter strips perform.

Research and operation sites exhibit certain characteristics that should be considered in planning a vegetated filter strip:

- Sheet flow must be maintained. Concentrated flow should be avoided unless low velocity grass waterways are used.
- Hydraulic loading must be carefully controlled to maintain desired depth of flow.
- Application of process-generated wastewater must be periodically carried out to allow rest periods for the vegetated filter strip. Storage of wastewater is essential for rest periods and for climatic influences.
- Unless infiltration occurs, removal of soluble constituents from the run-on water will be minimal.
- Removal of suspended solids and attached constituents from the run-on can be high, in the range of 60% to 80% for properly installed and maintained strips.
- Vegetated filter strips should not be used as a substitute for other appropriate structural and management practices. They generally are not a stand-alone practice.
- Maintenance, including proper care of the vegetation and removal of the accumulated solids, must be performed.
- Proper siting is essential to assure uniform slopes can be installed and maintained along and perpendicular to the flow path.

The criteria for planning, design, implementation, operation, and maintenance of vegetated filter strips for livestock operations and manure application sites are in Conservation Practice Standard 393, "Filter Strip."

5.3.5.4   Forestland for Animal Waste Treatment

Forestland provides an area for recycling animal waste. Wastewater effluent has been applied to some forest sites over extended periods of time with good nutrient removal efficiency and minimal impact on surface water or groundwater. On most sites, the soil is covered with layers, some several inches thick, of organic material. This material can efficiently remove sediment and phosphorus from the effluent. Nitrogen in the form of nitrates is partly removed from the wastewater in the top few feet of the soil, and the added fertility contributes to increased tree and understory growth. Caution must be taken not to over apply water that will leach nitrates out of the root zone and down toward the groundwater. Digested biosolids also has been applied to forest.

Trees take up considerable amounts of nutrients. Many of these nutrients are redeposited and recycled annually in the leaf litter. Leaves make up only 2% of the total dry weight of northern hardwoods. Harvesting trees with leaves on increases the removal of plant nutrients by the following percentages over that for trees without leaves:

Calcium   = 12%
Potassium   = 15%
Phosphorus = 4%
Nitrogen   = 19%

Whole-tree harvesting of hardwoods removes almost double the nutrients removed when only stemwood is taken. Stemwood, the usual harvested bole or log taken from the tree for lumber, makes up about 80% of the aboveground biomass (Hornbeck & Kropelin, 1982).

Riparian forest buffers are effective ecosystems between utilization areas and water bodies to control of contaminants from nonpoint sources (Lowrance et al., 1985). No specific literature has been reported on using these areas for utilization of nutrients in agricultural waste. These areas should be maintained to entrap nutrients in runoff and protect water bodies. They should not be used for waste spreading.

Key term: A *riparian buffer* is land next to streams, lakes, or wetlands that is managed for perennial vegetation (grass, shrubs, or trees) to enhance and protect aquatic resources from adverse impacts of agricultural practices.

Only 10% of the nitrogen in a 45-year Douglas fir forest ecosystem is in the trees. The greater part of the nutrient sink in a coniferous forest is in the tree roots and soil organic matter. Although nitrogen uptake in forests exceeds 100 lb per acre per year, less than 20% net is accumulated in eastern hardwood forest. The greater part of the assimilation is recycled from the soil and litter. Continued application rates of agricultural waste should be application rates of agricultural waste should be adjusted to meet the long-term sustainable need of the forest land, which generally is one-half to two-thirds that of the annual row crops (Keeney, 1980).

5.3.6   Nutrient Removal by Harvesting of Crops

The nutrient content of a plant depends on the amount of nutrients available to the plant and on the environmental growing condition. The critical level of nutrient concentration of the dry harvested material of the plant leaf is about 2% potassium, 0.25% phosphorus, and 1% potassium. Where nutrients are available in the soil in excess of plant sufficiency levels, the percentages can more than double.

In forage crops, the percent composition for nitrogen can range from 1.2% to 2.8%, averaging around 2% of the dry harvested material of the plant. The concentrations can reach as high as 4.5%, however, if the soil system has high levels of nitrogen (Walsh & Beaton, 1973).

The total uptake of nutrients by crops from agricultural waste applications increases as the crop yields increase, and crop yields for the most part increase with increasing soil nutrients, provided toxic levels are not reached and nutrient imbalances do not occur. The total nutrient uptake continues to increase with yield, but the relation does not remain a constant linear relationship.

Two important factors that affect nutrient uptake and removal by crop harvest are the percent nutrient composition in the plant tissue and the crop biomass yield. In general, grasses contain their highest percentage of nutrients, particularly nitrogen, during the rapid growth stage of stem elongation and leaf growth.

Nitrogen uptake in grasses like corn follows an S-shaped uptake curve with very low uptake in the first 30 days of growth, but uptake rises sharply until flowering and then decreases with maturity.

Harvesting the forage before it flowers would capture the plant's highest percent nutrient concentration. Multiple cuttings during the growing season maximize dry matter production. A system of two or three harvests per year at the time of grass heading optimizes the dry matter yield and plant tissue concentration, thus maximizing nutrient uptake and removal.

5.3.6.1 Nutrient Uptake Calculation

Table 5.13 can be used to calculate the approximate nutrient removal by agricultural crops. Typical crop yields are given only as default values and should be selected only in lieu of local information.

1. Select the crop or crops to be grown in the cropping sequence.
2. Determine the plant nutrient percentage of the crop to be harvested as a percentage of the dry or wet weight, depending on the crop value given in Table 5.13.
3. Determine the crop yield in pounds per acre. Weight to volume conversion is given.
4. Multiply the crop yield by the percentage of nutrient in the crop.

The solution is pound per acre of nutrients removed in the harvested crop.

TABLE 5.13

Plant Nutrient Uptake by Specified Crop and Removed in the Harvested Part of the Crop

Crop	Dry wt. lb/bu	Typical yield/acre plant part	Average concentration of nutrients (%)								
			N	P	K	Ca	Mg	S	Cu	Mn	Zn
Grain crops			% of the dry harvested material								
Barley	48	50 bu	1.82	0.34	0.43	0.05	0.10	0.16	0.0016	0.0016	0.0031
		1 T. straw	0.75	0.11	1.25	0.40	0.10	0.20	0.0005	0.0160	0.0025
Buckwheat	48	30 bu	1.65	0.31	0.45	0.09			0.0034	0.0034	
		0.5 T. straw	0.78	0.05	2.26	1.40	0.10				
Corn	56	120 bu	1.61	0.28	0.40	0.02	0.10	0.12	0.0007	0.0011	0.0018
		4.5 T. stover	1.11	0.20	1.34	0.29	0.22	0.16	0.0005	0.0166	0.0033
Oats	32	80 bu	1.95	0.34	0.49	0.08	0.12	0.20	0.0012	0.0047	0.0020
		2 T. straw	0.63	0.16	1.66	0.20	0.20	0.20	0.0008	0.0030	0.0072
Rice	45	5,500 lb	1.39	0.24	0.23	0.08	0.11	0.80	0.0030	0.0022	0.0019
		2.5 T. straw	0.60	0.09	1.16	0.18	0.10			0.0316	
Rye	56	30 bu	2.08	0.26	0.49	0.12	0.18	0.42	0.0012	0.0131	0.0018
		1.5 T. straw	0.50	0.12	0.69	0.27	0.07	0.10	0.0300	0.0047	0.0023
Sorghum	56	60 bu	1.67	0.36	0.42	0.13	0.17	0.17	0.0003	0.0113	0.0013
		3 T. stover	1.08	0.15	1.31	0.48	0.30	0.13		0.0116	
Wheat	60	40 bu	2.08	0.62	0.52	0.04	0.25	0.13	0.0013	0.0038	0.0058
		1.5 T. straw	0.67	0.07	0.97	0.20	0.10	0.17	0.0053	0.0053	0.0017

TABLE 5.13 (continued)
Plant Nutrient Uptake by Specified Crop and Removed in the Harvested Part of the Crop

Crop	Dry wt. lb/bu	Typical yield/acre plant part	Average concentration of nutrients (%)								
			N	P	K	Ca	Mg	S	Cu	Mn	Zn
Oil crops			% of the dry harvested material								
Flax	56	15 bu	4.09	0.55	0.84	0.23	0.43	0.25		0.0061	
		1.75 T. straw	1.24	0.11	1.75	0.72	0.31	0.27			
Oil palm	—	22,000 lb	1.13	0.26	0.16	0.19	0.09		0.0043	0.0225	
		5 T. fronds & stems	1.07	0.49	1.69		0.36				
Peanuts	22–30	2,800 lb	3.60	0.47	0.50	0.04	0.12	0.24	0.0008	0.0040	
		2.2 T. vines	2.33	0.24	1.75	1.00	0.38	0.36		0.0051	
Rapeseed	50	35 bu	3.60	0.79	0.76		0.66				
		3 T. straw	4.48	0.43	3.37	1.47	0.06	0.68	0.0001	0.0008	
Soybeans	60	35 bu	6.25	0.64	1.90	0.29	0.29	0.17	0.0017	0.0021	0.0017
		2 T. stover	2.25	0.22	1.04	1.00	0.45	0.25	0.0010	0.0115	0.0038
Sunflower	25	1,100 lb	3.57	1.71	1.11	0.18	0.34	0.17		0.0022	
		4 T. stover	1.50	0.18	2.92	1.73	0.09	0.04		0.0241	
Fiber crops			% of the dry harvested material								
Cotton		600 lb. Lint and 1,000 lb seeds	2.67	0.58	0.83	0.13	0.27	0.20	0.0040	0.0073	0.0213
		burs and stalks	1.75	0.22	1.45	1.40	0.40	0.75			
Pulpwood		98 cord	0.12	0.02	0.06		0.02				
		Bark, branches	0.12	0.02	0.06		0.02				
Forest			% of the dry harvested material								
Leaves			0.75	0.06	0.46						
Northern hardwoods		50 tons	0.20	0.02	0.10	0.29					
Douglas fir		76 tons	0.16								
Silage crops			% of the dry harvested material								
Alfalfa haylage	(50% dm)	10 wet/5 dry	2.79	0.33	2.32	0.97	0.33	0.36	0.0009	0.0052	
Corn silage	(35% dm)	20 wet/7 dry	1.10	0.25	1.09	0.36	0.18	0.15	0.0005	0.0070	
Forage sorghum	(30% dm)	20 wet/6 dry	1.44	0.19	1.02	0.37	0.31	0.11	0.0032	0.0045	
Oat haylage	(40% dm)	10 wet/4 dry	1.60	0.28	0.94	0.31	0.24	0.18			
Sorghum-sudan		10 wet/5 dry	1.36	0.16	1.45	0.43	0.34	0.04		0.0091	
			2.79	0.33	2.32	0.97	0.33	0.36	0.0009	0.0052	
Tobacco			% of the dry harvested material								
All types											
		2,100 lb	3.75	0.33	4.98	3.75	0.90	0.70	0.0015	0.0275	0.0035
Forage crops			% of the dry harvested material								
Alfalfa		4 tons	2.25	0.22	1.87	1.40	0.26	0.24	0.0008	0.0055	0.0053
Bahiagrass		3 tons	1.27	0.13	1.73	0.43	0.25	0.19			
Big bluestem		3 tons	0.99	0.85	1.75		0.20				
Birdsfoot trefoil		3 tons	2.49	0.22	1.82	1.75	0.40				

TABLE 5.13 (continued)
Plant Nutrient Uptake by Specified Crop and Removed in the Harvested Part of the Crop

Crop	Dry wt. lb/bu	Typical yield/acre plant part	Average concentration of nutrients (%)								
			N	P	K	Ca	Mg	S	Cu	Mn	Zn
Bluegrass-pastd.		2 tons	2.91	0.43	1.95	0.53	0.23	0.66	0.0014	0.0075	0.0020
Bromegrass		5 tons	1.87	0.21	2.55	0.47	0.19	0.19	0.0008	0.0052	
Clover-grass		6 tons	1.52	0.27	1.69	0.92	0.28	0.15	0.0008	0.0106	
Dallisgrass		3 tons	1.92	0.20	1.72	0.56	0.40				
Guineagrass		10 tons	1.25	0.44	1.89		0.43	0.20			
Bermudagrass		8 tons	1.88	0.19	1.40	0.37	0.15	0.22	0.0013		
Indiangrass		3 tons	1.00	0.85	1.20	0.15					
Lespedeza		3 tons	2.33	0.21	1.06	1.12	0.21	0.33		0.0152	
Little bluestem		3 tons	1.10	0.85	1.45		0.20				
Orchardgrass		6 ton	1.47	0.20	2.16	0.30	0.24	0.26	0.0017	0.0078	
Pangolagrass		10 tons	1.30	0.47	1.87		0.29	0.20			
Paragrass		10.5 tons	0.82	0.39	1.59	0.39	0.33	0.17			
Red clover		2.5 tons	2.00	0.22	1.66	1.38	0.34	0.14	0.0008	0.0108	0.0072
Reed canarygrass		6.5 tons	1.35	0.18		0.36					
Ryegrass		5 tons	1.67	0.27	1.42	0.65	0.35				
Switchgrass		3 tons	1.15	0.10	1.90	0.28	0.25				
Tall fescue		3.5 tons	1.97	0.20	2.00	0.30	0.19				
Timothy		2.5 tons	1.20	0.22	1.58	0.36	0.12	0.10	0.0006	0.0062	0.0040
Wheatgrass		1 ton	1.42	0.27	2.68	0.36	0.24	0.11			
Fruit crops			**% of the fresh harvested material**								
Apples		12 tons	0.13	0.02	0.16	0.03	0.02	0.04	0.0001	0.0001	0.0001
Bananas		9,900lb	0.19	0.02	0.54	0.23	0.30				
Cantaloupe		17,500 lb	0.22	0.09	0.46		0.34				
Coconuts		0.5 tons-dry copra	5.00	0.60	3.33	0.21	0.36	0.34	0.0010		0.0076
Grapes		12 tons	0.28	0.10	0.50		0.04				
Oranges		54,000 lb	0.20	0.02	0.21	0.06	0.02	0.02	0.0004	0.0001	0.0040
Peaches		15 tons	0.12	0.03	0.19	0.01	0.03	0.01			0.0010
Pineapple		17 tons	0.43	0.35	1.68	0.02	0.18	0.04			
Tomatoes		22 tons	0.30	0.04	0.33	0.02	0.03	0.04	0.0002	0.0003	0.0001
Sugar crops			**% of the fresh harvested material**								
Sugarcane		37 tons	0.16	0.04	0.37	0.05	0.04	0.04			
Sugar beets		20 tons	0.20	0.03	0.14	0.11	0.08	0.03	0.0001	0.0025	
tops			0.43	0.04	1.03	0.18	0.19	0.10	0.0002	0.0010	
Turf grass			**% of the dry harvested material**								
Bluegrass		2 tons	2.91	0.43	1.95	0.53	0.23	0.66	0.0014	0.0075	0.0020
Bentgrass		2.5 tons	3.10	0.41	2.21	0.65	0.27	0.21			
Bermudagrass		4 tons	1.88	0.19	1.40	0.37	0.15	0.22	0.0013		
Wetland plants			**% of the dry harvested material**								
Cattails		8 tons	1.02	0.18							
Rushes		1 ton	1.67								
Saltgrass		1 ton	1.44	0.27	0.62						

TABLE 5.13 (continued)
Plant Nutrient Uptake by Specified Crop and Removed in the Harvested Part of the Crop

Crop	Dry wt. lb/bu	Typical yield/acre plant part	Average concentration of nutrients (%)								
			N	P	K	Ca	Mg	S	Cu	Mn	Zn
Sedges	0.8 ton		1.79	0.26							
Water hyacinth			3.65	0.87	3.12						
Duckweed			3.36	1.00	2.13						
Arrowhead			2.74								
Phragmites			1.83	0.10	0.52						
Vegetable crops			**% of the fresh harvested material**								
Bell peppers	9 tons		0.40	0.12	0.49		0.04				
Beans, dry	0.5 ton		3.13	0.45	0.86	0.08	0.08	0.21	0.0008	0.0013	0.0025
Cabbage	20 tons		0.33	0.04	0.27	0.05	0.02	0.11	0.0001	0.0003	0.0002
Carrots	13 tons		0.19	0.04	0.25	0.05	0.02	0.02	0.0001	0.0004	
Cassava	7 tons		0.40	0.13	0.63	0.26	0.13				
Celery	27 tons		0.17	0.09	0.45						
Cucumbers	10 tons		0.20	0.07	0.33		0.02				
Lettuce (heads)	14 tons		0.23	0.08	0.46						
Onions	18 tons		0.30	0.06	0.22	0.07	0.01	0.12	0.0002	0.0050	0.0021
Peas	1.5 tons		3.68	0.40	0.90	0.08	0.24	0.24			
Potatoes	14.5 tons		0.33	0.06	0.52	0.01	0.03	0.03	0.0002	0.0004	0.0002
Snap beans	3 tons		0.88	0.26	0.96	0.05	0.10	0.11	0.0005	0.0009	
Sweet corn	5.5 tons		0.89	0.24	0.58		0.70	0.06			
Sweet potatoes	7 tons		0.30	0.04	0.42	0.03	0.06	0.04	0.0002	0.0004	0.0002
Table beets	15 tons		0.26	0.04	0.28	0.03	0.02	0.02	0.0001	0.0007	

Source: cals.arizona.edu/animalwaste/nrcstools/plantuptake.pdf. Accessed January 6, 2007.

5.3.6.2 Nutrient Uptake Example

Corn and alfalfa are grown in rotation and harvested as grain and silage corn and alfalfa hay. Follow the above steps to calculate the nutrient taken up and removed in the harvested crop.

1. Crops to be grown: Corn and alfalfa
2. Plant nutrient percentage in harvested crop (from Table 5.13):

    Corn grain: 1.61% nitrogen
    0.28% phosphorus
    0.40% potassium

    Corn silage: 1.10% nitrogen
    0.25% phosphorus
    1.09% potassium

    Alfalfa: 2.25% nitrogen
    0.22% phosphorus
    1.87% potassium

3. Crop yield taken from local database:

Corn grain: 130 bu/ac @56 lb/bu = 7,280 lb
Corn silage: 22 ton/ac @2,000 lb/ton@ 35% dm = 15,400 lb
Alfalfa hay: 6 ton/ac @2,000 lb/ton = 12,000 lb

4. Multiply the percent nutrients contained in the crop harvested by the dry matter yield:

Corn grain: 1.61% N × 7,280 lb = 117 lb N
                0.28% P × 7,280 lb = 20 lb P
                0.40% K × 7,280 lb = 29 lb K

Corn silage: 1.10% N × 15,400 lb = 169 lb N
                0.25% P × 15,400 lb = 39 lb P
                1.09% K × 15,400 lb = 168 lb K

Alfalfa: 2.25% N × 12,000 lb = 270 lb N
                0.22% P × 12,000 lb = 26 lb
                1.87% K × 12,000 lb = 224 lb

Nutrient values are given as elemental P and K. The conversion factors for phosphates and potash are:

$$\text{lb P} \times 2.3 = \text{lb P}_2\text{O}_5$$

$$\text{lb K} \times 1.2 = \text{lb K}_2\text{O}$$

Under alfalfa, nitrogen includes that fixed symbiotically from the air by alfalfa.

Table 5.13 shows nutrient concentrations that are average values derived from plant tissue analysis values, which can have considerable range because of climatic conditions, varietal differences, soil conditions, and soil fertility status. Where available, statewide or local data should be used in lieu of Table 5.13 values.

5.4 GEOLOGIC AND GROUNDWATER CONSIDERATIONS (NRCS/USDA, 1999)

Storing, treating, or using animal wastes and nutrients at or below the ground surface has the potential to contaminate groundwater (see Figure 5.2). Many animal waste management components can be installed on properly selected sites without any special treatment other than good construction procedures. The key is to be able to recognize and avoid potentially problematic site conditions early in the planning process. An appropriately conducted onsite investigation is essential to identify and evaluate geologic conditions, engineering constraints, and behavior of earth materials. The requirements for preliminary (planning) and detailed (design) investigations are explained in this section, which provides guidance in a wide variety of engineering geologic issues and water quality considerations that may be encountered in investigation and planning.

5.4.1 GEOLOGIC MATERIAL AND GROUNDWATER

The term *geologic material*, or *earth material*, covers all natural and processed soil and rock materials. Geologic material ranges on a broad continuum from loose granular soil or soft cohesive soil through extremely hard, unjointed rock. *Groundwater* refers to the supply of fresh water found beneath the earth's surface (usually in aquifers), often used for supplying wells and springs.

5.4.1.1 Geologic Material: Material Properties and Mass Properties

Material properties of soil or rock are either measured in the laboratory using representative samples or assessed in the field on in-place material. Common examples of material properties

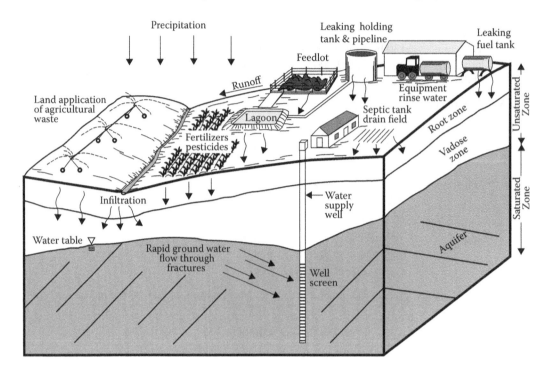

FIGURE 5.2   Agricultural sources of potential groundwater contamination. (Source: NRCS/USDA, 1999, p. 7-1.)

include mineral composition, grain size, consistency, color, hardness (strength), weathering condition, porosity, permeability, and unit weight. Some properties may be inferred by index tests of samples; for example, permeability may be roughly inferred in soils from their graduation and plasticity values.

Mass properties of geologic materials are large-scale features that can only be observed, measured, and documented in the field. They typically cannot be sampled. These properties include regional features, such as geologic structure or Karst topography. *Geologic structure* refers to the orientation and deformation characteristics, such as faults and joints. Karst topography is formed primarily in limestone terrain and characterized by solutionally widened joints, sinkholes, and caves. Mass properties also include discontinuities that are distinct breaks or abrupt changes in the mass. The two broad types of discontinuities are stratigraphic and structural, depending on mode of formation. The presence of discontinuities complicates the design of an animal waste management system.

Stratigraphic discontinuities originate when geologic material is formed under distinct changes in deposition or erosion. They are characterized by abrupt lateral or vertical changes in compositions or other material property, such as texture or hardness. These features apply to all stratified soil and rocks and can occur in many shapes described with common geologic terms, such as *blanket, tongue, shoestring,* or *lens.* Abrupt changes in composition or material property can result in contrasting engineering behavior of the adjacent geologic materials. A common example of a stratigraphic discontinuity is the soil-bedrock interface.

Structural discontinuities are extremely common in almost any geologic material. They include fractures of all types that develop sometime after a soil or rock mass has formed. Almost all types of bedrock are fractured near the Earth's surface. Forces acting on the mass that cause deformation include physical geologic stresses within the Earth's crust; biological stresses, such as animal burrows or tree roots; or artificial stresses, such as blasting. Fractures in rock materials may be systematically oriented, such as joint sets, fault zones, and bedding plane partings, or may be randomly oriented. In soil materials, fractures may include soil joints, desiccation cracks, and remnant structure from the parent bedrock in residual soils.

FIGURE 5.3 Karst areas in the United States. (Source: NRCS/USDA, 1999, p. 7-3.)

Many rural domestic wells, particularly in upland areas, derive water from fractures and joints in rock. These wells are at risk for contamination from waste impoundment facilities if rock occurs within the excavation limits, within feedlots or holding areas, and in waste utilization areas. Fractures in bedrock may convey contaminants directly from the site to the well. Discontinuities can, therefore, significantly affect water quality in a local aquifer. Although Karst topography (Figure 5.3) is well known as a problem because of its wide, interconnected fractures and open conduits, almost any near-surface rock type has fractures that can be problematic unless treated in design.

5.4.1.2 Groundwater

Many NRCS programs deal with the development, control, and protection of groundwater resources. The planner of animal waste management practices should be familiar with the principles of groundwater. NRCS references that include information on groundwater include *National Engineering Handbook* Section 16, Drainage of Agricultural Lands, and Section 18, Groundwater; *Engineering Field Handbook* Chapter 12, Springs and Wells, and Chapter 14, Drainage.

5.4.1.2.1 *Zones of Underground Water*

A portion of the precipitation that falls on land infiltrates the land surface, percolates downward through the soil under the force of gravity, and becomes groundwater. Groundwater, like surface water, is extremely important to the hydrologic cycle and to our water supplies. Almost half of the people in the United States drink public water from groundwater supplies. Overall, more water exists as groundwater than surface water in the United States, including the water in the Great Lakes. However, sometimes, pumping it to the surface is not economical and, in recent years, pollution of groundwater supplies from improper animal waste disposal has become a significant problem.

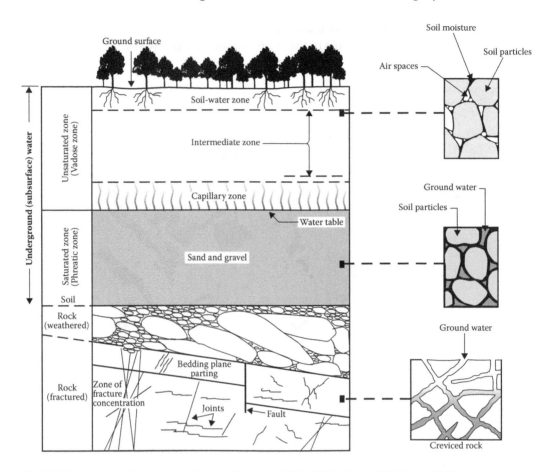

FIGURE 5.4 Zones of underground water. (Sources: AIPG, 1984; Heath, 1993; Todd, 1980.)

All water beneath the surface of the Earth is called *underground water,* or *subsurface water.* Underground water occurs in two primary zones: an upper zone of aeration called the *vadose,* or *unsaturated, zone,* and a lower zone of saturated called the *phreatic,* or *saturated, zone.* The vadose zone contains both air and water in the voids, and the saturated zone is where all interconnected voids are filled with water (Figure 5.4). Groundwater is the only underground water available for wells and springs.

The vadose zone includes the soil–water zone, the intermediate zone, and the capillary fringe. The soil–water zone extends from the ground surface to slightly below the depth of root penetration. Water in this zone is available for transpiration by plants or direct evaporation. This zone is usually at less than saturation except during rainfall or irrigation. Water held by surface tension moves by capillary action. Excess water percolates through the soil–water zone by gravity. An intermediate zone may separate the soil zone from the capillary fringe. An intermediate zone does not exist where the water table (described later) approaches the ground surface. Regions in the intermediate zone may be hundreds of feet thick. Water in the intermediate zone cannot move back up to the soil–water zone by capillary action. Intermediate zone water moves either downward under gravity or is held in place by surface tension.

Directly above the water table is a saturated zone, the capillary fringe. This zone occurs in fine- to medium-grained soils and in rocks with fractures less than 1/8 in. wide. Water in the capillary fringe is under less than atmospheric pressure. It rises from a few inches to more than 10 ft above the water table, depending on the earth materials (sand, low; clay, high). Surface tension and capillary action cause water in this zone to rise. Capillary rise increases as the pore spaces decrease.

In the saturated zone, water is under hydrostatic pressure and occupies all pore spaces. The upper surface of the saturated zone is called the *water table*. The elevation of the water table is at atmospheric pressure. The saturated zone extends from the plane of the water table down to impermeable geologic material.

5.4.1.3 Aquifers

Groundwater occurs in saturated layers called *aquifers* under the earth's surface. An aquifer is a geologic unit capable of storing and conveying usable amounts of groundwater to wells or springs. Three types of aquifers exist: unconfined, confined, and springs. Aquifers are made up of a combination of solid material, such as rock and gravel, and open spaces called *pores*. Regardless of the type of aquifer, the groundwater in an aquifer is in a constant state of motion. This motion is caused by gravity or by pumping. When siting any animal waste management component, you must know:

- What type(s) of aquifers may be present and at what depths
- What the aquifer use classification is, if any

Aquifers occur in many types of soil or rock material. Productive aquifers include sand and gravel alluvial deposits on the flood plains of perennial streams; glacial outwash; coarse-grained, highly porous, or weakly cemented sedimentary rocks (some sandstones and conglomerates); and Karst topography. An aquifer need not be highly productive to be an important resource. For example, millions of low-yielding (less than 10 gpm), private, domestic wells are in use throughout the country. In upland areas, often the only aquifer available for a groundwater source is fractured rock occurring near the surface (up to 300 ft deep).

Aquifers that lie just under the earth's surface in the zone of saturation are called *unconfined aquifers*. The top of the zone of saturation is the *water table*. An unconfined aquifer is only contained on the bottom and is dependent on local precipitation for recharge. This type of aquifer is often called a *water table aquifer*.

- *Important point:* The ability of an aquifer to allow water to infiltrate is called *permeability*.

Unconfined aquifers are a primary source of shallow well water (see Figure 5.5). These wells are not desirable as a public drinking water source. They are subject to local contamination from hazardous and toxic materials—fuel and oil, and septic tanks and agricultural runoff providing increased levels of nitrates and microorganisms. These wells may be classified as groundwater under the direct influence of surface water (GUDISW) and therefore require treatment for control of microorganisms.

Some unconfined aquifers result in flowing artesian wells. This occurs when the water table locally rises above the ground surface. Topography is the primary control on most flowing wells in major valley bottoms. The valleys serve as groundwater discharge areas. Because hydraulic potential increases with depth in valley bottoms, deep wells frequently tap a hydraulic head contour with a head value greater than that of the land surface and therefore will flow.

A *confined aquifer* is sandwiched (confined) between two impermeable layers that block the flow of water. The water in a confined aquifer is under hydrostatic pressure. It does not have a free water table (see Figure 5.6). The surface of groundwater under confined conditions is often subject to higher than atmospheric pressure because impermeable layers confine the aquifer.

Confined aquifers are called *artesian aquifers*. Wells drilled into artesian aquifers are called *artesian wells* and commonly yield large quantities of high quality water. Artesian wells are any well where the water in the well casing would rise above the saturated strata. Wells in confined

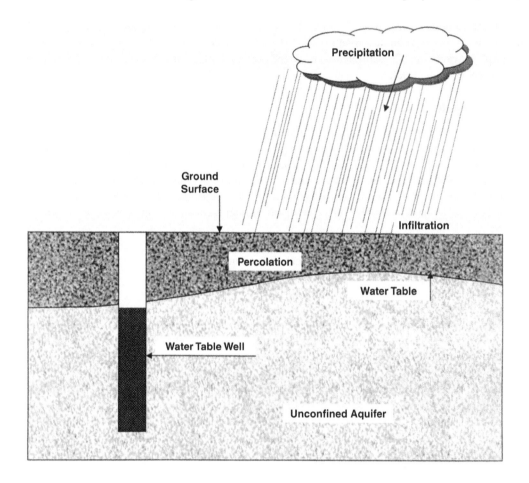

FIGURE 5.5   Unconfined aquifer.

aquifers are normally referred to as *deep wells* and are not generally affected by local hydrological events.

A confined aquifer is recharged by rain or snow in the mountains where the aquifer lies close to the surface of the earth. Because the recharge area is some distance from areas of possible contamination, the possibility of contamination is usually very low. However, once contaminated, confined aquifers may take centuries to recover.

- *Important point:* Groundwater naturally exits the earth's crust in areas called *springs*. The water in a spring can originate from a water table aquifer or from a confined aquifer. Only water from a confined spring is considered desirable for a public water system.

A *perched aquifer* is a local zone of unconfined groundwater occurring at some level above the regional water table. An unsaturated zone separates the perched aquifer from the regional water table. A perched aquifer generally is of limited lateral extent. It forms in the unsaturated zone where a relatively impermeable layer, called a *perching bed* (for example, clay), intercepts downward percolating water and causes it to accumulate above the bed. Perched aquifers can be permanent or temporary, depending on frequency and amount of recharge. Perched aquifers can present dewatering problems during construction if not discovered during investigation of the site.

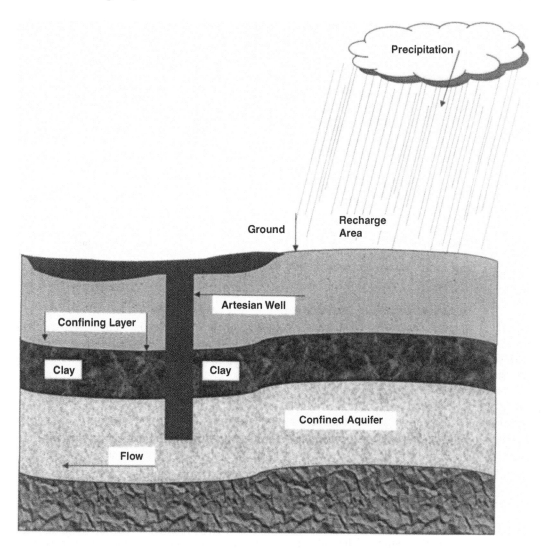

FIGURE 5.6 Confined aquifer.

5.4.1.4 Porosity

The actual amount of water in an aquifer depends upon the amount of space available between the various grains of material that make up the aquifer. The amount of space available is called *porosity;* it is defined as the ratio of the volume of voids to the total volume of a soil or rock mass, expressed as a percentage.

- *Important point:* The ease of movement through an aquifer depends on how well the pores are connected. For example, clay can hold a lot of water and has high porosity, but the pores are not connected, so water moves through the clay with difficulty.

The two main types of porosity are primary and secondary (see Figure 5.7). *Primary porosity* refers to openings formed at the same time the material was formed or deposited. An example of primary porosity is the voids between particles in a sand and gravel deposit. Primary porosity of soil depends on the range in grain size (sorting) and the shape of the grains. Porosity, however,

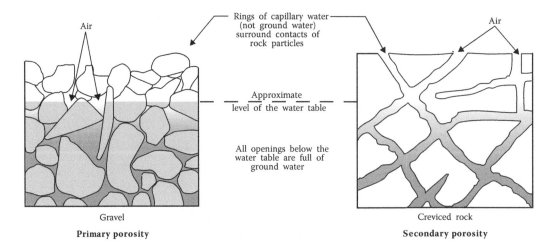

FIGURE 5.7 Porosity refers to how groundwater occurs in geologic materials. (Source: NRCS/USDA, 1999, p. 7-8.)

is independent of particle size. Thus, a bathtub full of bowling balls has the same porosity as the same tub full of baseballs. This assumes the arrangement (packing) is the same for bowling balls and baseballs. However, a tub full of a mixture of bowling balls and baseballs would have a lower porosity either the baseballs or the bowling balls.

Secondary porosity refers to openings formed after initial deposition or formation of a material. Processes that create secondary porosity include physical weathering (freezing-thawing, wetting and drying, heating and cooling), chemical or biological action, and other stresses that produce fractures and joints. Secondary porosity is extremely common in most geological materials near the earth's surface. This type of porosity enables contaminants to move with little attenuation (reduction) or filtration.

5.4.1.5   Specific Yield

Specific yield is the ratio of the volume of water that an unconfined aquifer (soil or rock) releases by gravity drainage to the volume of the soil or rock mass. A material with high porosity, such as clay, does not necessarily yield a high volume of water if the material also has low permeability. Such a material has low specific yield. See Table 5.14 for comparison of porosity and specific yield of some geologic materials.

5.4.1.6   Groundwater Quality

Generally, groundwater possesses high chemical, bacteriological, and physical quality. When pumped from an aquifer composed of a mixture of sand and gravel, if not directly influenced by surface water, groundwater is often used without filtration. It can also be used without disinfection if it has a low coliform count. However, as mentioned, groundwater can become contaminated. When septic systems fail, saltwater intrudes, improper disposal of animal wastes occurs, improperly stockpiled chemicals leach, underground storage tanks leak, hazardous materials spill, fertilizers and pesticides are misplaced, and mines are improperly abandoned, groundwater can become contaminated.

To understand how an underground aquifer becomes contaminated, you must understand what occurs when pumping is taking place within a well. When groundwater is removed from its underground source (for example, from the water-bearing stratum) via a well, water flows toward the center of the well. In a water table aquifer, this movement causes the water table to sag toward the well.

TABLE 5.14
Porosity and Specific Yield for Various Geologic Materials

Geologic material	Porosity (%)	Specific yield (%)
Soil		
Gravel (mix)	25–40	15–30
Sand (mix)	25–40	10–30
Silt	35–50	5–10
Clay	45–55	1–10
Sand, silt, clay mixes	25–55	5–15
Sand and gravel mixes	10–35	10–25
Rock		
Fractured or porous basalt	5–50	5–50
Fractured crystalline rock	0–10	0–10
Solid (unfractured)	0–1	0
Karst topography	5–50	5–50
Sandstone	5–30	5–15
Limestone, dolomite	1–20	0.5–5
Shale	0–10	0.5–5

Source: Adapted from Driscoll (1986); Johnson (1967).

This sag is called the *cone of depression.* The shape and size of the cone depends on the relationship between the pumping rate and the rate at which water can move toward the well. If the rate is high, the cone is shallow, and its growth stabilizes. The area that is included in the cone of depression is called the *cone of influence,* and any contamination in this zone will be drawn into the well.

The USEPA, under the provisions of the Safe Drinking Water Act, has the authority to designate sole source aquifers. A sole source aquifer is an aquifer that provides the principal or sole source of drinking water to an area. No federal funds can be committed to any project that the USEPA finds would contaminate the aquifer and cause a significant health hazard.

- *Important point:* A state may have designated use classifications to protect aquifers for future use by a municipality, for example. Some aquifers may be regulated against over-draft or groundwater mining.

5.4.2 ENGINEERING GEOLOGY CONSIDERATIONS

When dealing with nature's building material, soil, an engineer or planner should keep the following statement in mind:

Observe always that everything is the result of change, and get used to thinking that there is nothing Nature loves so well as to change existing forms and to make new ones like them.—Mediations, Marcus Aurelius

In this section, we provide guidance in determining what engineering geology considerations may need to be investigated for various animal waste management components listed as follows:

- Corrosivity
- Location of water table (uplift pressure)
- Depth to rock
- Stability for embankment and excavation cut banks

- Excavatability
- Seismic stability
- Dispersion
- Permeability
- Puncturability
- Settlement potential
- Shrink/swell
- Topography
- Availability and suitability of barrow material.

The significance of each consideration is briefly described, with some guidance on how to recognize it in the field. Keep in mind that most issues serve as signals or red flags that, if encountered, justify requesting assistance of a geologist or other technical expert.

5.4.2.1 Corrosivity

Soil is corrosive to many materials used in animal waste management system components. For design and corrosion risk assessment purposes, estimating the corrosivity of soils is desirable. Published soil surveys and the NRCS National Soil Characterization Database give corrosion potential for steel and concrete for soil map units. Note that data for map units normally apply only to the top 60 in of soil.

One of the simplest classifications is based on a single parameter, soil resistivity. The generally adopted corrosion severity ratings are:

Sandy soils are high up on the resistivity scale and therefore considered the least corrosive. Clay soils, especially those contaminated with saline water, are on the opposite end of the spectrum. The soil resistivity parameter is very widely used in practice and generally considered to be the dominant variable in the absence of microbial activity (Roberge, 1999).

5.4.2.2 Location of Water Table

Each soil unit has an estimated "depth to high water table" associated with it. Estimates of future high, medium, and low groundwater levels are needed for engineering and architectural design decisions and for appropriate selection of land uses.

The elevation and shape of the water table may vary throughout the year. High water tables and perched water tables in borrow areas can create access problems for heavy machinery. Rising water tables can also crack, split, and lift concrete slabs and rupture impoundment liners. The occurrence of a high water table may restrict the depth of excavation and require installation of relief or interceptor drainage systems to protect the practice from excessive uplift pressures.

Preliminary estimates of the depth to high water table can be obtained from published soils surveys and the NRCS National Soil Characterization Database. Site-specific groundwater depths may vary from values given in these sources. Stabilized water levels observed in soil borings or test pits provide the most accurate determination in the field. Seasonal variations in the water table also may be inferred from the logs of borings or pits. Recording soil color and mottling is particularly important. Mottling indicates seasonal changes in soil moisture. Perennially saturated soil is typically gray. Perennially aerated soil is typically various shades of red, brown, or yellow.

5.4.2.3 Depth of Rock

The selection of siting locations and components of an animal waste management system may be restricted by shallow depth to bedrock because of physical limitations or state and local regulations.

Soil resistivity (ohm cm)	Corrosivity Rating
> 20,000	Essentially non-corrosive
10,000 to 20,000	Mildly corrosive
5,000 to 10,000	Moderately corrosive
3,000 to 5,000	Corrosive
1,000 to 3,000	Highly corrosive
< 1,000	

The occurrence of hard, dense, massive, or crystalline rock at a shallow depth may require blasting or heavy excavators to achieve the designed grade. If the rock surface is highly irregular, differential settlement can be a hazard for steel tanks and monolithic structures, such as reinforced concrete tanks. Vegetative practices, such as filter strips, may be difficult to establish on shallow soil or exposed bedrock. Waste applied in areas of shallow or outcropping rock may contaminate groundwater because fractures and joints in the rock provide avenues for contaminants.

For animal waste impoundments, shallow bedrock generally is a serious condition requiring special design considerations. Bedrock of all types is nearly always jointed or fractured when considered as a unit greater than 0.5 to 10 acres in area. Fractures in any type of rock can convey contaminants from an unlined waste storage pond or treatment lagoon to an underlying aquifer. Fractures have relatively little surface area for attenuation of contaminants. In fact, many fractures are wide enough to allow rapid flow. Pathogens may survive the passage from the site to the well and thereby cause a health problem. Consider any rock type within 2 ft of the design grade to be a potential problem. The types of defensive design measures required to address shallow rock conditions depend on site conditions and economic factors. Design options include linings, waste storage tanks, or relocating to a site with favorable foundation conditions.

Sinkholes or caves in Karst topography or underground mines may disqualify a site for an animal waste storage pond or treatment lagoon. The physical hazard of ground collapse and the potential for groundwater contamination through the large voids are a severe limitation.

5.4.2.4 Stability for Embankment and Excavated Cut Slopes

Embankments and excavated cut slopes must remain stable throughout their design life. Control of groundwater prevents stability problems related to excessive pore pressure. Subsurface interceptor drains, relief drains, or open ditches may be needed to control excessive water pressure around structures. The foundation must be free-draining to prevent the increased loads caused by the static or dynamic weight of a component from causing downslope sling or slumping, especially for a clay foundation with low shear strength.

Key term: *Shear strength* refers to how well a member can withstand two equal forces acting in opposite directions.

Embankments and excavated cutbanks may be vulnerable to failure when wastewater is emptied or pumped out of a waste impoundment. Rapid drawdown of wastewater may saturate the soil in the bank above the liquid level, which may lead to bank caving. Designers must consider this in determining the stable side-slope of embankments and cutbanks and in designing the liner thickness. Consideration should be given to addressing the maximum rate that wastewater should be withdrawn from waste impoundments to minimize this problem in operation and maintenance plans.

5.4.2.5 Excavatability

Excavation characteristics of the geologic materials at the site determine the type and size of equipment needed and the class of excavation, either common or rock, for pay purposes (Table 5.15).

TABLE 5.15

Excavation Characteristics of Geologic Materials

Geologic material	Excavation characteristics	Equipment size flywheel horsepower
Very soft to very stiff cohesive soil or very dense granular soil	Hand pick and spade or light equipment (common excavation)	< 100
Very soft rock to moderately soft rock	Power tools or easy ripping (common excavation)	< 150
Moderately hard to hard rock	Hard to very hard ripping (rock excavation)	< 250
Very hard to extremely hard rock	Extremely hard ripping to blasting (rock excavation)	> 350 to blasting*

Source: Adapted from Kirsten (1987).
* Explosives may be an alternative to equipment.

Commonly available equipment may not be suitable in some situations. Blasting or specialized high horsepower ripping equipment may be required. Cemented pans, dense glacial till, boulders, an irregular bedrock surface, or a high water table can all increase the difficulty and cost of excavation.

5.4.2.6   Seismic Stability

Abrupt lateral or vertical changes in soil or rock materials may indicate faults (active or inactive) or bedrock structures, such as tight folds, shear zones, and vertical bedding. A foundation consisting of loose, saturated, fine-grained, relatively clean sand is most susceptible to seismic activity. Most well-compacted embankments and those foundations and embankments consisting of fine-grained soil with plasticity are inherently resistant to seismic shocks.

5.4.2.7   Dispersion

Dispersed soils are those in which the clay fraction is or may become deflocculated or disaggregated. The clay particles in these soils have minimal electrochemical attraction and are not tightly bonded when saturated, resulting in reduction of effective particle size and effective pore diameters. Dispersed soils are also characterized by high soluble sodium content. Whereas calcium ions promote flocculation, sodium ions enhance dispersion in clays. Dispersion tends to decrease permeability while flocculation tends to increase it. Dispersed soils occur in all regions of the United States. If dispersion is suspected, send representative soil samples to a laboratory for testing.

Key term: *Dispersed* is used in soil science to mean that soil particles do not cling together but rather separate into individual particles.

Typical characteristics of dispersed soils are:

- Relatively high content of soluble sodium and varying amounts of exchangeable sodium.
- Highly erodible. Clay and colloidal fractions go readily into suspension and remain there. Surface exposures have appearance of melted sugar. Gullying or rilling is extensive.
- Lower than normal shear strengths in CL, CH, and ML soils. Clay fraction goes into suspension within the pore fluid and reduces electrochemical attraction between particles.
- Generally, high shrink-swell potential, subjecting them to severe cracking when dried.
- Layers or lenses in a soil profile rather than extensive masses of a mappable soil series.

5.4.2.8 Permeability

Permeability, or *hydraulic conductivity*, refers to rate at which water flows through a material. The permeability of the underlying material is an important geologic planning consideration. For example, permeability of the soil material at the excavation limits of an animal waste impoundment is an important factor in determining the need for a liner. Permeability can also affect the attenuation of contaminants that are land applied in use of wastes. Soils with lower permeability may allow the time needed for transformation and plant uptake of nutrients, while soils with high permeability may leach contaminants. Permeability can be measured in the laboratory or estimated based on the characteristics of the material.

5.4.2.9 Puncturability

Puncturability is the ability of foundation materials to puncture a flexible membrane liner or steel tank. Angular rock particles greater than 3 in. in diameter in contact with a tank may cause denting or puncturing. Angular particles greater than 0.5 in. can be a puncture hazard to plastic and synthetic rubber membranes. Sharp irregularities in the bedrock surface itself also can cause punctures. Large angular particles can occur naturally or be created by excavation and construction activity.

5.4.2.10 Settlement Potential

Monolithic structures are designed to behave as a structural unit. Examples include poured-in-place reinforced concrete tanks and steel tanks. These structures are particularly vulnerable to settlement. Differential settlement occurs when the settlement is not even over the entire foundation. The potential for differential settlement can be an important design consideration in certain earthfill and concrete waste impoundment structures. Segmentally designed structures are built of structurally independent units, such as precast reinforced concrete retaining wall units. Although the potential of differential settlement may be less significant, some segmentally designed structures may be susceptive to settlement.

The six common geologic conditions that cause settlement to occur are:

1. Abrupt, contrasting soil boundaries—A foundation is susceptible to differential settlement if underlain by zones, lenses, or beds of widely different soil types with boundaries that change abruptly either laterally or vertically.
2. Compressible soil—Layers or zones greater than one foot thick consisting of soft clays and silts, peat and organic-rich soil (OL and OH in the Unified System), and loose sands may settle excessively when loaded by an embankment or concrete structure.
3. Weak foundations—Structures located in areas of active or abandoned underground mining or areas with a high rate of groundwater withdrawal can have problems resulting from settlement of the material.
4. Steep abutments—Differential settlement of embankments may occur on abutment slopes steeper than one horizontal to one vertical. Adequate compaction is difficult to achieve on steep slopes. Settlement cracks may occur in the fill in the area where the base of a steep abutment joins the flood plain.
5. Uneven rock surface—A foundation may settle if underlain by normally consolidated soil materials over a highly irregular, shallow bedrock surface or other uneven, unyielding material. As a rule, consider a foundation problematic if, in the foundation area, the difference between maximum and minimum thicknesses of the overlying compressible soil above an uneven rock surface divided by the maximum observed soil thickness is greater than 25%. This is expressed as "Problem Foundation" when [100 (max, depth − min. depth/max. depth)] > 25%.

6. Collapsible soil—This soil is common, especially in the western continental states. It has low density and low water content, and is formed in windblown silts and fine sands and rapidly deposited alluvial fans. This soil may undergo large, sudden settlement when it becomes saturated after loading by a structure built on it.

5.4.2.11   Shrink/Swell

Soil containing montmorllonite clay may undergo substantial changes in volume when saturated or dried. Some types of rock, such as gypsum and anhydrite, also change volume dramatically when wetted and dried. Soil with a high shrink/swell hazard is identified in published soil surveys or the NRCS National Soil Characterization database. Field investigations and previous experience in the area are often the only ways to foresee this problem.

5.4.2.12   Topography

Recognition of landforms and their associated problems is a valuable asset when planning a component for an animal waste management system. For example, flood plain sites generally have a higher water table compared to that of adjacent uplands, are subject to surface flooding, and can indicate presence of permeable soils.

Topography can indicate direction of regional groundwater flow. Uplands may serve as aquifer recharge areas, and valley bottoms, marshes, and lowlands as groundwater discharge areas.

Steep slopes restrict use for some structural and vegetative measures. Hazards include instability (landslide potential) and erosion.

Karst topography is formed on limestone, gypsum, or similar rocks by dissolution and is characterized by sinkholes, caves, and underground drainage. Common problems associated with Karst terrain include highly permeable foundations and the associated potential for groundwater contamination and collapsible ground. As such, its recognition is important in determining potential siting problems.

5.4.2.13   Availability and Suitability of Borrow Material

Borrow must meet gradation, plasticity, and permeability requirements for its intended use and be in sufficient quantity to build the component. Losses routinely occur during handling, transport, placement, and consolidation of fill materials. To compensate, as much as 150% of the design fill requirements should be located within an economical hauling distance. Conditions for the borrow area itself may limit the usefulness of borrow materials. Limitations may include such factors as moisture, thickness, location, access, land use, vegetation, and cultural resources.

5.4.2.14   Presence of Abandoned Wells and Other Relics of Past Use

The site and its history should be surveyed for evidence of past use that may require special design considerations or animal waste management system component site relocation. If an abandoned well exists on the site, special efforts are required to determine if the well was sealed according to local requirements. An improperly sealed well can be a direct pathway for contaminants to pollute an aquifer.

- *Important point:* Other remnants of human activity, such as old foundations, trash pits, or filled-in areas, require special animal waste management system design or site relocation.

5.4.3   Factors Affecting Groundwater Quality Considered in Planning

In the animal waste management system planning process, several groundwater quality factors should be considered, including:

- Attenuation potential of soil
- Groundwater flow direction
- Permeability of aquifer material
- Hydraulic conductivity
- Hydraulic head
- Hydraulic gradient
- Hydrogeologic settling
- Land topography
- Proximity to designated use aquifers, recharge areas, and wellhead protection areas
- Vadose zone material
- Type of aquifer

5.4.3.1 Attenuation Potential of Soil

The root zone of surface soils contains many biological, physical, and chemical processes that break down, lessen the potency, or otherwise reduce the volume of contaminants. These processes, collectively called *attenuation,* retard the movement of contaminants into deeper subsurface zones. The degree of attenuation depends on the time a contaminant is in contact with the material through which it travels. It also depends on the distance through which it passes and the total amount of surface area of particles making up the material. Thus, attenuation potential increases as clay content increases, the soil deepens, and distance increases between the contaminant source and the well or spring.

5.4.3.2 Clay Content

Increased clay content increases the opportunity for attenuation of contaminants because of its cation exchange capacity and its affect of reducing permeability. Clay particles hold a negative charge that gives them the capacity to interchange cations in solution. As such, clay can absorb contaminant ions and thus attenuate the movement of contaminants. Clay has a very low permeability. Therefore, the greater the amount of clay, the slower contaminants move and the greater the contact time that allows more opportunity for attenuation.

5.4.3.3 Depth of Soil

Deeper soil increases the contact time a contaminant has with the mineral and organic matter of the soil. The longer the contact time, the greater the opportunity for attenuation. Very shallow (thin to absent) soil provides little to no protection against groundwater contamination.

5.4.3.4 Distance between Contaminant Source and Groundwater Supply

Both the depth and the horizontal distance to a groundwater supply affect the attenuations of contaminants. Depth refers to the vertical distance through which a contaminant must pass to reach the top of an aquifer. Assuming all other factors remain constant, the greater this depth, and the greater the time of travel, the more opportunity for a contaminant to be in contact with the surrounding material for attenuation processes. Horizontal distance also affects attenuation of contaminants. The greater the horizontal distance between the source of the contamination and a well, spring, or other groundwater supply, the greater the travel time. The greater the travel time, the greater opportunity for attenuation of contaminants.

5.4.3.5 Groundwater Flow Direction

A desirable site for a waste storage pond or treatment lagoon is an area where groundwater is not flowing from the vicinity of the site toward a well, spring, or important underground water supply.

The direction of flow in a water table aquifer generally can be ascertained from the topography. In most cases, the slope of the land indicates the groundwater flow direction. In most humid regions, the shape of the water table is a subdued reflection of surface topography. Unconfined groundwater moves primarily from topographically higher recharge areas down gradient to withdrawal areas at lower elevations. Lower areas serve as discharge points where groundwater rises and emerges with perennial streams and ponds or flows as springs. However, radial flow paths and unusual subsurface geology can too often invalidate this assumption. Consider the case where secondary porosity governs the flow. A common example is rock in upland areas where the direction of groundwater flow is strongly controlled by the trend of prominent joint sets or fractures. Fracture patterns in the rock may not be parallel to the slope of the ground surface. Thus, assuming groundwater flow is parallel to the ground slope can be significantly misleading in terrain where flow is controlled by bedrock fractures.

5.4.3.6   Permeability of Aquifer Material

The ease or difficulty with which water flows through a material is controlled by the material's permeability. A material that is very permeable allows water to pass through easily. In contrast, water moves with more difficulty through a slightly permeable material. Permeability is determined by laboratory analysis but is also commonly determined as a mass property through field testing. The mass property is more accurately known as the aquifer's *hydraulic conductivity,* which integrates all of the aquifer's characteristics to conduct water.

The time available for attenuation in aquifer materials decreases as the permeability of the materials increases. Permeability may vary significantly among different types of materials or at different places within the same material. Permeability is commonly greater laterally than vertically. Ignored or undetected, a thin (0.5 in. or less) clay or shale seam in an otherwise uniform soil or rock aquifer can profoundly alter the outcome of mathematical analyses and design assumptions.

5.4.3.7   Hydraulic Conductivity

Hydraulic conductivity is a mass property of an aquifer that is determined through field-testing, such as pump tests or slug tests. Commonly known as permeability, hydraulic conductivity is the rate of flow (L/t) of water through an aquifer. Hydraulic conductivity reflects all of the aquifer's characteristics to transmit water.

- *Important point:* In most aquifers, the difference between vertical and horizontal conductivity rates is significant.

5.4.3.8   Hydraulic Head

Hydraulic head is the energy of a water mass produced mainly by difference in elevation, velocity, and pressure, expressed in units of length or pressure. Groundwater moves in the direction of decreasing hydraulic head. Hydraulic head in an aquifer is measured using piezometers.

Key term: *Head* is the measure of water pressure expressed as height of water in feet (1 psi = 2.31 ft of head). Stated another way, head is the equivalent distance water must be lifted to move from the supply tank or inlet to the discharge. Head can be divided into three components: static head, velocity head, and friction head.

Key term: A *piezometer* is an instrument that measures pressure head in a conduit or tank by determining the location of the free water surface.

5.4.3.9 Hydraulic Gradient

Hydraulic gradient is the change in hydraulic head per unit distance of flow in a given direction. It is expressed in units of length (elevation) per length (distance). Groundwater velocity is a function of the hydraulic gradient. Most water in an unconfined aquifer moves slowly in undeveloped aquifers. However, an action such as pumping water from a well can steepen local hydraulic gradients. This results in acceleration of flow in toward the well, carrying any dissolved contaminants with it into the well.

5.4.3.10 Hydrogeologic Setting

Hydrogeology is the study of the occurrence, movement, and quality of underground water. The hydrogeologic setting of an animal waste management system includes all the various geologic factors that influence the quality and quantity of underground water. Information on the hydrogeologic setting of a site is found in the following sources:

- State water quality management and assessment reports of surface and groundwater use designations and impairments
- Geologic maps showing rock types, faults, and similar information
- Regional water table maps and, if available, tables of static water levels in wells
- Groundwater vulnerability maps

5.4.3.11 Land Topography

The potential for groundwater contamination by infiltration is increased when topographic features contaminate runoff water. Example features include seasonal wetlands and level terraces. The hazard of surface water contamination from sediment increases as the slope and slope length increase.

5.4.3.12 Proximity to Designated Use Aquifers, Recharge Areas, and Wellhead Protection Areas

State water management and assessment reports and the following maps should be reviewed to ascertain the proximity of sensitive groundwater areas:

- Sole source of other types of aquifers whose uses have been designated by the state
- Important recharge areas
- Wellhead protection areas

5.4.3.13 Vadose Zone Material

The types of material in the vadose (unsaturated) zone affect the flow path and rate of flow of water and contaminants percolating through it. Flow rate is a function of the permeability of the material. Flow rate in the mass is greatly increased by macropores, such as soil joints. The time available for attenuation in this zone decreases as the permeability of the materials increases. Permeability rates may be inferred from the types of materials.

5.4.3.14 Type of Aquifer

Refer to Section 5.4.1 for details on confined and unconfined aquifers.

5.4.4 Planning and Design: Site Investigations

A site investigation should be done only after local regulations and permit requirements are known. The intensity of a field investigation is based on several factors, including:

- The quality of information that can be collected and studied beforehand
- Previous experience with conditions at similar sites
- Complexity of the animal waste management system or site

Clearly defined objectives for investigation are essential in this phase. For example, the objectives for investigating a site for a steel storage tank are significantly different from those for an earthen waste impoundment. The tank involves consideration of differential settlement of the foundation, while the earthen waste impoundment involves consideration of excavatability and permeability of foundation materials.

5.4.4.1 Preliminary Investigation

The purpose of a preliminary site investigation is to establish feasibility for planning purposes. A preliminary site investigation also helps determine what is needed in a detailed investigation. For many sites, the preliminary investigation and experience in the area are adequate to determine the geologic conditions, engineering constraints, and behavior of the geologic materials. Hand-auger borings and site examination often provide adequate subsurface information so that a detailed subsurface investigation is not required. A detailed investigation must be scheduled if reliable information for design cannot be obtained with the tools available during the preliminary investigation phase.

Make an initial evaluation of potential layouts of the component, access to the site, and location of active or abandoned wells, springs, and other such features. Farm "A" System worksheets and the Farm Bureau self-help water quality checklist are valuable tools in making initial site evaluations. These tools are not, however, suitable for making final design decisions.

All wells and well records near the site should be examined for proper construction. The condition of the concrete pad and, if possible, the annular seal or grout around the well casing should be examined.

Valuable background information about a proposed site is obtained from the following sources:

- Soil survey reports—These reports provide soil map units, photos of features near a site, information on seasonal flooding and the water table, and engineering interpretation and classification of soils.
- Topographic maps—USGS topographic quadrangles or existing survey data from the site provide information about slopes, location of forested areas, topographic relief, distances to identified resource features, such as wells, watercourses, houses, roads, and other cultural features.
- Aerial photos—These photos provide information on vegetation, surface runoff patterns, erosion conditions, proximity to cultural features, and other details.
- Local geologic maps and reports—These sources provide information on depth to and types of bedrock, bedrock structure, location of fault zones, characteristics of unconsolidated deposits, depth to water table, aquifer characteristics, and other geological and groundwater information.
- Conservation plans and associated logs.

5.4.4.2 Detailed Investigation

The purpose of a detailed geologic investigation is to determine geologic conditions at a site that will affect or be affected by design, construction, and operation of an animal waste management system component. The intensity of a detailed investigation is the joint responsibility of the designer and the person who has engineering job approval authority. Detailed investigations require application of individual judgment, use of pertinent technical references and state-of-the-art procedures, and timely consultation with other appropriate technical disciplines. Geologic characteristics are determined through digging or boring, logging the types and conditions of materials encountered, and securing and testing representative samples. An onsite investigation should always be conducted at a proposed waste impoundment location. State and local laws should be followed in all cases.

5.4.4.3 Investigation Tools

Soil probes, hand augers, shovels, backhoes, bulldozers, and power augers are used to allow direct observations for logging geologic materials, collection of samples, and access for field permeability testing. When logging soils with an auger, always consider that the augering process can obscure thin zones or mix soil layers. Test pits expose more of the foundation for detecting thin, but significant, lenses of permeable soil.

Geophysical methods are indirect techniques that employ geophysical equipment, such as electromagnetic induction meters, resistivity meters, refraction seismographs, and ground penetrating radar units, to evaluate the suitability of sites and the performance of the component. These techniques require trained, experienced specialists to operate the equipment and interpret results. Geophysical methods require correlation with test pits or borings for best results.

5.4.4.4 Logging Geologic Materials

During an investigation, all soil and rock materials at the site or in the borrow areas are identified and mapped. From an engineering standpoint, a mappable soil or rock unit is defined as a zone that is consistent in its mineral, structural, and hydraulic characteristics, and sufficiently homogeneous for descriptive and mapping purposes. A unit is referred to by formal name, such as Alford silt loan or Steele shale, or is a set in alphanumeric form, such as Sand Unit A-3.

- *Important point:* The NRCS classifies rock material using common rock type names as given in TR-71, Rock Material Field Classification Procedure; TR-78, The Characterization of Rock for Hydraulic Erodibility; and NEH part 628, Dams, Chapter 52, Field Procedure Guide for the Headcut Erodibility Index. Soils are classified for engineering purposes according to the Unified Soils Classification System, ASTM D-2488, Standard Practice for the Description and Identification of Soils, Visual-Manual Procedure.

When greater precision is needed, representative samples are analyzed in a soil mechanics laboratory. Laboratories commonly use ASTM D-2487, Standard Test Method for Classification of Soils for Engineering Purposes. Laboratory determinations of particle characteristics and Atterberg limits (liquid limit and plasticity index) are used to classify soils.

Use standard NRCS log sheets, such as NRCS-533, or the soil log sheet and checklist. Logs also may be recorded in a field notebook. Be methodical when logging soils. Identify and evaluate all applicable parameters according to criteria given in ASTM D-2488. Thorough logging requires only a few minutes on each boring or test pit and saves a trip back to the field to gather additional or overlooked information. Also, be prepared to preserve a test hole or pit to record the stabilized water elevation after 24 hours.

Each log sheet must contain the name of the project, location, date, investigator's name and title, and type of equipment used (back hoe), including make and model. For each soil type encountered in a test pit or drill hole, record the following information, as appropriate:

- Interval (depth range through which soil is consistent in observed parameters)
- Estimated particle size distribution (by weight, for fraction < 3 in.)
- Percent cobbles and boulders (by volume, for fraction > 3 in.)
- Angularity of coarse material
- Color of moist material, including presence of mottling (a possible indicator of the zone of water table fluctuation)
- Relative moisture content
- Structure
- Consistency (saturated fine-grained materials) or relative density (coarse-grained materials)
- Plasticity of fines
- Group name and USCS Symbol according to ASTM D-2488 flow charts
- Geologic origin and formal name, if known
- Sample (size, identification number, label, depth interval, date, location, name of investigator)
- Other remarks or notes (mineralogy of coarse material, presence of mica flakes, roots, odor, pH)
- Test hole or pit identification number
- Station and elevation of test hole or pit
- Depth (or elevation) of water table after stabilization (date measured and number of hours open)
- Depth to rock, refusal (limiting layer), or total depth drilled or dug

5.4.4.5 Samples

Samples of soil and rock materials collected for soil mechanics laboratory testing must meet minimum size requirements. Sample size varies according to test needs. Samples must be representative of the soil or rock unit from which they are taken. A geologist or engineer should help determine the tests to be conducted and may assist in preparing and handling samples for delivery to the lab. Test results are used in design to confirm field identification of materials and to develop interpretations of engineering behavior.

5.4.4.6 Guide to Investigation

For foundation of earthfill structures, use at least four test borings or pits on the proposed embankment centerline, or one every 100 ft, whichever is greater. If correlation of materials between these points is uncertain, use additional test borings or pits until correlation is reasonable. The depth to which subsurface information is obtained should be no less than equivalent maximum height of fill or to hard, unaltered rock or other significant limiting layer. For other types of waste storage structures, the depth should be to bedrock, dense sands or gravels, or hard fine-grained soils. Report unusual conditions to the responsible engineer or state specialist for evaluation.

For structures with a pool area, use at least five test holes or pits, or one per 10,000 ft^2 of pool area, whichever is greater. These holes or pits should be as evenly distributed as possible across the pool area. Use additional borings or pits, if needed, for complex sites. The borings or pits should be dug no less than 2 ft below proposed grade in the pool area or to refusal (limiting layer). Log all applicable parameters. Report unusual conditions to the responsible engineer or other specialist for further evaluation. Pay special attention to perched or high water tables and highly permeable materials in the pool area.

Borrow areas for embankment type structures and for clay liners should be located, described, and mapped. Locate suitable borrow to at least 150% of the required fill volume. Soil samples for natural water content determination should be obtained from proposed borrow and clay liner sources. Samples taken for testing should be maintained in moisture proof containers. Applicable parameters should be logged.

If a system requires a soil liner, consult soil survey reports and local surficial geologic maps to help identify potential borrow areas for investigation. Nearby clay-rich deposits for potential borrow sources should be located, mapped, and logged. Some designs may require bentonite or a chemically treated soil to reduce permeability. A qualified soil mechanics engineer should be consulted for guidance.

Depth to the water table in borrow areas is an important consideration. Dewatering a borrow area is usually impractical for such small components as waste structures. Installing drainage or excavating and spreading the materials for drying before placement generally is not cost-effective but may be necessary when suitable borrow is limited.

- *Important point:* Adhere to any state or local requirements for back filling investigation pits or plugging test holes.

5.5   SITING ANIMAL WASTE MANAGEMENT SYSTEMS

Planning and design options for arranging and integrating components of animal waste management systems into an existing or proposed farmstead are provided in this section. Application of waste products to the land is addressed only in terms of how it is affected by land use and site conditions. Although planning and design considerations vary depending on the type of waste and regional practices, the conservation planning process provides an essential framework for integrating the options presented in this section. A supplemental checklist is also included.

5.5.1   PROCESS AND PRINCIPLES

The animal waste management system siting process begins with defining the process and associated elements.

5.5.1.1   Landscape Elements

Manipulation of landscape elements, such as structures, landform, water, and vegetation, can improve the operations of an existing animal waste management system or help to integrate a new animal waste management system into the farmstead. Each farm can be viewed as a series of spaces used for different operations linked together by roads or paths. The arrangement of structures, landform, water, and vegetation within this system affects aesthetic quality, operational efficiency, energy consumption, runoff, and specific functions on the site. Manipulation of these elements can establish desirable views, buffer noise, determine circulation of animals and equipment, manage odor, modify air temperature, affect snow or windblown soil deposition, and optimize use of available space. In addition, proper placement can help reduce health and safety hazards and enhance the quality of life values.

5.5.1.2   Structures

Structures provide space for ongoing farm activities by creating enclosure. Existing barns, sheds, houses, fences, storage tanks, ponds, and silos are structural elements to be considered when siting components of an animal waste management system.

Planning for new animal waste management system components may give the decisonmaker an opportunity to update and reorganize farm structures and land uses between them. Existing operations and equipment may have indoor and outdoor spaces very different in size and shape than those currently needed. Structures also provide options for collecting runoff, channeling wind, controlling circulation of animals and equipment, and separating use areas.

5.5.1.3   Landform

Landform can be used as it occurs on the site or is modified to improve farm operations, direct or screen views, buffer incompatible uses, reduce massiveness of aboveground structures, control access, improve drainage, and influence microclimate. Existing landforms give each landscape its distinctive character. Landforms often provide a backdrop for an animal waste management system and serve as a model for designing new landforms, such as embankments, berms, and spoil disposal mounds.

Integrating aboveground animal waste management system components into flat landscapes is more difficult because structures often project above the horizon as prominent features. Many landform modifications can be employed to address this and other site conditions or land user objectives. Excavated soil, for example, can be used to build small landforms to reduce the prominence of new components. This effect is further enhanced through the addition of vegetation.

In excavating for a pond or lagoon, the shoreline can be irregularly shaped with smooth, curved edges to make the pond or lagoon appear natural. Operation and maintenance requirements of the structure need to be considered. Embankments can also be shaped to match the surround landform.

5.5.1.4   Water

Water has magnetic appeal. It can add to aesthetic quality, modify temperature, serve as a buffer between use areas, or divert attention from undesirable views.

5.5.1.5   Vegetation

Vegetation can also be used to organize space and circulation; establish desirable views; buffer noise, wind, or incompatible uses; reduce massiveness of aboveground structures; absorb particulates to reduce odor; cool air temperature; and reduce soil erosion and runoff. As with other elements, vegetation can be used to divert attention to other features.

Because native plants are often more hardy than introduced species of vegetation, they are recommended if compatible with the landscape setting. Existing vegetative patterns, such as hedgerows, stream corridors, and even-aged stands of trees or shrubs, can be expanded or duplicated with plantings to integrate a new animal waste management system into an existing landscape.

5.5.1.6   Siting the System

The process of placing animal waste management system components on the land is similar to that for integrating other conservation practices. The following process helps site the system as well as provide a means to document planning decisions.

5.5.1.7   Base Map

During the planning process, a topographic survey or aerial photograph is prepared (Figure 5.8). A conservation plan map may be sufficient for this purpose.

Although the decisionmaker's objectives will influence the scope and detail of the survey, the data to be obtained should include:

- Property lines, easements, and rights-of-way
- Names of adjacent parcel owners

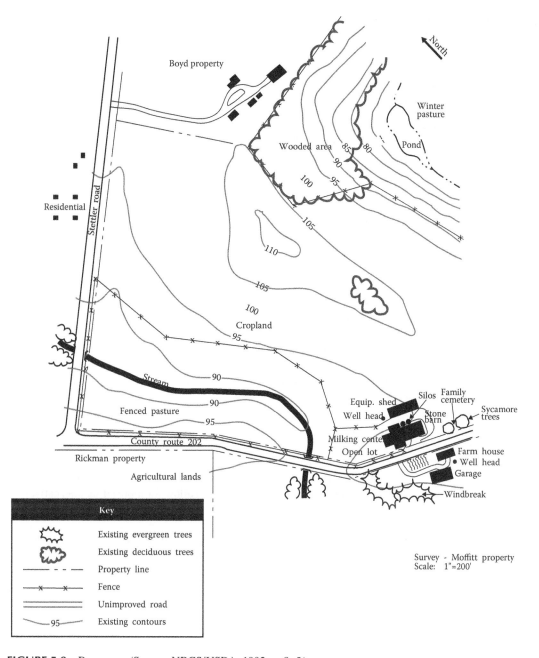

FIGURE 5.8 Base map. (Source: NRCS/USDA, 1992, p. 8–5.)

- Positions of buildings, wells, culverts, walls, fences, roads, gutters, and other paved areas
- Location, type, and size of existing utilities
- Location of wet areas, streams, and bodies of water
- Rock outcrops and other geological features
- Edges of wooded areas
- Elevations at contour intervals of 1 ft around anticipated storage/treatment areas and 2 to 5 ft around anticipated utilization areas
- Zoning ordinances and deed restrictions
- Land uses

- Geologic and soils data
- Climatic information
- Septic systems
- Wells

5.5.1.8   Site Analysis

One method of understanding site conditions and implementing the analysis of the resource data step is to prepare a site analysis diagram. This step of the process is the identification of problems and opportunities associated with installation of the animal waste management system. A topographic map, aerial photograph, or conservation plan map should be taken into the field where site conditions and observations can be noted. The site analysis should note such factors as:

- Land use patterns and their relationships
- Potential impacts to or from the proposed animal waste management system
- Existing or potential odor problems
- Existing or potential circulation (animals, equipment, and people) problems or opportunities
- Soil types and areas of erosion
- Water quality of streams and water bodies
- Vegetation to be preserved and removed
- Logical building locations, points of access, and areas for water utilization
- Good and poor views
- Sun diagram documenting the location of sunrise and sunset in winter and summer to determine sunny and shaded areas
- Slope aspect
- Prevailing summer and winter wind directions
- Frost pockets and heat sinks
- Areas where snow collects and other important microclimatic conditions
- Farmstead features with special cultural value or meaning to the decisionmaker
- Options for removal or relocation of existing buildings to allow for more siting alternatives for animal waste management system components

Figure 5.9 illustrates a site analysis for a 100-cow dairy on which the decisonmaker wishes to install an animal waste management system. The decisonmaker has requested an open view of the dairy operation and adjoining cropland from the residence and does not want views of the barn blocked. During summer, several neighbors downwind of the operation have complained of unpleasant odors. The site includes a family cemetery and some large sycamore trees with special meaning. The existing stone barn structure is unique to the area and is in good condition.

5.5.1.9   Concept Plan

As part of the formulation and evaluation alternative solutions steps, a concept plan or plans are developed to begin to evaluate alternative solutions (Figure 5.10). The area required for collection, storage, treatment, transfer, and use of waste is determined and first displayed at this step of the process. This and such related information as associated use areas, access ways, water management measures, vegetated buffer areas, and ancillary structures should be drawn freehand to approximate scale and configuration directly on the site analysis plan or an overlay.

In instances where several sites may satisfy the decisionmaker's objectives, propose the site that best considers cost differences, environmental impacts, legal ramifications, and operational capabilities. Continued analysis can further refine the location, size, shape, and arrangement of waste facilities. If the best area for a component requires a buffer, provide adequate space. If no site

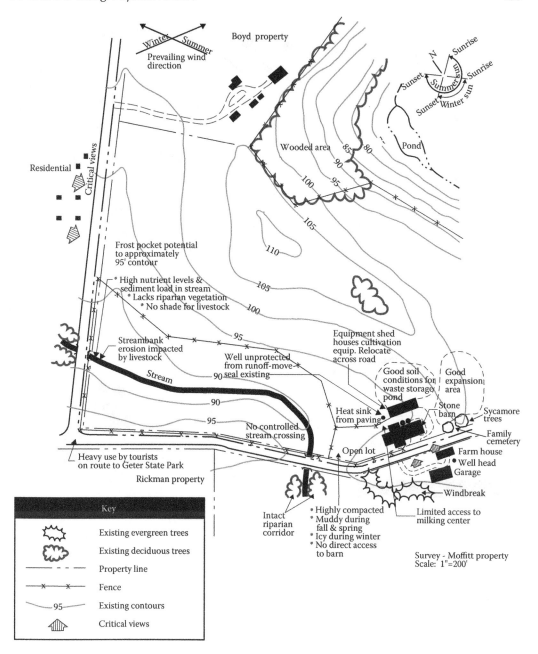

FIGURE 5.9 Site analysis diagram. (Source: NRCS/USDA, 1992, p. 8–7.)

seems viable, reassessment of the objectives in cooperation with the decisionmaker is appropriate. Generally, a minor adjustment in goals and objectives offers viable alternatives. Where a potential for major adverse effects exists, however, making significant adjustments in operations requiring a large economic commitment and attention to management may be necessary.

5.5.1.10 Site Plan

Completion of subsequent steps of the planning process results in the final site plan as preface to construction drawings and specifications (Figure 5.11). Final location and configuration of proposed

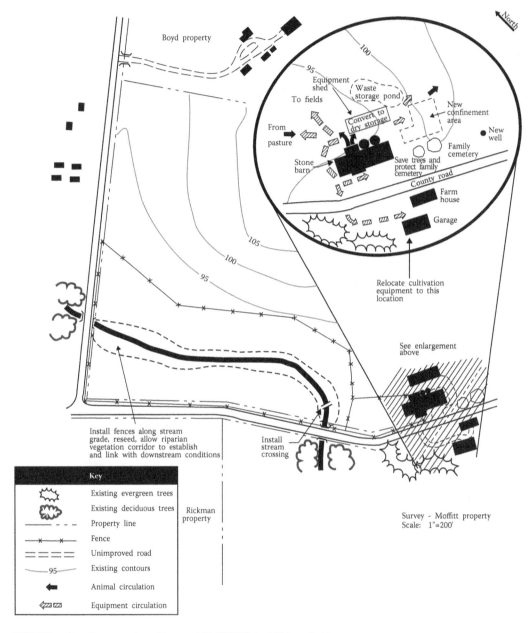

FIGURE 5.10   Concept plan. (Source: NRCS/USDA, 1992, p. 8–9.)

components and ancillary structures, finished elevations, construction materials and exterior finishes, suitable plant species and planting areas, circulation routes, utility corridors, and utilization areas are examples of information to be included. This plan is submitted to the decisonmaker for approval.

5.5.2 DESIGN OPTIONS

An animal waste management system should be designed to blend into the site and its surroundings with no adverse environmental effects. The following design options can aid the planner in achieving this objective.

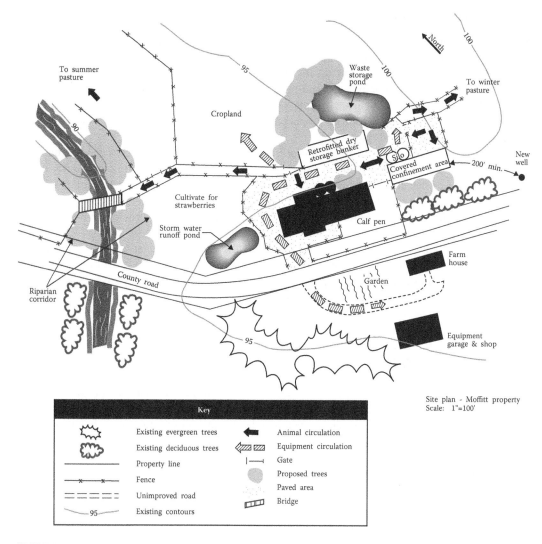

FIGURE 5.11   Site plan. (Source: NRCS/USDA, 1992, p. 8–10.)

5.5.2.1   Water Quality

The design of an animal waste management system must consider measures to improve and protect water quality. Water bodies in close proximity to the waste source are more susceptible to contamination. Relocating a pasture to an area further from a stream is often the best solution for preventing degraded streambanks and animal waste from entering the stream. Because this is not always possible, such measures as fencing, controlled stream crossings, and regarded and revegetated streambanks can aid in minimizing transport of contaminants in runoff directly entering the stream.

Developing a new animal waste management system for an existing system often presents an opportunity to improve runoff management. The addition of diversions, roof gutters to separate precipitation from waste sources, paved feedlots or loafing areas, drainage swales, and filter strips helps to minimize muddy areas and contaminated runoff. Landform mounds constructed from excess excavated material can be used to convey runoff and save the cost of hauling excess material to a disposal site. Either excess or imported soil can be used to fill depressions and improve drainage. Vegetation can serve many functions, including nutrient and sediment filtration, erosion control, moisture reduction, and temperature control.

5.5.2.2   Odor Reduction

The odor associated with the six functions of animal waste management often generates the most immediate response from the decisionmaker and adjacent residents—odor is not a problem until the neighbors complain! By anticipating the intensity, duration, and frequency of odors, animal waste management system components can be planned to reduce odors and for associated complaints. This includes areas of field application. Odor problems can be prevented or reduced through adequate drainage, runoff management, proper care to keep animals clean and dry, and appropriate waste removal, handling, and transport.

Locate waste management facilities and utilization areas as far as practical from neighboring residences, recreational areas, or other conflicting land uses. Avoid sites with radical shifts in air movement between day and night, such as those near large bodies of water or steep topography. A component's location in relation to surrounding topography may also strongly influence the transfer of odor because of daily changes in temperature and resulting airflow. To provide optimum conditions, prevailing winds should carry odors away from those who might object.

Providing conditions or design features that alter the microclimate around specific animal waste management system components can further mitigate odor. An abundance of sunlight and good ventilation, for example, helps keep livestock and poultry areas dry and relatively odor-free. A southern exposure with adequate slope to provide positive drainage for runoff is a preferred condition. Keeping waste aerated and at appropriate moisture and temperature levels slows the development of anaerobic conditions and reduces odor.

Odor-causing substances from waste material are frequently attracted to dust particles in the air. Collecting or limiting the transport of dust aids in reducing odor. Vegetation is very effective in trapping dust particles as is demonstrated by observing dust-covered trees and shrubs on the edges of unpaved roads and quarry sites. Surface features on leaves or needles, such as spines, hairs, and waxy or moist films, help trap particulates.

In addition to trapping dust particles, vegetation, landform, and structures can channel wind to carry odors away from sources of potential conflict (Figure 5.12).

5.5.2.3   Temperature and Moisture Control

Vegetation can alter microclimate and create lower temperatures. By shading the areas below them and through the process of evapotranspiration, trees and shrubs produce a cooling effect. They can also regulate temperature by reducing or increasing wind velocity. The placement of vegetation can help cool buildings in summer and allow heat-generating sunlight to penetrate in winter (Figure 5.13).

Key term: *Evapotranspiration* is an important part or process of the water cycle. More complex than precipitation, evapotranspiration is a land-atmosphere interface process whereby a major flow of moisture is transferred from ground level to the atmosphere. It returns moisture to the air, replenishing moisture by precipitation, and it also takes part in the global transfer of energy.

Dairy animals and other livestock seek streams or ponds and the shade of trees for their cooling effect. Where access to these features is removed, animals should be provided with other means of cooling.

The benefits and liabilities of sunlight, shade, and wind must be weighed in each geographic region. Cooler temperatures slow bacterial activity in waste treatment lagoons, which reduces necessary treatment of odor. Too much shade in a feedlot can allow an increase in snow or ice buildup and the amount of runoff during periods of thaw. It can also promote an increase in algae growth on paved surfaces, creating unsafe footing for animals and operators. Too little ventilation can cause the temperature and humidity to soar, while too much ventilation, especially in the form of winter winds, can create life-threatening conditions for animals.

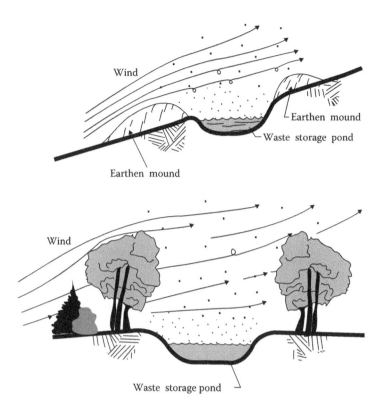

FIGURE 5.12 Topography and vegetation can uplift winds to disperse odor. (Source: NRCS/USDA, 1992, p. 8–13.)

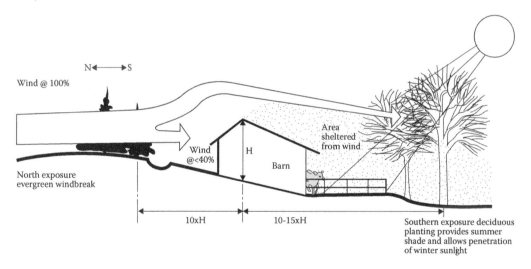

FIGURE 5.13 Vegetation modifies temperature in various ways. (Source: NRCS/USDA, 1992, p. 8–14.)

Structures can be located to influence internal temperatures (Figure 5.14). The central or long axis of new buildings can be oriented to regulate the angle and duration that sunlight strikes the roof and sides. In cool or temperate regions, for example, heat can be generated in buildings where drying of waste is needed by:

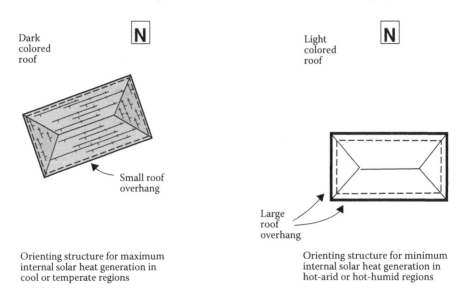

Dark colored roof

Light colored roof

Small roof overhang

Large roof overhang

Orienting structure for maximum internal solar heat generation in cool or temperate regions

Orienting structure for minimum internal solar heat generation in hot-arid or hot-humid regions

FIGURE 5.14 Orientation can influence the amount of internal sun-generated heat within buildings. (Source: NRCS/USDA, 1992, p. 8–15.)

- Orienting the long axis of the building in a northeast-southwest direction
- Constructing the roof with a small overhang to allow maximum sunlight to strike the sides of the building
- Locating the windows along the south and west walls
- Using dark roofing to enhance radiation adsorption

Where minimal internal heat is desired, such as in the hot, arid Southwest or the hot, humid Southeast, different building orientation and architecture are recommended. In these regions, minimizing the amount of sunlight on the sides of the building is best. Because the arc of the sun is higher in the sky, a minimum amount of sunlight can be expected to strike the south side of the building during midday. Therefore, the long axis of the building should be oriented in an east-west direction. The amount of wall and window area along the east and west walls should be minimized to reduce early morning and late afternoon exposure. The windows should be along the north and south walls. The roof should have wide overhangs and be finished in a light color.

If increased humidity is desirable, consider location storage ponds or treatment lagoons upwind of livestock or poultry confinement facilities. The air flowing over the pond or lagoon will pick up moisture and carry it through the confinement facilities. Care must be exercised, however, to avoid directing undesirable odor-bearing winds through the facilities. Orienting buildings to optimize prevailing winds, which can adversely effect upon the temperature or humidity within confinement facilities, can also enhance ventilation.

Temperature and moisture conditions greatly affect the presence of insects, rodents, and other pests, often a major concern of the decisionmaker and source of complaints from neighbors. Each type of livestock or poultry operation attracts specific species of insects that can affect not only the health and productivity of the animals but also the quality of the food product and the cost of production.

Several species of flies commonly breed in moist animal manure. Houseflies, which can impact areas up to 4 miles from their breeding location, are a major carrier of more than 100 human and animal pathogenic organisms. Other species of insects can range equal or further distances.

Because sanitation, including proper and timely manure handling procedures, has been reported to be the most important factor in reducing fly populations, the animal waste management system must be designed with this factor in mind. Avoid, for example, areas with odd shapes or corners, which prevent thorough scraping or other means of removing manure. Provide adequate drainage to aid in moisture control.

Many practices used for insect control also apply to rodents. Reducing nesting sites by careful selection and placement of vegetation around buildings and waste facilities helps to lower populations of insects and rodents. Many insect traps work best in full sunlight, one of many reasons to plot the course of sunlight through the farmstead.

5.5.2.4 Aesthetic Quality

Aesthetic quality is acknowledged as an integral part of daily life and underlies economic and other decisions about the land. Many land management decisions, including those related to planning and design of animal waste management systems, are made because of a decisionmaker's perception of what will enhance aesthetic quality and reflect a stewardship ethic to neighbors.

Highly visible animal waste management system components, such as storage tanks that are easily identified by their color and associated conservation practices, may be installed because they are attractive and show the decisonmaker cares about stewardship. Conversely, decisionmakers may be reluctant to install an animal waste management system that contradicts aesthetics norms for attractive or well-cared-for farmsteads and land.

5.5.2.5 Landscape Character

Patterns of land use and management, siting and design of structures, or field size and shape reflect cultural values with long guided farmstead planning and determined variations in landscape character. Landscapes are organized in response to surrounding environmental and cultural conditions and the decisionmaker's objectives.

The composition or structure of the site's surroundings must be understood so that waste management systems are designed to fit onto the landscape. To accomplish this objective, the patterns and linkages formed by farmsteads, riparian corridors, and similar features on the landscape should be examined.

Key term: A *riparian corridor* is a unique plant community consisting of the vegetation growing near a river or stream or other natural body of water. It serves a variety of functions important to people and the environment as a whole by:

- Preserving water quality by filtering sediment from runoff rivers and streams
- Protecting stream banks from erosion
- Providing a storage area for flood waters
- Providing food and habitat for fish and wildlife
- Preserving open space and aesthetic surroundings

Analyzing the compatibility of the proposed design alternatives with adjacent land uses helps to prevent potential conflicts. For example, where most residents are involved in poultry production, associated activities and impacts are expected and are therefore more likely to be accepted. The potential for incompatible land use is less likely in these situations than in those where isolated poultry operations are mixed with other uses.

Depending on objectives, components of the animal waste management system can be subdued or made prominent on the landscape. Generally, the components should blend with the surrounding landscape or be screened from view. The relationship of existing farmstead features to each other in terms of spacing, height, width, and orientation provides a clue to alternative siting locations. On a landscape divided into fields, hedgerows, and farmsteads, the animal waste management system components should be located where they will not disrupt existing relationship patterns.

Architectural style is a reflection of an area's cultural values. Unique structures, materials, or construction methods should be considered to avoid possible conflicts from proposed improvements. A historic barn, for example, can be diminished by placement of an adjacent aboveground waste storage tank, whereas a properly designed waste storage pond may serve the need and be less visually disruptive.

- *Important point:* Existing structures can often retain their original exterior appearance while their interiors are altered. The added expense may well be justified by the value of preserving an important cultural resource.

The architectural style (shape, height, and materials) of farmstead buildings should be analyzed to blend new structures into those existing. Modern, prefabricated buildings differ from traditional structures, which tend to be large and multistory and have dramatic visual line. The large floor space of traditional structures is balanced by height. Modern, prefabricated buildings generally have a lower profile, creating a greater horizontal appearance. Where possible, emulate the architectural style of existing farm buildings in the design of new structures.

The farm's layout and structures also should be discussed with the decisionmaker to identify special features. Long-established and enjoyed views from the farmhouse, large trees or windbreaks planted by ancestors, and an old springhouse or stonebase banked barn are just a few of the many possibilities that often provide a sense of place and have special meaning to the farm family or community.

5.5.2.6   Visibility

Important views to mountains and valleys, water bodies, or areas of special meaning to the decisionmaker should not be blocked with siting components unless other alternatives are not available.

Blending proposed as well as existing facilities with the surrounding landscape, while satisfying the decisionmaker's objectives, should be a primary consideration in designing an animal waste management system.

- *Important point:* The landform or vegetative patterns common to the existing landscape should be reproduced to screen an animal waste management system component.

In selecting new vegetation for screening, avoid plants that may later cause problems. Plants that are wrong for the available space, require frequent pruning, are poisonous to livestock, will not survive the ordinary growing conditions on the farm, or require more than normal maintenance should be avoided.

Reducing the visibility of an obtrusive facility is not accomplished by covering it with vegetation. To be effective, vegetation should be placed as an intervening feature between the viewer and the object being viewed. Generally, the closer the vegetation is to the viewer, the more effective it becomes in reducing visibility of the obtrusive facility.

Where vegetation is used to reduce visibility, the resulting effects upon available sunlight, air movement, snowdrift, freezing and thawing, and pest control should be considered. New plantings should be provided with the water and nutrients needed to become established.

- *Important point:* Structures can screen views of agricultural waste facilities. Roads and other landscape elements can also direct a viewer's attention away from animal waste management system components.

5.5.2.7   Compatibility

An important design consideration is restoring the site to a vegetated condition after construction is completed. Once reestablished, the newly planted trees will further enhance this effect.

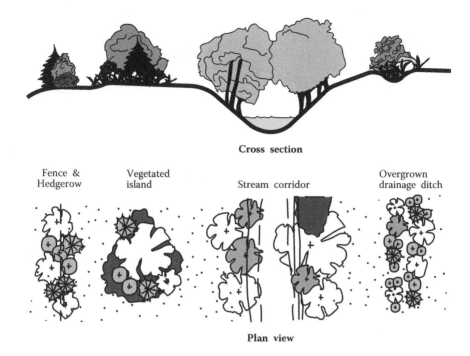

Cross section

Fence &
Hedgerow

Vegetated
island

Stream corridor

Overgrown
drainage ditch

Plan view

FIGURE 5.15   Common vegetative patterns. (Source: NRCS/USDA, 1992, p. 8–21.)

New plantings used to minimize scale or the geometric appearance of components should not attract attention by their color, texture, or form. Planting techniques include grouping plants in random arrangements to simulate natural patterns and using several sizes and species to duplicate natural vegetation. Figure 5.15 illustrates common vegetative patterns that can be used as models. The best guide, however, is to duplicate the vegetation patterns of the locality or region. Naturally occurring vegetation is more likely to be in irregular configurations rather than straight, geometric arrangements.

Whenever possible, existing vegetation should be used in siting components of the system. Fill or compaction by heavy equipment during construction or operation and maintenance can seriously reduce the amount of air available to the roots. Therefore, these activities should be avoided where the vegetation is to be saved.

Slope round and slope reduction (Figure 5.16) are two of many earth-grading and -shaping techniques that can reduce erosion and help to blend landforms into the landscape.

Coordinating colors of a new animal waste management system with colors and materials of the existing farm buildings reduces their visibility and preserves existing landscape character.

Large concrete surfaces of aboveground waste storage tanks or paved travel ways around below grade ponds can be textured or color-tinted (earth-tone colors based on surrounding soil conditions) to reduce contrast and reflectivity. Reflective metal can be painted or otherwise treated to harmonize with surroundings. Existing and planned facilities should be unified in style and materials.

5.5.2.8   Climatic Conditions

Snow and ice often hamper farm operations and cause critical runoff conditions during periods of melt. Where appropriate, the depth and location of snowdrift, as well as ice and other winter conditions, should be considered when siting an animal waste management system. Accumulation of snow on a waste storage pond or lagoon may not be desirable in areas where precipitation is abundant, especially as a waste storage pond nears capacity late in winter. Conversely, in more arid regions

Slope rounding

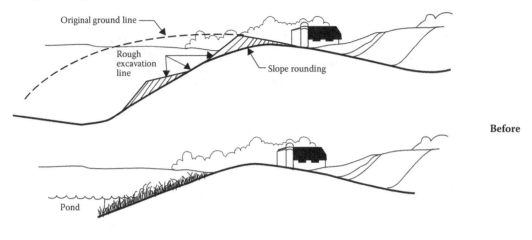

Before

After

Slope reduction

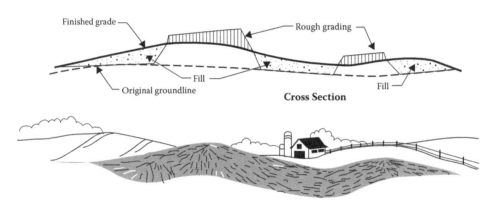

Cross Section

FIGURE 5.16   Slope rounding and reductions help to blend landforms into the landscape. (Source: NRCS/USDA, 1992, p. 8–23.)

or areas where most of the precipitation is received as snow, accumulation within the waste storage facility may be desirable. In both cases, vegetation and fences are effective in trapping snow.

The distance to which a fence or vegetative windbreak affects snow accumulation is dependent on its height and porosity and on the windspeed. A solid fence (0% porosity) causes most snow deposition to occur on the upwind (windward) side. However, its effective distance downwind (leeward) is so limited that it is not recommended for use with an animal waste management system. Fences with 15% to 25% porosity trap snow on the downwind side in an area that is as long as the fence and as wide as four or five times the fence's height. The standard snow fence is 4 ft high and 50% porous. Deposition occurs from the base of the fence to about 40 ft downwind. Figure 5.17 illustrates how fence porosity affects snow deposition patterns. As shown, a 50% porous barrier captures about four times as much snow as a 15% porous barrier. The same conditions are true for windblown soil in the more arid regions of the country.

Because of the additional height, vegetative windbreaks influence snow and windblown soil deposition over a greater distance than fences. Depending on location, they may provide additional benefits, including odor reduction, screening, temperature control, and wildlife habitat. Available

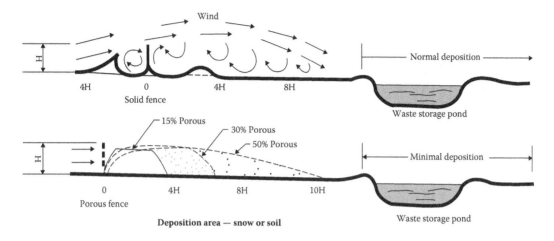

FIGURE 5.17 Fence porosity affects snow deposition. (Source: NRCS/USDA, 1992, p. 8-25.)

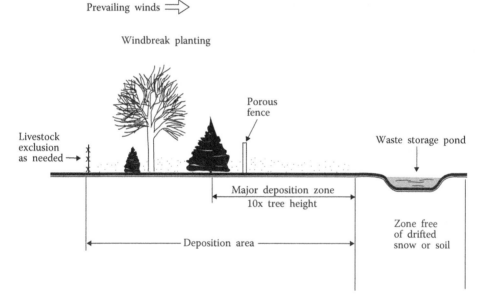

FIGURE 5.18 The combination of fence and windbreak plantings greatly enhances the pattern of snow and soil deposition. (Source: NRCS/USDA, 1992, p. 8–26.)

planting space and the amount of snow or soil deposition anticipated influence the location, width, and alignment of windbreaks.

When managing snow or soil deposition, the use of fences and vegetation should be combined whenever feasible. Fencing provides immediate results, wheras vegetation, which may require several years growing time, often provides additional multiple benefits. A second fence may be required near windbreaks to prevent livestock from damaging the vegetation. Figure 5.18 illustrates how a fence and multiple rows of vegetation with 50% porosity influence deposition.

- *Important point:* Agricultural waste facilities with the back wall protected from the wind, such as an open-front dry manure storage building, tend to have some snow accumulation just inside the front door. To prevent this, a 6- to 8-in. slot can be cut in the rear wall near the eaves to provide some wind penetration.

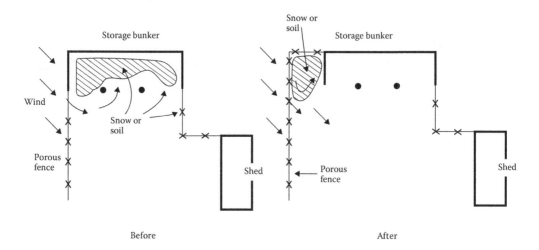

FIGURE 5.19   Fences affect snow and soil deposition around buildings. (Source: NRCS/USDA, 1992, p. 8–26.)

Ice buildup can be reduced by considering shade patterns of buildings and vegetation. Because deciduous trees shade only in summer and allow heat-generating sunlight in the winter, they are more effective than evergreens in regulating a microclimate affecting ice and snow accumulations. A mixture of deciduous trees and evergreen understory can minimize buildup.

Fences can also be located to deposit snow or windblown soil away from building openings (Figure 5.19).

5.5.2.9   Circulation

The circulation patterns of animals and equipment can be easily affected by installation of an animal waste management system. New roads and pathways are often required to ensure an efficient new system. Roads, pathways, and other forms of circulation should lead to their destination in an orderly and efficient manner. They should optimize the use of available area by providing adequate width, gradient, and turning space. In some cases, existing shortcuts must be abandoned and new circulation barriers must be used to accomplish this.

Alignment of roads and pathways should attempt to follow the existing contour of the land to prevent steep gradients and excessive cuts and fills. Sufficient drainage (0.5 to 0.75 in. per foot of slope for gravel surfaces and 0.25 to 0.5 in. per foot of slope for paved surfaces) should be provided. A minimum of 14 ft of vertical clearance should be allowed to accommodate equipment. Where feasible, existing roads, pathways, or parking areas can be eliminated or relocated to increase operation efficiency.

5.5.3   Checklist Approach to Siting Animal Waste Management System Components

The sample supplemental checklist provided in this section is designed to provide the animal waste management system decisionmaker a simplified data-generator from which prudent decisions can be made (NRCS/USDA, 1992, 1996).

5.5.3.1   Checklist of Siting Factors

Structures

1. Will the roofline, shape, materials, and color of proposed structures blend with existing structures?

2. Will proposed structures be located where their size and shape contribute to snow and ice management, wind reduction, cooling from shade, or windblown soil deposition?
3. Will outdoor lighting be installed at strategic spots, such as near steps or equipment areas, for safety and security?
4. Will signs be easily recognizable, legible, and uniform in appearance?
5. Will attaching signs to walls or other available structures reduce visual clutter? Can any signs be combined?
6. Can fences and walls be combined with plantings?
7. Will fences be uniform throughout the site to visually link discontinuous parts?
8. Will fences and walls be properly sited to prevent cold air pockets of snow, ice, and soil accumulation or to capture sun for maximum comfort levels?
9. Will fences and other linear components be located at existing landscape edges to enhance compatibility?
10. Will fencing be installed along ridges or the top of landforms where it is emphasized on the landscape? Could it be relocated at the bottom of the slope or below the horizon and still maintain its intended function?

Landforms

1. Will the plan consider highly erodible or ecologically important areas (steep slopes, areas with highly erodible soil, streambanks, natural areas, wetlands)?
2. Will disturbed areas be as small as possible?
3. Will established slopes be left undisturbed where possible?
4. Will grade changes be natural-appearing slopes that avoid abrupt transitions?
5. Will new construction fit elevations of existing landforms rather than requiring grading of the land to a continuous level, which may destroy its character?
6. Will grading and any new landforms allow successful runoff while assuring that the site is suitable for the animal waste management system?
7. Will excess excavated soil be used to create landforms to act as screens to buffer noise, wind, or incompatible facilities?

Vegetation

1. Will existing vegetation be retained to serve its important functions, such as screening, shading, wind control, erosion control, particulate control, and separation of incompatible uses?
2. Are roads of animal waste management system components designed to minimize disruption of vegetation?
3. Will roads, pathways, turnarounds, or other system components permit safe retention or introduction of vegetation?
4. Will required vegetative removal be staged to decrease the area and duration of exposure, thus reducing erosion and sedimentation potential?
5. Will removal of vegetation impact adjacent properties?
6. Will vegetation provide a buffer, visual barrier, and climatic and dust control for adjacent properties?
7. Will new vegetative species and patterns be based on those occurring naturally or appear compatible with those onsite and in the region?
8. Will measures be used during construction to protect trees or other vegetation and, if so, how successful will they be?
9. Will the survival rate of installed vegetation be acceptable? If not, what corrective measures can be used to guarantee establishment?

Water Quality

1. Will existing waterways be used and maintained for full value (open space, landscape character, and wildlife habitat)?
2. Will the design include measures to prevent runoff from draining across disturbed areas during construction?
3. Will the design preserve, restore, or enhance streambank vegetation?
4. Are slope changes designed for minimum slope length and gradient?
5. Will the design filter and deposit sediment onsite?
6. Where steeper slopes are unavoidable, will diversions be installed to intercept runoff before it reaches slopes?
7. Will retaining walls be used successfully to reduce slope gradients and improve aesthetic quality?
8. Will vegetative filter strips be retained or installed to slow down runoff, trap sediment, and reduce runoff volumes on slopes?
9. Will animals be provided with alternative water sources so they can be kept out of streams and ponds?
10. Can clean water be diverted to storage for such future uses as irrigation and stock watering?
11. If aquifer recharge is desired, will clean water runoff be directed to retention and infiltration facilities?
12. Where concentrated runoff leaves paved areas, will provisions be made for stabilized outlet points?
13. Will runoff be directed away from adjacent properties?
14. Will the design use paved watercourses where grassed swales would suffice?
15. Will roadways contribute to effective stormwater runoff management?

Odor Reduction

1. Will the design use wind control, fencing, or vegetation to reduce dust generation?
2. Is the animal waste facility sited downwind as far as practical from the farmhouse and neighbors?
3. Will the design provide maximum sunlight for biological decomposition?
4. Will the site of waste generation be designed to be as well drained as possible?
5. Will vegetation and water bodies be used to keep waste materials at optimum temperatures to control odor?
6. Will the design use landforms, vegetation, and structures to direct wind over or away from sources of odor?
7. Can equipment, work areas, storage areas, and livestock be kept as clean as practical?

Aesthetic Quality

1. Will the animal waste management system components retain or improve aesthetic quality of the farmstead and surrounding landscape?
2. Will the animal waste management system take full advantage of the natural features of the site?
3. Will the building materials and finishes be compatible with those existing?
4. Will color be used either to visually organize features on the site or to direct the eye away from undesirable views?
5. Will concrete and other building materials to textured or tinted to blend it into the landscape or reduce reflective surfaces?
6. Will roadways take advantage of desirable views?

7. Will the design allow for retention of landscape features with special meaning, such as specimen trees, exceptional views, or historic structures?

Temperature and Moisture Control

1. Will the species of pests on site be identified in order to control them at all stages of their development?
2. Has an integrated pest management plan been considered?
3. Will pest breeding sites be reduced by improving drainage, increasing sunlight and ventilation to manure generating sites?
4. Will vegetation placed around buildings and other animal waste management system components reduce pest breeding and nesting sites?
5. Will measures be installed for energy conservation (exposure to wind and sun, vegetation for shading)?
6. Will new structures be oriented and architecturally designed to benefit from or modify solar-generated heat and prevailing winds?

Compatibility

1. Will the measure adversely impact adjacent properties?
2. Will the reaction of community and nearby residents to the completed animal waste management system be positive or negative? What changes might obtain a more favorable design?
3. Will the measure be compatible with adjacent developments in terms of land use, density, scale, identity, and overall design?
4. Will structures, landform, water, and vegetation be used fully to buffer incompatible land uses?

Circulation

1. Will adequate pathways be provided for animals and humans?
2. Will paved walkways function to direct surface runoff?
3. Will drainage improvements interfere with vehicular, pedestrian, or animal circulation?
4. Will pedestrian, animal, and vehicular traffic be adequately separated?
5. Will maintenance access routes serve as pedestrian or animal walkways?
6. Will roads, pathways, and parking areas be designed to follow the shape of the land, thereby reducing costly grading and land disturbance?
7. Will roads, pathways, and parking areas be designed to allow for future expansion or change in size of equipment?
8. Will roads, pathways, and parking areas be designed to minimize disruption of vegetation and cropping practices?
9. Will roadways interrupt pedestrian and animal pathways?
10. Will sight distances be adequate for safe turning maneuvers?
11. Will access points onto highways be located at safe distances from intersections? Will warning signs, reflectors, or lane striping be installed as appropriate?
12. Will roads avoid wetlands, meadows, creeks, and other ecologically critical areas?
13. Will circulation routes be wide enough to accommodate anticipated traffic?

5.6 SUMMARY

Physical siting characteristics and considerations are essential to successful and environmentally sound AFO and CAFO operation.

CHAPTER REVIEW QUESTIONS

1. What physical factors are of concern for AFO/CAFO siting?
2. When should soil data be collected? What data is critical to site planning?
3. Where are soil maps available? What are they, why are they important, and for what are they used?
4. What are soil's three heterogeneous materials made up of?
5. What compounds make up soil, and what is the related concern?
6. What is critical to mineral and organic phases? Discuss why each is important.
7. What is critical about liquid/solution phase? Discuss the pertinent considerations.
8. What is critical about the gaseous phase? How does soil air differ from atmospheric air?
9. What are soil–animal waste interactions? Discuss the critical elements of filtration, biological degradation, and chemical reactions.
10. What are soil–animal waste mineralization concerns? Why are they important?
11. What effects does soil suitability have on land application of animal wastes? What soil properties are used to determine soil suitability and why?
12. Discuss the importance of available water capacity on soil suitability. What are the important limitations?
13. What is bulk density, and why is it important?
14. Why is cation exchange important to soil suitability?
15. What are the critical soil suitability concerns for depth to bedrock or cemented pan? Why?
16. Why is depth to high water table important?
17. What common concerns are caused by loading, and why?
18. What problems do rock fragments cause?
19. How does intake rate affect soil properties? Discuss its importance to soil suitability.
20. What's the difference between intake rate and soil permeability? What soil suitability concerns are related to permeability rates?
21. How and why does soil pH limit soil suitability?
22. What is the chief concern with ponding? Why?
23. What is the importance of soil salinity? How does it affect land application of animal wastes and crop selection? Why?
24. What concerns are related to slope and degree of slope? What incorporation practices apply, and why?
25. Discuss the objectives of a complete system approach to animal waste management design.
26. What benefits can properly designed animal waste management systems contribute?
27. Discuss the primary objective of applying animal waste to land in terms of plant nutrients.
28. Define, describe, and discuss the plant–soil system.
29. Define and discuss "nutrient budget." How is one calculated?
30. Describe the ways that waste material nutrient chemical compounds can be transformed. Why is this important?
31. Why are soil bacteria important?
32. Discuss cation-exchange differences between prairie and forest soils. Why is this important?
33. Discuss soil's function as a plant system support medium.
34. Define nutrient cycling. What is generally known about the process?
35. Discuss three elements essential to plant macronutrients and their roles in plant growth.
36. Discuss three trace elements commonly monitored in soil amendments.
37. How does nitrogen function as a limiting crop production factor?

38. Discuss phosphorus as a critical element in plant growth and phosphorus mobility in soil.
39. Describe and discuss how calcium, potassium, and magnesium can adversely affect plant fertility.
40. Where does soil sulphur come from, and how much is needed for plant growth?
41. What happens to trace elements (micronutrients) in soil?
42. What concerns are related to the presence of cadmium and lead in soils? Why?
43. What constraints to land application are related to synthetic organic compounds?
44. Discuss the balance of soil capacity to the amount of applied waste and residual.
45. Describe and discuss problems of plant nutrient deficiency in animal waste land application sites.
46. Describe and discuss problems of excessive plant nutrients on animal waste land application sites.
47. What is key to preventing the occurrence of soil toxicity problems?
48. What is the purpose of sidedressing?
49. How is the percent salt in waste estimated?
50. Discuss crop yields and soil salinity.
51. How are soils tested for salinity levels?
52. What trace elements are common on animal waste applied lands? How are these levels tested?
53. What are some common trace element toxicity symptoms?
54. What are three key steps to environmentally and economically sound animal waste application?
55. What field and forage crops are typically fertilized by animal waste application?
56. What are some of the factors that make a crop a good or poor choice for animal waste application? Be specific.
57. Define and discuss "grass tetany."
58. Discuss the advantages of perennial grasses as vegetative fillers.
59. Discuss the optimum times for land application.
60. Discuss land application of manure as it relates to ammonia toxicity.
61. What concerns and benefits are related to animal waste application on agricultural crops?
62. Discuss the USEPA, FDA, and USDA guidelines for biosolids application. What concerns do they raise?
63. Define, describe, and discuss "vegetative filter strips," their uses, and the processes involved. What site characteristics should be considered?
64. Describe and discuss how forest areas are used in animal waste recycling. What role do riparian buffers play in this process?
65. What percentage of taken-up nutrients may be redeposited in leaf litter?
66. Where, in a Douglas Fir forest ecosystem, is the greatest part of the nitrogen sink?
67. Describe and discuss the relationship between crop yield and nutrient removal. What are the critical factors?
68. How is nutrient uptake calculated?
69. What factors should a site investigation and evaluation address concerning local geology and groundwater?
70. What material properties of soil and rock are of importance and why?
71. Why are mass properties important? What effects can the two types of discontinuities have?
72. Describe and discuss underground water zones.
73. Describe and define the three types of aquifers.
74. What information on aquifers must be determined for siting an animal waste component?
75. What is porosity and how is it expressed?
76. How does soil type affect porosity?

77. How is specific yield expressed? What is it?
78. Discuss the issues connected with groundwater quality.
79. Many considerations of engineering geology related to different animal waste management components must be investigated. Describe and discuss the significance of each.
80. Describe the six common geological conditions that cause settlement.
81. What is the importance of knowing the history of the site?
82. Eleven different factors affect groundwater. Define and describe each.
83. What effect can increased clay content have on groundwater?
84. Discuss topography and direction of flow.
85. Discuss permeability and attenuation.
86. Define hydraulic conductivity.
87. Define and discuss hydraulic head.
88. How is hydraulic gradient measured?
89. Where can information on hydrogeologic setting be found?
90. What are the three types of sensitive groundwater areas and why?
91. What three factors affect the required intensity of a site investigation?
92. What role do clearly defined objectives have in site investigation issues?
93. What is the purpose of a preliminary investigation?
94. When is a detailed investigation needed?
95. What tools are useful for preliminary investigations?
96. What research resources are possible sources of site information?
97. What are detailed geologic investigations used for? Who determines the level of detail necessary? What factors affect the final outcome?
98. What additional tools can be used for detailed investigations?
99. What geophysical methods could be used?
100. How should geologic materials be logged?

THOUGHT-PROVOKING QUESTIONS

1. What are some of the scientific, emotional, and economic concerns for AFO/CAFO siting?
2. Define and discuss soil. How does it differ from its source materials? What are common soil source materials? What affects change in these materials?
3. What are the advantages and disadvantages of natural fertilizers (i.e., plant and animal wastes) versus chemical fertilizers? Discuss, including such topics as organic farming, land yield, physical quality, health benefits and risks related to each method, pollutants, and water quality.
4. Biosolids land application and animal waste land application have similarities and differences. Discuss, define, and describe them. What can AFO/CAFO operations learn from biosolids practices?
5. AFO/CAFO management may question the need for soil chemistry information. Discuss the need mastering this information.
6. Discuss and describe the interrelatedness of geology and groundwater.
7. Discuss "good neighbors" and aesthetics in site planning. How much effort and money should be expended on visuals? Why?

REFERENCES

Adam, R., Lagace, R., & Vallieres, M. 1986. Evaluation of beef feedlot runoff treatment by a vegetative filters. St. Joseph, MI: ASAE paper 86-208.
AIPG. 1984. Groundwater issues and answers. Arvada, CO: American Institute of Professional Geologists.

Brady, N.C., & Weil, R.R. 1996. *The Nature and Properties of Soils*, 11th ed. Upper Saddle River, NJ: Prentice-Hall.

Dillaha, T.A., Sherrard, H.H., Lee, D., Mostaghimi, S., & Shanholtz, V.O. 1988. Evaluation of vegetative filter strips as a best management practice for feed lots. *Journal WPCF* 60:1231–1238.

Doyle, R.C., & Stanton, G.S. 1977. Effectiveness of forest and grass buffer strips in improving the water quality of manure polluted runoff. St. Joseph, MI: ASAE paper 77-2501.

Driscoll, F.G. 1986. *Groundwater and Wells*, 2nd ed. St. Paul, MN: Johnson Div.

Firestone, M.K. 1982. Biological denitrification. In F.J. Stevenson (ed.). Nitrogen in agricultural soils. *Agronomy* 22:289–326.

Foth, H.D. 1990. *Fundamentals of Soil Science*, 8th ed. New York: Wiley & Sons.

Guyer, P.Q., et al. 1984. Grass tetany. University of Nebraska, NebGuide. http://www.ianr.unl.edu/pubs/animaldisease/g32.htm. Accessed February 17, 2005.

Heath, R.C. 1983. Basic ground-water hydrology. Washington, DC: U.S. Geol. Surv., Water Supply Paper 2220.

Hensler, R.R., Olson, R.J., & Attoe, O.J. 1970. Effects of soil pH and application rates of dairy cattle manure on yield and recovery of twelve plant nutrients by corn. *Agron. J.* 62:828–830.

Hornbeck, J.W., & Kropelin. 1982. Nutrient removal and leaching from a whole-tree harvest of northern hardwoods. *J. Environ. Qual.* 11:309–316.

Jenny, H. 1941. *Factors of Soil Formation*. New York: McGraw-Hill.

Johnson, A.L. 1967. Specific yield—compilation of specific yields for various materials. Washington, DC: U.S. Geol. Surv., Water-Supply Pap. 1662-D, US. Govt. Print. Off.

Kabata-Pendias, A., & Pendias, H. 1984. *Trace Elements in Soils and Plants*. Boca Raton, FL: CRC Press.

Keeney, D.R. 1980. Prediction of soil nitrogen availability in forest ecosystem: A literature review. *Forest Sci.* 26:159–171.

Kirkham, M.B. 1985. Agricultural use of phosphorus in sewage sludge. *Adv. Agron.* 35:129–161.

Kirsten, H.A.D. 1987. Case histories of groundmass characterization for excavatability. In *Rock Classification Systems for Engineering Purposes*, ASTM, STP-984, ed. L. Kirkdale, Philadelphia, PA: American Society for Testing and Materials.

Klausner, S.D., & Guest, R.W. 1981. Influence of NH_3 conversations from dairy cattle manure on the yield of corn. *Agron. J.* 73:720–723.

Lowrance, R., Leonard, R., & Sheridan, J. 1985. Managing riparian ecosystems to control nonpoint pollution. *J. Soil and Water Cons.* 40:87–91.

Martin, J.H., & Leonard, W.H. 1949. *Principles of Field Crop Production*. New York: Macmillan Company.

NRCS/USDA. 1992; 1996. Agricultural waste management field handbook. http://www.ftw.nrcs.usda.gov/awmfh.html.

NRCS/USDA. 1999. Geologic and groundwater considerations. http://www.ftw. nrcs.usda.gov/awmfh.html.

Proulx, A. 1992. *That Old Ace in the Hole*. New York: Scribner.

Roberge, P.R. 1999. *Handbook of Corrosion Engineering*. New York: McGraw-Hill.

USDA. 1998. Soil quality resource concerns: Available water capacity. http://soils.usda.gov. Last accessed February 7, 2005.

Sawyer, J.E., & Hoeft, R.G. 1990. Greenhouse evaluation of simulated injected liquid beef manure. *Agron. J.* 82:613–618.

Schertz, C.B., & Clausen, J.C. 1989. Vegetative filter treatment of dairy milkhouse wastewater. *J. Environ. Qual.* 18:446–451.

Schertz, D.I., & Miller, D.A. 1972. Nitrate-N accumulation in the soil profile under alfalfa. *Agron. J.* 64:660–664.

Schuman, B.A., & Elliott, L.F. 1978. Cropping an abandoned feedlot to prevent deep percolation of nitrate nitrogen. *Science* 126(4) 237–243.

Simpson, C.P. 1986. *Manure Handling*. Agricultural Waste Management Handbook. USDA/SC.

Simpson, K. 1998. *Fertilizers and Manures*. London: Longman Group Limited.

Spellman, F.R. 1996. *Stream Ecology and Self-Purification*. Lancaster, PA: Technomic Publishing Company.

Spellman, F.R. 1999. *The Science of Environmental Pollution*. Boca Raton, FL: CRC Press.

Stewart, B.A. 1974. *Selected Materials Relating to Role of Plants in Waste Management*. Bushland, TX: USDA Southwest Great Plains Res. Cent.

Sutton, A.L., Nelson, D.W., Hoff, J.D., & Mayrose, V.B. 1982. Effects of injection and surface application of liquid swine manure on crop yield and soil composition. *J. Environ. Qual.* 11:468–472.

Tisdale, S.L., Nelson, W.L., & Beaton, J.D. 1985. *Soil Fertility and Fertilizers*. New York: Macmillan.

Todd, D.K. 1980. *Groundwater Hydrology*, 2nd ed. New York: John Wiley & Sons.

USDA 1986. Utilization of sewage sludge compost as a soil conditioner and fertilizer for plant growth. Washington, DC: AIB 464, U.S. Govt. Print. Office.

USEPA. 1979. Animal waste utilization on cropland and pastureland. EPA-600/2-79-059. Washington, DC: U.S Govt. Print. Office.

USEPA. 1983. Land application of municipal sludge process design manual. Munic. Environ. Res. Lab., Cincinnati, OH: U.S. Govt. Print. Office, Wash., DC.

Walsh, LM. & Beaton, J.D. 1973. Soil testing and plant analysis. *Soil Sci. Soc. Amer.,* Madison, WI.

Wild, A. 1988. *Russell's Soil Conditions and Plant Growth.* New York: Longman Scientific & Technical, John Wiley & Sons.

Young, R.A., Huntrods, T., & Anderson, W. 1980. Effectiveness of vegetated buffer strips in controlling pollution from feedlot runoff. *J. Environ. Qual.* 9:483–487.

6 Animal Waste Pollutants

Headaches, sore throat, dizziness. Them hogs are pumped full a antibiotics and growth hormones. Eat that pork and it gets into you. Bacteria and viruses adapt to the antibiotics so the day is comin when if we get sick the antibiotics can't help.

(Proulx, 2002, p. 114)

6.1 INTRODUCTION

Pollutants most commonly associated with animal waste include nutrients (including ammonia), organic matter, solids, pathogens, and odorous compounds. Animal waste can also be a source of salts, various trace elements (including metals), pesticides, antibiotics, and hormones. These pollutants can be released into the environment through discharge or runoff if manure and associated wastewater are not properly handled and managed.

Pollutants in animal waste can enter the environment through a number of pathways. These include surface runoff and erosion, overflows from lagoons, spills and other dry-weather discharges, leaching into soils and groundwater, and volatilization (evaporation) of compounds (e.g., ammonia) and subsequent redeposit on the landscape. Pollutants from animal waste can be released from an operation's animal confinement area, treatment and storage lagoons, and manure stockpiles, and from cropland where manure is often land-applied (Federal Register [FR], 2003).

In this chapter, we present the pollutants associated with livestock and poultry operations (of which concentrated animal feeding operations [CAFOs] are a subset), the pathways by which the pollutants reach surface water, and their impacts on the environment and human health.

6.2 ANIMAL WASTE POLLUTANTS OF CONCERN

The primary pollutants associated with animal waste are nutrients (particularly nitrogen and phosphorus), ammonia, pathogens, and organic matter. Animal waste is also a source of salts, trace elements and, to a lesser extent, antibiotics, pesticides, and hormones. Each of these types of CAFO pollutants is discussed in the sections that follow. [*Note*: The estimates of manure pollutant production are based on average values reported in the scientific literature and compiled by the American Society of Agricultural Engineers (ASAE, 1999), U.S. Department of Agriculture (USDA)/National Resource Conservation Service (NRCS) (1996), and USDA/Agricultural Research Service (ARS) (1998)]. The actual composition of manure depends on the animal species, size, maturity, and health as well as on the composition (e.g., protein content) of animal feed (Phillips et al., 1992). After waste has been excreted, it may be altered further by the bedding and waste feed and may be diluted with water (Loehr, 1972; USDA, 1992).

- *Important point:* Ammonia is also a nutrient but is listed separately here because it exhibits additional environmental effects, such as aquatic toxicity and direct dissolved oxygen demand.

TABLE 6.1

Primary Nutrients in Both Livestock and Human Manures

	Animal Group						
	Swine	Layer	Broiler	Turkey	Beef	Dairy	Human
Mass of animal	135	4.00	2.0	15	800	1400	150
Nutrient	Pounds per 1,000 pounds live animal weight per day						
Nitrogen (total Kjeldahl)	0.52	0.84	1.1	0.62	0.34	0.45	0.20
Phosphorus (total)	0.18	0.30	0.30	0.23	0.092	0.094	0.02
Orthophosphorus	0.12	0.09	n/a	n/a	0.03	0.061	n/a
Potassium	0.29	0.30	0.40	0.24	0.21	0.29	0.07

Sources: Livestock data are "as excreted" and are from ASAE (1999); Human waste data are "as excreted" and are from USDA/NRCS (1996). Values rounded to two significant figures. n/a = not available.

6.2.1 NUTRIENTS

The three primary nutrients in manure are nitrogen, phosphorus, and potassium. Much of the past research on animal manure has focused on these constituents, given their importance as cropland fertilizers. The following discussion provides more detail on nitrogen and phosphorus characteristics and concentrations in manure. Scientific literature and policy statements commonly cite these two nutrients as key sources of water quality impairments. In the central United States, a 1995 estimate notes that 37% of all nitrogen and 65% of all phosphorus inputs to watersheds come from manure (U.S. Fish and Wildlife Service [USFWS], 2000). Actual or anticipated levels of potassium in groundwater and surface water are unlikely to pose hazards to human health or aquatic life (Wetzel, 1983). Potassium does contribute to salinity, however, and applications of high salinity manure are likely to decrease the fertility of the soil.

Table 6.1 presents the amounts of total Kjeldahl nitrogen, total phosphorus, orthophosphorus, and potassium generated per 1,000 lb live animal weight per day (ASAE, 1999). For comparison, Table 6.1 presents similar information for humans. The figures illustrate that per-pound nutrient output varies among animal types and is much higher for animals than for humans.

Key term: Total Kjeldahl nitrogen is the sum of organic nitrogen in the tri-negative oxidation state and ammonia.

6.2.1.1 Nitrogen

Nitrogen (N) is an essential nutrient required by all living organisms; ubiquitous in the environment, it accounts for 78% of the atmosphere as elemental nitrogen (N_2). This form of nitrogen is inert and does not impact environmental quality. It is also not bioavailable to most organisms and therefore has no fertilizer value. Nitrogen also forms other compounds that are bioavailable, mobile, and potentially harmful to the environment. The nitrogen cycle (Figure 6.1) shows the various forms of nitrogen and the processes by which they are transformed and lost to the environment.

- *Important point:* Nitrogen occurs in the environment in gaseous forms (elemental nitrogen, N_2; nitrogen oxide compounds, N_2O and NO_x; and ammonia, NH_3); water soluble forms (ammonia, NH_3; ammonium, NH_4^+; nitrite, NO_2^-; and nitrate, NO_3^-); and an organic nitrogen, bound up in the proteins of living organisms and decaying organic matter (Brady,

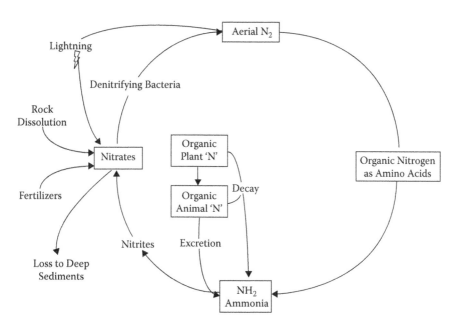

FIGURE 6.1 Nitrogen cycle. (Source: Spellman, 1996, p. 12).

1990). The transformation of the different forms of nitrogen among land, water, air, and living organisms is shown in Figure 6.1.

Manure nitrogen is primarily in organic form (organic nitrogen and ammonia nitrogen compounds (North Carolina Agricultural Extension Service [NCAES], 1982). Organic nitrogen in the solid content of animal feces is mostly in the form of complex molecules associated with digested food, whereas organic nitrogen in urine is mostly in the form of urea $((NH_2)_2CO)$ (USDA, 1992). In organic form, nitrogen is unavailable to plants. However, via microbial processes, organic nitrogen is transformed to ammonium (NH_4^+) and nitrate (NO_3^-) forms, which are bioavailable and therefore have fertilizer value.

These forms can also produce negative environmental impacts when they are transported to the environment.

- *Important point:* In an anaerobic lagoon, the nitrogen organic fraction is about 20% to 30% of total nitrogen (USDA, 1992).

Under aerobic conditions, ammonia can oxidize to nitrites and nitrates. Subsequent anaerobic conditions can result in denitrification (transformation of nitrates and nitrites to gaseous nitrogen forms). Overall, depending on the animal type and specific waste management practices, between 30% and 90% of nitrogen excreted in manure can be lost before use as a fertilizer (Vanderholm, 1975).

6.2.1.2 Phosphorus

Phosphorus exists in solid and dissolved phases, in both organic and inorganic forms. More than 70% of the phosphorus in animal manure is in organic form. Like nitrogen, the various forms of phosphorus are subject to transformation (Figure 6.2). Dissolved phosphorus in the soil environment consists of orthophosphates $(PO_4^{-3}, HPO_4^{-2}, or H_2PO_4^-)$, inorganic polyphosphates, and organic phosphorus (Poultry Water Quality Consortium, 1998). Solid phosphorus exists as organic phosphorus in dead and living materials; mineral phosphorus in soil components; adsorbed phosphorus

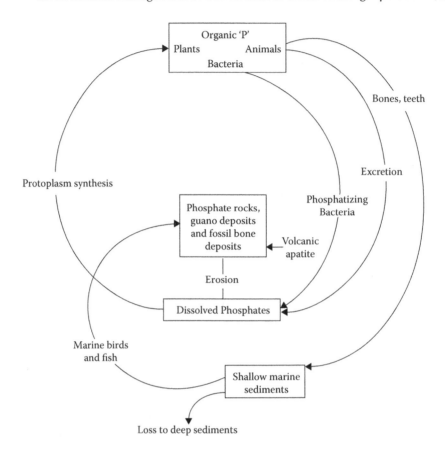

FIGURE 6.2   The phosphorous cycle. (Source: Spellman, 1996, p. 14).

on soil particles; and precipitate phosphorus, which forms upon reaction with soil cations such as iron, aluminum, and calcium (Poultry Water Quality Consortium, 1998). Orthophosphate species, both soluble and attached, are the predominant forms of phosphorus in the natural environment (Bodek et al., 1988). Soluble (available or dissolved) phosphorus generally accounts for a small percentage of total soil phosphorus. However, soils saturated with phosphorus can have significant occurrences of phosphorus leaching. Soluble phosphorus is the form used by plants and is subject to leaching. About 73% of the phosphorus in most types of fresh livestock waste is in the organic form (USDA, 1992). As animal waste ages, the organic phosphorus mineralizes to inorganic phosphate compounds and becomes available to plants.

- *Important point:* Inorganic phosphorus tends to adhere to soils and is less likely to leach into groundwater.

- *Important point:* Soil test data in the United States confirm that many soils in areas dominated by animal-based agriculture have elevated levels of phosphorus.

6.2.2   Ammonia

"Ammonia-nitrogen" includes the ionized form (ammonium, NH_4^+) and the unionized form (ammonia, NH_3). Ammonium is produced when microorganisms break down organic nitrogen products, such as urea and proteins in manure. This decomposition can occur in either aerobic or anaerobic environments. In solution, ammonium enters into an equilibrium reaction with ammonia, as shown in the following equation:

$$NH_4^+ \rightleftarrows NH_3 + N^+$$

As the equation indicates, higher pH levels (lower H^+ concentrations) favor the formation of ammonia, whereas lower pH levels (higher H^+ concentrations) favor the formation of ammonium. Both forms are toxic to aquatic life, although the unionized form (ammonia) is much more toxic.

- *Important point:* Fish kills from ammonia toxicity are a potential consequence of the direct discharge of animal wastes to surface waters. This is illustrated by a May 1997 incident in Wabasha County, Minnesota, in which ammonia in a dairy manure release killed 16,500 minnows and white suckers (Clean Water Action Alliance, 1998).

Up to 50% or more of the nitrogen in fresh manure may be in the ammonia form or convert to ammonia relatively quickly once manure is excreted (Vanderholm, 1975). Ammonia is very volatile, and much of it is emitted as a gas, although it may also be absorbed by or react with other substances.

Higher pH levels (lower H^+ concentrations) favor the formation of ammonia, whereas lower pH levels (higher H^+ concentrations) favor the formation of ammonium. The ammonia form is subject to volatilization.

The ammonia content of fresh manure varies in amount by animal species and changes as the manure ages. Ammonia content may increase as organic matter breaks down; it may decrease when volatilization occurs or when nitrate oxidizes to nitrite under aerobic conditions.

6.2.3 Pathogens

Pathogens are disease-causing organisms (bacteria, viruses, protozoa, fungi, and algae). Both manure and animal carcasses can be sources of pathogens in the environment (Juranek, 1995). Livestock manure may contain bacteria, viruses, fungi, helminthes, protozoa, and parasites, many of which are pathogenic (Jackson et al., 1987; USDA/ARS, 1998). For example, researchers have isolated pathogenic bacteria and viruses from feedlot wastes (Derbyshire et al., 1966; Derbyshire & Brown, 1978; Hrubant, 1973). In addition, the USFWS (2000) has shown fields receiving animal waste applications to have elevated levels of fecal coliforms and fecal streptococci. Specifically, bacteria such as *Escherichia coli* 0157:H7, *Salmonella* species, *Campylobacter jejuni*, *Listeria monocytogenes*, and *Leptospira* species are often found in livestock manure and have also been associated with waterborne disease. A recent study by the USDA revealed that about half the beef cattle presented for slaughter during July and August 1999 carried *Escherichia coli* 0157:H7 (Elder et al., 2000). Also, protozoa, including *Cryptosporidium parvum* and *Giardia* species (such as *Giardia lamblia*), may occur in animal waste. *Cryptosporidium parvum* is associated with cows in particular; newborn dairy calves are especially vulnerable to infection and excrete large numbers of infectious oocysts (USDA/ARS, 1998). Most pathogens are shed from host animals with active infections.

- *Important point:* Multiple species of pathogens may be transmitted directly from a host animal's manure to surface water, and pathogens already in surface water may increase in number from loadings of animal manure nutrients and organic matter.

Presence of bacteria (and other pathogens) is often measured by the level of fecal coliforms, *Escherichia coli*, or enterococci in manure (Bouzaher et al., 1993). Use of such indicator organisms has limitations, specifically, that no established relationships have been established between fecal coliform and pathogen contamination. However, indicators are still used because specific pathogen testing protocols are too time-consuming, expensive, or insensitive to be used for monitoring purposes (Shelton, 2000). Table 6.2 lists the number of total coliform bacteria, fecal coliform bacteria,

TABLE 6.2

Coliform Bacteria in Manure (Colonies per Cubic Foot of Manure, As Excreted)

Animal group	Total coliform bacteria	Fecal coliform bacteria	Fecal streptococcus bacteria
Swine	1.6×10^{11}	5.9×10^{10}	18×10^{11}
Poultry (layers)	4.7×10^{11}	3.2×10^{10}	0.69×10^{11}
Beef	3.2×10^{11}	14×10^{10}	1.5×10^{11}
Dairy	36×10^{11}	5.2×10^{10}	3.0×10^{11}

Source: ASA (1999).
Note: Values rounded to two significant figures.

and fecal streptococcus bacteria per cubic foot of manure for swine, poultry, beef, and dairy animals (ASAE, 1999).

- *Important point:* Over 150 pathogens found in livestock manure are associated with risks to humans.

- *Important point:* The Centers for Disease Control and Prevention (CDC) (1998) reported on an Iowa investigation of chemical and microbial contamination near large-scale swine operations. The investigation demonstrated the presence of pathogens not only in manure lagoons used to store swine waste before it is land applied but also in drainage ditches, agricultural drainage wells, tile line inlets and outlets, and an adjacent river.

6.2.4 Organic Matter

Livestock manures contain many carbon-based, biodegradable compounds. These compounds are of concern in surface water because dissolved oxygen is consumed as aquatic bacteria and other microorganisms decompose these compounds. This process reduces the amount of oxygen available for aquatic animals.

- *Important point:* Oxygen-depleting substances are the second leading stressor in estuaries. They are the fourth greatest stressor both in impaired rivers and streams and in impaired lakes, ponds, and reservoirs (Spellman, 1996). Biochemical oxygen demand (BOD) is an indirect measure of the concentration of biodegradable substances present in an aqueous solution. Alternatively, the chemical oxygen demand (COD) test uses a chemical oxidant. This test provides an approximation of the ultimate BOD and can be estimated more quickly than the 5 days required for the BOD test. If the waste contains only readily available organic bacterial food and no toxic matter, the COD values correlate with BOD values obtained from the same wastes (Dunne & Leopold, 1978).

Table 6.3 lists BOD and COD estimates for manure generated by swine, poultry, beef, and dairy animals and, for comparison, provides values for domestic sewage. Reported BOD values for various untreated animal manures range from 24,000 mg/L to 33,000 mg/L. COD values range from 25,000 mg/L to 260,000 mg/L. Dairy and beef cattle manure have BOD and COD values of similar magnitude. By comparison, the BOD value for raw domestic sewage ranges from 100 mg/L to 300 mg/L. Even after biological treatment in anaerobic lagoons, animal waste BOD concentrations (200 mg/L to 3,8000 mg/L) are much higher than those of municipal wastewater treated to the secondary level (about 20 mg/L) (USEA, 1992).

TABLE 6.3

Reported BOD and COD Concentrations for Manures and Domestic Sewage

Waste	BOD (mg/L)	COD (mg/L)
Swine manure		
Untreated	27,000 to 33,000	25,000 to 180,000
Anaerobic lagoon influent	13,000	n/a
Anaerobic lagoon effluent	300 to 3,600	n/a
Poultry manure		
Untreated (chicken)	24,000	100,000 to 260,000
Anaerobic lagoon influent (poultry)	9,800	n/a
Anaerobic lagoon effluent (poultry)	600 to 3,800	n/a
Dairy cattle manure		
Untreated	26,000	68,000 to 170,000
Anaerobic lagoon influent	6,000	n/a
Anaerobic lagoon effluent	200 to 1,200	n/a
Beef cattle manure		
Untreated	28,000	73,000 to 260,000
Anaerobic lagoon influent	6,700	n/a
Anaerobic lagoon effluent	200 to 2,500	n/a
Domestic sewage		
Untreated	100 to 300	400 to 600
After secondary treatment	20	n/a

Sources: Untreated values, except for beef manure BOD, are from NCAES (1982). The BOD value for beef manure is from ASAE (1997). Lagoon influent and effluent concentrations are USDA/NRCS (1996). Values rounded to two significant figures. n/a = not available

---

6.2.5   SALTS AND TRACE ELEMENTS

The salinity of animal manure is directly related to the presence of the nutrient potassium and dissolved mineral salts that pass through the animal. In particular, significant concentrations of soluble salts containing the cations sodium and potassium remain from undigested feed that passes unabsorbed through animals (NCAES, 1982). Other major cations contributing to salinity are calcium and magnesium; the major anions are chloride, sulfate, bicarbonate, carbonate, and nitrate (National Research Council [NRC], 1993). Salinity tends to increase as the volume of manure decreases during decomposition and evaporation (Gresham et al., 1990). Salt buildup deteriorates soil structure, reduces permeability, contaminates groundwater, and reduces crop yields.

- *Important point:* In fresh waters, increasing salinity can disrupt the balance of the ecosystem, making it difficult for resident species to remain viable. In laboratory settings, drinking water high in salt content has inhibited growth and slowed molting of mallard ducklings. Salts also contribute to degradation of drinking water supplies.

Trace elements in manure of environmental concern include arsenic, copper, selenium, zinc, cadmium, molybdenum, nickel, lead, iron, manganese, aluminum, and boron. Arsenic, copper, selenium, and zinc are often added to animal feed as growth stimulants or biocides (Sims, 1995). Trace elements may also end up in manure through use of pesticides, which farmers apply to livestock to suppress houseflies and other pests (USDA/ARS, 1998). Trace elements have been found in manure lagoons used to store swine waste before land application and in drainage ditches, agricultural drainage wells, and tile line inlets and outlets. They have also been found in rivers adjacent to hog and cattle operations.

- *Important point:* Spellman (1996) points out that metals are the fifth leading stressor in impaired rivers, the second leading stressor in impaired lakes, and the third leading stressor in impaired estuaries.

It is useful to compare trace element concentrations in manure to those in municipal sewage sludge, which is regulated by the USEPA's *Standards for the Use or Disposal of Sewage Sludge* promulgated under the Clean Water Act (CWA) and published in 40 CFR Part 503 (USEPA, 1993c). Regulated trace elements in sewage biosolids include arsenic, cadmium, chromium, copper, lead, mercury, molybdenum, nickel, selenium, and zinc. Sims (1995) has reported that total concentrations of trace elements in animal manures are comparable to those in some municipal biosolids, with typical values well below the maximum concentrations allowed by Part 503 for land-applied sewage biosolids.

6.2.6 ANTIBIOTICS

Antibiotics are used in animal feeding operations and can be expected to appear in animal wastes. The practice of feeding antibiotics to poultry, swine, and cattle evolved from the 1949 discovery that very low levels usually improved growth. Antibiotics are used both to treat illness and as feed additives to promote growth or to improve feed conversion efficiency. In 1991, farmers used an estimated 19 million pounds of antibiotics for disease prevention and growth promotion in animals. Between 60% to 80% of animals receive antibiotics during their productive life span (Tetra Tech, 2000a). Use as feed additives accounts for most of the mass of antibiotics used in both the swine and poultry industries and accounts for the presence of antibiotics in the resulting manure. Although antibiotic residues in beef and dairy manure are also a concern, the USEPA could not locate any literature on levels of antibiotics in manure. Estimated concentrations of the antibiotic chlortetracycline in the lagoon systems of a port produce in Nebraska range from 150 to 300 mg/L; that producer currently uses 16 different antibiotics as feed and drinking water additives (USFWS, 2000).

- *Important point:* Of greater concern than the presence of antibiotics in animal manure is the development of antibiotic-resistant pathogens. Use of antibiotics in raising animals, especially broad-spectrum antibiotics, is increasing. As a result, more strains of antibiotic-resistant pathogens are emerging, along with strains that are growing more resistant. Normally, about 2% of a bacterial population is resistant to a given antibiotic; however, up to 10% of bacterial populations from animals regularly exposed to antibiotics have been found to be resistant.

6.2.7 PESTICIDES AND HORMONES

Pesticides and hormones are compounds commonly used in animal feeding operations (AFOs) and can be expected to appear in animal wastes. Both of these types of pollutants have been linked with endocrine disruption.

Farmers may use pesticides on crops grown for animal consumption or directly in animal housing areas to control parasites (among other reasons). However, little information is available regarding the concentrations of pesticides in animal wastes or on their bioavailability in waste-amended soils.

- *Important point:* Pesticides are applied to livestock to suppress houseflies and other pests. Very little research has been performed on losses of pesticides in runoff from manured lands. Experience has shown that cyromazine losses (used to control flies in poultry litter) in runoff increase with the rate of poultry manure applied and the intensity of rainfall.

Hormones are chemical messengers that carry instructions to target cells throughout the body and are normally produced by the body's endocrine glands. Target cells read and follow the hormones' instructions, sometimes building a protein or releasing another hormone. These actions lead to many bodily responses, including a faster heartbeat or bone growth. Hormones include steroids (estrogen, progesterone, testosterone), peptides (antidiuretic hormone), polypeptides (insulin), amino acid derivatives (melatonin), and proteins (prolactin, growth hormone). Natural hormones are potent; only very small amounts are needed to cause an effect.

Specific hormones are administered to cattle to increase productivity in the beef and dairy industries, and several studies have shown that hormones are present in animal manures (Mulla et al., 1999). For example, poultry manure has been shown to contain about 30 ng/g of estrogen and about the same levels of testosterone (Shore et al., 1995). Estrogen was found in concentrations up to 20 ng/L in runoff from fields fertilized with chicken manure (Shore et al., 1995).

- *Important point:* In 1995, an irrigation pond and three streams in the Conestoga River watershed near the Chesapeake Bay had both estrogen and testosterone present. All of these sites were affected by fields receiving poultry litter.

6.2.8 OTHER POLLUTANTS OF CONCERN

CAFOs can also be a source of gas emissions and particulates. A general overview of each group of pollutants follows.

Gas emissions. The degradation of animal wastes by microorganisms produces a variety of gases. Sources of odor include animal confinement buildings, waste lagoons, and land application sites. In addition to ammonia (discussed earlier), the three main gases generated from manure are carbon dioxide, methane, and hydrogen sulfide. Aerobic conditions yield mainly carbon dioxide, and anaerobic conditions generate both methane and carbon dioxide. Anaerobic conditions, which dominate in typical, unaerated animal waste lagoons, also generate hydrogen sulfide and more than 150 other odorous compounds, including volatile fatty acids, phenols, mercaptans, aromatics, sulfides, and various esters, carbonyls, and amines (Bouzaher et al., 1993; O'Neill & Phillips, 1992; USDA, 1992).

Particulates. Sources of particulate emissions from CAFOs include dried manure, feed, epithelial cells, hair, and feathers. The airborne particles make up an organic dust, which includes endotoxin (the toxic protoplasm liberated when a microorganism dies and disintegrates), adsorbed gases, and possibly steroids. At least 50% of dust emissions from swine operations may be respirable (Thu, 1995).

6.3 MANURE POLLUTANTS: SURFACE WATER CONTAMINATION

Pollutants found in animal manure can reach surface water by several mechanisms. These can be characterized as either surface discharges or other discharges. Surface discharges can result from runoff, erosion, spills, and dry-weather discharges. In surface discharges, the pollutant travels overland or through drain tiles with surface inlets to a nearby stream, river, or lake. Direct contact between confined animals and surface waters is another means of surface discharge. For other types of discharges, the pollutant travels via another environmental medium (groundwater or air) to surface water.

- *Important point:* Animal agriculture is a common source of pollutants in watersheds, but it is never the only source. Indeed, the diverse and ubiquitous nature of pollutant forms in the environment introduces significant complexity to the increasingly important task of managing pollutants in watersheds.

6.3.1 Surface Discharges

Near the outset of this section, attempting a systematic quantification of pollutant sources in surface waters as a means of exploring the relative importance of animal agriculture's influence on pollutant control in aquatic ecosystems under different conditions is appropriate.

6.3.1.1 Runoff

Water that falls on manmade surfaces or soil and fails to be absorbed flows across the surface and is called *runoff*. Surface discharges of manure pollutants can originate from feedlots and from overland runoff at land applications. Runoff is especially likely at open-air feedlots, when rainfall occurs soon after application and when farmers over-apply or misapply manure. For example, experiments shown that for all animal wastes, the application rate has a significant effect on the runoff concentration (Daniel et al., 1995). Other factors that promote runoff to surface waters are steep land slope, high rainfall, low soil porosity or permeability, and close proximity to surface waters. In addition, manure applied to saturated or frozen soils is more likely to run off the soil surface (Mulla et al., 1999). Runoff of pollutants dissolved in rainwater is a significant transport mechanism for water soluble pollutants, including nitrate, nitrite, and organic forms of phosphorus.

Runoff of manure pollutants has been identified as a factor in a number of documented impacts for CAFOs. For example, in 1994, an environmental advocacy group noted multiple runoff problems for a swine operation in Minnesota (Clean Water Action Alliance, 1998), and in 1996, the State of Ohio identified runoff from manure spread on land at several Ohio operations that were feeding swine and chicken (Ohio Department of Natural Resources [ODNR], 1997). More discussion of runoff and its impacts on the environment and human health appears later in this section.

6.3.1.2 Erosion

In addition to runoff, surface discharges can occur by erosion, in which the soil surface is worn away by the action of water or wind. Erosion is a significant transport mechanism for land-applied pollutants, such as phosphorus, that are strongly sorbed to soils, of which phosphorus is one example (Gerritse & Zugec, 1977). In 1999, the ARS noted that phosphorus bound to eroded sediment particles makes up 60% to 90% of phosphorus transported in surface runoff from cultivated land. For this reason, most agricultural phosphorus control measures have focused on soil erosion control to limit transport of particulate phosphorus. However, soils do not have infinite adsorption capacity for phosphate or any other adsorbing pollutant, and dissolved pollutants, including phosphate, can still enter waterways via runoff and leachate even if soil erosion is controlled.

The NRCS reviewed manure production in a watershed in South Carolina. Agricultural activities in the project area are a major influence on the streams and ponds in the watershed and contribute to nutrient-related water quality problems in the headwaters of Lake Murray. The NRCS found that bacteria, nutrients, and sediment from soil erosion are the primary contaminants affecting the waters in this watershed. The NRCS has calculated that soil erosion, occurring on over 13,000 acres of cropland in the watershed, ranges from 9.6 to 41.5 tons per acre per year (USEPA, 1997).

6.3.1.3 Spills and Dry-Weather Discharges

Surface discharges can occur through spills or other discharges from lagoons. Catastrophic spills from large manure storage facilities can occur primarily through overflow following large storms or by intentional releases (Mulla et al., 1999). Other causes of spills include pump failures, malfunctions of manure irrigation guns, and breakage of pipes or retaining walls. Manure entering tile drains has a direct route to surface water. (Tile drains are a network of pipes buried in fields below the root zone of plants to remove subsurface drainage water from the root zone to a stream, drainage ditch, or evaporation pond.) In addition, spills can occur as a result of washouts from floodwaters when lagoons are sited on floodplains. Indications that discharges from siphoning lagoons occur deliberately as a means to reduce the volume in overfull lagoons have been recorded (Clean Water Action Alliance, 1998). An independent review of Indiana Department of Environmental Management records indicated that two common causes of waste releases in that state were intentional discharges and accidental discharges resulting from lack of operator knowledge (Hoosier Environmental Council, 1997).

Numerous such dry-weather discharges have been identified. For example, the ODNR documented chicken manure traveling through tile drains into a nearby stream in several instances occurring in 1994, 1995, and 1996 (ODNR, 1997). In 1995, a discharge of 25 million gallons of manure from swine farms in North Carolina was documented (Meadows, 1995; Warrick, 1995). Subsequent discharges of hundreds of thousands of gallons of manure were documented from swine operations in Iowa (1996), Illinois (1997), and Minnesota (1997) (Illinois Stewardship Alliance, 1997; Iowa Department of Natural Resources [IDNR], 1998; Macomb Journal, 1999; Clean Water Action Alliance, 1998). Between 1994 and 1996, half a dozen discharges from poultry operations in Ohio resulted when manure entered drain tiles (ODNR, 1997). In 1996, more than 40 animal waste spills occurred in Iowa, Minnesota, and Missouri alone (U.S. Senate, 1997). In 1998, a dairy feedlot in Minnesota discharged 125,000 gallons of manure (Clean Water Action Alliance, 1998). Acute discharges of this kind frequently result in dramatic fish kills. Fish kills were reported as a result of the North Carolina, Iowa, Minnesota, and Missouri discharges mentioned above.

6.3.1.4 Direct Contact between Confined Animals and Surface Water

Surface discharges can also occur as a result of direct contact between confined animals and the rivers, streams, or ponds located within their reach. Historically, people located their farms near waterways for both water access by animals and discharge of wastes. Certain animals, particularly cattle, wade into the waterbody, linger to drink, and often urinate and defecate in the water. This practice is now restricted for CAFOs; however, enforcement actions are the primary means for reducing direct access as described below (McFall, 2000).

In the more traditional farm production regions of the Midwest and Northeast, dairy barns and feedlots are often in close proximity to streams or other water sources. This close proximity to streams was formerly necessary to provide drinking water for the dairy cattle, to cool the animals in hot weather via direct access, and to cool milk prior to the widespread use of refrigeration. For CAFO-size facilities, this practice is now replaced with more efficient means of providing drinking water for the dairy herd. In addition, the use of freestall barns and modern milking centers minimizes the exposure of dairy cattle to the environment. For example, in New York, direct access of animals to surface water is more of a problem for smaller, traditional dairy farms that for older methods of housing animals. However, at these smaller facilities, direct access to surface water has relatively lower impact on surface water, compared to impacts associated with silage leachate and milkhouse waste (Dimura, 2000).

In the arid West, feedlots are typically located near waterbodies to allow for cheap and easy stock watering. Many existing lots were configured to allow the animals direct access to the water. The direct deposition of manure and urine contributes greatly to water quality problems. Environmental problems associated with allowing farm animals access to waters that are adjacent to the production area are well documented in the literature. USEPA Region X staff has documented dramatically elevated levels of *E. coli* in rivers downstream of CAFOs with direct access to surface water. Recent enforcement actions against direct access facilities have resulted in the assessment of tens of thousands of dollars in civil penalties (McFall, 2000).

6.3.2 Other Discharges to Surface Water

6.3.2.1 Leaching to Groundwater

Leaching of land-applied pollutants is a significant transport mechanism for water-soluble pollutants. In addition, leaking lagoons are a source of manure pollutants in groundwater. Although manure solids purportedly "self-seal" lagoons to prevent groundwater contamination, some studies have shown otherwise. A study for the Iowa legislature published in 1999 indicates that leaking is part of lagoon design standards and that all lagoons should be expected to leak (Iowa State University, 1999). A survey of swine and poultry lagoons in the Carolinas found that nearly two-thirds of the 36 lagoons sampled had leaked into the groundwater (Meadows, 1995). Even clay-lined lagoons have the potential to leak, since they can crack or break as they age and can be susceptible to burrowing worms. In a 3-year study of clay-lined swine lagoons on the Delmarva Peninsula, researchers found that leachate from lagoons located in well-drained loamy sand had a severe impact on groundwater quality (Ritter & Chirnside, 1990).

Pollutant transport to groundwater is also greater in areas with high soil permeability and shallow water tables. Percolating water can transport pollutants to groundwater, as well as to surface waters via interflow. Contaminated groundwater can deliver pollutants to surface waters through hydrologic connections. Nationally, about 40% of the average annual stream flow is from groundwater (USEPA, 1993b). In the Chesapeake Bay watershed, the U.S. Geological Survey (USGS) estimates that about half of the nitrogen loads from all sources to nontidal streams and rivers originate from groundwater (ASCE, 1998).

- *Important point:* Understanding the connection between groundwater and surface water is important when developing surface water protection strategies, because groundwater moves much more slowly than does surface water. For example, groundwater in the Chesapeake Bay region takes an average of 10 to 20 years to reach the bay; thus, it may take several decades to realize the full effect of pollutant additions or reductions (ASCE, 1998).

6.3.2.2   Discharge to Air and Subsequent Deposition

Atmospheric deposition can be a significant mechanism of transport to surface waters, as nitrogen emissions to air can return to terrestrial or aquatic environments in dry form or dissolved in precipitation (Agricultural Animal Waste Task Force, 1996). Discharges to air can occur as a result of volatilization of pollutants already present in manure and of pollutants generated as result of manure decomposition. Ammonia is very volatile and can have significant impacts on water quality through atmospheric deposition (Aneja et al., 1998). Ammonia losses from animal feeding operations can be considerable, rising from manure piles, storage lagoons, and land application fields. Other ways that manure pollutants can enter the air are from spray application methods for land applying manure and from particulates wind-borne in dust.

The degree of volatilization of manure pollutants is dependent on the manure management system. For example, losses are greater when manure remains on the land surface rather than being incorporated into the soil and are particularly high when farmers perform spray application. Environmental conditions, including soil acidity and moisture content, also affect the extent of volatilization—ammonia also readily volatizes from lagoons. Losses are reduced by the presence of growing plants (Follet, 1995).

Once airborne, pollutants can find their way into nearby streams, rivers, and lakes. The 1998 *National Water Quality Inventory* indicates that atmospheric deposition is the third largest cause of water quality impairment for estuaries and the fifth largest cause of water quality impairment for lakes, ponds, and reservoirs (USEPA, 2000a).

6.3.3   POLLUTANT-SPECIFIC TRANSPORT

6.3.3.1   Nitrogen Compounds

Livestock waste can contribute up to 37% of total nitrogen loads to surface water (Mulla et al., 1999). Nitrogen compounds and nitrates in manure can reach surface water through several pathways. As suggested by Follet (1995), agricultural nitrate contributions to surface water are primarily from groundwater connections and other subsurface flows. Although potentially less significant, overland runoff can also carry nitrate to surface waters. A recent Iowa investigation of chemical and microbial contamination near large-scale swine operations demonstrated the presence of nitrate and nitrite, not only in manure lagoons used to store swine waste before it is land applied but also in drainage ditches, agricultural drainage wells, tile line inlets and outlets, and an adjacent river (CDCP, 1998).

Studies of small geographical areas have revealed evidence of nitrate contamination in groundwater. As of 1988, 40% of wells in the Chino Basin, California, had nitrate levels in excess of the maximum containment level (MCL); the USEPA (1993b) identified dairy operations as the major source of contamination. This presents potentially widespread impacts, since water from the Chino Basin is used to recharge the primary source of drinking water for residents of heavily populated Orange County. On the Delmarva Peninsula, in Maryland, where poultry production is dominant, over 15% of wells were found to have nitrate levels exceeding the MCL. Wells located close to chicken houses contained the highest median nitrate concentrations (Ritter et al., 1989). Measured nitrate levels in groundwater beneath Delaware poultry houses are as high as 100 mg/L (Ritter et al., 1989).

- *Important point:* In 1994, the USGS analyzed nitrogen sources to 107 watersheds. Potential sources included manure (both point and nonpoint sources), fertilizers, point sources, and atmospheric deposition. The "manure" source estimates include waste from both confined and unconfined animals. As may be expected, the USGS found that proportions of nitrogen originating from various sources differ according to climate, hydrologic conditions, land use, population, and physical geography. Results of the analysis for selected watersheds for the 1987 base year show that, in some instances, manure nitrogen is a large portion of the total nitrogen added to the watershed. The study showed that, for the following nine watersheds, more than 25% of nitrogen originates from manure: Trinity River, Texas; White River, Arkansas; Apalachicola River, Florida; Altamaha River, Georgia; Potomac River, Washington, D.C.; Susquehanna River, Pennsylvania; Platte River, Nebraska; Snake river, Idaho; and San Joaquin River, California. Of these, California, Texas, Florida, Arkansas, and Idaho have large populations of confined animals.

Elevated nitrate levels can also exist in surface waters, although these impacts are typically less severe than groundwater impacts. In a historical assessment, the USGS (1997) found that nitrate levels in streams in agricultural areas were elevated compared to undeveloped areas. Nevertheless, the in-stream nitrate concentrations were generally less than those for groundwater in similar locations, and the drinking water MCL was rarely exceeded. The primary exception to this pattern was in the Midwest, where poorly drained soils restrict water percolation and artificial drainage provides a quick path for nutrient-rich runoff to reach streams (USGS, 1997).

- *Important point:* "Nitrate-N in streams originates from a variety of sources. Agricultural sources include nitrogen fertilizer, animal manure, mineralization of soil nitrogen and nitrogen-fixing crops. Other sources include human waste from sewage treatment plants, septic systems and landfills, and nitrogen produced as a waste or by-product or some industrial processes" (Rodecap, 2002, p. 1).

When farmers apply manure to land as fertilizer, risk of nitrate pollution generally increases rates of nitrogen application. Even when farmers land-apply manure at agronomic rates, nitrogen transport to surface water and groundwater can still occur for the following reasons: (1) nitrate is extremely mobile and may move below the plant root zone before being taken up; (2) ammonia may volatize and be redeposited in surface water; (3) the waste may be unevenly distributed, resulting in local "hot spots"; (4) obtaining a representative sample of the waste to determine the amount of mineralized (plant-available) nitrogen may be difficult; (5) uncertainties about the estimated rate of nitrogen mineralization in the applied waste are common; (6) transport is affected by the manure application method (for example, drip irrigation, spray irrigation, knifing, etc.); and (7) transport is affected by uncontrollable environmental factors such as rainfall and other local conditions (Follett, 1995).

6.3.3.2   Phosphorus Compounds

Phosphorus can reach surface waters via discharges directly into surface water, runoff of manure to surface water from feedlots, and runoff and erosion from land application sites. The organic phosphorus compounds in manure are generally water soluble and subject to leaching and dissolution in runoff (Gerritse & Zugec, 1977). Once in receiving waters, these compounds can undergo transformation and become available to aquatic plants. Overall, land-applied phosphorus is less mobile than nitrogen, since the mineralized (inorganic phosphate) form is easily adsorbed to soil particles. A report by the ARS noted that phosphorus bound to eroded sediment particles makes up 60% to 90% of phosphorus transported in surface runoff from cultivated land (USDA/ARS, 1999). For this reason, most agricultural phosphorus control measures have focused on soil erosion control to limit transport of particulate phosphorus. However, soils do not have infinite phosphate adsorption capacity, and dissolved inorganic phosphates can still enter waterways via runoff even if soil erosion is controlled (NRC, 1993).

- *Important point:* In the field of water quality chemistry, phosphorus is described by several terms. Some of these terms are chemistry-based (referring to chemically-based compounds), and others are methods-based (describing what is measured by a particular method).

Orthophosphate is a chemistry-based term that refers to the phosphate molecule all by itself. *Reactive phosphorus* is a corresponding method-based term that describes what is actually being measured when the test for orthophosphate is being performed. Because the lab procedure isn't quite perfect, mostly orthophosphate is obtained along with a small fraction of some other forms.

More complex inorganic phosphate compounds are referred to as *condensed phosphates* or *polyphosphates*. The method-based term for these forms is *acid hydrolysable* (Spellman & Drinan, 2000).

Livestock waste can contribute up to 65% of total phosphorus loads in surface waters (Mulla et al., 1999). Animal wastes typically have lower N:P ratios than crop N:P requirements, such that application of manure at a nitrogen-based agronomic rate can result in application of phosphorus at several times the agronomic rate (Sims, 1995). Summaries of soil test data in the United States confirm that many soils in areas dominated by animal-based agriculture have excessive levels of phosphorus (Sims, 1995). Research also indicates that there is a potential for phosphorus to leach into groundwater through sandy soils with already high phosphorus content (Citizens *Pfiesteria* Action Commission, 1997).

6.3.3.3   Ammonia

Ammonia can reach surface waters in a number of ways, including discharge directly to surface waters, leaching, dissolution in surface runoff, erosion, and atmospheric deposition. Leaching and runoff are generally not significant transport mechanisms for ammonia compounds in land-applied manure, because ammonium can be sorbed to soils (particularly those with high cation exchange capacities), incorporated (fixed) into clay or other soil complexes, or transformed into organic form by soil microbes (Follet, 1995). However, in these forms, erosion can transport nitrogen to surface waters. A recent Iowa investigation of chemical and microbial contamination near large-scale swine operations demonstrated the presence of ammonia not only in manure lagoons used to store swine waste before it is land applied but also to drainage ditches, agricultural drainage wells, tile line inlets and outlets, and an adjacent river (CDCP, 1998).

Ammonia losses from animal feeding operations to the air and subsequent deposition to surface waters can be considerable, arising from sources such as manure piles, storage lagoons, and land application fields. For example, in North Carolina, animal agriculture is responsible for over 90% of all ammonia emissions (Aneja et al., 1998). Ammonia composes more than 40% of the total estimated nitrogen emissions from all sources (Aneja et al., 1998). Data from Sampson County, North Carolina, indicates that ammonia levels in rain have increased with increases in the size of the pork industry. Levels more than doubled between 1985 and 1995 (Aneja et al., 1998). Based on USEPA estimates, swine operations in eastern North Carolina were responsible for emissions of 135 million pounds of nitrogen per year as of 1995. If deposited in a single basin, this would result in nitrogen loadings of almost 2.1 million pounds of nitrogen per year (Nowlin, 1997).

6.3.3.4 Pathogens

Sources of pathogen contamination from CAFOs include surface discharges and lagoon leachate. Surface runoff from land application fields can be a source of pathogen contamination, particularly if a rainfall event occurs soon after application or if the land is frozen or snow-covered (Mulla et al., 1999). Researchers have reported concentrations of bacteria in runoff water from fields treated with poultry litter at several orders of magnitude above contact standards (Giddens & Barnett, 1980; Coyne & Blevins, 1995).

A recent Iowa investigation of chemical and microbial contamination near large-scale swine operations demonstrated the presence of pathogens, not only in manure lagoons used to store swine waste before it is land applied but also in drainage ditches, agricultural drainage wells, tile line inlets and outlets, and an adjacent river (CDCP, 1998). Also, studies have reported that lands receiving fresh manure application can be the source of up to 80% of the fecal bacteria in surface waters (Mulla et al., 1999). Similarly, both *Cryptosporidium parvum* and *Giardia* species have also been found in over 80% of 66 surface water sites tested (LeChevallier et al., 1991). Since these protozoa do not multiply outside of the host, livestock animals are one potential source of this contamination. The bacterium *Erysipelothrix* spp., primarily a swine pathogen, has been isolated from many fish and avian species (USFWS, 2000).

- *Important point:* Waterborne disease outbreaks caused by microbial agents can be divided into three categories: "(1) Those associated with intestinal infection and feces from multiple species including humans such as Cryptosporidium parvum, Giardia species (sp), Escherichia coli 0157:H7, Campylobacter jejuni, and Salmonella sp.; (2) those associated with human intestinal infection and feces such as Shigella sp., Salmonella typhi and human intestinal viruses; and (3) those which live in the environment such as Pseudomonas and Legionella that are associated with a variety of human illnesses, including skin infections (dermatitis) and Legionaire's disease. Intestinal infections are the most common type of waterborne infection and affect the most number of people" (Stehman, 2000, p. 94).

High levels of indicator bacteria in surface water near CAFOs have been documented. For instance, Zirbser (1998) documented a report of fecal coliform counts of 3000/100 ml and fecal streptococci counts over 30,000/100 ml downstream from a swine waste lagoon site. (No sampling was performed upstream of the lagoon site.) Fecal coliform pollution from treated and partially treated sewage and stormwater runoff is often cited in beach closures and shell fish restrictions.

The natural filtering and adsorption action of soils typically causes a majority of the microorganisms in land-applied manure to be stranded at the soil surface (Crane et al., 1980). This phenomenon helps protect underlying groundwater but increases the likelihood of runoff losses to surface waters. Pathogens discharged to the water column can subsequently adsorb to sediments, presenting long-term health hazards. Benthic sediments harbor significantly higher concentrations of bacteria than the overlying water column (Mulla et al., 1999).

While surface waters are typically more prone to pathogen contamination than groundwaters, subsurface flows may also be a mechanism for pathogen transport depending on weather, site, and operating conditions. Groundwaters in areas of sandy soils, limestone formations, or sinkholes are particularly vulnerable. For example, the bacteria *Clostridium perfringens* was detected in the groundwater below plots of land treated with swine manure, and fecal coliform has been detected in groundwater beneath soils amended with poultry manure (Mulla et al., 1999). In 1998, *Campylobacter jejuni* was isolated from groundwater, and some of the strains were the same type as those from a dairy farm in the same hydrologic area (Stanley et al., 1998).

Other accounts of high levels of microorganisms in groundwater near feedlots are available. In cow pasture areas of Door County, Wisconsin, where a thin topsoil layer is underlain by fractured limestone bedrock, groundwater wells have commonly been shut down due to high bacteria levels (Behm, 1989). For example, a well at one rural household produced brown, manure-laden water (Behm, 1989). Private wells are more prone to contamination than public wells, since they tend to be shallower and therefore more susceptible to contaminants leaching from the surface. In a survey of drinking water standard violations in six states over a 4-year period, the U.S. General Accounting Office (USGAO, 1997) found that bacterial standard violations occurred in 3% to 6% of community water systems each year. By contrast, USGAO reported that some bacterial contamination occurred in 15% to 42% of private wells, according to statistically representative assessments performed by others.

- *Important point:* The USGAO reviewed compliance data from 1993 through 1996 for more than 17,000 community water systems in California, Illinois, Nebraska, New Hampshire, North Carolina, and Wisconsin.

Several factors affect the likelihood of disease transmission by pathogens in animal manure, including pathogen survivability in the environment. For example, *Salmonella* can survive in the environment for 9 months or more, providing for increased dissemination potential (USFWS, 2000); and *Campylobacter* can remain dormant, making water an important vehicle for campylobacteriosis (Altekruse, 1998). Recent studies are better characterizing the survivability and transport of pathogens in manure once it has been land applied. Several researchers (Dazzo et al., 1973; Himathongkham et al., 1999; Kudva et al., 1998; Maule, 1999; Van Donsel et al., 1967) have found that soil type, manure application rate, temperature, moisture level, aeration, soil pH, and the amount of time that manure is held before it is applied to pastureland are dominating factors in bacteria survival.

Experiments on land-applied poultry manure (Crane et al., 1980) indicated that the population of fecal organisms decreases rapidly as manure is heated, dried, and exposed to sunlight on the soil surface. However, regrowth of fecal organisms also occurred in these experiments. More recent research indicated that pathogens can survive in manure for 30 days or more (Himathongkham et al., 1999; Kudva et al., 1998; Maule, 1999). Kudva and colleagues (1998) found that *Escherichia coli* survived for 47 days in aerated cattle manure piles exposed to outdoor weather; drying the manure reduced the number of viable pathogens. Stehman (2000) also noted that *Escherichia coli* 0157:H7, *Cryptosporidium parvum*, and *Giardia* can survive and remain infectious in surface waters for a month or more.

The continued application of waste on a particular area could lead to extended pathogen survival and buildup (Dazzo et al., 1973). Additionally, repeated applications or high application rates increase the likelihood of runoff to surface water and transport to groundwater.

6.3.3.5 Organic Matter

Discharge and runoff of manure from feedlots cause large loadings of organic matter to surface waters. Numerous incidents of discharges from CAFOs directly to surface waters have occured nationwide. Discharges can also originate from land application sites when farmers overapply or misapply manure. Even if farmers apply manure with methods to ensure no concentrated discharge occurs, organic matter will be present in runoff from land application sites. As shown by Daniel et al. (1995), runoff or organic matter increases as application rate increases. For example, Daniel et al. (1995) reported that when the swine manure slurry application rate increased from 193 lb N/acre to 387 lb N/acre, COD levels in runoff (generated from a rainfall intensity of 2 in/hr) increased from 282 mg/L to 504 mg/L. By comparison, runoff from a control plot yielded 78 mg/L COD.

- *Important point:* In a series of experiments, Edwards and Daniel (1992b, 1993a, and 1993b, as reported by Daniel et al., 1995) measured runoff from fescue grass plots treated with poultry litter, poultry manure slurry, and swine manure slurry to determine how runoff quality is impacted by application rate and rain intensity. They found that, for all wastes, the application rate had a significant effect on the runoff concentration and mass loss of COD (as well as other constituents).

- *Important point:* The USEPA assumes that 175 lb N/acre is a typical requirement for a fescue crop in Arkansas, based on information from USDA extension agents (Tetra Tech, 2000b).

6.3.3.6 Salts and Trace Elements

Salts can reach surface waters via discharges from feedlots and runoff from land application sites. Salts can also leach into groundwater and subsequently reach surface water. Trace elements can also be transported by these mechanisms. A recent Iowa investigation showed that trace elements were present not only in manure lagoons used to store swine waste before land application but also in drainage ditches, agricultural drainage wells, tile line inlets and outlets, and an adjacent river (CDCP, 1998). Selenium concentrations have been detected in swine manure lagoons at up to 6 µg/L, copper has been detected in liquid swine manure prior to land application at 15 mg/L, and zinc has been detected in soils that receive applications of cattle manure at levels up to 9.5 mg/kg in the upper 60 cm of soil (USFWS, 2000).

6.3.3.7 Antibiotics

Little information is available regarding the fate and transport properties of antibiotics or the potential releases from animal waste compared to other sources, such as municipal and industrial wastewaters, septic tank leachate, runoff from land-applied sewage biosolids, crop runoff, and urban runoff. However, it is known that the primary mechanisms of eliminating antibiotics from livestock are through urine and bile. Also, essentially all of an antibiotic administered to an animal is eventually excreted, whether unchanged or in metabolite form (Tetra Tech, 2000a).

Although the presence of excreted antibiotics themselves may be of concern, the development of antibiotic-resistant pathogens due to exposure to environmental levels of antibiotics is generally of greater concern. The risk for development of antibiotic-resistant pathogens from this exposure is unknown.

6.3.3.8 Hormones

Hormones can reach surface waters through the same route as other manure pollutants, including runoff and erosion as well as direct contact of animals with the water. Estrogen is more likely to be lost by runoff than leaching, whereas testosterone is lost mainly through leaching (Shore et al., 1995).

Several sites have documented the presence of hormones in runoff and surface waters. Runoff from a field receiving poultry litter was found to contain estrogen. An irrigation pond and three streams in the Conestoga River watershed near the Chesapeake Bay had both estrogen and testosterone. Each of these sites was affected by fields receiving poultry litter (Shore et al., 1995). Runoff from fields with land-applied manure has been reported to contain estrogens, estradiol, progesterone, and testosterone, as well as their synthetic counterparts. Estrogens have also been found in runoff from heavily grazed land (Addis et al., 1999).

6.3.3.9   Other Pollutants

Very little research has been performed on losses of pesticides in runoff from manured lands. A 1999 literature review by the University of Minnesota discussed a 1994 study showing that losses of cyromazine (used to control flies in poultry litter) in runoff increased with the rate of poultry manure application and the intensity of rainfall. The 1999 literature review also includes a 1995 study documenting that about 1% of all pesticides enters surface water. However, the magnitude of the impacts of these losses on surface water is unknown (Mulla et al., 1999). In general, little information is available regarding the fate and transport of pesticides or their bioavailability in waste-amended soils. Furthermore, there is little information comparing potential releases of these compounds from animal waste to other sources, such as municipal and industrial wastewaters, septic tank leachate, runoff from land-applied biosolids, crop runoff, and urban runoff.

6.4   POTENTIAL HAZARDS FROM CAFO POLLUTANTS

As described in the previous section, AFOs are associated with a variety of pollutants, including nutrients (specifically nitrogen and phosphorus), ammonia, pathogens, organic matter, salts, trace elements, solids, antibiotics, hormones, gas and particulate emissions, and pesticides. These CAFO pollutants can produce multimedia impacts, including:

- *Surface water.* Impacts have been associated with surface discharges of waste, as well as leaching to groundwater and subsurface flow to surface water. Generally, states with high concentrations of feedlots experience 20 to 30 serious water quality pollution problems per year involving manure lagoon spills and feedlot runoff (Mulla et al., 1999). The waste's oxygen demand and ammonia content can result in fish kills and reduced biodiversity. Solids can increase turbidity and impact benthic organisms. Nutrients contribute to eutrophication and associated algae blooms. Algal decay and nighttime respiration can depress dissolved oxygen levels, potentially leading to fish kills and reduced biodiversity. Eutrophication is also a factor in blooms of toxic algae and other toxic microorganisms, such as *Pfiesteria piscicida*. Human and animal health imparts are primarily associated with drinking contaminated water (pathogens and nitrates), coming into contact with contaminated water (pathogens such as toxic algae and *Pfiesteria*), and consuming contaminated shellfish (pathogens such as toxic algae). Trace elements (e.g., arsenic, copper, selenium, and zinc) may also present human health and ecological risks. Salts contribute to salinization and disruption of ecosystem balance as well as degradation of drinking water supplies. Antibiotics, pesticides, and hormones may have low-level, long-term ecosystem effects.
- *Groundwater.* Impacts have been associated with pollutants leaching to groundwater. Human and animal health impacts are associated with pathogens and nitrates in drinking water. Leaching salts can increase health risks to salt-sensitive individuals and can make water unpalatable. Trace elements, antibiotics, pesticides, and hormones may also present human health and ecological risks through groundwater pathways.
- *Air.* Air impacts include human health effects from ammonia, hydrogen sulfide, other odor-causing compounds, particulates, and the contribution to global climate change due to methane emissions. In addition, volatilized ammonia can be redeposited on the earth and contribute to eutrophication.

- *Soil.* Trace elements and salts in animal manure can accumulate in soil and become toxic to plants. Salts also deteriorate soil quality, leading to reduced permeability and overall poor physical condition. Crops may provide a human and animal exposure pathway for trace elements and pathogens.

This section describes in greater detail the known or potential adverse human health and ecological effects of CAFO pollutants.

6.4.1 PRIMARY NUTRIENTS

In this section we review the hazards posed by primary nutrients in animal manure. We focus on nitrogen and phosphorus, which have received the greatest attention in the scientific literature. Actual or anticipated levels of potassium in groundwater and surface water are unlikely to pose hazards to aquatic life or human health (Wetzel, 1983). Potassium does contribute to salinity, however, and applications of high salinity manure are likely to decrease the fertility of the soil.

6.4.1.1 Ecology

6.4.1.1.1 *Eutrophication*
Eutrophication is the process in which phosphorus and nitrogen overenrich a waterbody and disrupt the balance of life in that waterbody. Perhaps the most documented impact of nutrient pollution is the increase in surface water eutrophication (nutrient enrichment) and its effects on aquatic ecosystems (Vallentyne, 1974). Although nutrients are essential for the growth of phytoplankton (free-floating algae), periphyton (attached algae), and aquatic plants, which form the base of the aquatic food web, the overabundance of nutrients can lead to harmful algal blooms and other adverse effects, such as:

- Increased biomass of phytoplankton
- Shifts in phytoplankton to bloom-forming species that may be toxic or inedible
- Changes in macrophyte species composition and biomass
- Death of coral reefs and loss of coral reef communities
- Decreases in water transparency
- Taste, odor, and water treatment problems
- Oxygen depletion
- Increased incidence of fish kills
- Loss of desirable fish species
- Reductions in harvestable fish and shellfish
- Decreases in aesthetic value of the waterbody (Carpenter et al., 1998)

The type of waterbody impacted may dictate which nutrient (nitrogen or phosphorus) will have the most impact. In estuaries and coastal marine waters, nitrogen is typically the limiting nutrient (i.e., in these waters, phosphorus levels are sufficiently high compared to nitrogen such that small changes in nitrogen concentrations have a greater effect on plant growth). In fresh waters, phosphorus is typically the limiting nutrient (Wendt & Corey, 1980; Robinson & Sharpley, 1995). Exceptions to this generalization can occur, however, especially in waterbodies with heavy pollutant loads. For example, estuarine systems may become phosphorus-limited when nitrogen concentrations are high. In such cases, excess phosphorus produces algal blooms (North Carolina's Nicholas School of the Environment's Agricultural Animal Waste Task Force, 1994). Thus, both nitrogen and phosphorus loads can contribute to eutrophication in either water type.

6.4.1.1.2 *Algae and Other Toxic Microorganisms*
Eutrophication causes the enhanced growth and subsequent decay of algae, which can lower dissolved oxygen content of a waterbody to levels insufficient to support fish and invertebrates. In some

cases, this situation can produce large areas devoid of life because of a lack of sufficient dissolved oxygen. One extreme example is the "Dead Zone," an area of hypoxic water larger than 10,000 km^2 that spreads off the Louisiana coast in the Gulf of Mexico each summer. The Dead Zone is believed to be caused by excess chemical fertilizer; however, nutrients from animal waste have also contributed to the problem. This condition has been attributed to excess nutrients delivered primarily by the Mississippi and Atchafalaya river systems (Atwood et al., 1994). The problem in the Gulf demonstrates that pollutant discharges can have far-reaching downstream impacts. In fact, the nutrient loadings to the Gulf originate from sources over a large land area covering approximately 41% of the contiguous United States (Goolsby et al., 1999).

Eutrophication can also affect phytoplankton and zooplankton population diversity, abundance, and biomass and can increase the mortality rates of aquatic species. For example, floating algal mats can prevent sunlight from reaching submerged aquatic vegetation, which serves as habitat for fish spawning, juvenile fish, and fish prey (e.g., aquatic insects). The resulting decrease in submerged aquatic vegetation adversely affects both fish and shellfish populations (USEPA, 2000a).

Another effect of eutrophication is increased incidence of harmful algal blooms, which release toxins as they die and can severely impact wildlife and humans. In marine ecosystems, blooms known as *red* or *brown tides* have caused significant mortality in marine mammals (Carpenter et al., 1998). In fresh water, cyanobacterial toxins have caused many incidents of poisoning of wild and domestic animals that have consumed impacted waters (Health Canada Environmental Health Programs, 1998). Published reports of wildlife poisoning from these blooms include amphibians, fish, snakes, waterfowl, raptors, and deer (USFWS, 2000).

Eutrophication is also associated with blooms of other toxic organisms, such as the estuarine dinoflagellate *Pfiesteria piscicida*. *Pfiesteria* (pronounced "Fee-steer-ee-ah") has been implicated as the primary causative agent of many major fish kills and fish disease events in North Carolina estuaries and coastal area (North Carolina State University [NCSU], 2000) as well as in Maryland and Virginia tributaries to the Chesapeake Bay (USEPA, 1997b). *Pfiesteria* (nicknamed "the cell from hell" because of its aggressive, flesh-eating nature) often lives as a nontoxic predatory animal, becoming toxic in response to human influences, including excessive nutrient enrichment (NCSU, 2000). While nutrient-enriched conditions are not required for toxic outbreaks to occur, excessive nutrient loadings are a concern because they help create an environment rich in microbial prey and organic matter that *Pfiesteria* uses as a food supply. By increasing the concentration of *Pfiesteria*, nutrient loads increase the likelihood of a toxic outbreak when adequate numbers of fish are present (Citizens *Pfiesteria* Action Commission, 1997). Researchers have documented stimulation of *Pfiesteria* growth by human sewage and swine effluent spills and have shown that the organism's growth can be highly stimulated by both inorganic and organic nitrogen and phosphorus enrichments (NCSU, 2000).

Increased algal growth can also raise the pH of waterbodies, as algae consume dissolved carbon dioxide to support photosynthesis. Many biological processes, including reproduction, cannot occur in water that is very acidic or alkaline (USEPA, 2000a).

6.4.1.1.3 Nitrites

Nitrites can also pose a risk to aquatic life, if sediments are enriched with nutrients, the concentrations of nitrites in the overlying water may be raised enough to cause nitrite poisoning or "brown blood disease" in fish (USDA, 1992). In addition, excess nitrogen can contribute to water quality decline by increasing the acidity of surface waters.

- *Important point:* Brown blood disease is named for the color of the blood of dead or dying fish, indicating that the hemoglobin has been converted to methemoglobin. According to Durborow and Crosby (2003), brown blood disease occurs in fish when water contains high nitrite concentrations. Nitrite enters a fish culture system after feed is digested by fish and the excess nitrogen is converted into ammonia, which is then excreted as waste into the water. Total ammonia nitrogen is then converted to nitrite that, under normal conditions, is quickly converted to nontoxic nitrate by naturally occurring bacteria. Uneaten (wasted) feed and other organic material also break down into ammonia, nitrite, and nitrate in a similar manner.

6.4.1.2   Human Health

6.4.1.2.1   Nitrates/Nitrites

The main hazard to human health from primary nutrients is elevated nitrate levels in drinking water. In particular, infants are at risk from nitrate poisoning (also referred to as *methemoglobinemia* or "blue baby syndrome"), which can be fatal. This poisoning results in oxygen starvation and is due to nitrite (a metabolite of nitrate), which is formed in the environment, foods, and the human digestive system. Unlike adults and older children, infants younger than age 6 months experience elevated nitrite production because their digestive systems have a higher concentration of nitrate-reducing bacteria. Nitrite oxidizes iron in the hemoglobin of red blood cells to form methemoglobin, which cannot carry sufficient oxygen to the body's cells and tissues. Although methemoglobin is continually produced in humans, an enzyme in the human body reduces methemoglobin back to hemoglobin. In most individuals, this conversion occurs rapidly. In infants, however, methemoglobin is not converted to hemoglobin as readily (Nebraska Cooperative Extension, 1995).

Because infants under six months have a higher concentration of digestive bacteria that reduce nitrates, and a lower concentration of methemoglobin-reducing enzyme, they are at higher risk for methemoglobinemia (Nebraska Cooperative Extension, 1995). To protect infant health, the USEPA set drinking water MCLs of 10 mg/L for nitrate-nitrogen and 1 mg/L for nitrite-nitrogen. MCLs are the maximum permissible levels of pollutants allowed in water delivered to public drinking water systems. Once a water source is contaminated, the costs of protecting consumers from nitrate exposure can be significant. Conventional drinking water treatment processes do not remove nitrate. Its removal requires additional, relatively expensive treatment units.

Although reported cases of methemoglobinemia are rare, the incidence of actual cases may be greater than the number reported. Studies in South Dakota and Nebraska have indicated that most cases of methemoglobinemia are not reported (Michel et al., 1996; Meyer, 1994). For example, in South Dakota between 1950 and 1980, only two cases were reported, but at least 80 were estimated to have occurred (Meyer, 1994). At least two reasons are responsible for this underreporting. First, methemoglobinemia can be difficult to detect in infants because its symptoms are similar to other conditions (Michel et al., 1996). In addition, doctors are not always required to report it (Michel et al., 1996).

In addition to blue baby syndrome, low blood oxygen due to methemoglobinemia has also been linked to birth defects, miscarriages, and general poor health in humans and animals. These effects are exacerbated by concurrent exposure to many species of bacteria in water (Integrated Risk Information System [IRIS], 2000). Studies in Australia found an increased risk of congenital malformations with consumption of high-nitrate groundwater (Bruning-Fann & Kaneene, 1993). Multigeneration animal studies have found decreases in birth weight, post-natal growth, and organ weights among mammals prenatally exposed to nitrite (IRIS, 2000). Nitrate- and nitrite-containing compounds may also cause hypotension or circulatory collapse (Bruning-Fann & Kaneene, 1993).

High nitrate levels in drinking water have also been implicated in higher rates of stomach and esophageal cancer, although a 1995 NRC report concludes that exposure to nitrate and nitrite concentrations in drinking water are unlikely to contribute to human cancer risks (NRC, 1995). However, nitrate metabolites such as N-nitroso compounds (especially nitrosamines) have been linked to severe human health effects, such as gastric cancer (Bruning-Fann & Kaneene, 1993). The formation of N-nitroso compounds occurs in the presence of catalytic bacteria (for example, those found in the stomach) or thiocyanate.

Generally, people drawing water from domestic wells are at greater risk for nitrate poisoning than those drawing from public wells (Nolan & Ruddy, 1996) because domestic wells are typically shallower and not subject to wellhead protection or monitoring requirements. Reported cases of methemoglobinemia are most often associated with wells that were privately dug and that may have been badly positioned in relation to the disposal of human and animal excreta (Addiscott et al., 1991). Because of water quality monitoring and treatment requirements, people served by public water systems are better protected even if the water becomes contaminated.

6.4.1.2.2   Phosphorus

Animal manure also contributes to increased phosphorus concentrations in water supplies. Previous evaluations of phosphorus have not identified significant adverse human health effects, but phosphate levels greater than 1 mg/L may interfere with coagulation in drinking water treatment plants and thereby increase treatment costs (North Carolina's Nicholas School of the Environment's Agricultural Animal Waste Task Force, 1994).

6.4.1.2.3   Eutrophication/Algal Blooms

To the extent that nitrogen and phosphorus contribute to algal blooms in surface water through accelerated eutrophication, these nutrients can reduce the aesthetic and recreational value of surface water resources. Algae can affect drinking water by clogging treatment plant intakes and by producing objectionable tastes and odors. Algae can also increase production of harmful chlorinated by-products (such as, trihalomethanes) by reacting with chlorine used to disinfect drinking water. These impacts can result in increased costs of drinking water treatment, reduced drinking water quality, and increased health risks.

Eutrophication can also affect human health by enhancing growth of harmful algal blooms that release toxins as they die. In marine ecosystems, harmful algal blooms such as red tides can result in human health impacts via shellfish poisoning and recreation contact (Thomann & Mueller, 1987). In fresh water, blooms of cyanobacteria (blue-green algae) may pose a serious health hazard to humans via water consumption. When cyanobacterial blooms die or are ingested, they release water-soluble compounds that are toxic to the nervous system and liver (Carpenter et al., 1998).

In addition, eutrophication is associated with blooms of a variety of other organisms toxic to humans, such as the estuarine dinoflagellate *Pfiesteria piscicida*. Although *Pfiesteria* is primarily associated with fish kills and fish disease events, the organism has also been linked with human health impacts through dermal or inhalation exposure. Researchers working with dilute toxic cultures of *Pfiesteria* exhibited symptoms such as skin sores, severe headaches, blurred vision, nausea and vomiting, sustained difficulty breathing, kidney and liver dysfunction, acute short-term memory loss, and severe cognitive impairment (NSCE, 2000). People with heavy environmental exposure have exhibited symptoms as well. In a 1998 study, such environmental exposure was definitively linked with cognitive impairment and less consistently linked with physical symptoms (Morris et al., 1998).

6.4.2   Ammonia

6.4.2.1   Ecology

Ammonia exerts a direct BOD on the receiving water. As ammonia is oxidized, dissolved oxygen is consumed. Moderate depressions of dissolved oxygen are associated with reduced species diversity, while more severe depressions can produce fish kills. In fact, ammonia is a leading cause of fish kills (USDA, 1992). Ammonia-induced fish kills are a potential consequence of the discharge of animal wastes directly to surface waters. For example, in a May 1997 incident in Wabasha County, Minnesota, ammonia in a dairy cattle manure discharge killed 16,5000 minnows and white suckers (Clean Water Action Alliance, 1998). Additionally, ammonia loadings can contribute to accelerated eutrophication of surface waters, which can significantly impact aquatic ecosystems in a number of ways, as noted.

6.4.2.2 Human Health

Ammonia is a nutrient form of nitrogen that can have several impacts. First, volatized ammonia is of concern because of direct localized impacts on air quality. Ammonia produces an objectionable odor and can cause nasal and respiratory irritation. In addition, ammonia contributes to eutrophication of surface waters. This phenomenon, as stated previously, is primarily a hazard to aquatic life but is also associated with human health impacts. As previously mentioned, eutrophication reduces the aesthetic and recreational value of water bodies. Additionally, the associated algae blooms can affect drinking water by clogging treatment plant intakes, producing objectionable tastes and odors and increasing production of harmful chlorinated by products. These impacts can result in increased drinking water treatment costs, reduced drinking water quality, and increased health risks. Eutrophication can also impact human health by enhancing the growth of toxic algae and other toxic organisms.

6.4.3 Pathogens

6.4.3.1 Ecology

Animal wastes carry pathogens, bacteria, and viruses, many of which have the potential to be harmful to wildlife (USDA, 1992; Jackson et al., 1987). Some bacteria in livestock waste cause avian botulism and avian cholera, which have killed thousands of migratory waterfowl in the past (USEPA, 1993b). Avian botulism is a form of food poisoning caused by ingestion of a neurotoxin produced by the bacterium Clostridium botulinum type C. and Salmonella spp., both of which naturally occur in the intestinal tracts of warm-blooded animals (USFWS, 2000).

Pathogens in surface water can adhere to the skin of fish or be taken up internally when present at high enough concentrations. In a controlled experiment, Fattal et al. (1992) detected significant bacterial concentrations in fish exposed to *Escherichia coli* and other microorganisms for up to 48 hours. The data suggest that harmful pathogens could be taken up by fish-eating carnivores feeding in contaminated surface waters.

Shellfish are filter feeders that pass large volumes of water over their gills. As a result, they can concentrate a broad range of microorganisms in their tissues (Chai et al., 1994). This provides a pathway for pathogen transmission to higher trophic organisms. However, little information is available to assess the health effects of contaminated shellfish on wildlife receptors.

6.4.3.2 Human Health

Pathogens may be transmitted to humans through contaminated surface water or groundwater used for drinking or by direct contact with contaminated surface water through recreational uses. By the year 2010, about 20% of the human population (especially infants, the elderly, and those with compromised immune systems) will be classified as particularly vulnerable to the health effects of pathogens (Mulla et al., 1999). Over 150 pathogens in livestock manure are associated with risks to humans (Council for Agricultural Science and Technology [CAST], 1992). Table 6.4 presents a list of several of these pathogens and their associated diseases, including salmonellosis, cryptosporidiosis, and giardiasis. Other pathogens associated with livestock waste include those that cause cholera, typhoid fever, and polio (USEPA, 1993b). Many of these pathogens are transmitted to humans via the fecal-oral route. In the water environment, humans may be exposed to pathogens through consumption of contaminated drinking water (although the USEPA assumes adequate drinking water treatment of public supplies) or by incidental ingestion during activities in contaminated waters.

Although a wide range of organisms may cause disease in humans, relatively few microbial agents are responsible for the majority of human disease outbreaks from water-based exposure routes. This point is illustrated by Table 6.5, which presents reports of waterborne disease outbreaks

TABLE 6.4

Selected Diseases and Parasites Transmittable to Humans from Animal Manure*

Disease	Responsible Organism	Symptoms
Bacteria		
Anthrax	*Bacillus anthracis*	Skin sores, fever, chills, lethargy, headaches, nausea, vomiting, shortness of breath, cough, nose and throat congestion, pneumonia, joint stiffness, joint pain
Brucellosis	*Brucella abortus, Brucella melitensis, Brucella suis*	Weakness, lethargy, fever, chills, sweating, headache
Colibaciliosis	*Escherichia coli* (some serotypes)	Diarrhea, abdominal gas
Coliform mastitismetritis	*Escherichia coli* (some serotypes)	Diarrhea, abdominal gas
Erysipelas	*Erysipelothrix rhusiopathiae*	Skin inflammation, rash, facial swelling, fever, chills, sweating, joint stiffness, muscle aches, headache, nausea, vomiting
Leptospirosis	*Leptospira pomona*	Abdominal pain, muscle pain, vomiting, fever
Listeriosis	*Listeria monocytogenes*	Fever, fatigue, nausea, vomiting, diarrhea
Salmonellosis	*Salmonella* species	Abdominal pain, diarrhea, nausea, chills, fever, headaches
Tetanus	*Clostridium tetani*	Violent muscle spasms, "lockjaw" spasms of jaw muscles, difficulty breathing
Tuberculosis	*Mycobacterium tuberculosis, Mycobacterium avium*	Cough; fatigue; fever; pain in chest, back, or kidneys
Rickettsia		
Q fever	*Coxiella burnetii*	Fever, headache, muscle pains, joint pain, dry cough, chest pain, abdominal pain, jaundice
Viruses		
Foot and mouth	virus	Rash, sore throat, fever
Swine cholera	virus	
New castle	virus	
Psittacosis	virus	Pneumonia
Fungi		
Coccidioidomycosis	*Coccidioides immitus*	Cough, chest pain, fever, chills, sweating, headache, muscle stiffness, joint stiffness, rash, wheezing
Histoplasmosis	*Histoplasma capsulatum*	Fever, chills, muscle ache, muscle stiffness, cough, rash, joint pain, joint stiffness
Ringworm	Various *Microsporum* and *Trichophyton*	Itching, rash
Protozoa		
Balantidiasis	*Balatidium coli*	Diarrhea, abdominal gas
Coccidiosis	*Eimeria* species	
Cryptosporidiosis	*Cryptosporidium parvum*	Watery diarrhea, dehydration, weakness, abdominal cramping
Giardiasis	*Giardia lamblia*	Diarrhea, abdominal pain, abdominal gas, nausea, vomiting, headache, fever
Toxoplamosis	*Toxoplasma* species	Headaches, lethargy, seizures, reduce cognitive function
Parasites/Metazoa		
Ascariasis	*Ascaris lumbricoides*	Worms in stool or vomit, fever, cough, abdominal pain, bloody sputum, wheezing, skin rash, shortness of breath

TABLE 6.4 (continued)
Selected Diseases and Parasites Transmittable to Humans from Animal Manure*

Disease	Responsible Organism	Symptoms
Sarcocystiasis	*Sarcosystis* species	Fever, diarrhea, abdominal pain

Sources: Diseases and organisms were compiled from USDA/NRCS (1996) and USEPA (1998). Symptom descriptions were obtained from various medical and public health service Internet sites.

* Pathogens in animal manure are a potential source of disease in humans and other animals. This list represents a sampling of disease that may be transmittable to humans.

TABLE 6.5
Etiology of Waterborne Disease Outbreaks Causing Gastroenteritis, 1989–1996

Etiologic agent	Total number of outbreaks	Outbreaks associated with drinking water		Outbreaks associated with recreational water	
		Surface	Ground	Natural	Pool/Park
Giardia spp.	27	12	6	4	5
Cryptosporidium parvum	21	4	4	2	11
Escherichia coli 0157:H7	11	—	3	7	1
Campylobacter jejuni	3	3	—	—	—
Salmonella typhimurium	1	—	1	—	—
Salmonella java	1	—	—	—	1
Leptospira grippotyphosa	1	—	—	1	—
Shigella sonnei	17	—	7	10	—
Shigella flexneri	2	—	1	1	—
Hepatitis A	3	—	—	—	3
Norwalk virus	1	—	1	—	—
Norwalk-like virus	1	—	—	—	1
Small round structured virus	1	1	—	—	—
Unidentified etiology	60	8	44	7	1
Cyanobacteria-like bodies	1	1	—	—	—

Source: Adapted from Stehman (2000).

and their causes (if known) in the United States for the period 1989–1996. Intestinal infections are the most common type of waterborne infection and affect the most people.

As presented in Table 6.5, most reported outbreaks were associated with protozoa and bacteria. As noted in Table 6.4, *Cryptosporidium parvum* can produce gastrointestinal illness, with symptoms such as severe diarrhea. Relatively low doses of both *Cryptosporidium parvum Giardia* species are needed to cause infection (Stehman, 2000). Although healthy people typically recover relatively quickly (within 2 to 10 days) from this type of illness, these diseases can be fatal in people with weakened immune systems. These individuals typically include children, the elderly, people with human immunodeficiencyvirus (HIV) infection, chemotherapy patients, and those taking medications that suppress the immune system.

Table 6.5 shows that infections caused by *Giardia* species and *Cryptosporidium parvum* (considered the two most important waterborne protozoa) were the leading causes of infectious waterborne disease outbreaks in which an agent was identified, both for total cases and for number of outbreaks (Mulla et al., 1999; Stehman, 2000). In 1993 in Milwaukee, Wisconsin, *Cryptosporidium parvum* (the bug that made Milwaukee famous) contamination of a public water supply caused more than 100 deaths and an estimated 403,000 illnesses (Smith, 1994; Casman, 1996). The outbreak cost an estimated $37 million in lost wages and productivity (Smith, 1994). The source of the oocysts was not identified, but speculated sources include runoff from cow manure application sites, wastewater from a slaughterhouse and meat packing plant, and municipal wastewater treatment plant effluent (Casman, 1996). Four documented cases of cryptosporidiosis occurring since 1984 have been linked to non-point-source agricultural pollution (Mulla et al., 1999). Two outbreaks of *Cryptosporidium parvum* were also traced to contamination of drinking water by cow manure in England (Stehman, 2000).

The mandated treatment of public water supplies helps reduce the risk of infection via drinking water, but the first step in providing safe drinking water is source water protection, especially because *Cryptosporidium parvum* is resistant to conventional treatment.

Escherichia coli is another important cause of bacterial waterborne infection in untreated and recreational water (Stehman, 2000). Infection can be life threatening, especially in the young and in the elderly. It can cause bloody diarrhea and, if not promptly treated, can result in kidney failure and death (Shelton, 2000). In particular, *Escherichia coli* 0157:H7 is emerging as the second most important cause of bacterial waterborne disease after *Shigella* species, which is associated with human feces. *Escherichia coli* 0157:H7 was unknown until 1982, when it was associated with a multistate outbreak of hemorrhagic colitis (Shelton, 2000). In 1999, an *Escherichia coli* outbreak occurred at the Washington Country Fair in New York State. This outbreak was possibly the largest waterborne outbreak of *Escherichia coli* 0157:H7 in U.S. history. It took the lives of two fair attendees and sent 71 others to the hospital. An investigation identified 781 persons with confirmed or suspected illness related to this outbreak. The outbreak is thought to have been caused by contamination of the Fair's Well 6, either by a dormitory septic system or manure runoff from the nearby Youth Cattle barn (New York State Department of Health [NYSDOH], 2000). More recently, in May 2000, an outbreak of *Escherichia coli* 0157.H7 in Walkerton, Ontario, resulted in at least seven deaths and 1,000 cases of intestinal problems; public health officials theorize that one possible cause was floodwaters washing manure contaminated with *Escherichia coli* into the town's drinking water well; an investigation is currently underway (Brook, 2000). An outbreak of *Escherichia coli* 0157:H7 was reported in Canada from well water potentially contaminated by manure runoff (Stehman, 2000).

- *Important point:* Researchers at Cornell University concluded that problems with *E. coli* 0157:H7 could be eliminated by feeding hay 3 days before slaughter.

- *Important point:* Cattle on hay diets defecate much more frequently, and the manure is much less viscous (Price, 1999).

Cow manure has specifically been implicated as a causative factor in the high bacteria levels and ensuing swimming restrictions on Tainter Lake, Wisconsin (Behm, 1989). Contact recreation can result in infections of the skin, eye, ear, nose, and throat (Juranek, 1995; Stehman, 2000). The USEPA's recommended ambient water quality standard for human health protection in contact-recreational fresh waters is either 120 *Escherichia coli* bacteria/100 ml or 33 enterococcus bacteria/100 ml. (This standard, finalized in 1986, replaces the previous standard of 200 fecal coliform bacteria/100 ml.) About 8% of U.S. outbreaks of *Escherichia coli* 0157:H7 between the years 1982 and 1996 occurred as a result of swimming (Griffin, 1998). Certain regions, in particular, may be adversely impacted. For example, pathogen impairment of surface waters is a great problem in most rural areas of southern Minnesota. This causes many rivers and lakes to be unsuitable for swimming (Mulla et al., 1999).

Most human infectious caused by bacteria such as *Escherichia coli* 0157:H7, *Salmonella* species, *Campylobacter jejuni,* and *Leptospira* species are spread by foodborne or direct contact (Stehman, 2000). Many pathogens are be transmitted through shellfish (Stelma & McCabe, 1992), which are filter feeders prone to accumulating bacteria and viruses. Others may be transmitted through inhalation. In particular, there is concern that pathogens may also be introduced to the air directly from animal feeding houses or during spray application of wastes. Flies and other vectors also present potential pathways for disease transmission.

A final concern is exposure to pathogens via consumption of raw foods improperly subjected to manure application. Cieslak et al. (1993) suggest that a 1993 *Escherichia coli* outbreak in Maine was the result of manure applications to a vegetable garden. Additionally, three *Escherichia coli* outbreaks (Montana in 1995, Illinois in 1996, and Connecticut in 1996) were traced to organic lettuce growers; the lettuces are suspected to have been contaminated by infected cattle manure (Nelson, 1997). In another incident in Maine, a few hundred children were sickened by *Cryptosporidium parvum.* The source was fresh-pressed apple cider made from apples gathered from a cattle pasture (Milliard et al., 1994). Although this exposure route can cause health problems, the proposed revisions to the USEPA regulations do not attempt to address it directly.

6.4.4 ORGANIC MATTER

6.4.4.1 Ecology

Increased organic matter loading to surface waters supports increased microbial population and activity; as these organisms aerobically degrade the organic matter, dissolved oxygen is consumed, reducing the amount available for aquatic organisms. This impact is exacerbated more in warm waters than in colder waters, because the dissolved oxygen saturation level is lower and because the higher temperatures support increased microbial metabolism.

As a result of dissolved oxygen depletion, aquatic species may suffocate (USEPA, 1993a) or be driven out of areas that lack sufficient oxygen. This phenomenon can occur rapidly, particularly with loadings of high-strength waste such as those that can result from catastrophic lagoon breaches (Goldman & Horne, 1983). Many examples of fish kills resulting from manure discharges from have occurred nationwide. In Nebraska in 1995, 50% of all agriculture-related fish kills investigated were because of livestock waste. In 1996, that percentage rose to 75%. In 1997 and 1998, 100% of agriculture-related fish kills were traced to livestock waste (USFWS, 2000).

Oxygen-stressed aquatic systems may also experience decreases in species richness or community structure as sensitive species are driven out or die off. Organisms living in borderline hypoxic (low oxygen) water are also likely to experience physiological stress, which can increase the potential for diseases, decrease feeding rates, or increase predation. Livestock has been widely reported to cause significant decreases in wildlife species and numbers (Mulla et al., 1999). For example, reduction in biodiversity due to AFOs has been documented in a study of three Indiana stream systems (Hoosier Environmental Council, 1997). That study shows that waters downstream of animal feedlots (mainly swine and dairy operations) contained fewer fish and a limited number of species of fish in comparison with reference sites. Excessive algal growth; altered oxygen content; and increased levels of ammonia, turbidity, pH, and total dissolved solids were also observed.

High oxygen depletion rates related to microbial activity have been reported in manure-amended agricultural soils as well. In soils, elevated microbial populations can affect crop growth by competing with plant roots for soil oxygen and nutrients (USDA, 1992).

6.4.4.2 Human Health

The release of organic matter to surface waters is a human health concern insofar as it can impact drinking water sources and recreational waters. As aquatic bacteria and other microorganisms degrade organic matter in manure, they consume dissolved oxygen. This can lead to foul odors and ecological impacts, reducing the water's value as a source of drinking water and recreation. Additionally, increased organic matter in drinking water sources can lead to excessive production of harmful chlorinated by-products, resulting in higher drinking water treatment costs and higher health risks. Pathogen growth is another concern, as large inflows of nutrient-rich organic matter, under the right environmental conditions, can cause rapid increases in microbial populations.

6.4.5 Salts and Trace Elements

6.4.5.1 Ecology

Salts in manure can impact water and soil environments. In fresh waters, increasing salinity can disrupt the balance of the ecosystem. Drinking water high in salt content was shown to inhibit growth and cause slowed molting in mallard ducklings (IEC, 1993). On land, salts can accumulate and become toxic to plants, reducing crop yields. Salts can damage soil quality by increasing acidity, reducing permeability and deteriorating soil structure. Salty soils cause plants to become nutrient deficient because they don't pick up enough of the nutrients they need, such as nitrogen and phosphorus (Spellman, 1998).

Trace elements in manure can impact plants, aquatic organisms, and terrestrial organisms. While many of the trace elements are essential nutrients at low concentrations, they can have significant ecotoxicological effects at elevated concentrations. For example, metals such as zinc (a feed additive) can accumulate in soil and become toxic to plants at high concentrations. Arsenic, copper, and selenium are other feed additives that can produce aquatic and terrestrial toxicity at elevated concentrations. Bottom-feeding birds can be quite susceptible to metal toxicity because they are attracted to shallow feedlot wastewater ponds and waters adjacent to feedlots. Metals can remain in aquatic ecosystems for long periods because of adsorption to suspended or bed sediments or uptake by aquatic biota.

Several of the trace elements in manure are regulated in treated municipal sewage biosolids (but not manure) by the CWA's Part 503 Rule. Total concentrations of trace elements in animal manures have been reported as comparable to those in some municipal biosolids, with typical values well below the maximum concentrations allowed by Part 503 for land-applied sewage sludge (Sims, 1995). Based on this information, trace elements in agronomically applied manures should pose little risk to human health and the environment. However, repeated application of manures above agronomic rates could result in exceedance of the cumulative metal loading rates established in Part 503, thereby potentially impacting human health and the environment (USFWS, 1991).

In 1991, the USFWS reported on suspected impacts from a large number of cattle feedlots on Tierra Blanca Creek, upstream of the Buffalo Lake National Wildlife Refuge in the Texas Panhandle. The USFWS found elevated concentrations of the feed additives copper and zinc in the creek sediment (as well as elevated aqueous concentrations of ammonia, chlorophyll a, coliform bacteria, chloride, conductivity, total Kjeldahl nitrogen, and volatile suspended solids). The relative contribution of these contaminants from various sources (e.g., runoff from facilities without containment lagoons, lagoon discharges, and lagoon leachate) was not assessed (USFWS, 1991).

In 1998, the USFWS found copper and zinc in wetlands fed by wastewater from a nearby swine production operation in Nebraska. Concentrations of copper exceeded both a proposed aquatic life criterion of 43 µg/L and the current least-protective criterion of 121 µg/L. Zinc concentration exceeded the concentrations recommended for the protection of aquatic life (USFWS, 2000).

6.4.5.2 Human Health

Salts from manure can impact surface and groundwater drinking water sources. Salt load into the Chino Basin from local dairies is over 1,500 tons per year, and the cost to remove that salt by drinking water treatment systems ranges from $320 to $690 for every ton (USEPA, 1993b). At lower levels, salts can increase blood pressure in salt-sensitive individuals, increasing the risk of stroke and heart attack. Salts can also make drinking water unpalatable and unsuitable for human consumption.

Some of the trace elements in manure are essential nutrients required for human physiology; however, they can induce toxicity at elevated concentrations. These include zinc, arsenic, copper, and selenium, which are feed additives (Sims, 1995). Although these elements are typically present in relatively low concentrations in manure, they are of concern because of their ability to persist in the environment and to bioconcentrate in plant and animal tissues. These elements could pose a hazard if manure is overapplied to land due to insufficient acreage available to accommodate manure from increasingly concentrated AFOs. Overapplied manure increases the likelihood of pollutants reaching surface water and ultimately being ingested.

Trace elements are associated with a variety of illnesses. For example, arsenic is carcinogenic to humans, based on evidence from human studies; some of these studies have found increased skin cancer and mortality from multiple internal organ cancers in populations that consumed drinking water with high levels of inorganic arsenic. Arsenic is also linked with noncancer health effects, including hyperpigmentation and possible vascular complications. Selenium is associated with liver dysfunction and loss of hair and nails, and zinc can result in changes in copper and iron balances, particularly copper deficiency anemia (IRIS, 2000).

6.4.6 SOLIDS

Excessive silting and sedimentation are prime agents responsible for the long-term degradation of rivers, streams, and lakes. Major sources of siltation include runoff from agricultural, urban, and forest lands and other non-point sources (USEPA, 1992b).

Solids entering surface water can degrade aquatic ecosystems to the point of nonviability. Suspended particles can reduce the depth to which sunlight can reach, decreasing photosynthetic activity (and the resulting oxygen production) by plants and phytoplankton. The increased turbidity also limits the growth of desirable aquatic plants that serve as critical habitat for fish, crabs, and other aquatic organisms. In addition, suspended particles can flog fish gills, degrade feeding areas, reduce visibility for sight feeders (Abt Associates, Inc., 1993), and disrupt migration by interfering with fish's ability to detect chemical communication signals in water (Goldman & Horne, 1983). Sediment can smother eggs, interrupt the reproductive process, and alter or destroy habitat for fish and benthic organisms.

Solids can also degrade drinking water sources, thereby increasing treatment costs. Furthermore, solids provide a medium for the accumulation, transport, and storage of other pollutants, including nutrients, pathogens, and trace elements. Sediment-bound pollutants often have a long history of interaction with the water column through cycles of deposition, resuspension, and redeposition.

6.4.7 ANTIBIOTICS AND ANTIBIOTIC RESISTANCE

Antibiotic-resistant strains of bacteria develop as a result of continual exposure to antibiotics. Use of antibiotics in raising animals, especially broad-spectrum antibiotics, is increasing. As a result, more strains of antibiotic-resistant pathogens are emerging, along with strains that are increasingly resistant (Mulla et al., 1999). Antibiotic-resistant forms of *Salmonella, Campylobacter, Escherichia coli,* and *Listeria* are known or suspected to exist. An antibiotic-resistant strain of the bacterium *Clostridium perfringens* was detected in the groundwater below plots of land treated with swine manure, while it was nearly absent beneath unmanured plots.

Antibiotic resistance poses a significant health threat. In April 2000, the *New England Journal of Medicine* published an article that discussed the case of a 12-year-old boy infected with a strain of *Salmonella* that was resistant to no fewer than 13 antimicrobial agents (Fey et al., 2000). The cause of the child's illness is believed to be exposure to the cattle on his family's Nebraska ranch.

The CDC, Food and Drug Administration, and National Institutes of Health issued a draft action plan in June 2000 to address the increase in antibiotic resistant diseases (CDCP, 2000). The plan is intended to combat antimicrobial resistance through survey, prevention and control activities, research, and product development. One of the action items involves conducting pilot studies to assess the impact of environmental contamination by antimicrobial drug residues and drug-resistant organisms that enter the soil or water from human and animal waste.

Case Study 6.1 (USGS, 2005)

In the following, we present a research project, for illustrative purposes, conducted under the auspices of USGS concerned with predicting sorption, mobility, accumulation, and degradation potential of antibiotics in Iowa's soil and water environment.

Research Problem

Approximately 32.6 billion pounds of antibiotics are used in the production of poultry (10.6 million pounds), hogs (10.3 million pounds), and cattle (3.7 million pounds) in the United States each year (Mellon et al., 2001). Over three-fourths of these antibiotics (24.6 million pounds) are given to healthy animals in low doses to promote growth (Levy, 1997). Most of the antibiotics given to farm animals are not metabolized in the body; rather they are excreted in the active form (Lee et al., 2000). The fate of antibiotics introduced into soil and aquatic environments with manure and other animal wastes is largely unknown. However, there is much concern that the presence and persistence of low levels of antibiotics in soil and aquatic environments could encourage the buildup of existing antibiotic-resistant bacterial populations and promote the development of new populations (Henry, 2000).

In Iowa, earthen waste storage structures (lagoons) are widely used for temporary storage of liquid animal wastes with the intent of protecting surface and groundwater from contamination and allowing farmers to use the wastes in a timely fashion. Liquid animal wastes are generally spread on agricultural soils both as a means of disposal of the wastes and as a nutrient source for crop production. The Iowa Department of Public Health found relatively high concentrations of chlortetracycline (11 to 540 µg/L) and erythromycin (10 to 275 µg/L) in such liquid animal wastes. The report also indicated that many of the 18 *E. coli* isolates, all three *Salmonella* species, and an isolate of *Enterococcus* demonstrated resistance to one or more of the antibiotics.

The antibiotics most commonly added to livestock feed as growth promoters (1 to 1000 mg per head per day) are chlorotetracycline (Aureomycin), oxytetracycline (Terramycin), and macrolides (such as erythromycin) (Sewell, 1993; FAC, 1998; Herman et al., 1995). The fate of these compounds in Iowa soils depends on sorption and desorption of the antibiotics on soils, leaching, and the rates of chemical, photochemical, and microbial decomposition of the antibiotics. The basic hypothesis of the study is that the fate (sorption/desorption, leaching, and decomposition) of antibiotics in soil environments is strongly influenced by the chemical reactions between the antibiotics and soils constituents.

Specific Objectives

1. Characterize three common Iowa soils and isolate and characterize reactive soil components (clay-humic complexes, clay minerals, and humic materials) from these soils.
2. Quantify sorption of tetracycline and chlorotetracycline on the soils and soil components.
3. Determine the effects of saturating cations (Ca vs. K) and ionic strength (1=0.05 and 1=0.005) on sorption of tetracycline and chlorotetracycline on the soils and soil components.
4. Quantify the influence of sorption on tetracycline and chlorotetracycline degradation rates.
5. Quantify mobility of tetracycline and chlorotetracycline in soil column.

Methodology

Soil samples, surface (0-15 cm) and subsurface (= 15 cm), were collected from three sites representing three different soil series and a range of soil physical and chemical properties. Both the studied soils and the general sampling locations had been previously characterized (McBride et al., 1987). Based on interviews with the landowners or operators, specific sampling sites that had never received manure applications were selected. The soils were characterized using standard analytical procedures to determine pH in $CaCl_2$, pH in KCl, pH in water, organic C, organic H, organic N, % sand, % coarse silt, % fine silt, % clay, and extractable cations (Ca, Mg, Na, and K).

Soil components were physically and chemically separated from the soils and prepared for the sorption and degradation studies. Clay-humic complexes were isolated from the soils by sedimentation (< 2 μm e.s.d.). Portions of the clay-humic complexes were K- or Ca-saturated by washing in 1M KCl or 0.5 M $CaCl_2$ and then dialyzed against distilled water and freeze dried. Other portions of the clay-humic complexes were treated with 30% H_2O_2 for removal of the humic materials before being K- and Ca-saturated, dialyzed against distilled water, and freeze-dried. Humic materials were separated from the three soils by hydrolyzing Na-saturated samples in 0.1 M NaOH under and N_2 purge. After hydrolysis, the humic materials were separated by centrifugation, neutralized to pH 7, K- or Ca-saturated, dialyzed and freeze-dried.

A batch equilibration technique has been designed and is being used to measure sorption of tetracycline and chlorotetracycline on the various soils and prepared soil components. High performance liquid chromatography (HPLC) is being used to quantify tetracycline and chlortetracycline in the supernatant solutions, and sorption is determined by difference. Variables being tested include soil components (clay-humic complexes, clay minerals, and humic substances), saturating cations (K vs. Ca), and ionic strength (1=0.05 and 1=0.005). Previous research has demonstrated the importance of pH; hence in this study pH is being carefully controlled at 6.5. The data will be used to prepare four point sorption isotherms with three replications for each point.

Key term: HPLC is used for separating, identifying, purifying, and quantifying various compounds.

After the sorption studies are complete, tetracycline will be incubated under both sterile and nonsterile conditions in aqueous controls and with soil components exhibiting both high and low sorption. Tetracycline will also be incubated under both sterile and nonsterile conditions with the soil exhibiting the highest sorption capacity. Degradation kinetics will be quantified for these systems by extracting tetracycline from the samples at various times during the incubations and quantifying parent and degradation products by HPLC.

The final stage of the research will be a column leaching study. Intact soil columns treated with tetracycline and chlorotetracycline will be leached with high and low ionic strength solutions with different ratios of K and Ca. The ionic strength and the K:Ca ratios of the leaching solutions will be selected to both encourage and discourage colloid mobility. Leachate will be analyzed by HPLC.

Principal Findings and Significance

The project is not yet far enough along to report findings and significance with respect to tetracycline sorption, degradation, and mobility in Iowa soils. Prior to Dr. Evangelou's death in March 2002, soil samples had been collected and initial analysis of the samples provided basic soil characterization data. Since March of 2002, the project has been reorganized and refocused. Specifically, we have conducted a literature review, refined the hypotheses being tested, developed new specific objectives, and designed three major sets of experiments focused on testing those hypotheses. From the literature review, it was apparent that the effects of pH on sorption and degradation of tetracyclines in soil environments have been carefully studied. However, little information was available to distinguish whether tetracyclines are dominantly sorbed on soil clays or soil humic materials and the

TABLE 6.6

Soil Sampled for the Study

Sample Site	Soil Series	Classification
Tama Co.	Fayette	Fine-silty mixed superactive mesic, Typic Hapludalfs
Boone Co.	Nicollet	Fine-loamy mixed superactive mesic, Aquic Hapludolls
Clarke Co.	Clarinda	Fine smectitic mesic, Vertic Agriaquolls

effects of saturating cation and ionic strength on sorption, degradation, and mobility of tetracyclines in soils. Therefore, the focus of the project has been targeted on filling these knowledge gaps.

The soils sampled for this study are listed in Table 6.6. Clay content of the sampled soils ranged from 19.2% in the Nicollet surface sample to 34.6% in the Clarinda subsoil sample. Organic C content ranged from 0.44% for the Fayette subsoil sample to 1.65 for the Fayette surface soil sample. Total exchangeable cations ranged from 13.6 $cmol_c$ kg^{-1} for the Nicollet surface soil to 19.7 $cmol_c$ kg^{-1} for the Fayette subsoil. The pH values in KCl ranged from 4.5 in the Clarinda subsoil sample to 6.5 in the Fayette surface soil sample. In general, the properties of the sampled soils are sufficiently diverse to allow a reasonable assessment of the influence of soil properties and soil components on the fate of antibiotics.

Preliminary chemical characterization of tetracycline, chlorotetracycline, and oxytetracycline were performed. UV-vis absorbance spectra of the antibiotics dissolved in water and various concentrations of KCl, $CaCl_2$, $MgCl_2$, and $AlCl_3$ were obtained using a UV-vis spectrophotometer (Varian Instruments, Cary 50 Bio model, Walnut Creek, CA, USA). Calibration curves for quantifying concentrations of the various antibiotics dissolved in water were developed for two wavelengths, near 270 nm (W1) and 370 nm (W2). Solubility of the oxytetracycline in water was measured by determining the concentration where the absorbance-concentration relationship deviates from Beer's law.

Potentiometric titrations indicate two and possibly three pKa's for the tetracyclines. The solubility of oxytetracycline was found to be approximately 300 mg L^{-1}. UV-VIS spectroscopy revealed two prominent absorption maxima near 280 and 360 nm for oxytetracycline and tetracycline and two prominent sorption maxima near 280 and 370 nm for chlorotetracycline. Absorption spectra for all three tetracyclines were only slightly affected by background $CaCl_2$ (0 to 50 meq L^{-1}) and $MgCl_2$ (0 to 40 meq L^{-1}) concentrations. By contrast, the presence of as little as 2 meq L^{-1} $AlCl_3$ substantially altered the absorbance spectra for all three tetracyclines. The cause of change in the absorbance spectra in the presence of $AlCl_3$ is not clear but may indicate either a pH effect or the formation Al-tetracycline complexes. More work is needed to resolve the cause of this effect. The results demonstrate that tetracycline, chlorotetracycline, and oxytetracycline concentrations in water and both $CaCl_2$ and $MgCl_2$ solutions can be quantified by UV-VIS spectroscopy with linear response for the 0 to 20 mg L^{-1} concentration range. The presence of Al in aqueous solution, however, many cause problems with spectrometric analysis.

Major accomplishments during the last year include the physical separation and chemical preparation of cation-saturated soil components (clay-humic complexes, clay minerals, and humic materials) from the studied soils and the development and testing of an HPLC method for quantification of tetracyclines. Considerable effort was expanded in developing the HPLC technique. The tetracyclines are not well behaved in HPLC because they have three ionizable moieties (i.e., portions of a molecular structure having some property of interest) and are zwitterions (i.e., compounds that carry both positive and negative charges in solution) over a large pH range. Several published HPLC methods performed poorly on our HPLC system, and considerable refinement of one of those methods was necessary to obtain high quality analytical data.

6.4.8 HORMONES AND ENDOCRINE DISRUPTION

The presence of estrogen and estrogen-like compounds in surface water has caused much concern. Their ultimate fate in the environment is unknown, although early studies indicate that no common soil or fecal bacteria can metabolize estrogen (Shore et al., 1995). When present in high concentrations, hormones in the environment are linked to reduced fertility, mutations, and the death of fish, and evidence is present that fish in some streams are experiencing endocrine disruptions (Shore et al., 1995; Mulla et al., 1999).

Estradiol, an estrogen hormone, was found in runoff from a field receiving poultry litter at concentrations up to 3.5 µg/L. Fish exposed to 0.25 µg/L of estradiol often have gender changes; exposures at levels above 10 µg/L can be fatal (Mulla et al., 1999). Estrogen levels of 10 µg/L have been shown to affect trout (Shore et al., 1995).

Endocrine disruptors have also been the subject of increasing concern because they alter hormone pathways that regulate reproductive processes in both human and animal populations. Estrogen hormones have been implicated in the drastic reduction in sperm counts among European and North American men (Sharpe & Skakkebaek, 1993) and widespread reproductive disorders in a variety of wildlife (Colburn et al., 1993). A number of agricultural chemicals have also been demonstrated to cause endocrine disruption as well, including pesticides (Shore et al., 1995). The effects of these chemicals on the environment and their impacts on human health through environmental exposures are not completely understood. They are currently being studied for neurobiological, developmental, reproductive, and carcinogenic effects (Tetra Tech, 2000a). The USEPA is not aware of any studies done on the human health impact of hormones from watersheds with impairment from animal manure.

6.4.9 OTHER POLLUTANTS OF CONCERN

6.4.9.1 Gas Emissions

Odor sources (see Chapter 7) include animal confinement buildings, waste lagoons, and land application sites. As animal waste decomposes, various gases are produced. The primary gases associated with aerobic decomposition include carbon dioxide and ammonia. Gases associated with anaerobic conditions, which dominate in typical, unaerated animal waste lagoons, include methane, carbon dioxide, ammonia, hydrogen sulfide, and over 150 other odourous compounds (USDA, 1992; Bouzaher et al., 1993; O'Neill & Phillips, 1992). These include volatile fatty acids, phenols, mercaptans, aromatics, sulfides, and various esters, carbonyls, and amines. The decomposition process is desirable because it reduces the BOD and pathogen content of the waste. However, many of the end products can produce negative impacts, including strong odors. Heavy odors are the most common complaint from neighbors of swine operations in particular (Agricultural Animal Waste Task Force, 1996).

Odor is itself a significant concern because of its documented effect on moods, such as increased tension, depression, and fatigue (Schiffman et al., 1995). Odor also has the potential for vector attraction and affects property values. Additionally, many of the odor-causing compounds can cause physical health impacts. For example, hydrogen sulfide is toxic, and ammonia gas is a nasal and respiratory irritant.

In 1996, the Minnesota Department of Health found elevated levels of hydrogen sulfide gas at residences near AFOs that were high enough to cause such symptoms as headaches, nausea, vomiting, eye irritation, respiratory problems (including shallow breathing and coughing), achy joints, dizziness, fatigue, sore throats, swollen glands, tightness in the chest, irritability, insomnia, and blackouts (Addis et al., 1999). In an Iowa study, neighbors within 2 miles of a 4,000-sow swine facility reported more physical and mental health symptoms than a control group (Thu, 1998). These symptoms included chronic bronchitis, hyperactive airways, mucus membrane irritation, headaches, nausea, tension, anger, fatigue, and confusion.

Methane and carbon dioxide are greenhouse gases that contribute to global warming. Methane also contributes to the formation of tropospheric ozone (a component of photochemical smog). Based on various USEPA estimates (USEPA, 1989, 1992a), methane emissions from U.S. animal wastes are a very small contributor to the global warming effects.

6.4.9.2   Particulates

Sources of particulate emissions from AFOs may include dried manure, feed, skin, hair, and possibly bedding. The airborne particles make up an organic dust, which includes endotoxin (the toxic protoplasm liberated when a microorganism dies and disintegrates), adsorbed gases, and possibly steroids (Thu, 1995). At least 50% of dust emissions from swine production facilities are believed to be respirable. The main impact downwind appears to be respiratory irritation due to the inhalation of organic dusts. Studies indicate that the associated microbes generally are not infectious but may induce inflammation (Thu, 1995).

6.4.9.3   Pesticides

Pesticides may pose risks to the environment, such as chronic aquatic toxicity, and human health effects, such as systemic toxicity. In a few studies, common herbicides have been shown to cause endocrine disruption. There is some evidence that fish in some streams are experiencing endocrine disruption and that contaminants including pesticides may be the cause (Mulla et al., 1999).

6.5   SUMMARY

Problems and concerns related to animal waste pollution are a highly important factor for AFO/CAFO operations. Other problems related to animal wastes and air quality are the topic of concern for Chapter 7.

CHAPTER REVIEW QUESTIONS

1. What pollutants are commonly associated with animal waste?
2. What are the common pathways for animal waste to enter the environment?
3. What pollutants present the biggest areas of concern? Why?
4. What percent of nitrogen and phosphorus come from manure?
5. Sketch the nitrogen cycle. What nitrogen components cause environmental problems?
6. Describe the chemical relationship between nitrogen and ammonia compounds.
7. What forms of phosphorus are of concern? Why?
8. What problems are related to ammonia toxicity?
9. What pathogens of concern may be present in animal wastes?
10. Why are increased levels of salinity problematic?
11. What could be learned by comparing trace element concentrations in manure to those in sewage biosolids?
12. What are the issues of concern that pertain to antibiotic use for animals in AFOs/CAFOs? Discuss.
13. What issues of concern are related to use of pesticides in AFOs/CAFOs? Discuss.
14. What issues of concern are related to use of hormones for animals in AFOs/CAFOs? Discuss.
15. What gas emissions and pollutants are of concern? Why?
16. What are the common pollution pathways between AFOs/CAFOs and surface water?
17. Describe and discuss pollution problems related to erosion.
18. Describe and discuss regional pollution issues related to direct contact between confined animals and surface water.

19. Describe and discuss the pollution issues related to groundwater.
20. Why is the connection between groundwater and surface water important? Be specific to AFO/CAFO operation and design.
21. Describe and discuss pollution issues related to discharge to the air and deposition.
22. What are the common pathways for nitrogen compounds to travel from AFO/CAFO sources to surface water and groundwater?
23. Nitrate pollution risk is elevated with land application of animal wastes, even at agronomic rates. Why?
24. How does phosphorus commonly reach surface waters? Why is phosphorus's limited mobility important?
25. What are the common pathways for ammonia compounds?
26. What are the common pathways for pathogens?
27. Discuss bacterial pathways and possible sources.
28. What are the three causal agents of waterborne disease outbreaks?
29. Discuss soil filtering and adsorption action as it relates to groundwater.
30. Discuss pathogen survivability and transmission.
31. Discuss organic matter related to discharge (intentional and accidental).
32. Does application rate have a significant effect on runoff accumulation? Discuss.
33. Discuss pathways for salts and trace elements to enter surface and groundwater.
34. Discuss potential pathways for antibiotics, hormones, and pesticides.
35. Describe and discuss potential hazards from AFO/CAFO pollutants in and across all environmental media.
36. What primary nutrients are of highest concern?
37. What is eutrophication, what causes it, and why is it of concern to AFO/CAFO management?
38. What adverse effects are related to eutrophication? What other problems are related to it?
39. Describe and discuss "the Dead Zone."
40. What is "brown blood disease"? What causes it?
41. Describe and discuss the human health issues from primary nutrients that are related to AFO/CAFO pollutants.
42. Describe and discuss the human health issues from AFO/CAFO pollutant-source eutrophication and algal blooms.
43. Discuss BOD and ammonia and receiving waters.
44. What problems and human health issues are related to ammonia?
45. What problems and human health issues are related to pathogens?
46. What problems and human health issues are related to organic matter?
47. What problems and human health issues are related to salts?
48. What problems and human health issues are related to trace elements?
49. What problems and human health issues are related to excessive levels of solids?
50. What problems and potential problems are related to antibiotic use for AFOs and CAFOs? What's being done to define and control the problems?
51. Discuss the objectives of Case Study 6.1's research hypothesis.
52. Discuss the ramifications of the effects of excessive estrogen and estrogen-like compounds on humans and both wild and domestic animals.
53. What are the problematic symptoms related to odors and gas emissions from AFOs/CAFO operations?
54. What other problems are related to gas emissions?
55. What is the chief problem related to particulate matter? Who does it affect most strongly? Why?
56. What problems are related to pesticide pollution?

THOUGHT-PROVOKING QUESTIONS

1. What problems are related to AFOs/CAFOs and outdated state and federal pollution regulations?
2. On a swine operations in Iowa, the CDC has found animal waste pathogens from operations far afield from waste storage locations—in drainage ditches, agricultural drainage wells, tile line inlets and outlets, and in an adjacent river. Discuss this from the standpoint of the AFO neighboring community, from the standpoint of AFO/CAFO ownership and management, and as a matter of public water quality and human health.
3. Odor as pollution is becoming a bigger issue as suburban development encroaches upon agricultural lands. Research and discuss both sides of this issue. Include a specific discussion of "NIMBY."
4. Problems related to antibiotic and growth hormone use for livestock are now coming to worldwide attention in several different ways. Examine this in light of the stance, "We do not know what we do not know."
5. What's problematic about the lack of pesticide pollution research from AFO/CAFO sources? Examine this in light of "We do not know what we do not know."

REFERENCES

Abt Associates, Inc. 1993. *Human Health Risk Assessment for the Use and Disposal of Sewage Sludge: Benefits of Regulation*. Prepared for Health and Ecological Criteria Division, Office of Science and Technology, Office of Water, U.S. Environmental Protection Agency. January.

Addis, P.G., Blaha, T., Crooker, B., Diez, F., Feirtag, J., Goyal, S., Greaves, I., Hathaway, M., Janni, K., Kirkhorn, S., Moon, R., Morse, D.E., Phillips, C., Reneau, J., Shutske, J., and Wells, S. 1999. *Generic Environmental Impact Statement on Animal Agriculture: A Summary of the Literature Related to the Effects of Animal Agriculture on Human Health (K)*. Prepared for the Environmental Quality Board by the University of Minnesota, College of Agriculture, Food, and Environmental Sciences.

Addiscott, T.M., Whitmore, A.P., & Powlson, D.S. 1991. *Farming, Fertilizers, and the Nitrate Problem*. Rothamsted Experimental Station. Oxon, United Kingdom: C-A-B International.

Agricultural Animal Waste Task Force. 1996. Policy recommendations for management of Agricultural Animal Waste in North Carolina: Report of the Agricultural Animal Waste Task Force.

Altekruse, S.F. 1998. *Campylobacter jejuni* in foods. *Journal of American Veterinary Medical Association* 213(12):1734–1735.

Aneja, V., Murray, G.C., & Southerland, J. 1998. Atmospheric nitrogen compounds: Emissions, transport, transformation, deposition, and assessment. *EM, Air & Waste Management Association's Magazine for Environmental Managers*, pp. 22–25.

ASAE. 1997. *Manure Production and Characteristics*. ASAE D384.1 (ASAE Standards).

ASAE 1999. *Manure Production and Characteristics*. ASAE D384.1 (ASAE Standards).

ASCE, 1998. *The Chesapeake Bay Experience*. Reiton, VA: American Society of Chemical Engineers.

Atwood, DK., Bratkovich, A., Gallagher, M., & Hitchcock G. (eds). 1994. Introduction to the dedicated issue. *Estuaries* 17(4):729–911.

Behm, D. 1989. Ill waters: The fouling of Wisconsin's lakes and streams. *The Milwaukee Journal Sentinel*. Special report: a series of articles published November 5–10.

Bodek, I., Lyman, W.J., Reehl, W.F., & Rosenblatt, D (eds.). 1988. *Environmental Inorganic Chemistry: Properties, Processes, and Estimation Methods*. New York: Pergamon Press.

Bouzaher, A., Lakshminarayan, P.G., Johnson, S.R., Jones, T., & Jones, R. 1993. The economic and environmental indicators for evaluating the national pilot project on livestock and the environment, Livestock Series Report 1. Center for Agricultural and Rural Development (CARD) at Iowa State University and Texas Institute for Applied Environmental Research at Tarleton State University. Staff Report 93-SR 64. October.

Brady, N. 1990. *The Nature and Properties of Soils*, 10th Edition. New York: Macmillan Publishing Company.

Brook, J. 2000. Few left untouched after deadly *E. coli* flows through an Ontario town's water. *The New York Times*. July 10.

Bruning-Fann, C.S., & Kaneene, J.B. 1993. The effects of nitrate, nitrite, and n-nitroso compounds on human health: A review. *Vet. Human. Toxicol.* 35(6). December.

Carpenter, S., Caraco, N.F., Correll, D.L., Howarth, R.W., Sharpley, A.N., & Smith, V.H. 1998. Nonpoint pollution of surface waters with phosphorus and nitrogen. *Issues in Ecology,* Number 3. Published by the Ecological Society of America, Washington, DC. Summer.

Casman, E.A. 1996. *Chemical and Microbiological Consequences of Anaerobic Digestion of Livestock Manure., A Literature Review.* Interstate Commission on the Potomac River Basin. ICPRB Report #96-6.

CAST. 1992. Water quality: Agriculture's role. Council for Agricultural Science and Technology Report 120. December.

CDCP, 1998. Iowa Lagoon, Surface Water Samples from Swine Water. Center for Disease Control & Prevention. Atlanta, GA.

Chai, T.J., Han, T., & Cockey, R.R. 1994. Microbiological quality of shellfish-growing waters in Chesapeake Bay. *J. Food Protect.* 57:229–234.

Cieslak, P.R., Barret, T.J., Griffin, P.M., Gensheimer, K.F., Beckett, G., Buffington, J., & Smith, M.G. 1993. *Escherichia coli* 0157:H7 infection from a manured garden. *The Lancet.* 342:367.

Citizens Pfiesteria Action Commission. 1997. Final report of the Citizens Pfiesteria Action Commission. Governor Harry R. Hughes, Commission Chairman. November.

Clean Water Action Alliance. 1998. Minnesota manure spills and runoff.

Colburn, T., von Saul, F.S., Soto, A.M., 1993. Developmental effects of endocrine-disrupting chemicals in wildlife and humans. *Environ Health Perspect* 101:378–384.

Coyne, M.S., & Blevins, R.L. 1995. Fecal bacteria in surface runoff from poultry manured fields. In Kenneth Steele (ed.), *Animal Waste and the Land-Water Interface*, Boca Raton, FL: CRC Lewis Publishers.

Crane, S.R., Westerman P.W., & Overcash, M.R. 1980. Die-off of fecal indicator organisms following land application of poultry manure. *Journal of Environmental Quality.* 9:531–537.

Daniel, T.C., Edwards, D.R., & Nichols, D.J. 1995. Edge-of-field losses of surface-applied animal manure. In Kenneth Steele (ed.), *Animal Waste and the Land-Water Interface*, Boca Raton: CRC Lewis Publishers.

Dazzo, F., Smith, P., & Hubbell, D. 1973. The influence of manure slurry irrigation on the survival of fecal organisms in Scranton fine sand. *Journal of Environmental Quality* 2:470–473.

Derbyshire, J.B., Clark, M.C., & Jessett, D.M. 1966. Observations on the fecal excretion of adenoviruses and enteroviruses in conventional and "minimal disease" pigs. *Vet. Record* 79: 595.

Derbyshire, J.B., & Brown, E.G. 1978. Isolation of animal viruses from farm livestock waste, soil, and water. *J. Hygiene* 81:295–302.

Dimura, J. 2000. New York State Department of Environmental Conservation. Personal communication with Patricia Harrigan, USEPA, Washington, DC on direct access of CAFO animals to surface water via e-mail. August 10.

Durborow, R., & Crosby M.D. 2003. *Brown blood disease.* BioFilter.Com, http://www.biofilter.com/MSU1390. htm. Accessed March 27, 2005.

Dunne, T., & Leopold, L.B. 1978. *Water in Environmental Planning.* San Francisco: W.H. Freeman and Company.

Edwards, D.R., & Daniel, T.C. 1992b. Potential runoff quality effects of poultry manure slurry applied to fescue plots. *Trans. ASA* 35(6):1827–1832. Cited in Daniel et al. (1995).

Edwards, D.R. & Daniel, T.C. 1993a. Effects of poultry litter application rate and rainfall intensity on quality of runoff from fescue grass plots. *Journal of Environmental Quality* 22(2):361–365. Cited in Daniel et al. (1995).

Edwards, D.R., & Daniel, T.C. 1993b. Runoff quality impacts of swine manure applied to fescue plots. *Trans ASAE* 36(1); 81–86. Cited in Daniel et al. (1995).

Elder, R.O., Keen, J.E., Siragusa, G.R., Barkocy-Gallagher, G.A., Koohmaraie, M., & Laegreid, W.W. 2000. Correlation of enterohemorrhagic *Escherichia coli* 0157 prevalence in feces, hides, and carcasses of beef cattle during processing. *PNAS* 97: 2999–3003.

FAC. 1998. *Feed additive compendium.* Sarah Muirhead, Ed. Minnetonka, MN: The Miller Publishing Co.

Fattal, B., Dotan, A., & Tchorsh, Y. 1992. Rates of experimental microbiological contamination of fish exposed to polluted water. *Water Resources Bulletin* 26(12):1621–1627.

Fey, P.D., Safranek, T.J., Rupp, M.E., Dunn, E. F., Ribot, E., Iwen, P.C., Bradford, P.A., Angulo, F.J., & Hinrichs, S.H. 2000. Ceftriaxone-resistant salmonella infection acquired by a child from cattle. *The New England Journal of Medicine* 342(17):1242–1249. April 27.

Follet, R.R. 1995. *Fate and transport of nutrients: nitrogen.* Working Paper No. 7. U.S. Department of Agriculture, Agricultural Research Service, Soil-Plant-Nutrient Research Unit. Fort Collins, CO. September.

FR. 2003. *CAFO Rules and Regulations.* Washington, DC: Federal Register, Vol. 68, No. 29.

Gerritse, R.F., & Zugec, I. 1977. The phosphorus cycle in pig slurry measured from $^{32}PO_4$ distribution rates. *J. Agric Sci.* 88(1):101–109.

Giddens, J., & Barnett, A.P. 1980. Soil loss and microbiological quality of runoff from land treated with poultry litter. *J. Environ. Qual.* 9:518–520.

Goldman, C., & Horne, A. 1983. *Limnology.* New York: McGraw-Hill Publishing Co.

Goolsby, D.A., Battaglin, W.A., Lawrence, G.B., Artz, R.S., Aulenbach, B.T., Hooper, R.P., Keeney, D.R., & Stensland, G.J. 1999. *Flux and sources of nutrients in the Mississippi – Atchafalaya River Basin.* Topic 3 Report of the Gulf of Mexico Hypoxia Assessment. Submitted to the White House Office of Science and Technology Policy Committee on Environment and Natural Resources Hypoxia Work Group. May. www.nos.noaa.gov/products/pubs_hypox.html.

Gresham, C.W., Janke, R.R., & Moyer, J. 1990. Composting of poultry litter, leaves, and newspaper. Rodale Research Center.

Griffin, P.M. 1998. Epidemiology of shiga toxin-producing Escherichia coli infections in humans in the United States. In *Escherichia coli 0157:H7 and Other Shiga Toxin-Producing Escherichia coli Strains*, edited by J.B. Kasper and O'Brien A.D. Washington, DC: ASM Press, pp. 15–22.

Health Canada Environmental Health Programs. 1998. *Blue-green algae (Cyanobacteria) and their toxins.* Publication of the Canadian federal government. http://www.hc-sc.gc.ca/ehp/ehd/catalogue/general/iyh/alea.htm.

Henry, C.M. 2000. Antibiotic resistance. *Chem. & Eng. News*, vol. 78, 10:41–58.

Herman, T., Baker, S., & Stokka, G.I. 1995. Medicated feed additives for beef cattle and calves. Cooperative Extension Service. Kansas State Univ. Publ. MF-2043.

Himathongkham, S., Bahari, S. Riemann, H., & Cliver, D. 1999. Survival of Escherichia coli 0157:H7 and Salmonella typhimurium in cow manure and cow manure slurry. FEMS Microbiology Letters. 178:251–257.

Hoosier Environmental Council. 1997. Internet home page. www.envirolink. Org/orgs/hecweb/monirotspring97/confined.htm.

Hrubant, G.R. 1973. Characterization of the dominant aerobic microorganism in cattle feedlot waste. *J. Appl. Microbiol.* 26:512–516.

IEC. 1993. *Irrigation return flow fee feasibility study.* Industrial Economics, Incorporated. Prepared for Office of Policy Analysis, Office of Policy, Planning and Evaluation, U.S. Environmental Protection Agency. March.

Illinois Stewardship Alliance. 1997. *The adverse impacts of CAFO's continue in Illinois without regulation.* July 10.

Iowa Department of Natural Resources (IDNR). 1998. Prohibited discharges at Iowa livestock operations Resulting in monetary penalties and/or restitution of fish kill being proposed, collected or pending—1992–present. June 3.

Iowa State University. 1999. *Earthen Waste Storage Structures in Iowa: A Study for the Iowa Legislature.* EDC-186. August.

IRIS. 2000. *Chemical files and Background documents and papers.* Washington, DC: Integrated Risk Information System. www.epa.gov/ngispgm3/iris/index.html.

Jackson, G., Keeney, D., Curwen, D., & Webendorfer, B. 1987. *Agricultural management practices to minimize groundwater contamination.* University of Wisconsin Environmental Resources Center. July.

Juranek, D.D. 1995. Cryptosporidiosis: Sources of infection and guidelines for prevention. *Clin Infect Dis.* 21(Suppl. 1):S57–61.

Kudva, I., Blanch, K., & Hovde, C. 1998. Analysis of *Escherichia coli* 0157:H7 survival in ovine or bovine manure and manure slurry. *Appl. Environ. Microbiol.* 64(9):3166–3174.

LeChevallier, M.W., Norton, W.D., & Lee, R.G. 1991. Occurrence of *Giardia* and *Cryptosporidium* spp. in surface water supplies. *Appl Environ. Microbiol.* 57(9):2610–2616.

Lee, W., Li, Z, Vakulenko, S., & Mobashery, S. 2000. A light-activated antibiotic. *J. Med. Chem.* 43:128–132.

Levy, S.B. 1997. Antibiotic resistance: An ecological imbalance. In Antibiotic Resistance: Origins, Evolution, Selection and Spread; Chadwick, D.J., Goode, F., Eds., Ciba Foundation Symposium 207; Wiley: Chichester: pp. 1–14.

Loehr, R.C. 1972. Animal waste management: Problems and guidelines for solutions. *Journal of Environmental Quality* 1(1): 71–78.

Macomb Journal. 1999. State settles with farmers in 1997 manure spill. August 8.

Maule, A. 1999. Survival of verocytotoxigenic *Escherichia coli* 0157 in soil, water and on surfaces. In *Society for Applied Microbiology Summer Conference Handbook*.

McBride, J.F., Horton, R., & Thompson, M.L. 1987. Evaluation of three Iowa soil materials as liners for hazardous-waste landfills. *Proc. Iowa Acad. Sci.*, pp. 1–14.

McFall, W. 2000. USEPA Region X. Personal communication with Patricia Harrigan, USEPA, Washington, DC, on direct access of CAFO animals to surface water via 2-mail. August 2.

Meadows, R. 1995. Livestock legacy. *Environmental Health Perspectives* 103(12):1096–1100.

Mellon, M., Benbrook, C., & Benbrook, K. 2001. Hogging it! Estimate of antimicrobial abuse in livestock. Union of Concerned Scientists, p. xiv, 109 or www.ussusa.org

Meyer, M. 1994. How common is methemoglobinemia from nitrate contaminated wells? A South Dakota perspective. Paper presented at the 39th Annual Midwest Groundwater Conference, Bismarck, ND. October. Cited in Michel et al. (1996).

Michel, K., Bacon, J.R., Gempesaw II, C.M., & Martin Jr., J.H. 1996. Nutrient management by Delmarva poultry growers: A survey of attitudes and practices. University of Delaware, College of Agricultural Sciences, Department of Food and Resource Economics. August.

Milliard, P.S., Gensheimer, K.F., Addis, D.G., Sosin, D.M., Beckett, G.A., Houch-Jankoski, A., & Hudson, A. 1994. An outbreak of cryptosporidiosis from fresh-pressed apple cider. *JAMA.* 272(20):1592–1596.

Morris, J. G. Jr., Matuszak, D.L., Taylor, J.L., Dickson, C., Benjamin, G.C., & Grattan, L.M. 1998. Focused issue: *Pfiesteria*—Beginning to unravel the mystery. *Maryland Medical Journal.* University of Maryland School of Medicine, Baltimore, Maryland. May. Cited in Brown (1998).

Mulla, D., Sekely, A., Birr, A., Perry, J., Vondracek, B., Bean, E., Macbeth, E. Goyal, S., Wheeler, B., Alexander, C., Randall, G., Sands, G., & Linn, J. 1999. Generic environmental impact statement on animal agriculture: A summary of the literature related to the effects of animal agriculture on water resources (G). Prepared for the Environmental Quality Board by the University of Minnesota, College of Agriculture, Food, and Environmental Sciences.

National Research Council. 1993. *Soil and Water Quality: An Agenda for Agriculture.* Washington, DC: National Academy Press.

National Research Council. 1995. *Nitrate and Nitrite in Drinking Water.* Washington, DC: National Academy Press.

NCAES 1982. Best management practices for agricultural nonpoint source control: Animal waste. Raleigh, NC: North Carolina State University, Biological and Agricultural Engineering Department. North Carolina Agricultural Extension Service.

NCSU. 2000. Aquatic Botany Laboratory *Pfiesteria piscicida* homepage. North Carolina State University.

Nebraska Cooperative Extension. 1995. Drinking water: Nitrate and methemoglobinemia ("blue baby" syndrome). NebGuide document G98-1369. July. www.ianr.unl. edu/PUBS/water/g1369.htm.

Nelson, H. 1997. The contamination of organic produce by human pathogens in animal manures. Ecological Agriculture Projects, McGill University, Canada. http://eap.megill.ca/_private/bl_head.htm.

Nolan, B.T., & Ruddy, B.C. 1996. Nitrate in ground waters of the United States – Assessing the risk. U.S. Geological Survey National Water Quality Assessment Program, Fact Sheet FS-092-96. http://water.usgs.gov/nawqa/FS-092-96.html.

North Carolina's Nicholas School of the Environment's Agricultural Animal Waste Task Force. 1994. Water resource characterization DSS-phosphorus and nitrate-nitrite. http://H2osparc.wq.nesu.edu/info.

Nowlin, M. 1997. Environmental implications of livestock production. *J. Soil Water Conservation,* 52.

NSCE, 2000. *Pfiesteria.* Washington, DC: National Council for Science and the Environment.

NYSDOH. 2000. Health commissioner releases *E. coli* outbreak report. March 31. New York State Department of Health. www.health.state.ny.us/nysdoh/commish /2000/ecoli.htm.

Ohio Department of Natural Resources. 1997. Division of Wildlife Pollution investigation report – Manure related spills and fish and wildlife kills. June 9.

O'Neill, D.H., & Phillips. V.R. 1992. A review of the control of odour nuisance from livestock buildings: Part 3, properties of the odourous substances which have been identified in livestock wastes or in the air around them. *Journal of Agricultural Engineering Research* 53:23–50.

Phillips, J.M., Scott, H.D., & Wolf, D.C. 1992. Environmental implications of animal waste application to pastures. In *Proceedings from the South Pasture Forage Crop Improvement Conference*. USDA/Agricultural Research Service. September. pp. 30–38.

Poultry Water Quality Consortium. 1998. Poultry waste management. In *Poultry Water Quality Handbook*, second edition. Chattanooga, TN. September. www.poultryegg.org/other.

Price, D. 1999. The latest on *E. coli* 0157:H7. *Beef Magazine*, November.

Proulx, A. 1992. *That Old Ace in the Hole*. New York: Scribner.

Ritter, W.F., & Chirnside, A.E.M. 1990. Impact of animal waste lagoons on groundwater quality. *Biologic Wastes* 34:39–54.

Ritter, W.R., Humenik, F.J., & Skaggs, R.W. 1989. Irrigated agriculture and water quality in east. *Journal of Irrigation and Drainage Engineering* 115(5):807–821.

Robinson, J.S., & Sharpley, A.N. 1995. Release of nitrogen and phosphorous from poultry litter. *Journal of Environmental Quality* 24(1):62–67.

Rodecap, J. 2002. Nitrate in surface water—trends and concerns. *Water Watch*. Iowa State University.

Schiffman, S.S., Sattely Miller, E.A., Suggs, M.S., & Graham, B.G. 1995. The effect of environmental odors emanating from commercial swine operations on the mood of nearby residents. *Brain Research Bulletin* 37:369–375.

Sewell, H.B. 1993. *Feed additives for beef cattle*. Agricultural Publication G02075, Dept. of Animal Sci., Univ. of Missouri. Columbia.

Sharpe, R.M., & Skakkebaek, N.E. 1993. Are oestrogens involved in falling sperm count and disorders of the male reproductive tract? *Lancet* 341:1392–1395.

Shelton, D.R. 2000. Sources of pathogens in a watershed: Humans, wildlife, farm animals? *Managing Nutrients and Pathogens from Animal Agriculture. Proceedings of a Conference for Nutrient Management Consultants, Extension Educators, and Producer Advisors*. Natural Resource, Agriculture, and Engineering Service, pp. 108–115. March 28–30.

Sherer, B.M., Miner, J.R., Moore, J.A., & Buckhouse, J.C. 1988. Resuspending organisms from a rangeland stream bottom. Trans. ASAE. 31:1217–1222.

Sherer, B.M., Miner, J.R., Moore, J.A., & Buckhouse, J.C. 1992. Indicator bacterial survival in stream sediments. *J. Environ. Qual.* 21:591–595.

Shore, L.S., Correll, D.L., & Chakraborty, P.K. 1995. Relationship of fertilization with chicken manure and concentrations of estrogens in small steams. In K. Steele (ed.), *Animal Waste and the Land-Water Interface* Boca Raton, FL: CRC Press/Lewis Publishers.

Sims, J.T. 1995. Characteristics of animal wastes and waste-amended soils: An overview of the agricultural and environmental issues. In K. Steele (ed.), *Animal Waste and the Land-Water Interface*, Boca Raton, FL: CRC Press/Lewis Publishers.

Smith, V. 1994. Disaster in Milwaukee: Complacency was the root cause. *EPA Journal*. 20:16–18.

Spellman, F.R. 1996. *Stream Ecology and Self-Purification*. Boca Raton: CRC Press.

Spellman, F.R. 1998. *The Science of Environmental Pollution*. Boca Raton: CRC Press.

Spellman, F.R. and Drinan, J., 2000. *Stream Ecology and Self-Purifications*. Boca Raton: CRC Press.

Stanley, K., Cunningham, R., & Jones, K. 1998. Isolation of *Campylobacter jejuni* from groundwater, *J. Appl. Microbiol.* 85:187–191.

Stehman, S.M. 2000. ag-related waterborne pathogens. In *Managing Nutrients and Pathogens from Animal Agriculture, Proceedings of a Conference for Nutrient Management Consultants, Extension Educators, and Produce Advisors*. Ithaca, NY: Natural Resource, Agriculture, and Engineering Service (NRAES).

Stelma, G.N., Jr., & McCabe, L.J. 1992. Nonpoint pollution from animal sources and shellfish sanitation. *Journal of Food Protection*. 55(8):649–656.

Tetra Tech. 2000a. Literature review and assessment of pathogens, heavy metals, and antibiotic content of waste and wastewater generated by CAFOs. EPA contract 68-C-99-263.

Tetra Tech. 2000b. Development of pollutant loading reductions from the implementation of nutrient management and best management practices. November 21.

Thomann, R.V., & Muller, J.A. 1987. *Principles of Surface Water Quality Modeling and Control*. New York: Harper Collins Publishers.

Thu, K. (ed.). 1995. *Understanding the Impacts of Large-Scale Swine Production*. Proceedings from an Interdisciplinary Scientific Workshop. Des Moines, Iowa. June 29–30.

U.S. Senate. 1997. *Animal Waste Pollution in America: An Emerging National Problem.* Report compiled by the Minority Staff of the United States Senate Committee on Agriculture, Nutrition and Forestry for Senator Tom Harkin.

USDA 1992. *Agricultural Waste Management Field Handbook.* 210-AWMFH, April.

USDA/ARS. 1998. Agricultural uses of municipal animal, and industrial byproducts. U.S. Department of Agriculture/Agricultural Research Service. Conservation Research Report Number 44, January.

USDA/ARS. 1999. Agricultural phosphorus and eutrophication. Department of Agriculture/Agricultural Research Service. ARS-149. July.

USDA/NRCS. 1996. *Agricultural Waste Management Field Handbook.* U.S. Department of Agriculture/ National Agricultural Statistics Service. 210-AWMFH.

USEPA. 1989. Policy options for stabilizing global climate. Draft Report to Congress. United States Environmental Protection Agency. Washington, DC. February. Cited in USEPA (1992a).

USEPA. 1992a. Global methane emissions from livestock and poultry manure. United States Environmental Protection Agency. Office of Air and Radiation. EPA/400/1-91/048.

USEPA. 1992b. Managing nonpoint source pollution: Final report to Congress. U.S. Environmental Protection Agency. Office of Water.

USEPA. 1993a. Guidance specifying management measures for sources of nonpoint pollution in coastal waters. Office of Water. 840-B-92-002. January.

USEPA. 1993b. The reports of the EPA/State Feedlot Workgroup. U.S. Environmental Protection Agency. Office of Wastewater Enforcement and Compliance.

USEPA. 1993c. Standards for the use of disposal of sewage sludge. 40 CFR 503.

USEPA. 1997b. Section 319 Success Stories: Volume II. Highlights of State and Tribal Nonpoint Source Programs. Washington, DC: United States Environmental Protection Agency Office of Water. Washington, DC: EPA-R-97-001. October. www.epa.gov/owow /NPS/Section 31911.

USEPA. 2000a. National water quality inventory: 1998 report to Congress. United States Environmental Protection Agency. EPA 841-R-00-001.

USEPA. 2001. Environmental assessment of proposed revisions to the national pollutant discharge elimination system regulations and the effluent guidelines for concentrated animal feeding operations, EPA-821-B-01-001. Washington, DC: United States Environmental Protection Agency.

USFWS. 1991. Contaminants in Buffalo Lake National Wildlife Refuge, Texas. U.S. Fish and Wildlife Service. Report by the Arlington Field Office. October.

USFWS. 2000. Environmental contaminants associated with confined animal feeding operations and their impacts to a service waterfowl production area. U.S. Fish and Wildlife Service. Report by Region 6. June.

USGAO. 1997. Drinking water: Information on the quality of water found at community water systems and private wells. Report to Congressional Requesters. GAO/RCED-97-123. www.gao.gov/AIndexFY97/ abstracts/re97123.htm. June.

USGS. 2005. Fate of Antibiotics in Soils. Washington, DC: United States Geological Survey.

USGS. 1997. Nutrients in the nation's waters—too much of a good thing? U.S. Geological Survey. National Water Quality Assessment Program. U.S. Geological Survey Circular 1136.

Vallentyne, J. 1974. The algal bowl: Lakes and man. Miscellaneous Special Publications 22. Dept. of the Environment Fisheries and Marine Service. Quebec: Canadian Government Publishing Center Supply and Services.

Van Donsel, D., Geldreich, E., & Clark, H. 1967. Seasonal variations in the survival of indicator bacteria in soil and their contribution to storm water pollution. *Appl Microbiol* 15:1362–1370.

Vanderholm, D.H. 1975. Nutrient losses from livestock waste during storage, treatment and handling. In *Managing Livestock Wastes.* Proceedings of 3rd Inter. Symp. on Livestock Wastes. Urbana-Champaign, Illinois: 21–24 apr. Am. Soc. Agric. Eng., St. Joseph, MI. 282–285.

Warrick, J. 1995b. Hog farm is fined $110,000 for spill. *The News and Observer.* August 23.

Wendt, R.C., & Corey, R.B. 1980. Phosphorus variations in surface runoff from agricultural lands as a function of land use. *Journal of Environmental Quality* 9(1):130–136.

Wetzel, R.G. 1983. *Limnology.* 2nd Edition. London: Saunders College Publishing.

Zirbser, K. 1998. Memo to administrative record, feedlots point source category study, re: bacterial count information from Dorothea Paul, February 5. (Mr. Zirbser reported results of independent monitoring of a stream flowing through Ms. Paul's property, downstream of a hog farm.)

7 Animal Waste and Air Quality Problems

Yes…we are next to the hog farm, and to tell you the truth, I don't know what in the world we are goin a do. It's not so bad now but when the wind changes and they turn on the fans it is very bad. My husband suffers from it a good deal. In the house we have nine special air conditions and six air purifiers runnin all the time, so it's not too awful, but outside, when the wind is right, your eyes just flame up and your throat hurts. That's why I only ask fifty dollars a month for the apartment. Otherwise it would be two hundred. So if you can stand the hog farm it's a good deal. Do you have a tendency to asthma?

(Proulx, 2002, p. 309)

Some years ago our family took a trip across the Midwest to visit relatives in Iowa, and for thousands of miles along the way we saw virtually no animal life except feedlots full of cattle—surely the most unappetizing sight and smell I've encountered in my life (and my life includes some years of intimacy with diaper pails). And we saw almost no plant life but the endless fields of corn and soybeans required to feed those pathetic penned beasts. Our kids kept asking, mile after mile, "What used to be here?"

(Kingsolver, 2002, p. 119–120)

7.1 INTRODUCTION

There are more than 1 million livestock and poultry farms in the United States. About one-third of these farms raise animals in confined areas, qualifying them as animal feed operations (AFOs) (USDA, 1999).

Airborne contaminants (air emissions) and odor have always been associated with livestock and poultry production. With the trend toward larger and more-concentrated production sites, however, gases, dust, and odors are rapidly becoming important issues for animal producers. "Even though very little evidence is available on the impact of airborne contaminants and odor from livestock operations on human health, the increasing intolerance of odors and the economic importance of animal agriculture have resulted in an urgent need for all stakeholders to find adequate solutions" (NWPS-18, 2002, P. 1). This chapter addresses the beef, dairy, swine, and poultry (broiler, laying hens, and turkey) sectors only. These animal sectors comprise the majority of animals raised in confinement in the United States; the smaller animal sectors (sheep, horses, goats, mules, rabbits, ducks, and geese) are not covered here because they do not generate emissions of the same magnitude as other animal sectors.

As previously mentioned, and as defined by the U.S. Environmental Protection Agency (USEPA, 40 CFR 122.23), an AFO is a facility where: (1) livestock or poultry are confined and fed for a total of 45 days or more in any 12-month period and (2) vegetative cover of any significance (crops, vegetative forage growth, or post harvest residues) is lacking. To be considered an AFO, the same animals do not have to be confined for 45 days, the 45 days do not have to be consecutive, and the 12-month period does not have to correspond to a calendar year. The stipulation of the absence of vegetative cover of any significance intentionally excludes operations where animals are maintained on pasture or rangeland. An AFO includes the confinement facility, manure management systems, and the manure application site.

Key term: The USEPA Office of Water uses the term *concentrated animal feeding operation* (CAFO) to designate AFOs that are point sources subject to the National Pollutant Discharge Elimination System (NPDES) permit system. Currently, 40 CFR 122.23 defines a CAFO as an AFO that confines 1,0000 animal units (AU) or more at any one time or that is designated as a CAFO on a case-by-case basis (according to 40 CFR 122.23).

Key term: *Animal unit* (AU) is a unit of measure used to compare different animal species. This text uses the definition of animal unit developed by the USEPA Office of Water (66 FR 2960-3138): 1 cattle excluding mature dairy and veal cattle; 0.7 mature dairy cattle; 2.5 swine weighing over 55 lb; 10 swine weighing 55 lb or less; 55 turkeys; 1000 chickens; and 1 veal calf.

7.2   AIR EMISSIONS FROM FEEDLOT OPERATIONS

AFOs emit gaseous and particulate substances. The primary mechanism for release of gaseous emissions is microbial decomposition of manure. The release of particulate matter is derived from the entrainment of feeds, dry manure, soil, and other material caused by movement of animals in both indoor and outdoor confinement. As previously mentioned, in this text, *manure* is defined as any combination of fecal matter, urine, and other materials that are mixed with manure (bedding material, waste feeds, wash water). Manure can be in a solid, slurry, or liquid state (surface liquids from storage facilities). Decomposition and the formation of these gaseous compounds begin immediately at excretion and continue until the manure is incorporated into the soil. Therefore, the substances generated and the subsequent rates of emission depend on a number of variables, including the species of animal, feeding practices, type of confinement facility, type of manure management system, and land application practices.

Animals also directly emit some of the gaseous substances mentioned above as a result of normal metabolic processes such as respiration. However, these emissions were not included in this text, given that they are uncontrollable (Alexander, 1977; Brock & Madigan, 1998; Tate, 1995).

7.2.1   GENERAL CHARACTERISTICS OF ANIMAL FEEDING OPERATIONS

An AFO has a confinement facility, a system for manure management (storage and in some cases stabilization), and a land application site. Because of different methods of confinement and associated manure management, no typical AFO is identified. Variances depending on animal type, regional climatic conditions, business practices, and preferences of the operator impact the design and operation of an AFO. However, the combinations of confinement and waste management systems that are most commonly used in each sector of animal agriculture are identified in this section. We present a general overview of AFOs below.

7.2.1.1   Confinement

A confinement facility may be a totally enclosed structure with full-time mechanical ventilation, a partially enclosed structure with or without mechanic ventilation, an open paved lot, or an open unpaved lot. Method of confinement, which varies among and within the animal species, probably is the most significant factor affecting emissions, because it influences ventilation and method of manure handling and disposal. Whether manure is handled as a solid, liquid, or slurry influences if microbial degradation occurs aerobically or anaerobically, and thus the substances generated.

Key term: *Aerobic* means occurring in the presence of free oxygen or capable of living or growing in the presence of free oxygen, such as aerobic bacteria.

Key term: *Anaerobic* means occurring in the absence of free or dissolved oxygen or capable of living and growing in the absence of oxygen, such as anaerobic bacteria.

7.2.1.2 Manure Management System

A manure storage facility may be an integral part of the confinement facility or located adjacent to the confinement facility. When manure is handled as a solid, storage may be within the confinement facility or in stockpiles that may or may not be covered. For liquid or slurry manure handling systems, manure may be stored in an integral tank, such as a storage tank under the floor of a confinement building, or flushed to an external facility, such as a pond or an anaerobic lagoon. Emissions from storage tanks and ponds differ those from anaerobic lagoons, which are designed for manure stabilization. *Stabilization* is the treatment of manure to reduce volatile solids and control odor prior to application to agricultural land. The use of the term "stabilization" rather than "treatment" is intended to avoid the implication that stabilized animal manure can be discharged to surface water or groundwater.

7.2.1.3 Land Application

Currently, almost all livestock and poultry manure is applied to cropland or pastures for ultimate disposal. The method of applying manure can affect emissions. Emissions from manure applied to the soil surface and not immediately incorporated are higher than those from manure that is immediately incorporated by disking or plowing. Injection, which is possible with manures handled as liquids or slurries, also reduces emissions. Conversely, the use of irrigation for the land application of liquid manure increases emissions of gaseous pollutants because of the increased opportunity for volatilization.

Key term: *Irrigation* is application of water and liquid wastes to land for agricultural purposes.

Key term: *Slurry* is manure with a total solids concentration of between approximately 5% to 15%. Slurries with a total solids concentration of less than 10% are pumpable. Above a total solids concentration of 10%, slurries are semisolids with a negligible angle of repose and can be scraped but not stacked for storage.

Table 7.1 presents an overview of the most common methods of confinement and manure management for large operations. These different combinations affect the relative magnitudes of emissions from each operation.

TABLE 7.1
Common Types of Animal Confinement and Manure Management Systems

Species	Animal confinement	Typical type of manure management systems
Broilers	Enclosed building	Integral with confinement, or open or covered stockpiles
Turkeys	Enclosed building	Integral with confinement, or open or covered stockpiles
Layers (dry manure)	Enclosed building	Integral with confinement
Layers (flush systems)	Enclosed building	Ponds and anaerobic lagoons
Swine	Enclosed building	Integral with confinement, or tanks, ponds, or anaerobic lagoons
Dairy	Enclosed building and open lots	Anaerobic lagoons, tanks and ponds, and uncovered stockpiles
Veal	Enclosed building	Integral with confinement, or tanks, ponds, or anaerobic lagoons
Beef	Open lots	Uncovered stockpiles

Source: USEPA (2001).

TABLE 7.2
Substances Potentially Emitted from Animal Feeding Operations

Animal sector	Operations	PM[a]	Hydrogen sulfide	Ammonia	Nitrous oxide	Methane	VOCs	CO_2
Broilers, turkeys, layers (dry)	Confinement	X		X				X
	Manure storage and treatment	X		X				X
	Land disposal	X		X	X			X
Layers (liquid)	Confinement	X	X	X	X	X	X	X
	Manure storage and treatment		X	X	X	X	X	X
	Land disposal		X	X	X		X	X
Swine (flush)	Confinement	X	X	X			X	X
	Manure storage and treatment		X	X		X	X	X
	Land disposal		X	X	X		X	X
Swine (other[1])	Confinement		X	X	X		X	X
	Manure storage and treatment		X	X	X	X	X	X
	Land disposal		X	X	X		X	X
Dairy (flush)	Confinement	X	X	X			X	X
	Manure storage and treatment		X	X		X	X	X
	Land disposal		X	X	X		X	X
Dairy (scrape)	Confinement	X	X	X			X	X
	Manure storage and treatment		X	X		X	X	X
	Land disposal		X	X	X		X	X
Dairy (drylot)	Confinement	X	X	X		X	X	X
	Manure storage and treatment	X	X	X		X	X	X
	Land disposal	X	X	X	X		X	X
Veal	Confinement	X	X	X			X	X
	Manure storage and treatment		X	X		X	X	X
	Land disposal	X	X	X	X		X	X
Beef	Confinement	X	X	X		X	X	X
	Manure storage and treatment	X	X	X		X	X	X
	Land disposal	X	X	X	X		X	X

Source: USEPA (2001).

Note: PM = particulate matter, as total suspended particulate; VOC = volatile organic compounds; CO_2 = carbon dioxide

[a] Other includes pit storage, pull plug pits, and pit recharge systems

7.2.2 SUBSTANCES EMITTED

A number of factors affect the emission of gases and particulate matter from AFOs. Most of the substances emitted are the products of microbial processes that decompose the complex organic constituents in manure. The microbial environment determines which substances are generated and at what rate. In this section, we describe the chemical and biological mechanisms that affect the formation and release of emissions.

Table 7.2 summarizes the substances that can be emitted from different operations with an AFO. Although all AFOs share the same three common elements (confinement faculties, manure management system, and land application site), the differences in production and manure management practices both among and within the different animal sectors result in different microbial environments and therefore different emission potentials. Several factors affect the emissions of ammonia, nitrous oxide, methane, carbon dioxide, volatile organic compounds (VOCs), hydrogen sulfide, particulate matter, and odors.

7.2.2.1 Ammonia

Ammonia is produced as a by-product of the microbial decomposition of the organic nitrogen compounds in manure. Nitrogen occurs as both unabsorbed nutrients in manure and as either urea (mammals) or uric acid (poultry) in urine. Urea and uric acid hydrolyze rapidly to form ammonia and are emitted soon after excretion. The formation of ammonia continues with the microbial breakdown of manure under both aerobic and anaerobic conditions. Because ammonia is highly soluble in water, ammonia accumulates in manures handled as liquids and semisolids or slurries but volatizes rapidly with drying from manures handled as solids. Therefore, the potential for ammonia volatization exists wherever manure is present, and ammonia is emitted from confinement buildings, open lots, stockpiles, anaerobic lagoons, and land application from both wet and dry handling systems.

Key term: *Anaerobic bacterium* are bacteria that do not require the presence of free or dissolved oxygen.

Key term: An *anaerobic lagoon* is a facility used to stabilize livestock or poultry manure using anaerobic microorganisms to reduce organic compounds to methane and carbon dioxide.

The volatilization of ammonia from any AFO can be highly variable depending on total ammonia concentration, temperature, pH, and storage time. Emissions depend on how much of the ammonia-nitrogen in solution reacts to form ammonia versus ionized ammonium (NH_4^+), which is nonvolatile. In solution, the partitioning of ammonia between the ionized (NH_4^+) and unionized (NH_3) species is controlled by pH and temperature. Under acidic conditions (pH values of less than 7.0), ammonium is the predominate species, and ammonia volatilization occurs at a lower rate than at higher pH values. However, some ammonia volatilization occurs even under moderately acidic conditions. Under acidic conditions, ammonia that is volatized is replenished because of the continual reestablishment of the equilibrium between the concentrations of the ionized and unionized species of ammonia in solution following volatilization. As pH increases above 7.0, the concentration of ammonia increases, as does the rate of ammonia volatilization. The pH of manures handled as solids can be in the range of 7.5 to 8.5, which results in fairly rapid ammonia volatilization. Manure handled as liquids or semisolids tend to have lower pH.

Because of its high solubility in water, the loss of ammonia to the atmosphere is more rapid when drying of manure occurs. However, little difference in total ammonia emissions occurs between solid and liquid manure handling systems if liquid manure is stored over extended periods prior to land application.

7.2.2.2 Nitrous Oxide

Nitrous oxide also can be produced from the microbial decomposition of organic nitrogen compounds in manure. Unlike ammonia, however, nitrous oxide is emitted only if nitrification occurs and is followed by denitrification. Nitrification is the microbial oxidation of ammonia to nitrites and nitrates and requires an aerobic environment. Denitrification most commonly is a microbially mediated process by which nitrites and nitrates are reduced under anaerobic conditions. The principal end product of denitrification is dinitrogen gas (N_2). However, small amounts of nitrous oxide as well as nitric oxide also can be generated under certain conditions. Therefore, for nitrous emissions to occur, the manure must first be handled aerobically (dry) and then anaerobically (wet).

Key term: *Denitrification* is the chemical or biological reduction of nitrate or nitrite with molecular nitrogen (N_2) as the primary end product. Other possible end products are nitrous oxide (N_2O) and nitric oxide (NO).

Key term: *Nitrification* is the microbially mediated biochemical transformation by oxidation of ammonium (NH_4^+) to nitrite (NO_2^-) or nitrate (NO_3^-).

Nitrous oxide emissions are most likely to occur from unpaved drylots for dairy and beef cattle and at land application sites, the sites most likely to have the necessary conditions for both nitrification and denitrification. At these sites, the ammonia nitrogen that is not lost by volatilization is adsorbed on soil particles and subsequently oxidized to nitrite and nitrate nitrogen. Emissions of nitrous oxide from these sites depends on two primary factors. The first is drainage. In poorly drained soils, the frequency of saturated conditions, and thus anaerobic conditions necessary for denitrification, are higher than for well-drained soils. Conversely, the opportunity for leaching of nitrite and nitrate nitrogen through the soil is higher in well-drained soils, and the conversion to nitrous oxide is less. Therefore, poorly drained soils enhance nitrous oxide emissions. The second factor is plant uptake of ammonia and nitrate nitrogen. Manure that is applied to cropland outside of the growing season has more available nitrogen for nitrous oxide emissions, as does manure that is applied at higher than agronomic rates.

Key term: *Drylots* are open feedlots sloped or graded from 4% to 6% to promote drainage away from the lot to provide consistently dry areas for cattle to rest. Drylots may be paved, unpaved, or partially paved.

At most operations, the manure application site is the principal source of nitrous oxide. However, if manure is applied correctly and at agronomic rates, little if any increase should occur in nitrous oxide emissions relative to emissions from application of inorganic commercial fertilizers.

7.2.2.3 Methane

Methane is a product of the microbial degradation of organic matter under anaerobic conditions. The microorganisms responsible, known collectively as *methanogens,* decompose and convert carbon (cellulose, sugars, proteins, fats) in manure and bedding materials into methane and carbon dioxide. Because anaerobic conditions are necessary, manures handled as a liquid or slurry emit methane. Manures handled as solids generally have a low enough moisture content to allow adequate diffusion of atmospheric oxygen to preclude anaerobic activity or permit the subsequent oxidation of any methane generated.

Methane is insoluble in water. Thus, methane volatilizes from solution as rapidly as it is generated. Concurrent with the generation of methane is the microbially mediated production of carbon dioxide, which is only sparingly soluble in water. Therefore, methane emissions are accompanied by carbon dioxide emissions. The mixture of these two gases is commonly referred to as *biogas.* The relative fractions of methane and carbon dioxide in biogas vary depending on the population of methanogens present. Under conditions favorable for the growth of methanogens, biogas normally are between 60% and 70% methane and 30% to 40% carbon dioxide. If, however, the growth of methanogens is inhibited, the methane fraction of biogas can be less than 30%.

Key term: *Biogas* is a combustible mixture of methane and carbon dioxide produced by the bacterial decomposition of organic wastes under anaerobic conditions. It may be used as a fuel.

The principal factors affecting methane emissions are the amount of manure produced and the portion of the manure that decomposes anaerobically, which depends on the biodegradability of the organic fraction and the management of manure. When manure is stored or handled as a liquid (anaerobic lagoons, ponds, tanks, or pits), it decomposes anaerobically and produces a significant quantity of methane. Anaerobic lagoons are designed to balance methanogenic microbial activity with organic loading and, therefore, produce more methane than ponds or tanks. The organic content of manure is measured as volatile solids. When manure is handled as a solid (in open feedlots or stockpiles), it tends to decompose aerobically and little or no methane is produced. Likewise, manure application sites are not likely sources of methane because the necessary anaerobic conditions generally do not exist, except when soils become saturated. In addition, because methane is insoluble in water, any methane generated during liquid storage or stabilization treatment is released immediately and is not present when manure is applied to cropland.

7.2.2.4 Carbon Dioxide

Carbon dioxide is a product of the microbial degradation of organic matter under both aerobic and anaerobic conditions. Under aerobic conditions, carbon dioxide and water are the end-products, with essentially all of the carbon emitted as carbon dioxide. Under anaerobic conditions, carbon dioxide is one of the products of the microbial decomposition of organic matter to methane. Under these conditions, carbon dioxide is formed as a by-product of the decomposition reactions involving complex organic compounds that contain oxygen. Thus, carbon dioxide is emitted under both aerobic and anaerobic conditions and occurs wherever manure is present. Land application sites emit carbon dioxide from the decomposition of manorial organic matter by soil microorganisms.

Although AFOs emit carbon dioxide, the emissions do not contribute to a net long-term increase in atmospheric carbon dioxide concentrations. The carbon dioxide from animal manures is a release of carbon sequestered by photosynthesis during the past 1 to 3 years at most. Thus, the carbon dioxide emitted is part of a cycling of carbon from the atmosphere to crops to animals and back into the atmosphere over a relatively short time period.

7.2.2.5 Volatile Organic Compounds

VOCs are formed as intermediate metabolites in the degradation of organic matter in manure. Under aerobic conditions, any VOC formed are rapidly oxidized to carbon dioxide and water. Under anaerobic conditions, complex organic compounds are degraded microbially to volatile organic acids and other VOCs, which in turn are converted to methane and carbon dioxide by methanogenic bacteria. When the activity of the methanogenic bacteria is not inhibited, virtually all VOCs are metabolized to simpler compounds, and the potential for VOC emissions is nominal. However, the inhibition of methane formation results in a buildup of VOCs in the manure and ultimate volatilization to the air. Inhibition of methane formation typically is caused by low temperatures or excessive loading rates of volatile solids in a liquid storage facility. Both of these conditions create an imbalance between populations of the microorganisms responsible for the formation of VOCs and methanogenic bacteria. Therefore, VOC emissions are minimal from properly designed and operated stabilization processes (such as anaerobic lagoons) and the associated manure application site. In contrast, VOC emissions are higher from storage tanks, ponds, overloaded anaerobic lagoons, and associated land application sites. The specific VOC emitted varies depending on the solubility of individual compounds and other factors (including temperature) that affect solubility.

7.2.2.6 Hydrogen Sulfide and Other Reduced Sulfur Compounds

Hydrogen sulfide and other reduced sulfur compounds are produced as manure decomposes anaerobically. The two primary sources of sulfur in animal manures are the sulfur amino acids contained in feed and inorganic sulfur compounds, such as copper sulfate and zinc sulfate, which are used as feed additives to supply trace minerals and serve as growth stimulants. Although sulfates are used as trace mineral carriers in all sectors of animal agriculture, their use is more extensive in the poultry and swine industries. A possible third source of sulfur in some locations is trace minerals in drinking water.

Hydrogen sulfide is the predominant reduced sulfur compound emitted from AFOs. Other emitted compounds include methyl mercaptans, dimethyl sulfide, dimethyl disulfide, and carbonyl sulfide. Small quantities of other reduced sulfur compounds are likely to be emitted as well.

Under anaerobic conditions, any excreted sulfur that is not in the form of hydrogen sulfide is reduced microbially to hydrogen sulfide. Therefore, manures managed as liquids or slurries are potential sources of hydrogen sulfide emissions. The magnitude of hydrogen sulfide emissions is a function of liquid phase concentration, temperature, and pH. Temperature and pH affect the solubility of hydrogen sulfide in water, which increases at pH values above 7. Therefore, as pH shifts from alkaline to acidic (pH < 7), the potential for hydrogen sulfide emissions increases (Snoeyink & Jenkins, 1980). Under anaerobic conditions, livestock and poultry manures are acidic, with pH values ranging from 5.5 to 6.5.

Key term: *pH* is the negative logarithm of the hydrogen ion concentration. The pH scale ranges from 0 to 14. Values below 7 are considered acidic; those above, alkaline.

Under aerobic conditions, any reduced sulfur compounds in manure are oxidized microbially to nonvolatile sulfate, and emissions of hydrogen sulfide are minimal. Therefore, emissions from confinement facilities with dry manure handling systems and dry manure stockpiles should be negligible, if adequate exposure to atmospheric oxygen to maintain aerobic conditions occurs. Any hydrogen sulfide that is generated in dry manure generally is oxidized as diffusion through aerobic areas occurs.

In summary, manure storage tanks, ponds, anaerobic lagoons, and land application sites are primary sources of hydrogen sulfide emissions whenever sulfur is present in manure. Confinement facilities with manure flushing systems that use supernatant from anaerobic lagoons also are sources of hydrogen sulfide emissions.

Key term: *Supernatant* is the liquid fraction above settled solids in a lagoon or storage tank.

7.2.2.7   Particulate Matter

In this text, the authors consider particulate matter (PM) as PM10 and PM2.5. PM 10 is commonly defined as airborne particles with aerodynamic equivalent diameters (AEDs) less than 10 μm. The number refers to the 50% cut diameter in a Federal Reference Method PM 10 sampler where particles of 10 μm AED are collected at 50% efficiency (62 Fed. Reg. 38651-38701). Similarly, PM2.5 refers to the particles that are collected in a Federal Reference Method PM2.5 sampler, which has a 50% cut diameter of 2.5 μm (62 Fed. Reg. 38651-38701).

Key term: *Particulate matter (PM)* is any airborne, finely divided solid or liquid matter with an aerodynamic diameter less than or equal to 100 μm. For this text, PM means total suspended particulate (TSP), except where noted specifically as PM 10.

Key term: *Aerodynamic equivalent diameter (AED)* is the diameter of a sphere (in μm) of unit density (1 g/cm^3) that has the same terminal settling velocity in air as the particle of interest (a 1 μm AED particle has 1000 times the volume of a 0.1 μm AED particle). The AED refers to an individual particle. The AED of PM is critical to its health and radiative effects. PM 2.5 can reach and be deposited in the smallest airways (alveoli) in the lungs, whereas larger particles tend to be deposited in the upper airways of the respiratory tract (National Research Council [NRC], 2002).

Sources of PM emissions include feed, bedding materials, dry manure, unpaved soil surfaces, animal dander, and poultry feathers. Therefore, confinement facilities, dry manure storage sites, and land application sites are potential PM emission sources. The relative significance of each source depends on three interrelated factors: (1) the type of animal being raised, (2) the design of the confinement facility being used, and (3) the method of manure handling.

The National Ambient Air Quality Standards (NAAQS, pronounced "knacks") currently regulate concentrations of PM with a mass median diameter of 10 μm of less (PM 10). Studies have shown that particles in the smaller size fractions contribute most to human health effects. The current PM 10 standard may be replaced by a standard for PM 2.5. A PM 2.5 standard was published in 1997, but has not been implemented pending the results of ongoing litigation.

The particle size distribution of PM emitted from AFOs has not been well-characterized. Virtually all of the emissions studies to date have measured total suspended particulate or did not report the test method used. Particle size distribution data was found only for beef feedlots. In one study, ambient measurements of PM 10 and PM 2.5 (using 5 hour sample collection periods) were taken downwind (15 to 61 m) of three cattle feedlots in the Southern Great Plains (Sweeten et al., 1998). In this study, PM 10 was measured as 20% to 40% of TSP (depending on the measurement method used), and PM 2.5 was 5% of TSP. No studies were found of particle size distribution from confinement buildings. Based on the emission mechanisms at AFOs, one would expect to find that: (1) PM from AFOs would have varying particle size distributions depending on the animal sector, method of confinement, and type of building ventilation used and (2) the PM emitted would include PM 10 and a lesser fraction of PM 2.5. In addition to direct emissions, PM 2.5 can be secondarily formed in the atmosphere from emissions of ammonia. If sulfur oxides or nitrogen oxides are present in the air, ammonia is converted to ammonium sulfate or ammonium nitrate, respectively. No information is available at this time to quantify the emissions of secondarily formed PM 2.5.

All confinement facilities are sources of PM emissions. However, the composition of these emissions vary. The only constant constituent is animal dander and feather particles from poultry. For poultry and swine, feed particles constitute a significant fraction of PM emissions because the dry, ground feed grains and other ingredients used to formulate these feeds are inherently dusty. Pelleting of feeds reduces, but does not eliminate, dust and PM emissions. Dried forages also generate PM, but most likely to a lesser degree. Silages, which have relatively high moisture content, tend to generate less PM than do other types of feed. Because veal calves are fed a liquid diet, feed does not contribute to particle emissions from veal operations.

The mass of PM emitted from totally or partially enclosed confinement facilities, as well as the particle size distribution, depend on type of ventilation and ventilation rate. Particulate matter emissions from naturally ventilated buildings are lower than those from mechanically ventilated buildings. Mechanically ventilated buildings emit more PM at higher ventilation rates. Therefore, confinement facilities located in warmer climates tend to emit more PM because of the higher ventilation rates needed for cooling.

While confinement facilities for dairy and beef cattle typically are all naturally ventilated, facilities for poultry, swine, and veal are mechanically ventilated for all or at least part of the year. When mechanical ventilation is used for only part of the year, it is used during the coldest and hottest months, with natural ventilation used during the remainder of the year.

Open feedlots and storage facilities for dry manure from broilers, turkeys, laying hens in high rise houses, dairy drylots, and beef cattle drylots also are potential sources of PM. These sites are intermittent sources of PM emissions, because of the variable nature of wind direction and speed and precipitation. Thus, the moisture content of the manure and the resulting emissions are highly variable. The PM emissions from covered manure storage facilities depend on the degree of exposure to wind.

Key term: A *feedlot* is a concentrated, confined animal or poultry growing operation for meat, milk, or egg production, or stabling, in pens or houses, wherein the animals or poultry are fed at the place of confinement and crop or forage growth or production is not sustained in the area of confined and is subject to 40 CFR 412.

Key term: *Broilers* are chickens of either sex specifically bred for meat production and marketed at approximately 7 weeks of age.

Key term: A *hen* is a mature female chicken.

7.2.2.8   Odors

Generally, odor is not a problem until the neighbors complain. In the not-too-distant past, agricultural activities were isolated, rural activities with few neighbors, and those neighbors were also agriculturally based. Thus, agricultural operations that produced odors in these settings, those having few neighbors, consequently had few, if any, complaints. With the expansion of urban centers (suburbia, exurbia), however, many formerly isolated agricultural operations have been squeezed from all directions by new construction and new neighbors. Odors that were not previously offensive because of lack of neighbors now have become irritating to the point of complaint—and complain these new, urban neighbors do.

Odor generated from an AFO is not the result of a distinct compound but rather is is the result of a large number of contributing compounds. Schiffman et al. (2001) identified 331 odor-causing compounds in swine manure. The principal compounds responsible for noxious odors are hydrogen sulfide, ammonia, and VOCs. The VOCs that contribute to odors are volatile acids (acetic, propionic, formic, butyric, and valeric), indole, phenols, volatile amines, methyl mercaptans, and skatole.

Most of the odorous compounds are products of anaerobic digestion or organic compounds. Therefore, the potential for odors is greater at operations with liquid manure management systems. In liquid systems, odors can be produced from storage pits, ponds, and land application. Properly designed and operated anaerobic lagoons should have relatively low odors, but odors can be produced under two conditions: (1) in the spring and fall, when sudden temperature changes can upset the microbial balance or (2) if the lagoon is overloaded with volatile solids. Drylots can produce odors whenever warm, wet conditions produce transient anaerobic conditions. Odors also can be caused by decaying animals, if the carcasses are stored too long prior to disposal.

- *Important point:* Land application of manure from livestock and poultry facilities is a frequent source of odor complaints from neighbors (the public).

7.2.3   SUMMARY OF FACTORS AFFECTING EMISSIONS

To summarize Section 7.2.2, emissions from AFOs depend on manure characteristics and manure management. Manure excreted by each type of animal has specific characteristics (nitrogen content, moisture content). These characteristics, however, can be altered depending on how the manure is collected, stored, and land applied. The potential for generating emissions from manure management systems used for the beef, dairy, swine, and poultry sectors depends on several factors. The potential for PM emissions depends on whether the manure is handled in a wet or dry state. The potential for gaseous emissions generally depends on: (1) the presence of an aerobic or anaerobic microbial environment, (2) the precursors present in the manure (for example, sulfur), (3) pH of the manure, and (4) time and temperature in storage, which primarily affect mass emitted. The effect of each of these factors on emission is summarized in Table 7.3 and described next.

TABLE 7.3
Pollutant Precursors

Substance emitted	Wet manure handling	Dry manure handling	pH	High temperature	Manure residence time	Precursors
Ammonia			> 7.0	X	X	Nitrogen
Nitrous oxide		X				Nitrogen
Hydrogen sulfide	X		< 7.0	X	X	Sulfur
Methane	X			X	X	Carbon
VOCs	X			X	X	Carbon
Particulate matter*		X				

Source: USEPA (2001a).

* Total suspended particulate. Fine particles (PM 2.5) in the form of ammonium sulfate and ammonium nitrate can be secondarily formed in the atmosphere from ammonia emissions, if sulfur oxides or nitrogen oxides are present in the air.

Wet/dry manure management system. To form hydrogen sulfide (and other reduced sulfur compounds), methane, and VOCs requires an anaerobic environment. Therefore, the potential to emit these substances is greatest when manure is handled as a liquid or slurry. Ammonia is generated in both wet and dry manure. Nitrous oxide is formed only when manure handled in a dry state becomes saturated (thus forming transient anaerobic conditions).

pH. Emissions of ammonia and hydrogen sulfide are influenced by pH. Manure pH affects the partitioning between these compounds and their ionized forms (NH_4^+ and HS^-), which are nonvolatile.

Temperature. Temperature has two effects: (1) Temperature affects gas phase vapor pressure and, therefore, volatility. For substances that are soluble in water (ammonia, some VOCs, hydrogen sulfide, and other reduced sulfur compounds), emissions are greater at higher temperatures. Emission rates of these substances is greater in warmer climates and in the summer than in colder climates and winter. Methane is insoluble in water and, at any temperature, is emitted very quickly after formation. (2) Higher temperature favors the microbial processes that generate methane and other substances.

Time in storage. Long periods of manure residence time in confinement, storage, or stabilization facilities provide greater opportunities for anaerobic breakdown and volatilization to the air. Also, masses emitted increase with time.

Precursors. The amount of sulfur ingested by an animal affects the potential for hydrogen sulfide production in manure. Sulfur can be present in feed additives and, in some cases, from water supplies. The amount of nitrogen in feed (proteins and amino acids) affects ammonia and nitrous oxide emission potential. The amount of carbon affects methane and carbon dioxide potential. Ensuring that the composition of feedstuffs does not exceed the nutritional needs of the animals reduces emissions.

7.3 EMISSIONS FROM LAND APPLICATION

Applying animal manure from AFOs to cropland generates air emissions. These emissions result primarily from the volatilization of ammonia at the point the material is applied to land (Anderson, 1994). Additional emissions of nitrous oxide are released from cropland when nitrogen applied to the soil undergoes nitrification and denitrification. Loss through denitrification depends on the oxygen levels of the soil to which manure is applied. Low oxygen levels, resulting from wet, compacted, or warm soil, increase the amount of nitrate-nitrogen released into the air as nitrogen gas or nitrous oxide (Ohio State University Extension [OSUE], 2000). However, a study by Sharpe and Harper (1997), which compared losses of ammonia and nitrous oxide from the sprinkler irrigation of swine effluent, concluded that ammonia emissions contributed more to airborne nitrogen losses. This analysis of air emissions from land application activities focuses on the volatilization of nitrogen as both ammonia and nitrous oxide and quantified both on- and off-site emissions.

Note that because of the numerous variables affecting the nature and emission rates of ammonia, PM, nitrous oxide, hydrogen sulfide, methane, and VOCs, even generally quantifying emissions of these substances from land application sites is difficult. Adding to this problem is the effect of emissions of these substances prior to land application. For example, a high rate of ammonia loss from an anaerobic lagoon because of warm summer temperatures translates into lower emissions from the land application site. Conversely, a low rate of ammonia loss from an anaerobic lagoon translates into a higher loss during land application. Thus, the lack of consistent estimates of emissions from land application sites found in the literature is understandable.

Emissions from land application occur in two phases. The first phase occurs during and immediately following application. These short-term emissions are influenced by the type of manure application method used. The second phase is the release from the soil that occurs over a longer period as a result of the microbial breakdown of substances in the applied manure.

7.3.1 Short-Term Emissions

7.3.1.1 Particulate Matter

If manure is handled as a solid and has a relatively low moisture content, PM emissions occur during the spreading process and also may occur immediately after spreading as the result of wind action. The duration of PM emissions from wind action after spreading depends on weather conditions and is highly variable. For example, a precipitation event occurring immediately after spreading can essentially eliminate PM postspreading emissions. Irrigation, obviously, has the same effect. Conversely, a period of windy, dry weather after spreading increases PM emissions.

7.3.1.2 Nitrogen Compounds, Hydrogen Sulfide, and VOCs

If ammonia, hydrogen sulfide, or VOCs are present in the manure being spread, emissions occur by volatilization to the air. The magnitudes of these emissions primarily depends on whether the manure is incorporated into the soil by disking, plowing, or direct injection. Theoretically, injection should be the most effective technique for minimizing the emissions of these compounds because it prevents exposure to the atmosphere. Efficiency depends to a degree, however, on subsequent closure of the channel or slit in the soil formed by the injector. With disking and plowing, efficiency depends on the time between spreading and incorporation. Plowing is more effective than disking in reducing emissions, because disking leaves some manure exposed to the atmosphere. Precipitation or irrigation immediately following manure spreading also reduces emissions of ammonia, hydrogen sulfide, and VOCs by the transport of these water-soluble compounds into the soil. In the short-term, nitrification, and consequently nitrous oxide emissions, do not occur (Alexander, 1977; Brock & Madigan, 1988; Tate, 1995).

7.3.1.3 Methane

Little or no methane is emitted in the short-term because methane is essentially insoluble in water. Only methane in manure has volatilized prior to land application. Therefore, any short-term methane emissions from land applications sites are limited to small amounts that are formed immediately following application of manure slurries and liquid manure. Aerobic conditions limit additional formation of methane to negligible amounts during application.

7.3.2 Long-Term Emissions

Land application sites used for the disposal of livestock and poultry manure are potential short-term sources of emissions of PM, ammonia, hydrogen sulfide, and VOCs. Given the number of variables with the potential to influence the magnitude of actual emissions, developing typical emission factors is problematic. Long-term emissions should be limited to possibly some nitrous oxide emissions. However, these emissions should not be substantially different from those resulting from the use of inorganic nitrogen fertilizers.

Cropland soils are generally aerobic microbial environments except for transient periods of saturation associated with precipitation and possibly irrigation events. Therefore, manorial ammonia, hydrogen sulfide, and VOCs not lost by volatilization during or immediately after manure spreading and entering the soil profile should be oxidized microbially to nitrate, sulfate, and carbon dioxide and water, respectively. The nitrogen, sulfur, and carbon in organic compounds subsequently mineralized are also are oxidized.

7.3.2.1 Nitrogen Compounds

Under transient periods of saturation and anaerobic conditions, any nitrate remaining after plant uptake and leaching to groundwater may undergo microbially mediated denitrification. As previously discussed, the principal end product of denitrification is dinitrogen gas. However, small amounts of nitrous oxide and nitric oxide also may be emitted under certain environmental conditions. Therefore, land used for manure disposal can be considered a potential source of nitrous oxide emissions. However, nitrous oxide is also generated when denitrification follows the application of inorganic nitrogen fertilizer materials. Thus, it appears nitrous oxide emissions would be no greater than if commercial fertilizers are used if nitrogen (in manure) application rates are based on crop requirements. However, application rates in excess of crop requirements result in higher emissions.

7.3.2.2 Hydrogen Sulfide

Hydrogen sulfide is oxidized to sulfate in the soil but subsequently may be reduced back to hydrogen sulfide during transient saturated soil conditions. The high solubility of hydrogen sulfide and other reduced sulfur compounds, however, should preclude any significant emissions. Reoxidation occurs following the return to aerobic conditions (Alexander, 1977; Brock & Madigan, 1988; Tate, 1995).

7.3.2.3 Methane and VOCs

Under transient saturated conditions, any remaining organic compounds in manure may be reduced to VOCs and methane. However, any VOCs formed are oxidized to carbon dioxide when aerobic conditions are reestablished. Given that methanogenic bacteria are obligate anaerobes (microorganisms that do not grow in the presence of oxygen), the presence of a population sufficient to generate any significant quantity of methane under transient anaerobic conditions is highly unlikely. In addition, if methane is formed, a population of methanotrophic (methane-oxidizing) microorganisms capable of oxidizing methane to carbon dioxide may be present (Alexander, 1977; Brock & Madigan, 1988; Tate, 1995).

7.4 EMISSIONS FROM VIRTUAL FARMS

This section explains USEPA (2001a) methods used to estimate emissions from model farms. The model farms reflect combinations of different confinement facilities, manure collection systems, and manure storage practices. For this section, we have modified and adapted USEPA models and changed the format from model application to virtual format in accordance with current digital practice. In addition, emission factors were developed for each element of a virtual farm (for example, drylot, storage pond). The estimated emissions for the entire virtual farm were then calculated by summing the emissions from each element. The following approaches were employed to develop emission factors:

- Emissions factors were gathered from the literature or derived from emission measurements data found in the literature (Section 7.4.1).
- If emission data were not available from the literature, an emission factor developed for one animal species was translated to another species, when justifiable (Section 7.4.2).
- If emission factors were unavailable from the literature and could not be translated from one species to another, an emission factor was derived based on the quantity of precursors in the manure, where appropriate (for example, nitrogen content of manure was used to estimate ammonia and nitrous oxide emissions in some cases) (Section 7.4.2). The method for estimating the quantity of precursors in manure is explained in Section 7.4.3.
- Where no emission factors or estimation methods were identified, no emissions were estimated, but the results identified elements of the model farm where emissions are expected. This judgment was based on knowledge of fundamental microbial and emission mechanisms.

Section 7.4.4 presents the emission factors and the annual emissions from the virtual farms. To provide a perspective on these results, Section 7.4.5 compares the virtual farm emissions to the amount of volatile solids, sulfur, and nitrogen in manure (the upper limits for transformation into gaseous substances).

7.4.1 DEVELOPMENT OF EMISSION FACTORS FROM LITERATURE SOURCES

The first step in developing emission factors was a literature search to locate published information about emissions. Included in this search were relevant peer-reviewed journals and published conference proceedings and research reports available as of May 2001. The AGRICOLA (Agricultural Online Access) bibliographic data base was used (Eastern Research Group [ERG], 2000). A total of 481 seemingly applicable references were identified, obtained, and reviewed. [*Note:* For the reader who wants a detailed listing of all reference material used by USEPA (2001a) for their model farms presentation, the original USEPA document includes several appendices listing sources with descriptions.]

7.4.1.1 Emission Data Review

In the review of each publication, the principal objective was to find emission factors or measurements data to allow derivation of emission factors by the individual elements of the virtual farms. Each publication was reviewed to ensure that the information presented was representative of expected emissions from the virtual farms defined herein. Studies that could not be partitioned to estimate emission factors for individual elements of the virtual farms were not used. Accordingly, the studies were screened to identify emission data that could be related to the following parameters:

- Animal species
- Number of animals present
- Type of confinement facility
- Type of manure handling and storage system
- Phase of production (e.g., finishing operation)
- Specific emission points tested
- Units of measure that could be converted to mass per year

When the publications were screened, that many of these articles did not contain the necessary information to develop emission factors became evident. Some of the articles provided only concentration or flux measurements without any background information, such as confinement capacity, number and age or size of animals present, or characterization of any accumulated manure present to allow translation of the measured values reported into an emissions factor. For example, a reference might provide a measurement of ammonia emissions but not indicate what size farm or number of animals were associated with the emissions measurements. Some studies provided concentration measurements at confinement houses without indication of the volumetric flow rates needed to convert concentrations to an emission rate. Some emission factors were expressed in units of measure that could not be converted to a mass per year per AU basis (e.g., mass per kilogram of litter per day). Some references presented data from laboratory studies and novel manure management techniques that were unlikely to be representative of typical U.S. operations. In addition, some of the references did not have adequate documentation of the emission points measured. For example, a reference might not indicate if emissions were measured from a flush house, anaerobic lagoon, or a combination of both. Other articles provided emission factors for the entire farm or from several emission sources combined.

TABLE 7.4
References Identified with Useful Emission Information

Animal type	Number of references*
Beef	6
Dairy	6
Veal	0
Broiler	8
Layer	7
Swine	24

* References are identified in Section 7.4.4.

That no approach was being employed to enable the direct comparison of emission factors on a standard basis became apparent as well. One of the more commonly used approaches was a per unit confinement capacity per year basis (e.g., mass emitted per number of broilers confined in a year). However, approaches were encountered such as mass per area confined per hour, mass per pig place, and mass per animal lifetime.

Table 7.4 tabulates the number of references identified with useful emission information to develop emission factors for each animal type. These references account for approximately 6% of the publications reviewed.

7.4.1.2   Emission Factor Development

From these emission data, emission factors were developed on the basis of mass per year per AU (lb/year-AU). An AU is a standard basis for comparing the size of AFOs across different species. While different definitions of AU are in use, this study used the definition by the USEPA Office of Water. The proposed revisions to the NPDES regulations and Effluent Limitation Guidelines and Standards for Concentrated Animal Feeding Operations (66 FR 2960) define an AU (for the purpose of reader review) as the capacity to confine:

- 1 cattle, excluding mature dairy and veal cattle
- 0.7 mature dairy cattle
- 2.5 swine weighing over 55 lb
- 10 swine weighing 55 lb or less
- 55 turkeys
- 100 chickens
- 1 veal

An annual basis was used to adequately reflect the differences in production cycles, feed consumption, and manure production among various species of animals. Thus, emission data or factors that were expressed on another basis (confinement capacity or time period) were converted to an annual basis using typical values for live weight, lengths of production cycles, and number of production cycles per year. The values used to make these conversions are described in Section 7.4.3.2.

In many cases, the emission factors were based on only one or two references. Where valid emission factors were available from more than one study, a mean emission factor was calculated for that particular pollutant and element of the virtual farm. In some instances, a reference contained results from emission measurements during different seasons of the year or at different geographic locations. Where ranges of emission values were reported in a study, the mean of the values reported was used to develop the emission factor.

TABLE 7.5
Summary of Emission Estimation Methods

Substance	Emission factors	Translated from one animal type to another	Factors based on precursor generation
Ammonia	X	X*	
Nitrous oxide			X**
Hydrogen sulfide	X	X***	
VOCs			X**
Particulate matter	X		

Source: USEPA (2001).
* Flush dairy barns, dairy lagoons, poultry lagoons, turkey barns.
** All emission factors.
*** Poultry lagoons, dairy lagoons.

7.4.2 OTHER METHODS USED TO CALCULATE EMISSIONS

In the absence of emission factor estimates based on measured values, two alternative approaches were employed. The first approach was to translate emission factors from one animal species to another by adjusting for differences in the quantity and composition of manure. The second approach was to calculate emissions based on precursors in the manure (nitrogen, sulfur, and volatile solids). These approaches were used when a rational basis and sufficient data were available to support the alternate approach. The option of using theoretical models, especially for estimating ammonia emissions, was considered. However, these models would have required an extensive degree of validation that was outside the scope of this text. The alternative approaches used for each pollutant (when emissions data were not available) are summarized in Table 7.5 and described in the following sections.

Key term: *Volatile solids* are solids lost upon ignition at 550°C (using Method 2540 E of the American Public Health Association). Volatile solids provide an approximation of organic matter (carbon) present.

7.4.2.1 Ammonia

For most emission sources, ammonia emission factors were found in the literature. However, no emission factors were found for dairy freestall barns with flush systems or anaerobic lagoons for dairy and laying hen manure. For these sources, translating ammonia emission factors from the swine sector-developed emission factors. Although manure characteristics differ significantly from one animal species to another, the mechanism by which ammonia is formed and the chemistry of ammonia in solution should not be different (Alexander, 1977; Brock & Madigan, 1988; Tate, 1995). Therefore, for these sources, emission factors developed for one species could be reasonably well translated to another by adjusting to reflect differences in excretion rates. Accordingly, emission factors from swine lagoons were applied to anaerobic lagoons in the laying hen and dairy virtual farms. Emission factors for swine flush houses were translated to dairy flush barns.

These translations were done by assuming that the ratio of ammonia emitted to the nitrogen in manure is the same for swine, poultry, and dairy cows. The following equation illustrates the translation for anaerobic lagoons for laying hens:

$$EF_{am, h} = \frac{EF_{am, s}}{M_{ns}} \times M_{n, h}$$

where

$EF_{am, h}$ = Emission factor for ammonia from anaerobic lagoons for laying hens (lb/year).
$EF_{am, s}$ = Emission factor for ammonia from an anaerobic lagoon for swine (lb/year).
$M_{n, s}$ = Nitrogen excretion rate in swine manure (lb/year).
$M_{n, h}$ = Nitrogen excretion rate in laying hen manure (lb/year).

7.4.2.1.1 Calculation of Emission Factors

In the absence of ammonia and hydrogen sulfide emissions data for an animal species, an approach was developed to estimate emissions based on translating emissions information from another animal species. The approach involved adjusting emissions based on the nitrogen and sulfur excretion rates of different animal species.

Emissions information was only translated from one species to another if: (1) no emissions information was available from the literature review and (2) the operation is expected to have similar emission mechanisms regardless of the animal type (for example, anaerobic microbes at a dairy lagoon act similar to those in a swine lagoon).

Consequently, ammonia and hydrogen sulfide emission factors for dairy flush houses, dairy anaerobic lagoons, and layer anaerobic lagoons were developed by translating emissions information from swine model farms. Ammonia emission factors for turkey houses were also calculated using information from broiler houses.

Section 7.4.2.1 discusses the methodology used to develop emission factors using this approach. Example calculations of the methodology are presented in this section.

7.4.2.1.2 Calculation of Dairy Ammonia Flush House Emissions Factor from Swine Flush House Information

1. The fraction of excreted nitrogen (N) (or sulfur (S)) emitted from the operation/animal type for which emissions factors were translated from (i.e., the source) was calculated.

A finisher pig excretes 0.42 lb N/day-1000 lb live weight (LW)-day, average LW of finisher pig is 154 lb, with a 119 day cycle and 2.8 cycles per year, 2.5 pigs per AU. Therefore:

$$\text{Excreted\_N} = \frac{0.42 \text{ lb N}}{\text{day} \cdot 1000 \text{ lb LW}} \times \frac{154 \text{ lb LW}}{\text{pig}} \times \frac{119 \text{ days}}{\text{cycle}} \times \frac{2.8 \text{ cycles}}{\text{yr}} \times \frac{2.5 \text{ pig}}{\text{AU}} = 53.9 \text{ lb N/AU-yr}$$

The emission factor from the literature is 10.3 lb NH_3/AU-yr, which converts to 8.5 lb N/AU-yr. The resultant fraction is:

$$\text{Fraction\_emitted} = \frac{\text{Emission\_factor}}{\text{Excreted\_N}} = \frac{8.5 \text{ lb N/AU-yr}}{53.9 \text{ lb N/AU-yr}}$$

2. The fraction emitted from the source animal type was multiplied by the annual nitrogen excretion in the target animal type.

In this example, given an excretion rate of 0.45 lb N/day-1000lb LW-day, an average LW of 1350 lb, 335 day cycle, one cycle per year, and that one AU is equal to 0.7 cows, dairy cows excrete 142 lb N/AU-yr. (Dry cows are not included in N and S excretion to flush freestall barns or anaerobic lagoons; these barns would be filled to capacity with lactating cows in the herd.)

Therefore, the dairy flush house ammonia emission factor is calculated as follows:

$$\text{Dairy\_emissions} = \text{Excreted\_N} \times \text{Fraction\_emitted} = \frac{142 \text{ lb N}}{\text{AU-yr}} \times 0.16 = 23 \text{ lb N/AU-yr}$$

This converts to an emission factor for dairy flush houses of 28 lb NH_3/AU yr.

The other instances where emissions information were translated from one animal species to another are shown in the following calculations.

7.4.2.1.2.1 Dairy Anaerobic Lagoon Ammonia from Swine Anaerobic Lagoon Ammonia

Swine N excretion is calculated as follows:

$$\text{Excreted\_N} = \frac{0.42 \text{ lb N}}{\text{day} \cdot 1000 \text{ lb LW}} \times \frac{154 \text{ lb LW}}{\text{pig}} \times \frac{119 \text{ days}}{\text{cycle}} \times \frac{2.8 \text{ cycles}}{\text{yr}} \times \frac{2.5 \text{ pig}}{\text{AU}} = 539 \text{ lb N/AU-yr}$$

The emission factor from the literature is 15.1 lb NH_3/AU-yr, which converts to 12.4 lb N/AU-yr.

The resultant fraction is:

$$\text{Fraction\_emitted} = \frac{\text{Emission\_factor}}{\text{Excreted\_N}} \times \frac{12.4 \text{ lb N/AU-yr}}{53.9 \text{ lb N/AU-yr}} = 0.23$$

Dairy cows excrete 142 lb N/AU-yr (calculated above). Dairy ammonia emissions from anaerobic lagoons are calculated as follows:

$$\text{Dairy\_emmissions} = \text{Excreted\_N} \times \text{Fraction\_emitted} = \frac{142 \text{ lb N}}{\text{AU-yr}} \times 0.23 \times \frac{17}{14} = 40 \text{ lb } NH_3/\text{AU-yr}$$

7.4.2.1.2.2 Layer Anaerobic Lagoon Ammonia from Swine Anaerobic Lagoon Ammonia

Swine N excretion is 53.9 lb N/AU-yr. The emission factor for swine anaerobic lagoon is 12.4 lb N/AU-yr, which results in a fraction emitted of 0.23 (calculated above). Layer N excretion rate is calculated as follows:

$$\text{Excreted\_N} = \frac{0.83 \text{ lb N}}{\text{day} \cdot 1000 \text{ lb LW}} \times \frac{3.97 \text{ lb LW}}{\text{hen}} \times \frac{350 \text{ days}}{\text{cycle}} \times \frac{1 \text{ cycle}}{\text{yr}} \times \frac{100 \text{ hen}}{\text{AU}} = 115.4 \text{ lb N/AU-yr}$$

Therefore, emissions from layer anaerobic lagoons are calculated according to the following:

$$\text{Lay\_emmissions} = \text{Excreted\_N} \times \text{Fraction\_emitted} = \frac{115.4 \text{ lb N}}{\text{AU-yr}} \times 0.23 \times \frac{17}{14} = 32.2 \text{ lb } NH_3/\text{AU-yr}$$

7.4.2.1.3 Daily Anaerobic Lagoon Hydrogen Sulfide

Swine S excretion is calculated as follows:

$$\text{Excreted\_S} = \frac{0.078 \text{ lb S}}{\text{day} \cdot 1000 \text{ lb LW}} \times \frac{154 \text{ lb LW}}{\text{pig}} \times \frac{119 \text{ days}}{\text{cycle}} \times \frac{2.8 \text{ cycles}}{\text{yr}} \times \frac{2.5 \text{ pig}}{\text{AU}} = 10.0 \text{ lb S/AU-yr}$$

Emission factors for anaerobic lagoons following flush houses and nonflush houses are 9.8 and 2.6 lb H_2S/AU-yr, respectively. These result in S emissions of 9.2 and 2.4 lbs S/AU-yr. The fraction emitted for anaerobic lagoons following flush houses is calculated by:

$$\text{Fraction\_emitted} = \frac{\text{Emission\_factor}}{\text{Excreted\_S}} = \frac{9.2 \text{ lb } S/\text{AU-yr}}{10 \text{ lb } S/\text{AU-yr}} = 0.92$$

Using the same method, an S fraction emitted of 0.24 for anaerobic lagoons following nonflush houses is calculated. Dairy cow S excretion is calculated as:

$$\text{Excreted\_S} = \frac{0.051 \text{ lb } S}{\text{day} \cdot 1000 \text{ lb LW}} \times \frac{1350 \text{ lb LW}}{\text{pig}} \times \frac{335 \text{ days}}{\text{cycle}} \times \frac{1 \text{ cycle}}{\text{yr}} \times \frac{0.7 \text{ cow}}{\text{AU}} = 16.1 \text{ lb } S/\text{AU-yr}$$

Emissions are calculated as follows for lagoons following flush operations in dairies:

$$\text{Dairy\_emissions} = \text{Excreted\_S} \times \text{Fraction\_emitted} = \frac{16.1 \text{ lb } S}{\text{AU-yr}} \times 0.92 \times \frac{34}{32} = 15.7 \text{ lb } H_2S$$

Following the same logic, an emission factor of 4.1 lb H$_2$S/AU-yr was calculated for lagoons at nonflush dairy operations.

7.4.2.1.4 Layer Anaerobic Lagoon Hydrogen Sulfide from Layer Anaerobic Lagoon Hydrogen Sulfide

Swine S excretion rate (calculated above) is 10 lb S/AU-yr. For anaerobic lagoons following nonflush houses (there is no model farm for layer flush houses), the fraction emitted as H$_2$S is 0.24. Layer S excretion is given as follows:

$$\text{Excreted\_S} = \frac{0.14 \text{ lb } S}{\text{day} \cdot 1000 \text{ lb LW}} \times \frac{3.97 \text{ lb LW}}{\text{hen}} \times \frac{350 \text{ days}}{\text{cycle}} \times \frac{1 \text{ cycle}}{\text{yr}} \times \frac{100 \text{ hen}}{\text{AU}} = 19.4 \text{ lb } S/\text{AU-yr}$$

The hydrogen sulfide emission factor is calculated by:

$$\text{Layer\_emissions} = \text{Excreted\_S} \times \text{Fraction\_emitted} = \frac{19.4 \text{ lb } S}{\text{AU-yr}} \times 0.24 \times \frac{34}{32} = 4.9 \text{ lb } H_2S/\text{AU-yr}$$

7.4.2.1.5 Turkey House Ammonia from Broiler House Ammonia

$$\text{Broiler\_N} = \frac{1.10 \text{ lb } N}{\text{day} \cdot 1000 \text{ lb LW}} \times \frac{26 \text{ lb LW}}{\text{broiler}} \times \frac{49 \text{ days}}{\text{cycle}} \times \frac{5.5 \text{ cycles}}{\text{yr}} \times \frac{100 \text{ broilers}}{\text{AU}} = 77 \text{ lb } N/\text{AU-yr}$$

The emission factor for broiler confinement is 24.4 lb NH$_3$/AU-yr, which translates into 20.0 lb N/AU-yr. The fraction N emitted as NH$_3$ is given by:

$$\text{Fraction\_emitted} = \frac{\text{Emission\_factor}}{\text{Excreted\_N}} = \frac{20.0 \text{ lb } N/\text{AU-yr}}{77 \text{ lb } N/\text{AU-yr}} = 0.26$$

Since hens and toms have differing production characteristics, nitrogen excretion for both were calculated and then averaged to produce one annual N excretion value for turkeys. For hens:

$$\text{Excreted\_N} = \frac{0.74 \text{ lb N}}{\text{day} \cdot 1000 \text{ lb LW}} \times \frac{11.5 \text{ lb LW}}{\text{turkey hen}} \times \frac{105 \text{ days}}{\text{cycle}} \times \frac{2 \text{ cycles}}{\text{yr}} \times \frac{55 \text{ turkeys}}{\text{AU}} = 98.3 \text{ lb N/AU-yr}$$

For toms:

$$\text{Excreted\_N} = \frac{0.74 \text{ lb N}}{\text{day} \cdot 1000 \text{ lb LW}} \times \frac{16.8 \text{ lb LW}}{\text{turkey tom}} \times \frac{133 \text{ days}}{\text{cycle}} \times \frac{2 \text{ cycles}}{\text{yr}} \times \frac{55 \text{ turkeys}}{\text{AU}} = 182 \text{ lb N/AU-yr}$$

The average of toms and hens is 140 lb N/AU-yr. The emission factor for turkey confinement is then calculated by:

$$\text{Turkey\_emissions} = \text{Excreted\_N} \times \text{Fraction\_emitted} = \frac{140 \text{ lb N}}{\text{AU-yr}} \times 0.26 \times \frac{17}{14} = 44 \text{ lb NH}_3/\text{AU-yr}$$

The calculation of nitrogen excretion rates is discussed in Section 7.4.3. Table 7.6 summarizes the basis for the ammonia emission factor used.

TABLE 7.6
Sources of Ammonia Emission Factors

Animal Type	Source with Emission Factor Available	Source of Emission Factor
Beef	Drylot	Literature review
	Stockpile	Literature review
Veal	None	None
Dairy	Freestall barn (flush)	Translated from swine flush
	Freestall barn (scrape)	Literature review
	Drylot	Literature review
	Liquid manure application	Literature review
	Solids storage	Literature review
	Anaerobic lagoon	Translated from swine anaerobic lagoon emissions
Swine	Flush house	Literature review
	House with pit recharge	Literature review
	House with pull plug pit	Literature review
	House with pit storage	Literature review
	Anaerobic lagoon	Literature review
	Liquid land application	Literature review
Broilers	House	Literature review
	Storage (cake and litter)	Literature review
	Solid manure land application	Literature review
Layers	Flush house	Literature review
	High rise house	Literature review
	Manure land application (solid and liquid)	Literature review
	Anaerobic lagoon	Translated from swine anaerobic lagoon emissions
Turkeys	House	Translated from broiler house emission
	Storage (cake and litter)	Literature review
	Solid manure land application	Literature review

Source: USEPA (2001a).

TABLE 7.7
Nitrous Oxide (MF$_{N_2O}$) Factors

Source	(MF$_{N_2O}$)*
Anaerobic lagoon	0.001
Deep pit	0.001
Drylot	0.02
Poultry manure with bedding	0.02
Poultry manure without bedding	0.005
Stacked solids	0.02
Storage pond	0.001

Source: USEPA (2001b).

* MF$_{N_2O}$ = Factor relating N$_2$O emissions as nitrogen in manure,
   pound N$_2$O-N emitted per pound nitrogen in manure.

7.4.2.2 Nitrous Oxide (N$_2$O)

Emission factors for nitrous oxide were not found in the literature. In all cases, therefore, nitrous oxide emissions were based on the nitrogen content of manure. Factors relating the emission of nitrous oxide (as nitrogen) to the amount of nitrogen in the manure (MF$_{N_2O}$) were provided for several emission points (USEPA, 2001b). The factors are listed in Table 7.7.

Nitrous oxide emissions were estimated using Table 7.7:

$$E_{N_2O} = 1.57 \, M_N \times MF_{N_2O}$$

where
E_{N_2O} = N$_2$O emissions, lb/yr.
M_N = Nitrogen excretion rate for a 500 AU farm, lb/year.
MF_{N_2O} = Nitrous oxide factor (Table 7.7), lb N$_2$O-N.

The value 1.57 is the conversion factor to express the emission estimate on a nitrous oxide rather than a nitrous oxide-nitrogen basis (USEPA, 2001b). The method for estimating nitrogen excretion is explained in Section 7.4.3. Nitrous oxide emission factors in Section 7.4.4 were calculated by dividing the nitrous oxide emissions by 500 AU and converting tons to pounds.

While these factors are the best available for nitrous oxide emissions, they were used with the qualification that they may overestimate emissions for some elements of the virtual farm. The basis of this conclusion is the absence of the necessary microbial environment (without inhibitory conditions) for nitrification to occur prior to land application. Except for operations with drylots, it is highly probable that manure application sites are the principal source of nitrous oxide emissions. The following paragraphs explain why nitrification is unlikely to occur at liquid storage sites or poultry confinement houses.

Anaerobic lagoons, deep pit storage tanks, and ponds. Given the high carbonaceous oxygen demand of animal manures and the low solubility of oxygen in water, any oxygen transferred from the atmosphere is rapidly used by the facultative heterotrophic microorganisms present. Thus, the oxygen necessary for nitrification is not available. The presence of nitrite or nitrate nitrogen in livestock or poultry manure as excreted is highly unlikely for two reasons. One is the toxicity of these compounds, which makes ultilization of feedstuffs containing more than trace concentrations of the ions undesirable. Plants such as corn only accumulate these ions under stressed growth conditions, such as drought conditions. Normally, most plants reduce nitrites and nitrates enzymatically to ammonia before or during uptake to provide the necessary precursor for amine acid synthesis. Secondly, any nitrate nitrogen consumed will be reduced due to the anaerobic microbial environment of the gastrointestinal tract; if any nitrous oxide is formed, it should be emitted upon excretion and not subsequently from anaerobic lagoons or manure storage tanks or ponds (Alexander, 1977; Brock & Madigan, 1988; Tate, 1995).

Poultry Confinement Houses. The aerobic environment in dry poultry confinement facilities suggests that nitrification and subsequent denitrification with nitrous oxide emissions is possible. However, the high ammonia-nitrogen emissions that have been measured from broiler and turkey litters suggest the absence of any significant nitrifying activity (Anderson et al., 1964; Carlile, 1984; Caveny and Quarles, 1978; Deaton et al., 1984; Valentine, 1964). Although the factor for these sources, 0.02, appears small, it suggests significant nitrifying activity, if dinitrogen gas is the principal product of denitrification. Given the alkaline environment present, this determination appears to be a reasonable assumption since it is well established that acidic environments are more conducive to the formation of nitrous oxide as a product of denitrification (Alexander, 1977; Tate, 1995). If, hypothetically, 5% of the nitrogen gases produced by denitrification is nitrous oxide, the poultry factor of 0.02 in Table 7.7 translates into the nitrification of 40% of the nitrogen excreted. If a population of nitrifying bacteria capable of this level of nitrification is present, it is probable that complete nitrification would occur, and the high level of ambient air ammonia concentrations that have been measured in broiler and turkey production facilities would not exist. Although the reason or reasons for the lack of nitrification are not clear, free ammonia inhibition is a possible explanation (Anthonisen et al., 1976).

The substantial difference, a factor of four, between the nitrous oxide emission factor for poultry manure with and without bedding (litter) appears questionable if the latter category applies to laying hen manure produced in high-rise type facilities. The rate of microbial heat production necessary for successful operation of high-rise houses indicates the necessity of an aerobic environment (Martin & Loehr, 1977). If nitrification occurs in poultry manure with bedding, it seems logical to also assume that nitrification also occurs in high-rise facilities from laying hens and use the same default emission factor value. Conversely, the default value of 0.005 for laying hen manure handled as a liquid or slurry due to the anaerobic microbial environment is suspect. A possible explanation for the value of 0.0005 is some distribution of total bird numbers between high-rise type facilities and facilities handling manure as a slurry or liquid.

7.4.2.3 Hydrogen Sulfide

Hydrogen sulfide emission factors were available for swine operations but not for poultry, dairy, and veal. For these animal sectors, hydrogen sulfide emission factors for anaerobic lagoons were calculated by translating hydrogen sulfide emission factors from the swine sector. Although manure characteristics differ significantly from one animal species to another, the rates of hydrogen sulfide formation from the various sulfur compounds contained in livestock and poultry manures under anaerobic conditions and the chemistry of hydrogen sulfide in solutions (for example, pH levels) should not be different (Alexander, 1977; Brock & Madigan, 1988; Tate, 1995). Therefore, for anaerobic lagoons, emission factors developed for one species could be translated to another by adjusting to reflect differences in excretion rates. The swine emission factor was adjusted to reflect different manure characteristics using the same methodology described for ammonia.

Hydrogen sulfide emissions for other AFOs either could not be calculated because of lack of information or were not expected because of aerobic conditions. For beef and veal, lack of information about typical hydrogen sulfide concentrations and concurrent pH levels in manure-holding tanks in confinement facilities, storage tanks and ponds, and anaerobic lagoons precluded the development of a theoretical model to predict hydrogen sulfide emissions. Under aerobic conditions, such as those present in dry manure collection and storage facilities, sulfur excreted should be oxidized to nonvolatile sulfate. Even if transient anaerobic conditions leading to hydrogen sulfide formation

TABLE 7.8
Sources of Hydrogen Sulfide Emission Factors

Animal type	Operation with emission factor available	Source of emission factor
Beef	None	None
Veal	None	None
Dairy	Anaerobic lagoon	Translated from swine anaerobic lagoon emissions
Swine	House with pit storage	Literature review
	Anaerobic lagoon	Literature review
	Liquid land application	Literature review
Broilers	None	None
Layers	Anaerobic lagoon	Translated from swine anaerobic lagoon emissions
Turkeys	None	None

Source: USEPA (2001a).

occur, subsequent oxidation to sulfate is probable. Thus, hydrogen sulfide emissions from broiler and turkey confinement facilities, high-rise type confinement facilities for laying hens, and drylots for beef and dairy cattle were considered to be insignificant (Alexander, 1977; Brock & Madigan, 1988; Tate, 1995). Table 7.8 indicates animal types and operations for which hydrogen sulfide emission factors have been developed.

7.4.2.4 Methane

Methane emissions were not estimated for the virtual farms. Methane emissions are a function of the mass of volatile solids present in manure, the method of manure handling, and the temperature and moisture of the manure. Temperature is an important variable because microbial decomposition decreases at low temperatures and ceases at the freezing point. Because temperature varies by geographic region and season, it was not practical within the scope of this text to incorporate a temperature variable into the virtual farms. Excluding virtual farm emissions, this section explains the methods used currently by USEPA to estimate methane emissions for the U.S. greenhouse gas inventory (USEPA, 2001b).

This methodology can be applied to individual farms. As an example, methane emissions were estimated from anaerobic lagoons for swine based on the 1999 monthly temperature profiles at two locations (North Carolina and Iowa). For a 500 AU farm in North Carolina, emissions from the anaerobic lagoon were estimated at 42 tons per year (38 Mg/year). For Iowa, emission estimates were 38 tons per year (35 Mg/year). Swine lagoons were chosen because they generally represent the largest methane emission source at AFOs. The emission calculations for these two virtual farms are shown below, along with an explanation of the methodology.

The USEPA methodology is based on the following equation. Emissions are a function of the mass of volatile solids excreted, the methane-producing capacity of manure from different animals, the type of waste management system, and the temperature of the manure.

$$\text{Methane Emissions (per head)} = \text{VS}_{\text{excreted}} \times \text{B}_{\text{o}} \times 0.67 \ \text{kg}/\text{m}^3 \times \text{MCF}$$

where
$\text{VS}_{\text{excreted}}$ = Volatile solids excreted (kg/yr).
B_{o} = Maximum methane-producing capacity (m^3CH_4/kg VS).
MCF = Methane conversion factor based on the waste minimization system (%).
0.67 = Methane density at 20°C, 1 atmosphere (kg/m^3).

TABLE 7.9

Methane Production Potential from Livestock and Poultry Manure

Animal type	Bo (m3CH4/kg VS excreted)	Reference
Mature dairy cow	0.24	Morris, 1976
Heifer	0.17	Bryant et al., 1976
Calf	0.17	Bryant et al., 1976
Beef (high energy diet)	0.33	Hashimoto et al., 1981
Broilers	0.36	Hill, 1984
Turkeys	0.36	Hill, 1984
Laying hens	0.39	Hill, 1982
Swine (grow-finish)	0.48	Hashimoto, 1984
Swine (farrow to finish)	0.48	Hashimoto, 1984

Source: USEPA (2001a).

TABLE 7.10

Methane Conversion Factors for Various Livestock and Poultry Manure Management System Components

Manure management system	Methane conversion factor (%) by climate		
	Cool*	Temperate**	Warm***
Anaerobic lagoon	0–100	0–100	0–100
Composting	0.5	0.5	0.5
Deep pit (< 1 month)	0	0	30
Deep pit (> 1 month)	39	45	72
Drylot	1	1.5	5
Poultry manure with bedding	1.5	1.5	1.5
Poultry manure without bedding	1.5	1.5	1.5
Stacked solids	1	1.5	5
Manure storage pond	39	45	72

Source: USEPA (2001b).
*   Temperatures are less than 15°C.
**   Temperatures are between 15°C and 25°C.
*** Temperatures are greater than 25°C.

The calculation of volatile solids excreted is discussed in Section 7.4.3.

The methane-production potential of animal waste (B_o) is the maximum quantity of methane (m^3CH_4) that can be produced per kilogram of volatile solids (VS) in the manure. Values for B_o are available from literature and are based on the animal species and diet. Table 7.9 presents the values for B_o that have been used in developing the USEPA's greenhouse gas inventory and other USEPA studies (USEPA, 2001b).

The methane conversion factor (MCF) is an estimate of the fraction of volatile solids that are converted to methane in a given type of manure management system at a specific temperature. The MCFs used in the greenhouse gas inventory for various livestock and poultry manure management options are listed in Table 7.10. Because the rate of reduction of VS to methane is a direct function of process temperature, MCFs vary with climate and season of the year.

The USEPA inventory method uses the MCF values in Table 7.10 for dry manure handling systems (composting, drylots, poultry manure, and stacked solids). For wet systems (anaerobic lagoon, deep pit, and storage ponds), the method uses the Van't Hoff–Arrhenius equation to estimate MCF. The Van't Hoff–Arrhenius equation allows a more precise estimate of the effect of local temperature variations on the biological conversion to methane.

$$f - \exp\left[\frac{E(T2 - T1)}{RT1T2}\right]$$

where

f = Temperature adjustment factor, substituting for MCF, dimensionless.
T1 = 303.16 K.
R = Ideal gas constant (1.987 cal/K mol).
E = Activation energy constant (15,175 cal/mol).
T2 = Ambient temperature for a geographic region (K).

For deep pits and manure storage ponds, the USEPA bases the value of "f" on annual average temperature in each state. The annual average state temperatures are based on the counties where the specific animal population resides (the temperatures were weighted based on the percent of animals located in each county). The approach used for anaerobic lagoons is also based on the Van't Hoff–Arrhenius equation but is calculated on a monthly basis instead of yearly to account for the longer retention time and associated build-up of VS in these systems.

Virtual Example: Methane Emissions Methodology
In the following example, the methane emission methodology is used to calculate emissions from an anaerobic lagoon at a 500-AU swine virtual farm in Iowa in January 1999.

1. Monthly temperatures are calculated by using county-level temperature and population data. The weighted-average temperature for a state is calculated using the population estimates and average monthly temperature in each county. Table 7.11 presents the monthly average temperatures from Iowa and North Carolina in 1999 from the USEPA's greenhouse gas inventory (USEPA, 2001b).
2. Monthly temperatures are used to calculate a monthly Van't Hoff–Arrhenius "f" factor.

For January 1999, in Iowa, f is calculated to be:

$$f = \exp\left[\frac{15,175\ (264.2 - 303.16)}{(1,987)(264.2)(303.16)}\right] = 0.0243$$

3. Monthly production of VS is calculated based on the number of animals present. Table 7.15 provides the annual production of VS for a 500-AU swine virtual farm, 173 tons/yr. On a per-day basis, this converts to 0.47 tons/day or 430.37 kg/day. On a monthly basis for January (31 days), this converts to 13,341 kg.
4. Monthly production of VS that are added to the system are adjusted using a management and design practices factor. This factor accounts for other mechanisms by which VS are removed from the management system prior to conversion to methane, such as solids being removed from the lagoon for application to cropland. This factor, equal to 0.8, was estimated in USEPA's greenhouse gas inventory using currently available methane measurement data from anaerobic lagoon systems in the United States (USEPA, 2001b).

TABLE 7.11

Calculation of Methane Emission from 500 AU Swine Virtual Farm in Iowa in 1999

Month	Average monthly temperature[a]				Volatile solids (kg)				Methane Emitted (kg)[g]
	(K)	(C)	(F)	f[b]	Produced[c]	Adjusted Production[d]	Cumulative Produced[e]	Consumed[f]	
October	284.4	11.3	52.3	0.19	13,341	10,673	10,673	2,028	973
November	277.6	4.4	39.9	0.10	12,911	10,329	18,974	1,856	891
December	271.3	(1.8)	28.7	0.05	13,341	10,673	27,791	1,445	694
January	264.2	(8.9)	15.9	0.02	13,341	10,673	37,019	904	434
February	273.1	(0.0)	32.0	0.06	12,050	9,640	45,756	2,869	1,377
March	275.3	2.1	35.8	0.08	13,341	10,673	53,560	4,165	1,999
April	282.8	9.6	49.3	0.16	12,911	10,329	59,723	9.720	4,666
May	288.9	15.7	60.3	0.29	13,341	10,673	60,676	17,412	8,358
June	293.4	20.3	68.5	0.43	12,911	10,329	53,593	23,230	11,150
July	297.8	24.7	76.4	0.64	13,341	10,673	41,036	26,113	12,534
August	294.0	20.9	69.6	0.46	13,341	10,673	25,596	11,710	5,621
September	289.1	15.9	60.7	0.29	12,911	10,329	24,215	7,111	3,413
October	283.2	10.0	50.1	0.17	13,341	10,673	10,673	1,807	867
November	279.7	6.5	43.7	0.12	12,911	10,329	19,195	2,310	1,109
December	270.6	(2.6)	27.4	0.05	13,341	10,673	27,558	1,327	637
Sum[h]					157,084	25,667	458,600	108,679	52,166

[a] From EPA's greenhouse gas inventory (USEPA, 2001b).

[b] Calculated using Van't Hoff–Arrhenius equation (Step 2).

[c] From volatile solids (VS) in swine manure in Table 7.15 and converting to a monthly basis (Step 3).

[d] Adjusted VS produced using a management and design practices factor of 0.8 (USEPA, 2001) (Step 4).

[e] Cumulative VS from previous month and current month minus VS consumed in the previous month (Step 5).

[f] Calculated by multiplying by monthly "f" factor (Step 6).

[g] Calculated from VS consumed multiplied by methane potential of waste, B0. For swine, B0 is equal to 0.48 m3/kg VS. Volume of methane was converted to mass (kg) using a density of methane of 0.67 kg/m3 from Step 8.

[h] Sums for January through December of 1999.

$$\text{Adjusted volatile solids} = (0.8)(13,341) = 10,673 \text{ kg for January}$$

5. The amount of VS available for conversion to methane is set equal to the adjusted amount of VS produced during the month (from Step 4) plus VS that may remain in the system from the previous month (VS produced in the previous month minus the VS consumed in the previous month). To account for the carry-over of VS from the year prior to the inventory year for which estimates are calculated, it is assumed in the methane calculation for lagoons that a portion of the VS from October, November, and December of the year prior to the inventory year are available in the lagoon system starting January of the inventory year. From Table 7.11, the VS remaining from the previous month were calculated at 26,346 kg (i.e., 27,791 − 1445 = 26,346). The total VS in January is calculated at:

$$26,346 + 10,673 = 37,019 \text{ kg}$$

6. The amount of VS consumed during the month is equal to the amount available for conversion multiplied by the "f" factor.

TABLE 7.12

Calculation of Methane Emission from 500 AU Swine Virtual Farm in North Carolina in 1999

Month	Average monthly temperature[a]				Volatile solids (kg)				Methane Emitted (kg)[g]
	(K)	(C)	(F)	f[b]	Produced[c]	Adjusted Production[d]	Cumulative Produced[e]	Consumed[f]	
October	290.4	17.2	63.0	0.33	13,352	10,681	10,681	3,521	1,690
November	285.7	12.6	54.6	0.21	12,921	10,337	17,497	3,758	1,804
December	283.2	10.1	50.2	0.17	13,352	10,681	24,421	4,153	1,993
January	282.7	9.0	48.2	0.15	13,352	10,681	30,949	4,751	2,280
February	281.8	8.6	47.5	0.15	12,060	9,648	35,846	5,292	2,540
March	282.7	9.5	49.1	0.16	13,352	10,681	41,236	6,637	3,186
April	290.3	17.1	62.8	0.33	12,921	10,337	44,936	14,702	7,057
May	292.8	19.6	67.3	0.41	13,352	10,681	40,915	16,703	8,018
June	300.2	23.6	74.5	0.58	12,921	10,337	34,548	20,059	9,628
July	299.9	27.1	80.7	0.78	13,352	10,681	25,170	19,648	9,431
August	294.8	26.8	80.2	0.76	13,352	10,681	16,204	12,335	5,921
September	289.5	21.7	71.0	0.49	12,921	10,337	14,205	6,963	3,342
October	289.5	16.3	61.3	0.30	13,352	10,681	10,681	3,239	1,555
November	287.1	14.0	57.1	0.24	12,921	10,337	17,779	4,349	2,087
December	281.3	8.2	46.7	0.14	13,352	10,681	24,112	3,405	1,635
Sum[h]					157,206	125,764	336,583	118,083	56,680

[a] From EPA's greenhouse gas inventory (USEPA, 2001b).

[b] Calculated using Van't Hoff–Arrhenius equation (Step 2).

[c] From volatile solids in swine manure in Table 7.15 and converting to a monthly basis (Step 3).

[d] Adjusted volatile solids produced using a management and design practices factor of 0.8 (USEPA, 2001) (Step 4).

[e] Cumulative volatile solids from previous month and current month minus volatile solids consumed in the previous month (Step 5).

[f] Calculated by multiplying by monthly "f" factor (Step 6).

[g] Calculated from volatile solids consumed multiplied by methane potential of waste, B_0. For swine, B_0 is equal to 0.48 m^3/kg volatile solids. Volume of methane was converted to mass (kg) using a density of methane of 0.67 kg/m^3 from Step 8.

[h] Sums for January through December of 1999.

$$(37,019)(0.0243) = 900 \text{ kg}$$

7. The amount of VS carried over from 1 month to the next is equal to the amount available for conversion minus the amount consumed.

$$37,019 - 900 = 36,119 \text{ kg}$$

8. The estimated amount of methane generated during the month is equal to the monthly VS consumed multiplied by the maximum methane potential of the waste (B_0). For swine, B_0 is equal to 0.48 m^3 methane/kg VS (Table 7.9).

= 900 kg VS consumed multiplied by 0.48 m^3 methane/kg VS

= 432 m^3 methane

= 289 kg methane (assuming a density of 0.67 kg/m^3

Tables 7.11 and 7.12 show the calculations for 500-AU swine farms in Iowa and North Carolina, respectively, in 1999. Numbers in the example may not exactly match the tables due to rounding.

7.4.2.5   Volatile Organic Compounds

A variety of volatile organic compounds may be present in livestock and poultry manures. Many of these compounds are present in freshly excreted manure but also may be formed subsequently when the manure is stored under anaerobic conditions. Under anaerobic conditions, the organic carbon in manure is converted to methane and carbon in a complex set of reactions in which VOCs are created and then consumed as intermediates. When the microbial reduction of the carbon to methane and carbon dioxide is inhibited (for example, by cold temperatures or bacterial imbalances), VOCs accumulate and may be emitted (Alexander, 1977; Brock & Madigan, 1988; Tate, 1995).

Under aerobic conditions, such as found in the broiler industry, carbon is degraded to carbon dioxide and water, and no VOCs are emitted. Thus, emissions of VOCs from broiler and turkey production facilities, high-rise-type confinement facilities for laying hens, and drylots for beef and dairy cattle should be minimal in comparison to facilities used for liquid manure storage and anaerobic stabilization.

Emissions from anaerobic lagoons for swine, laying hens, and dairy cattle manure also should be minimal, except when low temperatures reduce the rate of conversion of organic carbon to methane and carbon dioxide. However, VOC emissions occur from anaerobic lagoons located in colder climates when lagoon temperatures increase in the spring and the balance between the heterotrophic microorganisms (capable of producing these complex organic compounds) and methanogenic bacteria become reestablished.

The literature review did not produce any emission factor data for VOCs. However, based on the recognition that no biological process is 100% efficient, some nominal level of VOCs should be emitted from anaerobic lagoons and a somewhat higher level from storage ponds. To provide some sense of the possible magnitude of VOC emissions, the VOC emissions for anaerobic lagoons were calculated, based on professional judgement, as 1% of the methane production potential of these manures. The 1% value was used for anaerobic lagoons for swine, dairy, and wet layer manures. The methane-producing capacity of animal manure is discussed in Section 7.4.2.4. VOC emissions were calculated:

$$\text{VOC}_{\text{emitted}} = \text{VS}_{\text{excreted}} \times B_o \times 0.67 \times 0.01$$

where

$\text{VOC}_{\text{emitted}}$ = VOC emitted (kg/AU-year).
$\text{VS}_{\text{excreted}}$ = VS excreted (kg/AU-year).
B_o = Methane production potential (m^3 CH$_4$/kg VS).
0.67   = Methane density at 20°C, 1 atmosphere (kg/m^3).
0.01   = Fraction of the methane production potential emitted as VOC.

Clearly, VOC is emitted in more significant quantities from confinement facilities (especially those with integral manure storage tanks), manure storage tanks and ponds, solids manure storage facilities, and manure application sites. However, any attempt to estimate possible VOC emissions from these sources is difficult because of the absence of any reasonable basis for estimating methane production potential. The approach for anaerobic lagoons was based on the judgement that the destruction of readily biodegradable VS is essentially complete. For potential sources of VOCs other than anaerobic lagoons, that assumption would not be valid because stabilization is not an objective of these manure storage facilities. The degree of biodegradable VS destruction occurring could vary significantly among these sources, given differences in times of storage and other factors. Thus, no defensible estimates of emissions from these sources were thought possible.

7.4.3 ESTIMATION OF NITROGEN, SULFUR, AND VOLATILE SOLIDS IN MANURE

The development of some emission factors required an estimate of the mass of precursors in manure. The maximum possible levels of ammonia, nitrous oxide, and hydrogen sulfide emissions from animal manures are limited by the quantities of nitrogen, VS (carbon), and sulfur that are available for microbial transformation (precursors, for example). Estimates of excretion rates of these precursors were used to compute emissions directly, convert units of measure, or translate an emission factor from one animal sector to another. The average rates of nitrogen, sulfur, and VS for each animal type are discussed in Section 7.4.3.1. Section 7.4.3.2 explains how the daily rates were converted to annual rates for a virtual farm based on the production practices of the different animal sectors.

7.4.3.1 Daily Nitrogen, Sulfur, and Volatile Solids Excretion Rates

The characteristics of livestock and poultry manures differ significantly, reflecting differences in nutritional requirements and feeding programs designed to satisfy these requirements. These differences exist not only among species but also within individual species maintained for different purposes. For example, concentrations of nitrogen, sulfur, and organic carbon estimated using VS as a surrogate differ significantly between broiler-type chickens and layer hens. Even within the same species and breed or genetic strain maintained for the same purpose, manure characteristics may differ significantly, because of differences in diet, climate, or physiology. These differences in feed conversion efficiency are a reflection of both genetic potential and animal management practices.

To estimate the amount of nitrogen, sulfur, and VS excreted annually, assumptions about typical rates of excretion were necessary. The two primary sources of such information are the American Society of Agricultural Engineers (ASAE) and the Natural Resources Conservation Service (NRCS). While general agreement is evident among these sources, whether either represents typical excretion rates is not clear. For example, the background documentation for the estimates presented in both sources was not available; therefore, the values reflecting current production practices could not be determined. Additionally, no information was available on the number of point estimates included. For some parameters, the standard deviations, and therefore the coefficients of variation, are substantial (e.g., 20% for sulfur content of dairy cow manure). Given the lack of background information, the source of variation is unclear. It could be due to changes in feeding practices with time, a reflection of a limited data base with one or more outliers skewing the mean, or the factors previously discussed in this section. Despite concerns about their representativeness, the ASAE and NRCS data were used for this virtual study because no other information was available. The ASAE and NRCS data were assumed to be derived from point estimates that are normally distributed and that they would provide reasonable estimates of daily excretion rates per unit of live weight. Waste streams other than manure (e.g., waste waters) were considered to be nominal sources and were not estimated.

The NRCS (USDA, 1992) database was used to estimate nitrogen and VS excretion rates because it allowed estimates for different stages in swine and dairy production cycles. Because no sulfur excretion rates are available in the NRCS database, the ASAE (ASAE, 1999) values were used for sulfur. The excretion rates are listed in Table 7.13.

7.4.3.2 Calculation of Nitrogen, Sulfur, and Volatile Solids Excreted Annually

The mass of nitrogen, sulfur and VS excreted annually was computed for each animal section:

$$M_{s,a} = LW \times R_{a,au} \times ER_s \times P \times T$$

where
$M_{s,a}$ = Quantity of substance S excreted from animal A (lb/AU-yr).
LW = Average live weight of animal (lb/animal).

TABLE 7.13

Rates of Nitrogen, Volatile Solids, and Sulfur Excretion by Livestock and Poultry, lb per day per 1,000 lb live weight

Species	Nitrogen**	Volatile Solids**	Sulfur***
Poultry			
Broilers	1.10	15.00	0.085
Laying hens	0.83	10.80	0.14
Turkeys	0.74	9.70	ND*
Swine			
Feeder pigs****	0.42	5.40	0.078
Nursery pigs	0.60	8.80	ND
Gestating sows	0.19	2.13	ND
Lactating sows	0.47	5.40	ND
Gilts	0.24	2.92	ND
Boars	0.15	1.70	ND
Dairy Cattle			
Lactating	0.45	8.50	0.051
Dry	0.36	8.10	ND
Replacements	0.31	7.77	ND
Veal Calves	0.20	0.85	ND
Beef Cattle			
Feeder	0.30	5.44	0.046

*   No data
**   USDA, 1992
***   ASAE, 1999
**** For grow-finish operations.

ER_s = Excretion rate of substance S (lb/lb LW-day).
P   = Number of production cycles per year.
T   = Days per production cycle.
$R_{a,au}$ = Number of animals per AU.

The values used for average live weights (i.e., the average weight of an animal over the period of its confinement), lengths of production cycles, and numbers of production cycles per year are presented in Table 7.14 (Ensminger & Olentine, 1978; North & Bell, 1990; USEPA, 2000).

The excretion rates over a 1-year period for a 500-AU confinement facility are summarized in Table 7.15.

For turkey virtual farms, reflecting the differences between male (tom) and female (hen) turkeys in average LW and lengths of production cycles was necessary (Table 7.14). The values for males and females were calculated separately and then averaged based on the assumption of equal numbers of males and females in a flock.

The computational process for dairy cattle was more complex. Because of differences in feeding programs, the generation rate of manure constituents had to be calculated separately for mature cows ad replacements (heifers—young cows that have not given birth to calves) and then combined.

TABLE 7.14
Typical Animal Live Weights and Production Cycles

Animal species (subtypes)	Average live weight lb/animal-day	Length of production cycle, days	Number of production cycles per year
Broilers	2.60	49	5.5
Layer Hens	3.97	350	1
Turkeys			
Hens	11.5	105	2
Toms	16.8	133	2
Swine			
Feeder pig*	154	119	2.8
Nursery pigs	37	35	1.7 to 8–9**
Gestating sows	452	185	1.7
Lactating sows	496	30	1.7
Gilts	249	190	1
Boars	396	365	1
Dairy cattle			
Lactating (> 24 mo)	1,350	335	1
Dry (> 24 mo)	1,350	30	1
Replacements (0 to 24 mo)	634	365	1
Veal calves	139	56	6
Beef cattle			
Feeder (6 to 12 mo)	815	180	2

Sources: Ensminger and Olentine (1978); North & Bell (1990); USEPA (2000a).

* For grow-finish operations.

** Eight to nine production cycles per year for standalone nursery operations.

TABLE 7.15
Quantities of Volatile Solids, Nitrogen, and Sulfur Excreted Per 500 Animal Unit Virtual Farm

Animal	Composition of animal manure as excreted (tons/yr)		
	Volatile solids	Nitrogen	Sulfur
Beef	399	22	3
Veal	10	3	0
Dairy*	705	35	4
Swine-feeder pigs	173	14	3
Poultry-broiler	262	20	1.5
Poultry-layer	374	28	4
Poultry-turkey**	375	29	0

Source: USEPA (2001a).

* Based on replacing 25% of mature cow population each year.

** 50% of population are toms, and 50% are hens.

Similarly, for mature cows, the generation rates for lactating cows and dry cows were calculated separately and then combined. A 500-AU virtual dairy farm will have the equivalent of 350 mature cows. Approximately 25% of mature cows are replaced each year, resulting in 280 mature cows and 70 replacements (expressed as mature cows). Table 7.14 shows that the average LW of a replacement is approximately half that of a mature cow, indicating that one mature cow is equivalent to two replacements. For a 500-AU virtual farm, this results in 140 replacements and 280 mature cows, or 420 total animals. A typical period of lactation for mature cows of 335 days per year followed by a dry period of 30 days was used (Ensminger & Olentine, 1978; Van Horn, 1998). Calculations were based on no difference in LW between the two periods (Table 7.14). Because a new period of lactation typically begins every 12 months, there is one production cycle per year.

For swine, significant differences occur in the rates of excretion of nitrogen and VS between gestating sows, lactating sows, and nursery pigs (Table 7.13). Due to the complexity and variety of configurations of swine farrow-to-finish and nursery operations, an accurate distribution of the different pig subtypes in a model swine farm could not be determined. Therefore, swine model farms were designed to represent grow-finish operations, and the information for feeder pigs was used. The other swine subtypes are shown only for informational purposes.

7.4.4 EMISSION FACTORS AND ESTIMATES FROM VIRTUAL FARMS

This section presents the emission factors and estimated emissions for each virtual farm. The virtual farms are summarized in Table 7.16. Emissions were estimated only from emission sources that are related to manure management and animal related activities (e.g., feeding, housing). Emissions from trucks, tractors, and other farm equipment as well as those related to the generation of electricity were not included.

Emissions were estimated for NH_3, N_2O, H_2S, VOCs, and PM. In this virtual presentation, PM represents TSP, except where specifically noted at PM 10. Information was not available to quantify emissions of odor-causing compounds other than H_2S and VOCs.

Emissions were computed for virtual farms with a confinement capacity of 500 AU, the maximum number of animals that could be confined at one time. Based on the USEPA Office of Water definition, 500 AUs are equal to:

- 500 cattle, excluding mature dairy cattle and veal
- 350 mature dairy cattle
- 500 veal
- 1,250 swine each weighing over 55 lb
- 5,000 immature swine each weighing less than 55 lb
- 27,500 turkeys
- 50,000 chickens

The virtual study results are presented in two tables for each animal type. The first table summarizes the emission factors used for each emission point and indicates the range of emission factors from the literature, the number of emission factors, and the average and median of the emission factors found. Median values are provided as an indication of how normally the data points were distributed (i.e., a median significantly different than the average would indicate the presence of "outliers" in the data used for the emission factor). The table also identifies the references for each emission factor and the methodology used to estimate emissions where emission factors were not available. The second table presents the annual emission estimates for each virtual farm, calculated by multiplying the average emission factor (lb/year-AU) by 500 (AU capacity of the virtual farms) and correcting to tons per year. Where emission factors are not presented, the table indicates elements where (1) emissions are expected to be negligible and (2) emissions are expected but could not be estimated because of a lack of useable data.

TABLE 7.16
Summary of Virtual Farms

Animal	Virtual Farm ID	Confinement and Manure Collection System	Components of Virtual Farms		
			Solids Separation Activities	Manure Storage and/or Stabilization	Land Application
Beef	B1A	Drylot (scraped)	Solids separation for runoff (using a settling basin)	Storage pond (wet manure) and stockpile (dry manure)	Liquid manure application; solid manure application
	B1B		No solids separation		
Veal	V1	Enclosed house (flush)	None	Anaerobic lagoon	Liquid manure application
	V2	Enclosed house with pit storage	None	None	Liquid manure application
Dairy	D1A	Freestall barn (flush); milking center (flush); drylot (scraped)	Solids separation	Anaerobic lagoon (wet manure) and stockpile (dry manure)	Liquid manure application; solid manure application
	D1B		No solids separation		
	D2A	Freestall barn (scrape); milking center (flush); drylot (scraped)	Solids separation	Anaerobic lagoon (wet manure) and stockpile (dry manure)	Liquid manure application; solid manure application
	D2B		No solids separation		
	D3A	Milking center (flush); drylot (scraped)	Solids separation	Storage pond (wet manure) and stockpile (dry manure)	Liquid manure application; solid manure application
	D3B		No solids separation		
	D4A	Drylot feed alley (flush); milking center (flush); drylot (scraped)	Solids separation	Anaerobic lagoon (wet manure) and stockpile (dry manure)	Liquid manure application; solid manure application
	D4B		No solids separation		
Swine	S1	Enclosed house (flush)	None	Anaerobic lagoon	Liquid manure application
	S2	Enclosed house (pit recharge)	None	Anaerobic lagoon	Liquid manure application
	S3A	Enclosed house (pull plug pit)	None	Anaerobic lagoon	Liquid manure application
	S3B			External storage tank or pond	Liquid manure application
	S4	Enclosed house with pit storage	None	None	Liquid manure application
Poultry–broilers	C1A	Broiler house with bedding	None	Covered storage of cake; open litter storage	Solid manure application
	C1B		None	Covered storage of cake	Solid manure application
Poultry–layers	C2	Caged layer high rise house	None	None	Solid manure application
	C3	Caged layer house (flush)	None	Anaerobic lagoon	Liquid manure application
Poultry–turkeys	T1A	Turkey house with bedding	None	Covered storage of cake; open litter storage	Solid manure application
	T1B			Covered storage of cake	

Source: USEPA (2001a).

TABLE 7.17
Summary of Beef Emission Factors

Emission Source	Substance	Emission Factor Range (lb/yr-AU)	Number of Emission Factors	Average/Median Emission Factor (lb/yr-AU)	Reference
Drylot	NH_3	9.7– 41.4	3	22.0/25.6	European Environmental Agency, 1999; Grelinger, 1997; Hutchinson et al., 1982
Drylot	N_2O	—	—	2.8	a
Drylot	PM 10	5.4–20.0	2	12.7/12.7	USDA, 2000; Grelinger, 1997
Storage pond	N_2O	—	—	0.14	a
Stockpile	NH_3	4.2	1	4.2/4.2	European Environmental Agency, 1999
Stockpile	N_2O	—	—	2.8	a
Solid manure spreader	NH_3	8.0–38.2	5	18.8/23.1	USEPA, 1999; Van der Hoek, 1998

Source: USEPA (2001a).
[a] Calculated using a nitrogen in manure-to-nitrous oxide conversion factor. See Section 7.4.2.2.

For land application, the emission estimates represent short-term releases that occur from the application of manure to land. No information was found for estimating residual emissions from manure application sites over the long term following application (i.e., soils releases). Typically, the most prominent soil release is N_2O. However, if manure is applied at agronomic rates, N_2O emissions should be the same as if inorganic commercial fertilizers are applied. Conditions on farm land generally do not favor the formation of methane, water, or VOCs, except under transient conditions (e.g., extended rainfall) when saturated soil and warm temperatures promote microbial activity.

7.4.4.1 Beef Virtual Farms

Beef cattle emission factors and emission estimates for the two beef cattle virtual farms are summarized in Table 7.17 and Table 7.18, respectively. Emission factors from the literature search were used to estimate ammonia emissions from drylots, stockpiles, and solid manure land application activities. Emission factors also were found for PM 10 emissions from drylots. Nitrous oxide emissions from drylots, storage ponds, and stockpiles were calculated assuming that a fraction of the nitrogen in manure would be emitted as nitrous oxide, using the methodology and information presented in Section 7.4.2.2.

7.4.4.2 Veal Virtual Farms

No emission factors were identified for veal operations from the literature search. Emission factors for nitrous oxide and VOCs from anaerobic lagoons (Table 7.19) were derived based on a fraction emitted of the nitrogen and VS in the manure, using the methodologies in Sections 7.4.2.2 and 7.4.2.5. Estimates of hydrogen sulfide emissions could have been made by translating emissions from anaerobic lagoons at swine virtual farms, as discussed in Section 7.4.2.3, but no information on the sulfur content of veal manure was available to apply the hydrogen sulfide ratios to veal. Table 7.20 summarizes the emission estimate for the two veal virtual farms.

7.4.4.3 Dairy Virtual Farms

Dairy cattle emission factors and emission estimates for the eight dairy virtual farms are summarized in Tables 7.21 and 7.22, respectively.

TABLE 7.18

Summary of Emissions from Beef Virtual Farms (tons/yr-500-AU farm)

Virtual ID	Emission Source	NH_3	N_2O	H_2S	VOCs	PM10
B1	Drylot	5.5	0.7	Neg.[a]	Neg.[a]	3.2
	Solids separation	Neg.[a]	Neg.[a]	Neg.[a]	Neg.[a]	Neg.[a]
	Storage pond	Neg.[a]	Neg.[a]	[b]	[b]	Neg.[a]
	Liquid manure land application	[b]	[b]	[b]	[b]	Neg.[a]
	Stockpile	1.0	0.7	[c]	[c]	[c]
	Solid manure land application	4.7	[b]	[c]	[c]	[c]
	Total	**11.2**	**1.4**	**[b]**	**[b]**	**3.2**
B2	Drylot	5.5	0.7	Neg.[a]	Neg.[a]	3.2
	Storage pond	Neg.[a]	Neg.[a]	[b]	[b]	Neg.[a]
	Liquid manure land application	Neg.[a]	Neg.[a]	[b]	Neg.[a]	Neg.[a]
	Stockpile	1.0	0.7	[c]	[c]	[c]
	Solid manure land application	4.7	[b]	[c]	[c]	[c]
	Total	**11.2**	**1.4**	**[b]**	**[b]**	**3.2**

Source: USEPA (2001a).

[a] No emissions or negligible emissions are expected from this emission source.

[b] Emissions are expected from this source, but information is not available for estimation.

[c] Emissions may occur from this source depending on whether manure is dry (PM, N2O) or wet (NH3, H2S, VOCs). Information is not available to estimate emissions.

TABLE 7.19

Summary of Veal Emissions Factors

Emission source	Substance	Emission factor range (lb/yr-AU)	Number of emission factors	Average emission factor (lb/yr-AU)	References
Anaerobic lagoon	N_2O	—	—	0.02	[a]
	VOC	—	—	0.08	[b]

Source: USEPA (2001a).

[a] Calculated using a nitrogen in manure-to-nitrous oxide conversion factor. See Section 7.4.2.2.

[b] Calculated using a volatile solids-to-VOCs conversion factor. See Section 7.4.2.5.

Emission factors were developed from literature sources for ammonia emissions from scrape freestall barns, drylots, liquid manure land application activities, and solids storage. Emission factors also were found for PM emissions from drylots.

Ammonia emissions from flush barns and anaerobic lagoons were derived by translating emissions from comparable swine operations, using the methodology and assumptions presented in Section 7.4.4.1. The hydrogen sulfide emission factor for anaerobic lagoons was derived by the same method.

Nitrous oxide emissions from drylots, storage ponds, and anaerobic lagoons were estimated by calculating a fraction of the nitrogen in manure that would be emitted as nitrous oxide, using the methodology and information presented in Section 7.4.2.2. VOC emissions were estimated for anaerobic lagoons based on a fraction of the potential methane emissions being converted to VOCs, using the methodology and information presented in Section 7.4.2.5.

TABLE 7.20
Summary of Emissions from Veal Virtual Farms (tons/year-500 AU farm)

Virtual ID	Emission source	NH₃	N₂O	H₂S	VOC	PM
V1	Confinement (flush)	Neg.[a]	Neg.[a]	Neg.[a]	Neg.[a]	Neg.[a]
	Anaerobic lagoon	b	0.005	b	0.02	Neg.[a]
	Liquid manure land application	b	b	b	b	Neg.[a]
	Total	b	**0.005**	b	**0.02**	—
V2	Confinement w/pit storage	b	Neg.[a]	b	b	Neg.[a]
	Liquid manure land application	b	b	b	b	Neg.[a]
	Total	b	b	b	b	**Neg.[a]**

Source: USEPA (2001a).

[a] No emissions or negligible emissions are expected from this emission point.

[b] Emissions are expected, but information is not available to estimate emissions.

TABLE 7.21
Summary of Dairy Emission Factors

Emission source	Substance	Emission factor range (lb/yr-AU)	Number of emission factors	Average emission factor (lb/yr-AU)	References
Freestall barn (flush)	NH₃	—	—	28	d
Freestall barn (scrape)	NH₃	15.2–16.8	2	16.0/16.0	Demmers et al. (2001); University of Minnesota (1999)
	NH₃	4.5–13.4	3	10.2/9.0	Bouwman et al. (1997); Misselbrook et al. (1998); Van der Hoek (1998)
Drylot	N₂O	—	—	4.4	a
	PM	2.3	1	2.3/2.3	USDA (2000)
	NH₃	—	—	40	b
	N₂O	—	—	0.22	a
Anaerobic lagoon	H₂S	—	—	15.7[e], 4.1[f]	b
	VOC	—	—	4.5	c
Storage pond	N₂O	—	—	0.22	a
Liquid manure land application	NH₃	18.7	1	18.7/18.7	Van der Hoek (1998)
Solids storage	NH₃	5.9	1	5.9/5.9	Van der Hoek (1998)
	N₂O	—	—	4.4	a

Source: USEPA (2001a).

[a] Calculated using a nitrogen in manure to nitrous oxide conversion factor. See Section 7.4.2.2.

[b] Calculated by transferring emissions from swine anaerobic lagoons. See Sections 7.4.2.1, 7.4.2.3.

[c] Calculated using a volatile solids to VOC conversion factor. See Section 7.4.2.5.

[d] Calculated by transferring emissions from swine flush houses.

[e] Used for virtual farms D1A, D1B, D4A and D4B; data was transferred from anaerobic lagoons following flush houses for swine.

[f] Used for model farms and D2A and S2B; data was transferred from anaerobic lagoons not following flush houses for swine.

TABLE 7.22

Summary of Emissions from Dairy Virtual Farms (tons/yr-500-AU farm)

Virtual ID	Emission source	NH$_3$	N$_2$O	H$_2$S	VOCs	PM
D1A	Drylot	2.5	1.1	Neg.[a]	Neg.[a]	0.6
	Freestall barn (flush)	7.0	Neg.[a]	Neg.[a]	Neg.[a]	Neg.[a]
	Milking center	Neg.[a]	Neg.[a]	Neg.[a]	Neg.[a]	Neg.[a]
	Solids separation	Neg.[a]	Neg.[a]	Neg.[a]	Neg.[a]	Neg.[a]
	Anaerobic lagoon	10.0	0.1	3.9	1.1	Neg.[a]
	Liquid manure land application	4.7	[b]	[b]	[b]	Neg.[a]
	Stockpile	1.5	1.1	[c]	[c]	[c]
	Solid manure land application	[b]	[b]	[c]	[c]	[c]
	Total	**26**	**2.3**	**3.9**	**1.1**	**0.6**
D1B	Drylot	2.5	1.1	Neg.[a]	Neg.[a]	0.6
	Freestall barn (flush)	7.0	Neg.[a]	Neg.[a]	Neg.[a]	Neg.[a]
	Milking center	Neg.[a]	Neg.[a]	Neg.[a]	Neg.[a]	Neg.[a]
	Anaerobic lagoon	10.0	0.1	3.9	1.1	Neg.[a]
	Liquid manure land application	4.7	[b]	[b]	[b]	Neg.[a]
	Stockpile	1.5	1.1	[c]	[c]	[c]
	Solid manure land application	[b]	[b]	[c]	[c]	[c]
	Total	**26**	**2.3**	**3.9**	**1.1**	**0.6**
D2A	Drylot	2.5	1.1	Neg.[a]	Neg.[a]	0.6
	Freestall barn (scrape)	4.0	Neg.[a]	Neg.[a]	Neg.[a]	Neg.[a]
	Milking center	Neg.[a]	Neg.[a]	Neg.[a]	Neg.[a]	Neg.[a]
	Solids separation	Neg.[a]	Neg.[a]	Neg.[a]	Neg.[a]	Neg.[a]
	Anaerobic lagoon	10.0	0.1	1.0	1.1	Neg.[a]
	Liquid manure land application	4.7	[b]	[b]	[b]	Neg.[a]
	Stockpile	1.5	1.1	[c]	[c]	[c]
	Solid manure land application	[b]	[b]	[c]	[c]	[c]
	Total	**23**	**2.3**	**1.0**	**1.1**	**0.6**
D2B	Drylot	2.5	1.1	Neg.[a]	Neg.[a]	0.6
	Freestall barn (scrape)	4.0	Neg.[a]	Neg.[a]	Neg.[a]	Neg.[a]
	Milking center	Neg.[a]	Neg.[a]	Neg.[a]	Neg.[a]	Neg.[a]
	Anaerobic lagoon	10.0	0.1	1.0	1.1	Neg.[a]
	Liquid manure land application	4.7	[b]	[b]	[b]	Neg.[a]
	Stockpile	1.5	1.1	[c]	[c]	[c]
	Solid manure land application	[b]	[b]	[c]	[c]	[c]
	Total	**23**	**2.3**	**1.0**	**1.1**	**0.6**
D3A	Drylot	2.5	1.1	Neg.[a]	Neg.[a]	0.6
	Milking center	Neg.[a]	Neg.[a]	Neg.[a]	Neg.[a]	Neg.[a]
	Solids separation	Neg.[a]	Neg.[a]	Neg.[a]	Neg.[a]	Neg.[a]
	Liquid manure land application	4.7	[b]	[b]	[b]	Neg.[a]
	Stockpile	1.5	1.1	[c]	[c]	[c]
	Solid manure land application	[b]	[b]	[c]	[c]	[c]
	Total	**8.7**	**2.3**	**[b]**	**[b]**	**0.6**

TABLE 7.22 (continued)
Summary of Emissions from Dairy Virtual Farms (tons/yr-500-AU farm)

Virtual ID	Emission source	NH$_3$	N$_2$O	H$_2$S	VOCs	PM
D3B	Drylot	2.5	1.1	Neg.[a]	Neg.[a]	0.6
	Milking center	Neg.[a]	Neg.[a]	Neg.[a]	Neg.[a]	Neg.[a]
	Storage pond	[b]	0.1	[b]	[b]	Neg.[a]
	Liquid manure land application	4.7	[b]	[b]	[b]	Neg.[a]
	Stockpile	1.5	1.1	[c]	[c]	[c]
	Solid manure land application	[b]	[b]	[c]	[c]	[c]
	Total	**8.7**	**2.3**	**[b]**	**[b]**	**0.6**
D4A	Drylot	2.5	1.1	Neg.[a]	Neg.[a]	0.6
	Drylot feed alley (flush)	[b]	Neg.[a]	Neg.[a]	Neg.[a]	Neg.[a]
	Milking center	Neg.[a]	Neg.[a]	Neg.[a]	Neg.[a]	Neg.[a]
	Solids separation	Neg.[a]	Neg.[a]	Neg.[a]	Neg.[a]	Neg.[a]
	Anaerobic lagoon	10.0	0.1	3.9	1.1	Neg.[a]
	Liquid manure land application	4.7	[b]	[b]	[b]	Neg.[a]
	Stockpile	1.5	1.1	[c]	[c]	[c]
	Solid manure land application	[b]	[b]	[c]	[c]	[c]
	Total	**19**	**2.3**	**3.9**	**1.1**	**0.6**
D4B	Drylot	2.5	1.1	Neg.[a]	Neg.[a]	0.6
	Drylot feed alley (flush)	[b]	Neg.[a]	Neg.[a]	Neg.[a]	Neg.[a]
	Milking center	Neg.[a]	Neg.[a]	Neg.[a]	Neg.[a]	Neg.[a]
	Anaerobic lagoon	10.0	0.1	3.9	1.1	Neg.[a]
	Liquid manure land application	4.7	[b]	[b]	[b]	Neg.[a]
	Stockpile	1.5	1.1	[c]	[c]	[c]
	Solid manure land application	[b]	[b]	[c]	[c]	[c]
	Total	**19**	**2.3**	**3.9**	**1.1**	**0.6**

Source: USEPA (2001a).

[a] No emissions or negligible emissions are expected.

[b] Emissions are expected from this operation, but information is not available to estimate them.

[c] Emissions may occur from this operation depending on whether manure is dry (PM, N2O) or wet (NH3, H2S, VOCs). Information is not available to estimate emissions.

7.4.4.4 Swine Virtual Farms

Swine emission factors developed for this study and emission estimates for the five virtual swine farms are summarized in Table 7.23 and Table 7.24, respectively. Emission factors were developed from literature sources for ammonia, PM, and hydrogen sulfide. Emission factors from the literature search were used to estimate ammonia emissions from flush houses, houses with pit storage, houses with pull plug pits, houses using pit recharges, anaerobic lagoons, and liquid land application activities.

The same PM emission factor was used for each of the different swine confinement houses because the majority of PM would come from feed handling, which would be the same for all swine confinement houses. Hydrogen sulfide emission factors were developed from the literature for the house with pit storage, anaerobic lagoon, and liquid land application activities.

Nitrous oxide emissions from the anaerobic lagoon and external storage were calculated based on a fraction of the nitrogen in manure being emitted as nitrous oxide, using the methodology and information presented in Section 7.4.2.2. VOC emissions were estimated for anaerobic lagoons

TABLE 7.23

Summary of Swine Emissions Factors

Emission source	Substance	Emission factor range (lb/yr-AU)	Number of emission factors	Average/median emission factor (lb/yr-AU)	Reference
Flush house	NH_3	6.4–17.1	3	10.3/11.8	Hoeksma & Monteny, 1993; Oosthoek et al., 1991
	PM	4.6–13.0	3	8.0/8.8	Grelinger & Page, 1999, Takai et al., 1998
	NH_3	10.8–17.1	2	14.0/14.0	Oosthoek et al., 1991; University of Minnesota, 1999
House w/pit recharge	PM	4.6–13.0	3	8.0/8.8	Grelinger & Page, 1999; Takai et al., 1998
	NH_3	9.1–16.5	3	13.7/12.8	Andersson, 1998; Hoeksma & Monteny, 1993; Oosthoek et al., 1991
House w/pull plug pit	PM	4.6–13.0	3	8.0/8.8	Grelinger & Page, 1999; Takai et al., 1998
	NH_3	0.6–44.6	15	17.2/22.6	Andersson, 1998; Hoeksma & Monteny, 1993; Ni et al., 2000; Oosthoek et al., 1991; Secrest, 2000; USDA, 2000; USEPA, 1994; Zhu et al., 2000
House w/pit storage	PM	4.6–13.0	3	8.0/8.8	Grelinger & Page, 1999; Takai et al., 1998
	NH_3	2.8–39.4[a]	9	15.1/21.1	Aneja et al., 2000; Cure et al., 1999; Harper & Sharp, 1998; Martin, 2000[g]; NCDENR[f]
Anaerobic lagoon	N_2O	—	—	0.085	[d]
	H2S	0.8–9.8	5	9.8/9.8[b]; 2.6/2.9[c]	Grelinger & Page, 1999; Secrest, 2000
	VOC	—	—	2.4	[e]
Liquid land application	NH_3	20.9–44.3	5	29.4/32.6	USEPA, 1994; Van der Hoek, 1998
	H_2S	0.6	1	0.6/0.6	Grelinger & Page, 1999
External storage	N_2O	—	—	0.085	[d]

Source: USEPA (2001a).

[a] Three of the emissions factors were reported as nitrogen and converted to ammonia, assuming all nitrogen was ammonia.

[b] Used for virtual farm S1 because emission factor is for anaerobic lagoon following a flush house.

[c] Used for virtual farm S2 and S3A because emission factors were representative of anaerobic lagoons not following a flush house.

[d] Calculated using nitrogen in manure to nitrous oxide conversion factor. See Section 7.4.2.2.

[e] Calculated using a volatile solids in manure to VOC conversion factor. See Section 7.4.2.5.

[f] Report did not provide background test data.

[g] Based on a mass balance completed from tests of an anaerobic lagoon, showing 63.6% loss of total, Kjeldahl nitrogen input.

based on a fraction of the potential methane emissions being converted to VOCs, using the methodology and information presented in Section 7.4.2.5.

7.4.4.5 Poultry Virtual Farms

Poultry emission factors developed for this virtual study and emission estimates for the six poultry virtual farms are summarized in Table 7.25 and Table 7.26, respectively. Emission factors from the literature search were used to estimate ammonia emissions from (1) broiler and turkey housing, manure storage, and solid manure land application and (2) layer flush houses, high-rise houses, solid manure land applications, and liquid manure land applications. The ammonia emission factor for

TABLE 7.24

Summary of Emissions from Swine Virtual Farms (tons/yr-500-AU farms)

Virtual ID	Emission source	NH$_3$	N$_2$O	H$_2$S	VOC	PM
S1	Flush house	2.6	Neg.[a]	Neg.[a]	Neg.[a]	2.0
	Anaerobic lagoon	4.6	0.021	2.4	0.6	Neg.[a]
	Liquid manure land application	7.3	[b]	0.15	[b]	Neg.[a]
	Total	**1.5**	**0.021**	**2.6**	**0.6**	**2.0**
S2	House with pit recharge	3.5	Neg.[a]	[b]	[b]	2.0
	Anaerobic lagoon	4.6	0.021	0.7	0.6	Neg.[a]
	Liquid manure land application	7.3	[b]	0.15	[b]	Neg.[a]
	Total	**15**	**0.021**	**0.9**	**0.6**	**2.0**
S3A	House w/pull plug pit	3.4	Neg.[a]	[b]	[b]	2.0
	Anaerobic lagoon	4.6	0.021	0.7	0.6	Neg.[a]
	Liquid manure land application	7.3	[b]	0.15	[b]	Neg.[a]
	Total	**15**	**0.021**	**0.9**	**0.6**	**2.0**
S3B	House with pull plug pit	3.4	Neg.[a]	[b]	[b]	2.0
	External storage	[b]	0.021	[b]	[b]	Neg.[a]
	Liquid manure land application	7.3	[b]	[b]	[b]	Neg.[a]
	Total	**11**	**0.021**	**[b]**	**[b]**	**2.0**
S4	House with pit storage	4.3	0.021	0.3	[b]	2.0
	Liquid manure land application	7.3	Neg.[a]	[b]	[b]	Neg.[a]
	Total	**12**	**0.021**	**0.3**	**[b]**	**2.0**

[a] No emissions or negligible emissions are expected.
[b] Emissions are expected, but information is not available to estimate emissions.

broiler houses was used for turkey houses because of the similarity in houses, manure, and manure handling activities. The references that provided ammonia emission factors for cake and litter storage did not distinguish between the covered storage of cake and the open storage of litter. Given that no basis occurred by which the emission factor could be partitioned, equal amounts of ammonia were assumed to be emitted from both types of storage. Therefore, the emission factors for covered storage and open storage were multiplied by 50%. Emission factors from the literature were also used for PM emissions from broiler and turkey houses.

Nitrous oxide emissions in all cases were calculated based on a fraction of the nitrogen in manure being emitted as nitrous oxide, using the methodology and information presenting in Section 7.4.2.2. VOC emissions were estimated for anaerobic lagoons based on a fraction of the potential methane emissions being converted to VOCs, as explained in Section 7.4.2.5. Ammonia and hydrogen sulfide emissions from anaerobic lagoons were derived from swine anaerobic lagoons, using the methodology presented in Section 7.4.2.1 and Section 7.4.2.3, respectively.

7.4.5 Comparison of Emission Estimates to Manure Characteristics

Table 7.27 compares the annual emission estimates for the virtual farm to the quantities of VS, nitrogen, and sulfur compounds that are excreted annually, which defines the theoretical upper limit of ammonia and hydrogen sulfide emissions, respectively, if 100% mineralization occurs. The VS excreted annually define the theoretical upper limit for combined emissions of methane and VOCs, if all of the VS excreted are biodegraded. Obviously, only a fraction of these excreted compounds will be emitted as ammonia, hydrogen sulfide, methane, and VOCs. However, this comparison provides a method to assess the general validity of the various emissions estimates.

As shown in Table 7.27, the amount of excreted nitrogen that is emitted as the sum of ammonia-nitrogen and nitrous oxide-nitrogen rates from about 25% for drylot dairies to 94% for turkeys (cake

TABLE 7.25
Summary of Poultry Emissions Factors

Animal feeding operation	Substance	Emission factor range (lb/yr-AU)	Number of emission factors	Average/median emission factor (lb/yr-AU)	Reference
Broiler house w/bedding	NH_3	10–51	8	24.3/31	Groot Koerkamp et al., 1998; Kroodsma et al., 1988; Tamminga, 1992; USEPA, 1994; Van der Hoek, 1998; Zhu et al., 2000
	N_2O	—	—	2.4	a
	PM	2.9–14	2	8.2/8.2	Grub et al., 1965; Takai et al., 1998
	NH_3	2.2	1	2.2/2.2	Van der Hoek, 1998[c]
Broiler covered storage of cake	N_2O	—	—	2.4	a
	NH_3	2.2	1	2.2/2.2	Van der Hoek, 1998[c]
Broiler open litter storage	N_2O	—	—	2.4	a
Broiler solid manure land application	NH_3	22–24	2	23/23	Van der Hoek, 1998; USEPA, 1994; Groot Koerkamp et al., 1998
Caged layer flush house	NH_3	16.5–44	6	32.8/30.3	Kroodsma et al., 1988; Tamminga, 1992; Van der Hoek, 1998; USEPA, 1994
	N_2O	—	—	3.6	a
Layer high-rise house	NH_3	13.1–44	8	28.5/28.6	Groot Koerkamp et al., 1998; Hartung & Phillips, 1994; Kroodsma et al., 1988; Tamminga, 1992; USEPA, 1994; Valli et al., 1991; Van der Hoek, 1998
	N_2O	—	—	3.6	a
Layer solid manure land application	NH_3	11.1–36	4	24/24	USEPA, 1994; Van der Hoek, 1998; Witter, 1991
	NH_3	—	—	32	b
Layer anaerobic lagoon	N_2O	—	—	0.02	a
	H_2S	—	—	4.9	b
	VOC	—	—	4	c
Layer liquid manure	NH_3	11.1–36	4	24/24	USEPA, 1994; Van der Hoek, 1998; Witter, 1991
	NH_3	—	—	44	d
	N_2O	—	—	3.6	a
Turkey w/bedding	PM	1.4–36	2	18.7/18.7	Grub et al., 1965; Takai et al., 1998
	NH_3	7	1	7	Van der Hoek, 1998
Turkey covered storage	N_2O	—	—	3.6	a
	NH_3	7	1	7	Van der Hoek, 1998
Turkey open litter storage	N_2O	—	—	3.6	a
Turkey solid manure land application	NH_3	46–65	2	55/55	USEPA, 1994; Van der Hoek, 1998

[a] Calculated using a nitrogen in manure-to-nitrous oxide conversion factor. See Section 7.4.2.2.

[b] Calculated by transferring emissions from swine virtual models. See Sections 7.4.2.1 and 7.4.2.3.

[c] Calculated using a volatile solids in manure-to-VOC conversion factor. See Section 7.4.2.5.

[d] Calculated by transferring emission factors from broiler house. See Section 7.4.2.1.

[e] References provided emission factors for cake and litter storage but did not distinguish between the covered storage of cake and the open storage of litter. Given that there was no basis to partition the emission factors, it was judged that equal amounts of ammonia would be emitted from both types of storage. Half the emission factor was assigned to covered storage and half to open litter storage.

TABLE 7.26

Summary of Emissions from Poultry Virtual Farms (tons/yr-5000-AU farm)

Virtual ID	Emission source	NH$_3$	N$_2$O	H$_2$S	VOCs	PM
C1A	Broiler house with bedding	6.1	0.60	Neg.[a]	Neg.[a]	2.1
	Covered storage of cake	0.55	0.60	Neg.[a]	Neg.[a]	Neg.[a]
	Open litter storage	0.55	0.60	c	c	c
	Solid manure land application	5.8	b	c	c	c
	Total	**13.0**	**1.8**	c	c	**2.1**
C1B	Broiler house with bedding	6.1	0.60	Neg.[a]	Neg.[a]	2.1
	Covered storage of cake	0.55	0.60	Neg.[a]	Neg.[a]	Neg.[a]
	Solid manure land application	5.8	b	c	c	c
	Total	**13**	**1.2**	c	c	**2.1**
C2	Caged layer high rise house	7.1	0.90	Neg.[a]	Neg.[a]	b
	Solid manure land application	5.9	b	Neg.[a]	Neg.[a]	Neg.[a]
	Total	**13**	**0.90**	**Neg.[a]**	**Neg.[a]**	b
C3	Caged layer flush house	8.2	0.046	b	b	b
	Anaerobic lagoon	8.0	0.046	1.2	0.98	Neg.[a]
	Liquid manure land application	5.9	Neg.[a]	b	b	Neg.[a]
	Total	**22**	**0.092**	**1.2**	**0.98**	b
T1A	Turkey house with/bedding	11	0.90	Neg.[a]	Neg.[a]	4.7
	Covered storage of cake	0.9	0.90	Neg.[a]	Neg.[a]	Neg.[a]
	Open litter storage	0.9	0.90	c	c	c
	Solid manure land application	14	b	c	c	c
	Total	**27**	**2.7**	c	c	**4.7**
T1B	Turkey house with bedding	11	0.90	Neg.[a]	Neg.[a]	4.7
	Covered storage of cake	0.9	0.90	Neg.[a]	Neg.[a]	Neg.[a]
	Solid manure land application	14	b	c	c	c
	Total	**26**	**1.8**	c	c	**4.7**

Source: USEPA (2001a).

[a] No emissions or negligible emissions are expected.

[b] Emissions are expected, but information is not available to estimate emissions.

[c] Emissions may occur depending on whether manure is dry (PM, N$_2$O) or wet (NH$_2$, H$_2$S, VOCs). Information is not available to estimate emissions.

and litter storage virtual model T1A). The amount of excreted sulfur that is emitted as hydrogen sulfide-sulfur ranges from 10% (swine pit storage mode S4%) to 80% (swine flush house model S1). For the most part, these values appear to be reasonable ranges; however, the nitrogen emissions for turkeys (T1A and T1B) are probably significant overestimates. Nitrogen emissions as a percentage of excreted nitrogen from the broiler and turkey virtual farms should be similar, based on the understanding that typically no more than 80% of manurial nitrogen is readily mineralized. As a result, emission estimates of greater than 80% of excreted nitrogen are unrealistic. The loss of 80% of excreted sulfur as hydrogen sulfide for swine virtual farm S1 also appears to be unrealistically high.

TABLE 7.27

Comparison of Nitrogen, Sulfur, and Volatile Solids in Substances Emitted to Manure Loading

Virtual model	Manure loading (tons/500 AU-yr)			Emissions (tons/500 AU-yr)				
	N	S	VS	NH3-N	N2O-N	N-Total	H2S-S	VOC-C*
B1	22	3	399	9.2	0.9	10.1	—	—
B2	22	3	399	9.2	0.9	10.1	—	—
V1	3	—	10	—	0.003	0.003	—	0.01
V2	3	—	10	—	—	—	—	—
D1A	35	4	705	21.4	1.5	22.9	3.7	0.6
D1B	35	4	705	21.4	1.5	22.9	3.7	0.6
D2A	35	4	705	18.9	1.5	20.4	0.9	0.6
D2B	35	4	705	18.9	1.5	20.4	0.9	0.6
D3A	35	4	705	7.2	1.5	8.7	—	—
D3B	35	4	705	7.2	1.5	8.7	—	—
D4A	35	4	705	15.6	1.5	17.1	3.7	0.6
S1	14	3	173	12.4	0.01	12.4	2.4	0.3
S2	14	3	173	12.4	0.01	12.4	0.8	0.3
S3A	14	3	173	12.4	0.01	12.4	0.8	0.3
S3B	14	3	173	9.1	0.01	9.1	—	—
S4	14	3	173	9.9	0.01	9.9	0.3	—
C1A	20	2	262	10.7	1.1	11.8	—	—
C1B	20	2	262	10.3	0.8	11.1	—	—
C2	28	4	374	10.7	0.6	11.3	—	—
C3	28	4	374	18.1	0.06	18.2	1.1	0.6
T1A	29	—	375	22.2	1.7	23.9	—	—
T1B	29	—	375	21.4	1.1	22.5	—	—

Source: USEPA (2001a).

* Assumes VOCs consist of equal parts butyric acid, methylamine, and phenol.

7.5 EMISSION CONTROL METHODS

This section summarizes the possible control methods for reducing air emissions from CAFOs. The information assembled for this section was obtained by a review of the available literature (ERG, 2000). This section does not, however, specifically discuss the control of odors. Odor is not the result of the formation and emission of a distinct compound but rather is an indicator of the presence of one or more of the compounds (ammonia, hydrogen sulfide, and VOCs) that collectively contribute to odor. Although methods for reducing odor emissions are not specifically addressed in this section, the methods identified for reducing emissions of ammonia, hydrogen sulfide, and VOCs can also be used to reduce odor.

Emissions can be controlled by preventing or inhibiting the formation of emitted substances, suppressing emissions of substances once formed, or capturing and controlling a substance that is emitted. Inhibition techniques either reduce the amount of nitrogen and sulfur available to form ammonia and hydrogen sulfide or remove the conditions that favor formation. Suppression techniques prevent the release of substances once they have been generated. Because the substances are not physically altered or destroyed, they can be emitted at a later time or at another location (for

TABLE 7.28

Summary of Control and Suppression Techniques for Particulate Matter Emissions[a]

Description	Outdoor confinement	Indoor confinement	Manure storage and stabilization	Land application	Carcass handling
Suppression techniques					
Water application	X			X	
Oil application		X (60–80%)			
Modification of feed handling delivery systems		X (35–70%)			
Covering of manure stockpiles			X		
Capture and control techniques					
Filtration		X (50–60%)			
Ionization		X (40–60%)			
Wet scrubbing		X (≤ 90%)			

Source: USEPA (2001a).
[a] When available, percent reductions from literature provided.

example, covering a manure storage pond or lagoon will contain ammonia but will not prevent emission during subsequent land application if manure is surface-applied). Control techniques reduce emissions by capturing airborne emissions or altering the chemical composition of compounds to another form (for example, converting ammonia to nitrate).

Table 7.28 through Table 7.32 summarize the control methods found for PM, ammonia, hydrogen sulfide, methane, and VOCs, respectively. The tables categorize the control methods by inhibition, suppression, and control; indicate the parts of the farm (e.g., confinement, manure management, etc.) to which the method applies; and provide available information on control efficiency.

The remainder of this section summarizes the information obtained from the literature review. The controls for PM, gaseous emissions, and land application are presented in Sections 7.5.1, 7.5.2, and 7.5.3, respectively. Each section briefly describes the control methods and the emission control mechanisms and presents the information found on control efficiency, costs, and secondary environmental impacts.

7.5.1 Particulate Matter Emission Controls

PM is emitted from outdoor and indoor confinement facilities as well as stockpiles of manure solids. PM emissions from outdoor confinement facilities and manure stockpiles consist primarily of dry manure particles and soil. Low moisture feedstuffs, such as hay, also can be sources of PM emissions. Wind and movement of animals and vehicles generate the emissions of PM to the atmosphere.

With indoor confinement facilities, the primary sources of PM emissions are dried manure, feedstuffs, litter (bedding), and animal dander. Feathers from poultry also are a source of PM emissions. PM suspension is caused by movement of animals and by air circulation from natural or mechanical ventilation. The amount of PM generated from dried manure depends on the method of manure handling used in the indoor confinement facility. For example, manure is a significant fraction of the PM emissions from broiler and turkey production facilities as well as high-rise-type houses for laying hens, because the manure is handled as a dry solid. Conversely, manure that is handled as semisolid, slurry, or liquid, such as swine and dairy cow manure, is not a source of PM emissions.

TABLE 7.29

Summary of Inhibition, Suppression, and Control Techniques for Ammonia Emissions[a]

Description	Outdoor confinement	Indoor confinement	Manure storage and stabilization	Land application	Carcass handling
Inhibition techniques					
Design and operating methods	X	X			
Diet manipulations	X	X (28–53%)	X	X	
Manure additives			X		
Suppression techniques					
Acidification of manure		X	X		
Covers				X	
Rapid incorporation				X (> 87%)	
Direct injection					
Capture and control techniques					
Biofiltration		X (50–80%)			
Bioscrubbing		X (≤ 89%)			
Gas absorption			X		
Covering of anaerobic lagoons with biogas		X (≤ 53%)			
Anaerobic digestion			X		
Chemical oxidants			X		
Ozonation		X (15%)[c]			
Incineration					X
Composting			X[b]		X[c]

Source: USEPA (2001a).

[a] Where available, percent reductions from literature provided.

[b] The performance of this technique has not been consistently reproduced.

[c] Assumes adequate aeration to maintain predominantly aerobic conditions.

PM emissions associated with feedstuffs are primarily associated with handling, such as transfer into storage and delivery to animals. Finely ground feedstuffs for poultry and swine, which may be fed in pelletized form, are significant sources of PM.

This section discusses the following control methods for reducing PM emissions from animal confinement: water application, oil application, modifications to feed handling and delivery systems, filtration, ionization, wet scrubbing, and covering of manure stockpiles. Although descriptions of these techniques were found in the literature review, full-scale evaluations and demonstrations are lacking.

7.5.1.1 Water Application

7.5.1.1.1 Description and Applicability of Technique

To suppress PM emissions from outdoor feedlots, water sprays or sprinkler systems can be used to prevent the confinement surface (e.g., manure and soil) from becoming too dry. In practice, tanker trucks are used to dispense water over the confinement area surface. However, the suppression technique may only be practical for small operations, since large amounts of water are needed. Sweeten et al. (2000) cited the amount of water for suppressing dust was similar to the cattle drinking water requirements during the dry season (0.1 to 0.25 in per day). No discussion of using water sprays

TABLE 7.30

Summary of Control and Suppression Techniques for Hydrogen Sulfide Emissions[a]

Description	Outdoor confinement	Indoor confinement	Manure storage and stabilization	Land application	Carcass handling
Suppression techniques					
Diet manipulation	X	X	X	X	
Manure additives		X	X[b]		
Covers			X (> 95%)		
Prompt removal to disposal		X			X
Capture and control techniques					
Biofiltration		X (80–86%)			
Biocovers		X			
Gas absorption		X			
Aerobic treatment	X	X			
Covering of anaerobic lagoons with biogas control			X		
Anaerobic digestion			X		
Ozonation		X			
Incineration					X
Composting			X[c]		X[c]

Source: USEPA (2001a).

[a] Where available, percent reductions from literature are provided.

[b] The performance of this technique has not been consistently reproduced.

[c] Assumes adequate aeration to maintain predominantly aerobic conditions.

for indoor confinement for PM emissions suppression were found in the literature review, although increasing the humidity level indoors (e.g., using water sprays) should reduce the suspended PM concentration. Misting systems are used in indoor confinement facilities for broilers, turkeys, and swine. However, these systems are typically only used during hot weather for evaporative cooling.

7.5.1.1.2   Summary of Performance and Cost Data

No data are available to characterize the effectiveness of water sprays on reducing PM emissions from outdoor or indoor operations. However, increasing the moisture content of outdoor confinement soil or litter in broiler and turkey production facilities may increase other emissions because microbial activity in the manure is stimulated. The controls costs include the delivery system (e.g., tanker truck, misting system), water availability, and labor and management costs.

7.5.1.2   Oil Application

7.5.1.2.1   Description and Applicability of Technique

Suppression of PM from confinement housing has been achieved by applying vegetable oil on interior building surfaces (using handheld sprayers or sprinklers systems) and by applying oil to the skin of swine (using rollers or scratching posts that dispense oil on contact). However, the oil can be a safety hazard (i.e., slippery floors) for both personnel and animals. Also, the oily surfaces can increase building clean-out times between production cycles and may contribute to gaseous emissions as the residue undergoes microbial decomposition.

TABLE 7.31

Summary of Control and Suppression Techniques for Methane Emissions[a]

Description	Outdoor confinement	Indoor confinement	Manure storage and stabilization	Land application	Carcass handling
Suppression techniques					
Manure additives		X[b]			
Covers			X		
Prompt removal to disposal		X			X
Capture and control techniques					
Biocovers			X		
Covering of anaerobic lagoons w/biogas control			X		
Anaerobic digestion			X		
Ozonation		X			
Incineration					X
Composting			X[c]		X[c]

Source: USEPA (2001a).

[a] Where available, percent reductions from literature are provided.

[b] The performance of this technique has not been consistently reproduced.

[c] Assumes adequate aeration to maintain predominantly aerobic conditions.

TABLE 7.32

Summary of Control and Suppression Techniques for Volatile Organic Compound Emissions[a]

Description	Outdoor confinement	Indoor confinement	Manure storage and stabilization	Land application	Carcass handling
Suppression techniques					
Manure additives		X[b]	X[b]		
Covers			X		
Prompt removal to disposal		X			
Capture and control techniques					
Biofiltration		X			
Covering of anaerobic lagoons with biogas control			X		
Anaerobic digestion			X		
Vent gas capture/control		X	X		
Incineration					X
Composting			X[c]		X[c]

Source: USEPA, 2001a.

[a] Where available, percent reductions from literature are provided.

[b] The performance of this technique has not been consistently reproduced.

[c] Assumes adequate aeration to maintain predominantly aerobic conditions.

7.5.1.2.2   Summary of Performance and Cost Data

Several studies (Mankell et al., 1995; Takai et al., 1993; Zhang et al., 1996a) discussed reducing indoor PM concentrations using oil sprays. One study (Takai et al., 1993) achieved from 60% to 80% reduction in suspended PM concentrations using oil sprays in a swine confinement building. No secondary impacts related to this suppression technique have been reported in the literature.

- Important point: No basis to estimate the cost of this suppression technique or the potential increase in cleaning cost was found. However, the control costs would include the delivery system (e.g., portable sprayer), oil, and labor and management costs.

7.5.1.3   Modification of Feed Handling and Delivery System

7.5.1.3.1   Description and Applicability of Technique

PM emissions generated by the feed handling and delivery system can be reduced by the following modifications to the system:

- Mixing vegetable oil or animal fats with the feed
- Using totally enclosed delivery systems and covered feeders (except poultry feeders)
- Using pelletized feed

These modifications generally are applicable only to grain-based poultry and swine feeds that are fed directly after grinding or following pelleting.

Oils and fats commonly are added to poultry and swine rations as sources of metabolizable or digestable energy with use depending in part on the cost of other sources of energy. They also are used as a binder for pelleting. One drawback of adding fats or oils to feeds is the possibility of spoilage and the possible development of a rancid flavor, reducing feed consumption.

Options to control PM from feed handling systems generally are limited to the capture of dust generated when feeds are transferred to storage bins. Capturing PM emitted from feed bin vents when bins are filled with feed can do this. For swine operations, automatically closing feeder covers may reduce PM emissions to some degree by reducing the air movement over the feed.

Pelleting of animal feeds is also a control technique for PM emissions. However, this technique is not applicable to some feeds, such as starter rations for broilers and turkeys, which cannot be pelletized.

7.5.1.3.2   Summary of Performance and Cost Data

Several studies (Chiba et al., 1987; Heber & Martin, 1988; Takai et al., 1996) reported that PM reductions in air concentrations ranging from 35 to 70% have been achieved by adding fats or oils (1% to 4%) to the feed of indoor confinement housing (primarily swine and poultry). However, pelleting can reduce the digestibility of swine ad poultry rations.

- *Important point:* Using fat or oils for PM suppression could result in increased feed costs.

7.5.1.4   Filtration

7.5.1.4.1   Description and Applicability of Technique

Filters remove PM by impaction of entrained particulates on the filter media as air is passed through the filter. Filtration of indoor air can reduce PM emissions from confinement housing. Filters are not a feasible control option for outdoor confinement because the contaminated air cannot practically be captured and conveyed to the control device. Filtration can be applied to building exhaust ventilation air, where mechanical ventilation is used, to reduce dust emissions from totally or partially enclosed confinement housing. Filters also can be integrated into an air recirculation system that does not vent to the atmosphere.

Commercially available units using synthetic filter media could be used to reduce PM emissions from indoor confinement housing. Also, systems have been fabricated using natural material (e.g., straw and other crop residues) as the filter media. In these systems, building exhaust is routed through a structure containing the crop residue.

Over time, the filter media becomes clogged with PM and the media must either be cleaned or replaced. Filters made from synthetic materials typically are reused after cleaning, whereas natural filter media are replaced with new material.

7.5.1.4.2   Summary of Performance and Cost Data

Data on the performance of filters in reducing PM concentration was reported in only one study (Carpenter & Fryer, 1990). In that study, a synthetic filter achieved reductions in indoor PM concentrations from swine confinement ranging from 50% to 60%. The filter was a two-stage system that consisted of a coarse pre-filter and a fine filter, in series.

The secondary impacts associated with using filters is the emissions from the generation of the additional electricity needed for fans used to convey the contaminated air through filters. Also, the filters themselves can generate waste streams, depending on type of cleaning mechanism used (i.e., solid waste if the spent filter media is disposed of; liquid waste if the media is washed).

- *Important point:* Capital costs include duct work for routing building exhaust air, the filter housing, and filter media. Annual costs include maintenance, labor, and management costs and any additional costs of electricity used for powering duct work fans, if needed.

7.5.1.5   Ionization

7.5.1.5.1   Description and Applicability of Technique

Ionization is a potential method for reducing PM emissions from indoor confinement housing, although evaluation of its applicability to CAFOs has been limited. In ionization, gas molecules (e.g., oxygen) acquire a charge from high-energy electrons created by an electrically generated corona field. The ionized gases adhere to particulates, which then move to the nearest grounded surface (e.g., building surface, grounded collection). This collection mechanism is used by electrostatic precipitators (ESPs) in other industries, such as utilities.

For CAFO applications, commercially available room ionizers have been used to charge the indoor air molecules. Building surfaces have been used to collect PM (separate collection plates were not used).

7.5.1.5.2   Summary of Performance and Cost Data

Although ionization (i.e., ESPs) has been demonstrated to achieve PM removal efficiencies of 99% or greater in other industries, ionization has been shown to reduce PM emissions by only 40% to 60% in agricultural applications, based on the results of three separate studies (Bundy, 1984; Bundy, 1994; Moller, n.d.). No explanation for the lower PM removal efficiencies of ionization used for agricultural applications was found in the studies. However, high moisture content of the air stream may have been a factor.

The secondary impacts associated with using ionization include the emissions from the generation of the electricity needed to convey the contaminated air and to generate the corona field. Ionization also produces ozone and nitrous oxide. As with filters, the material collected using ionization requires disposal.

No data were found for estimating the costs of ionization for the reduction of PM emissions from indoor confinement facilities.

7.5.1.6   Wet Scrubbing

7.5.1.6.1   Description and Applicability of Technique

Wet scrubbing is a potential control technique for reducing PM emissions from confinement housing ventilation exhaust. A wet scrubber is typically an enclosed tower (with or without packing material) or wetted pad where a particulate-laden gas stream flows counter current to the flow of water. Particulates are removed by direct impaction and interception with or diffusion into water droplets.

7.5.1.6.2   Summary of Performance and Cost Data

The evaluation of wet scrubbers in the CAFO industry has been limited. One study (Pearson, 1989) showed a PM reduction of up to 90% using wet scrubbing. The secondary impacts associated with using a wet scrubber include the emissions from the generation of the electricity needed to convey the contaminated air to the scrubber and the electricity needed to run the scrubber pumps. Wet scrubbers also generate a liquid waste stream (i.e., scrubber effluent).

The capital costs for wet scrubbers applied to indoor confinement include the cost of the scrubber (or wetted pad), pumps for circulating scrubbing media, electric fans for moving confined housing air, and any duct work needed to convey building air to the scrubber. Annual operations costs include the electricity for pumps and fans and labor and management costs.

7.5.1.7   Covering of Manure Stockpiles

7.5.1.7.1   Description and Applicability of Technique

The potential for direct PM emissions from manure storage facilities obviously is limited to those used to handle manure as a solid with wind being the mechanism responsible for PM suspension and transport. Thus, covering stacked manure with sheet plastic or tarpaulins or use of windbreaks reduces PM emissions from these storage facilities.

7.5.1.7.2   Summary of Performance

Covering stored manure can potentially create anaerobic conditions that could initiate or increase ammonia, hydrogen sulfide, methane, and VOC emissions.

7.5.2   Gaseous Emissions Controls

Gaseous compounds are generated by microbial decomposition of animal manure in confinement and manure storage and stabilization facilities. Gaseous compounds are also generated by microbial decomposition of animal carcasses. The presence of aerobic versus anaerobic conditions determines the nature of gaseous compounds formed.

Under aerobic conditions, the principal gaseous emissions are carbon dioxide and ammonia. The carbon in organic compounds is oxidized to carbon dioxide, and nitrogen is mineralized to ammonia. Also, any reduced forms of sulfur, including hydrogen sulfide, are oxidized to nonvolatile sulfate. Aerobic conditions are typically associated with storage and stabilization of manure solids. The potential for aerobic conditions is limited to low moisture content manures, such as broiler and turkey manures and other manures handled as solid.

Under anaerobic conditions, the carbon in organic carbon compounds is reduced primarily to methane and various VOCs with some formation of carbon dioxide also occurring. Nitrogen and sulfur are reduced to ammonia and hydrogen sulfide, respectively. Because oxygen only is sparingly soluble in water, resulting in a very slow rate of natural diffusion, conditions exist when manure is handled as a liquid or slurry, unless external aeration is provided.

Gaseous emission control techniques include techniques for inhibiting and suppressing gaseous emissions and for altering the chemical composition of gaseous compounds (e.g., converting reduced compounds to oxidized compounds). With the exception of covering anaerobic lagoons (with and without biogas collection), anaerobic digestion, and composting, full-scale evaluation and demonstration under commercial conditions of the control methods described in this section generally have been lacking.

7.5.2.1 Confinement Facility Design and Operating Methods

7.5.2.1.1 Description and Applicability of Technique

Confinement facility design and operating practices can inhibit the generation of reduced gaseous compounds or suppress emissions once they have been generated. However, suppression techniques may only transfer the point of emissions to another CAFO process (e.g., to manure storage or a land application site). Also, because of their nature, these design and operating practices may be applicable only to new facilities.

On outdoor feedlots, moist conditions lead to anaerobic decomposition of manure. Faster drying of manure and frequent removal of manure from confinement areas can achieve suppressed or reduced emission of gaseous compounds. Sloping of the feedlot surface (4% to 6%) toward the south to southeast direction ensures that the feedlot receives the most insolation and that the accumulated manure dries more quickly. Ammonia and other gaseous emissions can also be reduced by removing solid manure frequently (every 7 days or less). However, manual removal tends to transfer ammonia and other gaseous emissions to manure storage and stabilization processes.

With slurry systems, frequent flushing or scraping to remove manure from partially or totally enclosed facilities also reduces the potential for gaseous compound emissions from the confinement facility. A smooth floor surface increases the effectiveness of frequent removal by both flushing and scraping. Ideas such as flow-through partitions and under floor ventilation have been proposed to enhance manure drying in partially or totally enclosed confinement facilities, but both their effectiveness and practicality seem questionable.

In facilities where manure is collected in shallow or deep pits, which typically are located under slatted floors, filling the pit with enough water so that all of the accumulating manure solids are submerged may reduce ammonia, hydrogen sulfide, and VOC emissions to some degree. Both ammonia and hydrogen sulfide are highly soluble in water, as are some VOCs. If these pits are not ventilated and have little natural air movement, a decrease will occur in the concentration gradient across the interface between the liquid and gas phases, with the consequence of decreased rates of mass transfer. Because methane is essentially insoluble in water, methane emissions will not be decreased.

7.5.2.2 Acidification of Manure in Confinement Housing

7.5.2.2.1 Description and Applicability of Technique

As discussed earlier, ammonia volatilization is inhibited under acidic conditions. At a pH of approximately 4.5 or lower, virtually all of the ammonia present exists as nonvolatile ammonium ion (NH_4^+). Consequently, ammonia emissions can be suppressed by acidification of solid and liquid manure. However, decreasing manure pH increases the potential for volatilization of hydrogen sulfide.

Acidification is used extensively to reduce ammonia emissions during the initial stage of broiler and turkey grow-out cycles to decrease the incidence of ammonia-induced respiratory problems and blindness in young birds. For many years, phosphoric acid was used as the acidifying agent, but concern about high phosphorus concentrations in land-applied manures has resulted in a shift to other materials, such as sodium bisulfate and aluminum sulfate. Usually ammonia volatilization is suppressed only for about 2 weeks because buffering agents (calcium and magnesium carbonates) are continually added in freshly excreted manure. Repeat applications of an acidifying agent can prolong the period of suppression but may only delay emissions to storage or land application processes.

This technique is also applicable to manure collection in confinement housing for swine and dairy operations that use flushing systems. Using low-pH liquid with flushing systems can decrease the rate of ammonia volatilization.

- *Interesting point:* In theory, ammonia emissions from manures handled as liquids or slurries or manure accumulations on open lots could also be reduced using acidification.

7.5.2.2.2   Summary of Performance and Cost Data

For acidification of manure, no data were found during the literature review to estimate the decrease in emissions of reduced gaseous compounds achieved with this technique. With regard to flushing systems, Heber et al. (1999) reported that those flushing swine confinement areas with low-pH liquid one to two times daily achieved approximately 70% reduction in ammonia emissions.

Because acidification is a suppression technique, the potential exists for ammonia to be volatilized from downstream processes (e.g., storage or land application) if the pH increases above 4.5. Also, the chemistry of hydrogen sulfide suggests that acidifying manures with an anaerobic microbial environment will increase hydrogen sulfide emissions.

The use of acids may not be economical since sophisticated application systems are typically required because of their dangerous and corrosive nature. Although using base-precipitating salts is less expensive and hazardous than acidifying agents, the reduction in manure slurry pH is more transient and more frequent applications would be required to maintain a low pH.

No information for quantifying the cost of flushing with low-pH liquid was found in the literature review. Because of the higher buffer capacity of livestock and poultry manures, it appears reasonable to conclude that the cost of acidification would be significant.

7.5.2.3   Biofiltration of Confinement Housing Exhaust

7.5.2.3.1   Description and Applicability of Technique

Biofilters use microbial action in all aerobic environments to oxidize the reduced compounds generated by indoor confinement into carbon dioxide, water, salts, and biomass. In biofiltration, building air from the ventilation system exhaust is passed through a filter bed with an established, diverse population of aerobic microorganisms. As the air stream flows through the filter media, oxidation of the gaseous compounds occurs.

A typical biofilter consists of a piping system for distributing the contaminated air throughout the filter bed. The filter media is usually organic (soil, compost, wood chips) with sufficient bulk to allow the air stream to pass and to prevent anaerobic conditions. Additionally, biofilters must have a drainage system (either active or passive) to remove excess condensate and precipitation. Although some moisture (50% to 60%) in the filter bed is needed to maintain microbial activity, excess moisture can lead to anaerobic conditions and failure of the biofilter. A filtration system upstream of the biofilter may be needed in some cases to remove PM, since accumulated dust can clog the filter over time. The filter bed must be rodent- and weed-free to avoid channeling of gases through the filter media and a loss of performance.

Because biofilters rely on microbial activity, performance is affected by ambient conditions (lower temperatures slow microbial activity) and variations in the pollutant concentrations in the contaminated air stream. The activity rate of microorganisms in the filter increases with increasing temperature. Consequently, the performance of biofilters varies seasonally unless provisions are made to preheat the incoming air stream during cold weather. Excessive variation in pollutant concentrations also can cause performance variability.

7.5.2.3.2   Summary of Performance Data

Although biofilters have been successfully used in other industries, in a few reported cases, biofilters have been shown to be economically viable when applied to CAFOs (Zahn et al., 2001). However, various pilot studies (University of Minnesota, 1999), primarily with swine operations, have shown that biofilters can reduce ammonia emissions by 50% to 80% and hydrogen sulfide emissions by 80% to 86%.

Biofilters can also be a source of nitrous oxide emissions because of denitrification following the oxidation of ammonia to nitrate and nitrate nitrogen. Periodically, the filter media must be replaced because of decomposition and compaction that occurs over time. This material is a potential source of solid waste. However, most organic media could be disposed by land application.

7.5.2.3.3 Summary of Cost Data

One article (Boyette, 1998) summarizing general biofilter performance reported that the operating and maintenance expenses for a biofilter range from \$2 to \$14/ft^3 of air treated. Another article (Leson & Winer, 1991) summarized the general design and performance data for biofilters used in other industries. This article presented ranges of capital cost estimates for open swing-bed filters of \$55 to \$90/ft^2 of filter area and \$90 to 500/ft^2 for enclosed systems.

7.5.2.4 Gas Absorption of Confinement Housing Exhaust

The operation of a gas absorber for removing gases, primarily ammonia and hydrogen sulfide, is very similar to that of a wet scrubber used for removing PM. However, the mechanism for removing gaseous compounds differs.

In a gas absorber, building air is collected and passed through an enclosed (typically packed) tower with the absorption media (e.g., caustic solution) flowing countercurrent to the incoming air stream. Gases in the air stream diffuse into and are absorbed by the media.

Although water is used as the scrubbing medium in many applications, the absorption of the gases can be enhanced using chemical reactions between the target gases and the absorbing medium, such as using caustic solution to remove acid gases.

7.5.2.4.1 Summary of Performance Data

Although no performance data was located during the literature review for absorbers applied to gaseous emissions from animal housing, one study (University of Minnesota, 1999) reported the ammonia removal achieved by a washing wall (a water curtain intended to remove PM as the building air passes through it, using the same removal mechanism [i.e., impaction] as a wet scrubber) at a swine facility. Because of ammonia's solubility in water, the washing wall was shown to reduce ammonia emissions up to 53%.

The secondary impacts associated with using a gas absorber include the emissions from the generation of the electricity needed to convey the contaminated air to the scrubber and the electricity needed to run the scrubber pumps. The effluent from a gas absorber is also a potential waste stream. If a caustic solution is used to remove acidic compounds, such as hydrogen sulfide, or an acidic solution is used to remove basic compounds, such as ammonia, from the air stream, the salts formed, such as sodium sulfate and ammonium phosphate, are removed from the scrubber as precipitates. If water is used as the scrubbing media, ammonia and hydrogen sulfide go into solution. Because only ionization occurs, the ammonia and hydrogen sulfide removed from the air stream can revolatilize from the scrubber effluent (e.g., if saturated effluent is exposed to the atmosphere).

7.5.2.4.2 Summary of Cost Data

No cost data for gas absorbers were found in the literature. However, one study (NCSU, 1998) noted that the installation cost of a washing wall system was approximately \$6 per unit of pig production capacity.

7.5.2.5 Bioscrubbing of Confinement Housing Exhaust

The concept behind a bioscrubber is similar to that of biofiltration, except that the microorganisms are housed in an enclosed packed tower with water circulated counter-current to the incoming building air, instead of in a filter bed. As contaminated air is passed through the scrubber, water-soluble compounds (ammonia, hydrogen sulfide) are absorbed by the water and oxidized microbially. Some scrubber designs contain a vessel that is used as a biological reactor. Effluent from the scrubber is routed to the vessel, where additional retention time is provided for microbial oxidation. No information was found in the literature review regarding the ultimate disposal of the effluent from bioscrubbers; however, this stream probably could be land-applied. Periodically, the filter media (especially organic media) must be replaced because of decomposition and compaction that occurs over time.

The rate of microbial oxidation in a bioscrubber is affected by temperature and variations in pollutant concentrations. However, bioscrubbers are unaffected by PM in the incoming gas stream. Periodically, the filter media (especially organic media) must be replaced because of decomposition and compaction that occurs over time.

7.5.2.5.1 Summary of Performance Data
A study of three bioscrubbers at swine operations showed that reductions of ammonia emissions up to 89% could be achieved (Lais et al., 1997). The secondary impacts from using biofilters include those associated with generation of the electricity needed to power fans and pumps. Although not specifically identified in the literature review, biofilters can be a source of nitrous oxide emissions if denitrification of the nitrified ammonia capture occurs. Bioscrubbers also are a source of solid waste (spend filter media) and wastewater (effluent from the scrubber).

The capital cost estimates of the three bioscrubbers at swine operations ranged from $9 to $17 per pig finished (Lais et al., 1997). No estimates of bioscrubber operating costs were found in the literature, but they would include the cost of electricity for pumps, maintenance, labor, and management.

7.5.2.6 Ozonation of Confinement Housing Air

7.5.2.6.1 Description and Applicability of Technique
Ozone (O_3) is a strong oxidant that reacts with most organic materials, including organic compounds and microorganisms. Although ozone has been used in treating drinking water, limited work has been conducted in evaluating the use of ozone to oxidize reduced gaseous compounds (ammonia and hydrogen sulfide) from CAFOs. Because the half-life of ozone is very short (10 to 30 minutes), it cannot be stored and must therefore be generated on-site. Typically, ozone is created by passing air through an electric field generated by a corona discharge cell.

7.5.2.6.2 Summary of Performance Data
One study (Priem, 1977) found that releasing ozone into the swine confinement building reduced ammonia levels in the air by 15% and 50% during the summer and winter ventilation conditions. The lower reduction was achieved during the summer months, which reflects the increased air circulation rate through the building for cooling.

The secondary impacts include the emissions from generation of the electricity needed to power fans for moving building air and for generating the corona discharge. Additionally, ozone usage has the potential for generation of nitrous oxide and sulfur oxides as by products.

7.5.2.6.3 Summary of Cost Data
One study (NCSU, 1998) estimated that ozonation of indoor air cost approximately $6 to $11 per unit of pig production capacity (the study did not specify if the cost estimate was for capital or annual costs).

7.5.2.7 Chemical Oxidation of Liquid Manure Storage

7.5.2.7.1 Description and Applicability of Technique
Oxidation of liquid manures by aerating storage basins or lagoons can reduce emissions of ammonia, hydrogen sulfide, methane, and VOCs. In aerobic stabilization, organic matter (containing carbon, hydrogen, oxygen, nitrogen, and sulfur) is microbially oxidized to carbon dioxide, water, and nitrate and sulfate ions. However, high-rate aeration, as used in the treatment of municipal and industrial wastewaters, is energy intensive, with high utility costs. Consequently, aeration of liquid manures is not typically practiced. Control of gaseous emissions is achieved by using chemical oxidants and biological treatment.

Chemical oxidants can be applied in liquid form to stored manure to oxidize ammonia, hydrogen sulfide, methane, and VOCs. Agents such as potassium permanganate and hydrogen peroxide can be applied to the manure surface to reduce emissions. However, a large amount of these types of additives is typically required because of the high level of organic matter content of animal manures. The emissions reduction achieved by these additives also appears to be short-term, requiring frequent applications to consistently reduce gaseous emissions. Ozone has been used to reduce gaseous emissions from manure slurries by bubbling or diffusing it through the slurry. However, ozone must be produced on-site, which requires costly generation and application systems (McCrory & Hobbs, 2001).

7.5.2.7.2   Summary of Performance and Cost Data
No characterization of chemical oxidant performance or identification of secondary impacts was found in the literature review. Based on the results of a laboratory study (Ritter et al., 1975), estimated costs of chemical oxidants for reducing hydrogen sulfide emissions from liquid dairy manure ranged from $0.06 to $12/10 m^3 of manure. These cost estimates were for a single application, with no indication of the required frequency of repeat applications.

7.5.2.8   Manure Additives

7.5.2.8.1   Description and Applicability of Technique
Manure additives include commercially available products intended to reduce ammonia volatilization from manure. The additives are typically mixed with water and poured evenly into the manure slurry. Also included are digestive additives (for example, select microorganisms, enzymes) intended to enhance the biodegradation of manure. Additives for absorbing ammonia and ammonium have also been used (McCrory & Hobbs, 2001).

7.5.2.8.2   Summary of Performance and Cost Data
No quantitative characterizations of the performance of manure additives or identification of possible secondary impacts were found in the literature reviewed. However, if absorbents are used, ammonia may be released during land application.

One study (Johnson, 1997) evaluated the effectiveness of eight manure additives from various suppliers. For all the additives tested, the cost was less than $0.65 per pig. However, the cost estimate did not include the labor required to apply the additives.

7.5.2.9   Covering of Liquid Manure Storage Tanks and Ponds

7.5.2.9.1   Description and Applicability of Technique
Liquid manure from swine and dairy operations is stored under anaerobic conditions in tanks or ponds or in anaerobic lagoons. Storage ponds and lagoons are large earthen impoundments operated under ambient conditions (no external heating). Anaerobic lagoons can be either single-cell or two-cell systems. Either a single basin (i.e., cell) is used for stabilization and storage, or the first cell is used exclusively for stabilization and the second cell is used as an effluent storage pond for two cell systems.

Liquid manure storage tanks and ponds and lagoons are sources of ammonia, hydrogen sulfide, methane, and VOC emissions. The population of methanogenic bacteria present determines the relative amounts of methane and VOCs emitted. Undersized lagoons emit greater quantities of VOCs, but even properly sized lagoons emit significant quantities of VOCs following extended periods of cold weather as the population of methanogenic bacteria becomes reestablished.

Where feasible, covering liquid manure storage tanks and ponds and anaerobic lagoons can suppress gaseous emissions of ammonia, hydrogen sulfide, and VOCs by reducing the air circulation above the manure surface, thus providing a barrier to diffusion from solution. However, covers that are not sealed do not suppress methane emissions because the primary constituents of biogas, methane, and carbon dioxide are essentially insoluble in water. Thus, escape of methane to the atmosphere occurs via some path of least resistance as biogas accumulates under an unsealed cover. (Sealed covers for anaerobic lagoons are discussed in the next section.) Although a wide range of covers is available, they can generally be categorized into two types: those that are self-supporting and those that are supported by the manure surface (i.e., floating covers).

Generally, self-supporting covers are made from such materials as wood, plastic, and concrete. These covers typically are fabricated on-site. Additionally, certain covers, depending on design, may require a drainage system for removing accumulated precipitation to prevent damage. Permanent covers are largely unaffected by ambient conditions, although some problems have been encountered with inflatable covers (a plastic membrane supported by captured biogas) under high wind conditions.

Floating covers can be permanent (e.g., polymer sheeting, polystyrene blocks) or temporary (e.g., surface crust, straw). Permanent floating covers are usually less expensive than self-supporting covers and provide greater emission reductions than temporary floating covers. Because they are typically attached to the tank or lagoon perimeter, permanent floating covers are less likely to be affected by wind. However, because they are attached, permanent floating covers may not be applicable to cases where the level of the manure surface fluctuates appreciably. Similar to self-supporting covers, permanent floating covers made from continuous materials (e.g., plastic sheeting) may require a drainage system for removing accumulated precipitation.

To form temporary floating covers, the covering materials (e.g., chopped straw) are applied directly to the manure surface, although in some cases, a crust naturally forms on the manure surface. Rather than provide an impermeable barrier, these covers reduce emissions by slowing the rate of diffusion and volatilization of gaseous compounds. Although they are the least expensive type of covering, they also achieve the lowest emission reduction relative to permanent floating and self-supporting covers. Channeling of gases can occur if holes or cracks develop in the cover, and natural covers can be disturbed by weather conditions (e.g., high winds), thereby reducing the effectiveness of the cover. Some temporary covering materials can become saturated and sink into the stored manure, potentially clogging the pumping system.

7.5.2.9.2   Summary of Performance Data

Permanent covers (made from plastic or concrete) were shown to suppress ammonia emissions by 80% at a swine facility (Sommer et al., 1993). Inflatable covers have been shown to suppress ammonia and hydrogen sulfide emissions by greater than 95% (Mannebeck, 1985; Zhang & Gaakeer, 1996) when applied to manure storage at swine facilities. Floating covers made of polystyrene or polyvinyl chloride and rubber have achieved suppression of gaseous emissions from swine manure by 90% or more (Clanton et al., 1999). No performance data were found in the literature for temporary covers made from natural materials.

No secondary impacts are associated with the use of covers unless electricity is used to power drainage system pumps. However, the suppressed emissions are released from the impoundment when the cover is removed and when the stored manure is land applied. Additionally, covers deteriorate over time because of temperature fluctuations and sunlight and must be periodically replaced.

7.5.2.9.3   Summary of Cost Data

The cost of covers is dependent on the material of construction and the surface area to be covered. Floating covers made from synthetic materials range in capital cost from $20 to $40/100 ft^2, depending on the type of material (Mannebeck, 1985). One study estimated the capital cost ($6,000) of an inflatable cover installed on an anaerobic stabilization lagoon sized for 200 sows at a farrow-to-finish facility (Zhang & Gaakeer, 1996b). This same study stated that a large concrete cover for the same size lagoon (i.e., 200 sows) could cost up to $50,000 (no design specifics were cited for the concrete cover).

7.5.2.10   Covering of Anaerobic Lagoons with Biogas Collection and Combustion

7.5.2.10.1   Description and Applicability of Technique

Although unsealed covers can suppress emissions of ammonia, hydrogen sulfide, and VOC emissions from manure storage tanks, ponds, and anaerobic lagoons, these gases can be emitted when the cover is removed or during land application of the biologically stabilized manure. However, sealed covers not only suppress emissions of ammonia, hydrogen sulfide, and VOCs but also capture the methane produced for disposal by flaring or use as a fuel. Given the relatively low rate of methane production from manure storage tanks and ponds, use of sealed covers with biogas collection only can be economically justified with anaerobic lagoons, which are designed to reduce VS to methane for waste stabilization. Although covered lagoons are not used extensively in the management of animal manures, a small number of full-scale covered anaerobic lagoons are in use for swine and dairy manures.

- *Interesting point:* Manure is partially digested feed. The remaining partially degraded and unused materials continue to decompose upon leaving the animal. Bacterial decomposition begins in any manure containment and continues until the manure is removed or is stabilized. The process can be simplified into two steps: Step 1 – Anaerobic bacteria degrade wet, unfrozen manure into the odiferous compounds associated with "that nasty manure smell"; Step 2 – Methane bacteria consume Step 1 compounds, given adequate time at a temperature above freezing, substantially eliminating the odors (Wilkie, et al. 1995; Roos et al., n.d.).

7.5.2.10.2   Summary of Performance Data

Although the performance data for covered anaerobic lagoons with biogas capture and use were not found in the literature review, reductions of ammonia, hydrogen sulfide, VOC, and methane emissions from the covered lagoon should approach 100%. However, subsequent emissions of ammonia, hydrogen sulfide, and VOCs from effluent storage ponds with two cell systems probably equal those from uncovered lagoons.

Because the collected biogas is sent to a combustion device (i.e., oxidized), the combustion device would be an emission source of carbon dioxide, nitrogen oxides, sulfur dioxide, and products of incomplete combustion. If, however, the captured biogas is used as a boiler fuel or electricity generation, these emissions would be in place of those resulting from the combustion of fossil fuels replaced.

7.5.2.10.3   Summary of Cost Data

One article (Roos et al., 1999) summarized cost estimates from eight vendors of lagoon covers designed for biogas collection. The installed cost (including cover components, labor, and shipping) ranged from $0.37 to $5.81/ft^3 of lagoon surface area. The range of costs was attributed the differences in cover materials, warranties, and installations. The cost estimates did not include the cost of the gas collection system (e.g., duct work, fans) or the combustion device.

Another article (USEPA, 2000b) summarized the installation costs for 11 covered lagoons with biogas collection and combustion. Detailed cost breakdowns were not provided in the article; the cost estimates did include the costs of cover components and combustion devices (e.g., flare, boiler). The surface areas of the lagoons covered were not provided in the article, though an estimate of the costs can be obtained by dividing the installed farrow-to-finish swine facility, which ranged from $133 to $158 per pig. The installed cost for swine nursery operations ranged from $5 to $73 per pig. For dairy operations, the installed costs ranged from $34 to $750 per cow.

7.5.2.11   Anaerobic Digestion

7.5.2.11.1   Description and Applicability of Technique

A small number of full-scale anaerobic digesters are in operation at commercial dairy and swine farms. Anaerobic digesters use the same microbial processes for stabilizing animal (swine and dairy) manure as anaerobic lagoons. However, an anaerobic digester is a closed reactor that is heated and possibly mixed to optimize the production of methane from the anaerobic decomposition process.

The main components of an anaerobic digester are the digester, effluent storage, and biogas collection and use equipment. Anaerobic digesters for animal manures may be either completely mixed or plug flow reactors with continuous or semicontinuous flow. The biogas produced contains about 60% to 70% methane, about 30% to 40% carbon dioxide, and trace amounts of hydrogen sulfide, VOCs, and moisture. The captured biogas is used either as a boiler fuel for space or water heating or used to fuel engine-generator sets to produce electricity. A fraction of the biogas energy is used for digester heating.

The benefits of anaerobic digestion are reduced emissions of methane, VOCs, hydrogen sulfide, and ammonia. However, ammonia and hydrogen sulfide emissions may only be delayed, depending on how the effluent is managed. The capital and annual operating costs of anaerobic digesters can be high but are at least partially offset by the value of the energy recovered. Digested fiber from dairy manure can be used as bedding material or sold.

7.5.2.11.2   Summary of Performance Data

No information was found in the literature review regarding the quantitative emissions reductions achieved by anaerobic digesters. However, because the digester is completely enclosed and the collected biogas is combusted, the percent destruction efficiency for gaseous pollutants would be similar to the performance of VOC incinerators (98%).

However, because the biogas is sent to a combustion device (either for energy recovery or control of emissions), the combustion device would be an emission source of carbon dioxide, nitrogen oxides, and sulfur dioxide and products of incomplete combusting. Again, an emissions off-set occurs with the replacement of fossil fuel combustion.

7.5.2.11.3   Summary of Cost Data

The costs of installing and operating an anaerobic digester vary, depending on the system design, location, and contractors. One report (USEPA, 1000b) summarized the installation costs of the various anaerobic digester systems operating in the United States. For complete mix digesters, the installed costs ranged from $18 to $325 per unit of confinement capacity (for swine facilities) and $750 to $1,852 per unit (for dairy operations). The high-end cost estimate for the dairy facilities included other costs associated with the operation's manure management system (e.g., storage tanks, scraper system). For plug-flow systems at dairy operations, the installed digester costs $200 to $1100 (the high-end cost estimate included other costs associated with manure management systems). The installed costs for plug-flow digesters at a swine facility and a poultry facility were $133 and $3, respectively.

The information found in the literature regarding operating costs of anaerobic digesters was limited. One report (USEPA, 2000b) presented long-term annual operating costs (electricity, maintenance) of approximately $2000 for a digester installed at an 8,600-head swine finishing operation. The report did not specify if this operating cost estimate included the benefits of biogas energy recovery; the report did summarize the estimated benefits (electricity, hot water, digested dairy fiber) associated with digester operations.

For dairy operations, the annual cost benefits (electricity and hot water offsets) ranged from $24 to $34 per cow. The value of the digested dairy solids ranged from $22 to $30 per cow. For swine operations, the annual cost benefits ranged from $12 to $27 per pig.

7.5.2.12   Biocovers for Liquid Manure Storage and Anaerobic Lagoons

7.5.2.12.1   Description and Applicability of Technique

In general, a biocover is a permeable cover made from natural (for example, chopped straw) or synthetic materials that floats on the surface on a storage or stabilization basin. The biocover provides a boundary layer between the surface of the manure and the atmosphere and a substrate for the growth of aerobic bacteria. As the reduced compounds (e.g., ammonia, hydrogen sulfide) diffuse through the cover, they are microbially oxidized.

7.5.2.12.2 Summary of Performance Data
No quantitative performance data were found for biocovers applied to manure storage or stabilization processes.

7.5.2.12.3 Summary of Cost Data
One study (Zahn et al., 2001) at a single facility estimated the capital and labor costs for a biocover (interlocked, perforated panels constructed of polymeric and geotextile materials) to be $2.37/m² of surface area ($1.14 per finisher pig). This study also cited a capital cost of $1.62/m² for a biocover made of a single layer of geotextile material. A life expectancy of 3 years was cited in the study as a conservative estimate.

7.5.2.13 Composting of Manure Solids

7.5.2.13.1 Description and Applicability of Technique
Composting is a predominantly aerobic biological waste stabilization process characterized by a significant elevation in temperature from microbial heat production. When properly operated, organic compounds are degraded with the oxidation of organic carbon to carbon dioxide to provide energy for cell maintenance and growth. In addition, any reduced sulfur compounds are oxidized to sulfates. Some methane and VOCs may be generated if localized anaerobic conditions occur but should be subsequently oxidized. Compost piles either are aerated continuously using air forced upward through the pile or tilled or turned periodically (typically daily) to ensure predominantly aerobic conditions. Bulking agents such as straw can be used to aid in maintaining aerobic conditions.

The magnitude of ammonia emissions during manure composting depends on the ratio of carbon to nitrogen. Without the addition of a supplemental source of carbon, ammonia emissions during manure composting are high. Because of the elevated temperature, which may reach 50° to 60°C (122° to 140°F), nitrification does not occur. However, studies confirm that the use of a sufficiently high initial carbon-to-nitrogen ration in the composted material (for example, achieved by adding high carbon–low nitrogen bulking agents such as straw) can minimize emissions of ammonia, as well as hydrogen sulfide, methane, and VOCs.

Manure can be composted in open piles or in open or enclosed structures. An impermeable surface is desirable to avoid groundwater contamination. With open piles, excess moisture from precipitation can lead to the development of anaerobic conditions and generate contaminated runoff requiring collection, storage, and disposal to avoid impairment of adjacent surface waters.

7.5.2.13.2 Summary of Performance and Cost Data
Composting capital costs include construction of composting bins and any equipment needed to till or turn the compost. Operating costs include maintenance, labor, and management costs.

If the conditions in the compost become predominately anaerobic, emissions of hydrogen sulfide, methane, and VOCs will occur. The secondary impacts of composting are associated with the use of energy for aeration and mixing.

7.5.2.14 Diet Manipulation

7.5.2.14.1 Description and Applicability of Technique
Recent studies, primarily involving swine and poultry, have demonstrated the potential for reducing gaseous emissions (e.g., ammonia) from manure by diet manipulation. The manipulation methods focus on improving nutrition. However, additional research is needed to fully evaluate the effectiveness of diet manipulation techniques since the digestion process is highly complex and the analytical results have not been consistent.

Improving nutrient utilization by animals (and consequently the reduction of nitrogen and sulfur excreted) has been shown to reduce emissions. Excess protein not used by the animal is excreted and contributes to ammonia emissions from manure. Several studies have shown that reducing dietary crude protein can reduce emissions of ammonia. Since proteins contain nitrogen, reducing the amount of protein that passes through an animal results in lower potential ammonia emissions. Zeolites and charcoal have been added to swine feeds in an attempt to bind ammonia and thereby reduce emissions. The enzyme phytase has been added to poultry and swine feeds to decrease the amount of excreted phosphorus. Phytase also appears to increase protein utilization.

Other additives (calcium salts, calcium benzoate) have been tried to reduce the pH (i.e., reduce the volatilization potential of ammonia) of excreted urine and manure. Research has also been conducted to evaluate the effectiveness of feeding specific substrates (e.g., polysaccharides, tea polyphenols) or microbial cultures to animals to alter the microflora contained in their digestive tracts.

7.5.2.14.2   Summary of Performance Data
One report (James et al., 1999) showed a 28% reduction in ammonia emissions from dairy cows fed a diet containing 9.5% crude protein. Another study (Whitney et al., 1999) showed that reducing the amount of sulfur in feeds and water reduced the amount of hydrogen sulfide and odor emissions from manure. Decreasing the digestive tract pH by increasing the level of calcium benzoate in sow diets achieved a reduction in ammonia emissions of up to 53% (Mroz et al., 1998). One study (Sutton et al., 1992) showed a 56% decrease in ammonia emissions from manure from swine fed the yucca extract.

7.5.2.14.3   Summary of Cost Data
Dietary manipulation has the potential of reducing feed costs. Additional research is needed to determine if diet manipulation adversely affects the animal's health or the productivity of the operation.

7.5.2.15   Carcass Disposal

7.5.2.15.1   Description and Applicability of Technique
In all livestock and poultry CAFOs, premature animal deaths occur. Decomposition of animal carcasses can emit reduced gases (ammonia, hydrogen sulfide, methane, and VOCs) and pathogenic bacteria to the atmosphere if the carcasses are not disposed of in a timely and proper manner. Chicken, turkey, and swine carcasses through the nursery stage of production are most commonly disposed of on-site either by composing, burial, or incineration. If disposal cannot be achieved within 24 hours, carcasses can be refrigerated to slow the decomposition process and thus minimize gaseous emissions.

Dairy, beef cattle, and feeder pig carcasses usually are disposed of by rendering off-site. In this section, only the techniques for on-site carcass disposal are addressed because emissions from rendering occur off-site.

7.5.2.15.2   Summary of Performance and Cost Data
Carcass incineration has the potential for generating emissions of particulates and other air pollutants (carbon dioxide, nitrogen oxides, and sulfur dioxide emissions and products of incomplete combustion). In many states, incinerators for animal carcass disposal are subject to regulations under state air quality statues with the requirements of operating permits that specify limits for PM emissions and other air pollutants. With carcass composting, PM emissions are limited to land application of composted residue and then only if the composted carcasses have a low moisture content.

The cost of a carcass composting facility for 25,000 birds of turkey confinement capacity is approximately $3,500 (Carter et al., 1993).

7.5.3   Land Application

As discussed earlier, the majority of animal manure (both solid and liquid) generated by CAFOs is applied to cropland or pasture for ultimate disposal. PM emissions associated with land application depend on the manure moisture content. Land application of manure handled as a solid, such as broiler and turkey litter, can be a significant source of PM emissions during and after land application. If present, ammonia, hydrogen sulfide, methane, and VOCs are also emitted during and following land application. The magnitudes of these emissions depend on: (1) the method of application and (2) the time of direct exposure of the applied manure to the atmosphere.

Solid manure is always applied to the soil surface while slurry and liquid manures can be either applied to the soil surface or injected into the soil. Both tractor-drawn and truck-mounted spreaders are used for application of manure to cropland pasture. Irrigation also is used for the disposal of liquid manure. Liquid manure from spreaders may be discharged under pressure using a splash plate to achieve a uniform spray pattern or distributed on the soil surface using devices such as band spreaders. The objective of using band spreaders, which distribute manure at ground level, is to reduce the surface area of manure exposed to the atmosphere during and after spreading. Equipment for injection of liquid or slurry manures has been available for several decades. Several different types of direct injection techniques are possible (e.g., shallow, deep), but the common characteristic is that they produce channels or holes for accepting the manure, which are subsequently closed by using a wheel or disc.

7.5.3.1   Particulate Matter Emissions from Land Application

7.5.3.1.1   Description and Applicability of Technique
Suppression of PM emissions during and after land application of dry manure could be achieved by either increasing manure moisture content before spreading or by using water sprays during or after spreading or both. However, neither can be considered a practical option. Increasing moisture content before spreading would require thorough mixing to insure uniform moisture distribution, and the volume of water required for water sprays would be prohibitive. However, a minimal degree of irrigation during and after spreading is a seemingly feasible option if sprinkler irrigation is available. However, most cropland and pastures used for manure disposal are not irrigated. Another feasible control option is avoiding the spreading of dry manure during windy conditions to reduce entrainment of PM.

7.5.3.2   Gaseous Emissions from Land Application

7.5.3.2.1   Description and Applicability of Technique
Suppression of gaseous emissions can be achieved by reducing the amount of time that the applied manure is exposed to the atmosphere. This can be accomplished by rapidly incorporating the applied manure into the soil. In general, a technique that applies and incorporates the manure in a single step has lower emissions than a technique that requires several steps. For example, applying manure using direct injection methods reduces emissions when compared to band spreading followed by disking or plowing, since direct injection applies and covers the manure in a single pass of the machinery. When manure is incorporated into the soil, ammonia, VOCs, and hydrogen sulfide are absorbed onto soil particles, providing the opportunity for oxidation by soil microorganisms to nitrates, sulfates, carbon dioxide, and water.

7.5.3.2.2   Summary of Performance Data
Land application of liquid manure using band spreaders with rapid incorporation into the soil (e.g., disking) has been shown to reduce gaseous emissions by 55% to 60%, compared to conventional broadcasting application using splash plate spreaders (Ministry of Agriculture FAF, 1992). One

study (Burton, 1997) that summarized the available European data from 1992 to 1997 showed that land application using a drag shoe for direct incorporation achieved reductions of 63% to 73% (depending on the type of land receiving the manure), compared to conventional broadcasting application.

Higher reductions of gaseous emissions have been reported using direct injection of the manure slurry into the soil. Studies have shown that ammonia reductions from 87% to 98% (Burton, 1997) can be achieved using direct injection (at various depths). Additionally, acidification of the manure slurry just prior to land application has been shown (Berg & Hornig, 1997; Burton, 1997) to achieve reductions of ammonia, but no quantitative reductions were given in these studies.

No secondary impacts are expected with these suppression techniques, other than the gaseous emissions from additional fuel combusted in the vehicles used to incorporate the manure, relative the amount of fuel needed to apply the manure.

7.5.3.2.3 Summary of Cost Data

Lazarus (1999) found that disk harrows, used for incorporating liquid and solid manures, ranged in price from $5,600 to $34,000 depending on their size and functionality. However, a disk harrow is a standard piece of tillage equipment on most farms engaged in crop production. Annual operation and maintenance costs were estimated to be 2% of the capital cost ($400 annually) plus an additional $30/hr for tractor operation and $10 for labor. Another study (USEPA, 1998) reported that the capital cost of a 4,200-gallon tank with injectors was about $20,000. One study (Wright, 1997) reported that tanker spreaders without injectors cost between $9,000 and $18,500, depending on the size; a 4,500-gallon tanker costs $14,000.

CHAPTER REVIEW QUESTIONS

1. What is the primary mechanism of release for gaseous emissions?
2. What are the chief sources of PM?
3. When does decomposition of manure begin and end?
4. What is a confinement facility? How does method of confinement affect emissions?
5. Does the method of land application effect emissions? How and why or why not?
6. What factors affect emissions of gases and PM?
7. How does ammonia form, and what are the design factors that affect accumulation and volatilization?
8. How does extended storage time affect ammonia formation? What chemical processes are responsible?
9. What conditions must exist for nitrous oxide emission? Where and when are they most likely to occur?
10. What is the principal source of nitrous oxide for most operations?
11. How is methane generated on AFOs and CAFOs?
12. Describe and discuss the relationship between methane volatilization and carbon dioxide emissions.
13. What are the principal factors affecting methane production?
14. How is carbon dioxide produced?
15. Do AFOs and CAFOs contribute to a net long-term increase in carbon dioxide? Explain.
16. How are VOCs formed?
17. What is the result of methane formation inhibition?
18. How are hydrogen sulfide and other reduced sulfur compounds produced?
19. What reduced sulfur compounds other than hydrogen sulfide are commonly emitted by AFOs and CAFOs?

20. What happens to excreted sulfur under anaerobic conditions? Under aerobic conditions?
21. Define and describe PM 10 and PM 2.5 particulate matter. What are the principal PM sources for AFOs and CAFOs?
22. What problems are presented by the data for PM emissions?
23. What are the usual ways PM can be reduced?
24. How does type and rate of ventilation affect PM mass? Discuss.
25. How have odor problems evolved over the gradual change from family farms to large-scale agribusinesses?
26. What are the principal compounds responsible for odor problems?
27. Where and how are these compounds produced?
28. Discuss odor problems associated with anaerobic lagoons.
29. What chief controlling elements affect emissions?
30. Describe and discuss the potential conditions for gaseous emissions.
31. Describe and discuss problems related to emissions from land application of animal wastes.
32. Describe and discuss the possible variables that can affect land application emissions of PM.
33. Discuss the short-term effects from emissions of PM.
34. Discuss the short-term effects from emissions of nitrogen compounds.
35. Discuss the short-term effects from emissions of hydrogen sulfides and VOCs.
36. Discuss the short-term effects from emissions of methane.
37. Describe and discuss general long-term emission problems associated with land applications.
38. Describe and discuss long-term nitrogen compound emission problems associated with land applications.
39. Describe and discuss long-term hydrogen sulfide and VOC emission problems associated with land applications.
40. Describe and discuss long-term methane emission problems associated with land applications.
41. Describe and discuss the approaches used to develop emissions factors for virtual farms.
42. Where can original literature sources for these virtual farms be found?
43. What parameters are related to the emissions data?
44. What problems were reflected in the source material?
45. How do the proposed NPDES permit regulations and standards for CAFOs define an AU?
46. What two alternative approaches were employed to develop emissions data?
47. What was the process used to put the reference material into a standard format?
48. What alternative methods were used to define emissions factors for ammonia? How were they calculated?
49. What alternative methods were used to define emissions factors for hydrogen? How were they calculated?
50. What alternative methods were used to define emissions factors for nitrous oxide? How were they calculated?
51. What qualifications have been placed on nitrous oxide emissions factors for the virtual farms?
52. Why is nitrification unlikely to occur at liquid manure storage sites?
53. Why is nitrification unlikely to occur at poultry confinement houses?
54. What justifies the use of swine hydrogen sulfide data for poultry operations?
55. Why aren't hydrogen sulfide emissions considered a factor for broiler and turkey confinement facilities?

56. Why isn't methane included?
57. What methods are used for methane emission estimation?
58. Describe and discuss the USEPA inventory method.
59. What is the methane conversion factor? Why is it important?
60. Calculate emissions from an aerobic lagoon at a 1000-AU swine virtual farm in North Carolina in July.
61. What is the difference in VOC production between anaerobic and aerobic manure storage systems? How does this affect virtual farm emission estimation?
62. How are VOC emissions estimated?
63. Under what storage conditions are more VOCs emitted? What factors make estimation difficult?
64. What steps are involved in estimating the mass of precursors in manure? Why is this necessary?
65. What factors cause significant differences in manure characteristics, even within the same species?
66. What uncertainties does the literature present for nitrogen sulfide and VS estimation?
67. What difficulties do dairy cattle estimation computations for nitrogen sulfide and VS pose? How are they solved?
68. What is the standardized confinement capacity of these virtual farms?
69. How can emissions be controlled? Describe and discuss various methods for the following emissions: general emissions, PM, ammonia, hydrogen sulfide, methane, and VOCs.
70. How are water applications used to control PM? Describe and discuss.
71. How are oil applications used to control PM? Describe and discuss.
72. How are feed handling and delivery modifications used to control PM? Describe and discuss.
73. How are filtration methods used to control PM? Describe and discuss.
74. How are ionization methods used to control PM? Describe and discuss.
75. How are wet scrubbing technologies used to control PM? Describe and discuss.
76. How is covering stockpiles used to control PM? Describe and discuss.
77. Describe and discuss gaseous emission control techniques and applications. Cover the following methods: confinement facility design and operation methods, acidification of manure in confinement housing, biofiltration, gas absorption, bioscrubbing, ozonation, chemical oxidation, manure additives, covering liquid manure storage, covering anaerobic lagoons and biogas collection and consumption, anaerobic digestion, biocovers, composting of manure solids, diet manipulation, and carcass disposal.
78. Describe and discuss land application emission control techniques for PM.
79. Describe and discuss land application emission control techniques for gaseous emissions.

THOUGHT-PROVOKING QUESTIONS

1. Air quality problems can present particularly difficult neighbor relationship problems. Why is this so, how can these problems develop, and what can be done about them?
2. What are the possible human health effects of inhaled PM from AFOs and CAFOs?
3. Discuss the problems that can be created by "fixing" another problem: for example, using water applications to control PM transfers at least part of the pollutant to another environmental media, where the pollutant presents other problems.

REFERENCES

Alexander, M. 1977. *Introduction to Soil Microbiology*, 2nd ed. New York: John Wiley and Sons.
Anderson, D.P., Beard, C.W., & Hanson, R.P. 1964. The adverse effects of ammonia on chickens including resistance to infection with newcastle disease virus. *Avian Diseases* 8:369–379.

Anderson, B. 1994. Animal manure as a plant resource. http://www.agcom. Purdue.edu/AfCom/Pubs/ID/ID-101.html. Accessed April 17, 2005.

Andersson, M. 1998. Reducing ammonia emissions by cooling of manure in manure culverts. *Nutrient Cycling in Agroecosystems* 51: 73–79.

Aneja, V.P., Chauhan, J.P., & Walker, J.T. 2000. Characterization of atmospheric ammonia emissions from swine waste storage and treatment lagoons. *Journal of Geophysical Research* 105: 11535–11545.

Anthonisen, A.C., Loehr, R.C., Prakasam, T.B.S., & Srinath, E.G. 1976. Inhibition of nitrification by ammonia and nitrous acid. *Journal of Water Pollution Control Federation* 48:835–852.

ASAE. 1999. *1999 ASAE Standards, Engineering Practices and Data,* 46th ed. American Society of Agricultural Engineers.

Berg, W. & Hornig, G. 1997. Emission reduction by acidification of slurry – Investigations and assessment. J.A.M. Voermans, G. Monteny (Ed.). Procs. of the Intl. Symp. on Ammonia and Odour Control from Animal Production Facilities Vinkeloord, The Netherlands. Rosmalen, The Netherlands: NVTL. 2: 459–466.

Bouwman, A.F., Lee, D.S., Asman, W.A.H, Dentener, F.J., Van der Hoek, K.W., & Olivier, J.G.J. 1997. A global high-resolution emission inventory for ammonia. *Global Biogeochemical Cycles* 11(4):561–587.

Boyette, R.A. 1998. Getting down to (biofilter) basics. *Biocycle* 39(5):58–62.

Brock, T.D., & Madigan, M.T. 1988. *Biology of Microorganisms*, 5th ed. New York: John Wiley and Sons.

Bryant, M.P., Schlegel, H.G., & Barnea, J. 1976. Microbial energy conversions, pp. 399–412, Erich Gottze KG, Gottingen, W. Germany.

Bundy, D.S. 1984. Rate of dust decay as affected by relative humidity, ionization and air movement [in animal confinement buildings]. *Transactions of the American Society of Agricultural Engineers* (ASAE). 27(3):865–870.

Bundy, D.S. 1991. Electrical charge plays role in dust-collection system. *Feedstuffs* 63(12):30.

Burton, C.H. (ed.) 1997. *Manure Management – Treatment Strategies for Sustainable Agriculture.* Bedford, United Kingdom: Silsoe Research Institute.

Carlile, F.S. 1984. Ammonia in poultry houses: A literature review. *World's Poultry Science Journal* 40:99–113.

Carpenter, G.A. & Fryer, J.T. 1990. Air filtration in a piggery: Filter design and dust mass balance. *Journal of Agricultural Engineering Research* 46(3):171–186.

Carter, T.A., Anderson, K.E., Arends, J., Barker, J.C., Bunton, S.S., Hawkins, B., Parsons, J., Scheideler, S.E., Stringham, S.M., & Winel, M.J. 1993. Composting poultry mortality: Poultry science and technology guide. North Carolina State University, North Carolina Cooperative Extension Service, Raleigh, North Carolina. December 1993.

Caveny, D.D. & Quarles, C.L. 1978. The effect of atmospheric ammonia stress on broiler performance and carcass quality. *Journal of Poultry Science* 57:1124–1125.

Chiba, L. I., Peo Jr., E.R., & Lewis, A.J. 1987. Use dietary fat to reduce dust, aerial ammonia and bacterial colony forming particle concentrations in swine confinement buildings. *Transactions of the American Society of Agricultural Engineers.* 30(2):464–468.

Clanton, F.J., Schmidt, D.R., Jacobson, L.D., Nicolai, R.E., Goodrich, P.R., & Janni, K.A. 1999. Swine manure storage covers for odor control. *Applied Engineering in Agriculture* (In press).

Cure, W., McCulloch, R.B., & Robarge, W. 1999. Nitrogen emissions in North Carolina. Air and Waste Management Association Conference, October 26–28, 1999.

Deaton, J.W., Reese, F.N., & Lott, B.D. 1984. Effect of atmospheric ammonia on pullets at point of lay. *Journal of Poultry Science* 63:384–385.

Demmers, T.G.M., Phillips, V.R., Short, L.S., Burgess, L.R., Hoxey, R.P., & Wathes, C.M. 2001. Validation of ventilation rate measurement methods and the ammonia emission from naturally ventilated dairy and beef buildings in the United Kingdom. *Journal of Agricultural Engineering Research*, pp. 1–10. November.

Ensminger, M.E. & Olentine, Jr., C.C. 1978. *Feeds and Nutrition*, 1st ed. Clovis, CA: Ensminger.

ERG. 2000. Summary of literature search and review to characterize AFO air emissions. Memorandum from Eastern Research Group, Inc. (ERG) to U.S. Environmental Protection Agency (USEPA). October 27, 2000.

European Environment Agency. 1999. *EMEP CORINAIR Atmospheric Emission Inventory Guidebook for Agriculture.*

Grelinger, M.A. 1997. Improved emission factors for cattle feedlots. Emission Inventory: Planning for the future, Proceedings of Air and Waste Management Association, U.S. Environmental Protection Agency Conference. Volume 1, pp. 515–524. October 28–30.

Grelinger, MA. & Page, A. 1999. Air pollutant emission factors for swine facilities. Air and Waste Management Conference Proceedings, pp. 398–408. October 26–28.

Groot Koerkamp, P.W.G., Metz, J.H.M., Uenk, G.H., Phillips, V.R., Holden, M.R., Sneath, R.W., Short, J.L., White, R.B., Hartung, J., Seedorf, J., Schroder, M., Linkert, K.H., Pederson, S., Takai, H., Johnson, J.O., & Wathes, C.M. Concentrations and emissions of ammonia in livestock buildings in Northern Europe. *Journal of Agricultural Engineering Research* 70:79–95.

Grub, W., Rollo, C.A., & Howes, J.R. 1965. Dust problems in poultry environments. Transactions of the American Society of Agricultural Engineers, pp. 338–339, 352.

Harper, L. & Sharpe, R. 1998. Ammonia emissions from swine waste lagoons in the Southeastern U.S. Coastal Plains. North Carolina Department of Environment and Natural Resources Report, USDA-ARS Agreement No. 58-6612-7M-022.

Hartung, J. & Phillips, V.R. 1994. Control of gaseous emissions from livestock buildings and manure stores. *Journal of Agricultural Engineering Research* 57:173–189.

Hashimoto, A.G. 1984. Methane from swine manure: Effect of temperature and influent substrate composition on kinetic parameter (k). *Agricultural Wastes* 9:299–308.

Hashimoto, A.G., Varel, V.H., & Chen, Y.R. 1981. Ultimate methane yield from beef cattle manure: Effect of temperature, ration constituents, antibiotics, and manure age. *Agricultural Wastes* 3: 241–256.

Heber, A.J. & Martin, C.R. 1988. Effect of additives on aerodynamic segregation of dust from swine feed. *Transactions of the American Society of Agricultural Engineers* 31(2):558–563.

Heber, A., Jones, D., & Sutton, A. 1999. *Methods and Practices to Reduce Odor from Swine Faculties.* Purdue University Cooperative Extension Service. West Lafayette, IN.

Hill, D.T. 1982. Design of digestion systems for maximum methane production. *Transactions of the American Society of Agricultural Engineers* 25 (1):226–230.

Hill, D.T. 1984. Methane productivity of the major animal types. *Transactions of the American Society of Agricultural Engineers* 27 (2) 530–540.

Hoeksma, P., Verdoes, N., & Monteny, G.J. 1993. Two options for manure treatment to reduce ammonia emission from pig housing. Proceedings of the First International Symposium on Nitrogen Flow in Pig Production and Environmental Consequences. Wageningen, The Netherlands. 69:301–306.

Hutchinson, G.L., Mosier, A.R., & Adre, C.E. 1982. Ammonia and amine emissions from a large cattle feedlot. *Journal of Environmental Quality* 11(2): 288–293.

Jacobson, L. et al. 1999. *Odor and Gas Emissions from Animal Manure Storage Units and Buildings.* American Society of Agricultural Engineers Annual International Meeting. July 18–22, 1999. Toronto, Ontario, Canada.

James, T., Meyer, D., Esparza, E., Depeters, E., & Perez-Monti, H. 1999. Effects of dietary nitrogen manipulation on ammonia volatilization from manure from Holstein heifers. *Journal of Dairy Science* 82(11): 2430–2439.

Johnson, J. 1997. *Final Report: Evaluation of Commercial Manure Additives.* Agricultural Utilization Research Institute. October 1, 1997.

Kingsolver, B. 2002. *Small Wonder.* New York: Perennial, Harper Collins.

Kroodsma, W., Scholtens, R., & Huis in't Veld, J. 1988. Ammonia emissions from poultry housing systems volatile emissions from livestock farming and sewage operations. *Proceedings of CIGR Seminar Storing, Handing and Spreading of Manure and Municipal Waste,* September 20–22, Uppsala, Sweden. Volume 2:7.1–7.13.

Lais, S., Hartung, E., & Jungbluth, T. 1997. Reduction of ammonia and odour emissions by bioscrubbers. Voermans JAM, Monteny G, editors. Proceedings of the International Symposium on Ammonia and Odour Control from Animal Production Faculties Vinkeloord. The Netherlands. Rosmalen, The Netherlands: NVTL. 2:533–536.

Lazarus, W.F. 1999. Farm machinery economic costs for 1999: Minnesota estimates with adjustments for use in Canada. Staff Paper p. 99–95. University of Minnesota, Department of Applied Economics, St. Paul, MN.

Leson, G. & Winer, A.M. 1991. Biofiltration: An innovative air pollution control technology for VOC emissions. *Journal of the Air and Waste Management Association* 41(8):1045–1054.

Mankell, K.O., Janni, K.A., Walker, R.D., Wilson, M.E., Pettigrew, J.L., Jacobson, L.D., & Wilcke, W.F. 1995. Dust suppression in swine feed using soybean oil. *Journal of Animal Science* 72(4):981–985.

Mannebeck, H. 1985. Covering manure storing tanks to control odour. V.C. Nielsen, J.H. Voorburg & P. L'Hermite (eds.), *Odour Prevention and Control of Organic Sludge and Livestock Farming.* London: Elsevier Applied Science, pp. 188–193.

Martin, J.H., & Loehr, R.C. 1977. *Poultry Waste Management Alternatives: A Design and Application Manual.* EPA-600/2-77-204. U.S. Environmental Protection Agency.

Martin, J.H. 2000. *A comparison of the performance of three swine waste stabilization systems.* Prepared by Resource Conservation Management for Eastern Research Group, Inc., Lexington, MA.

McCrory, D.F. & Hobbs, P.J. 2001. Additives to reduce ammonia and odor emissions from livestock wastes: A review. *Journal of Environmental Quality* 30:345–355.

Ministry of Agriculture FaF. 1992. *Code of Good Agricultural Practice for the Protection of Air.* London, United Kingdom: MAFF Publications.

Misselbrook, T.H., Pain, B.F., & Headon, D.M. 1998. Estimates of ammonia emission from dairy cow collecting yards. *Journal of Agricultural Engineering Research* 71:127–135.

Mollen, F. (n.d.). Stovreduktion I Stalde ved ionisering. (Dust Reduction by Ionisering). SJF orienteering nr 74. Bygholm, 8700Horsens, Denmark: National Institute of Agricultural Engineering.

Morris, G.R. 1976. Anaerobic fermentation of animal wastes: A kinetic and empirical design evaluation. Unpublished M.S. Thesis, Cornell University, Ithaca, NY.

Mroz, Z., Krasucki, W., & Grela, E. 1998. Prevention of bacteriuria and ammonia emission by adding sodium benzoate to diets for pregnant sows. Proc. Annual Mtg. EAAP Vienna, Austria.

MWPS-18. 2002. *Outdoor Air Quality.* Ames, IA: MidWest Plan Service.

NCDENR. 1999. Status report on emissions and deposition of atmospheric nitrogen compounds from animal production in North Carolina. North Carolina Department of Environment and Natural Resources.

NCSU. 1998. Control of odors from animal operations. North Carolina Agricultural Research Service, North Carolina State University, Swine Odor Task Force. Raleigh, NC.

Ni, J., Heber, A.J., Diehl, C.A., Lim, T.T. 2000. Ammonia, hydrogen sulphide and carbon dioxide release from pig manure in under-floor deep pits. *Journal of Agricultural Engineering Research* 77:53–66.

North, M.O. & Bell, D.D. 1990. *Commercial Chicken Production Manual*, 4th ed. New York: Chapman and Hall.

NRC. 2002. *The Airliner Cabin Environment and Health of Passengers and Crew.* National Research Council. Washington, DC: National Academy Press.

Oosthoek, J., Kroodsma, W., & Hoeksma, P. 1991. Ammonia emission from dairy and pig housing systems. *Odor and Ammonia Emissions from Livestock Farms.* Elsevier Applied Science.

OSUE. 2000. Selecting forms of nitrogen fertilizer. http://www2.ag.ohio-state.edu/~ohioline/agf-fact/0205.html. Ohio State University Extension. April 15, 2005.

Pearson, C.C. 1989. Air cleaning with wet scrubbers. *Farm Buildings and Engineering* 6(2):36–9.

Pedersen, S., et al. 2000. Dust in pig buildings. *Journal of Agriculture Safety and Health* 6(4):261–274.

Priem, R. 1977. Deodorization by means of ozone. *Agriculture and Environment* 3(2/3):227–37.

Proulx, A. 1992. *That Old Ace in the Hole.* New York: Scribner.

Ritter, W.F., Collins, Jr., N.E. & Eastburn, R.P. 1975. Chemical treatment of liquid dairy manure to reduce malodors. *Managing Livestock Wastes.* pp. 381–84.

Roos, K.F., & Moser, M.A., (eds.). n.d. *The AgSTAR Handbook,* USEPA, EPA-430-B-97-015.

Roos, K.F., Moser, M.A., & Martin, A.G. 1999. AgSTAR charter farm Program: Experience with five floating lagoon covers. Presented at Fourth Biomass Conference of the Americas, Oakland, California, August 29 – September 2, 1999.

Schiffman, S.S., Bennett, J.L., & Raymer, J.H. 2001. Quantification of odors, and odorants from swine operations in North Carolina. *Agricultural Forest Meteorology* 108:213–240.

Secrest, C. 2000. Field measurement of air pollutants near swine confined animal feeding operations using UV DOAS and FTIR. Unpublished report.

Sharpe, R.R., & Harper, L.A. 1997. Ammonia and nitrous oxide emissions from sprinkler irrigation applications of swine effluent. *Journal of Environmental Quality* 26:1703–1706.

Snoeyink, V. & Jenkins, D. 1980. *Water Chemistry.* New York: John Wiley and Sons.

Sommer, S.G., Christensen, B.T., Nielsen, N.E., & Schjorring, J.K. 1993. Ammonia volatilization during storage of cattle and pig slurry: Effect of surface cover. *Journal of Agricultural Science* 121(pt. 1):63–71.

Sutton, A.L., Goodall, S.R., Patterson, J.A., Mathew, A.G., Kelly, D.T., & Meyerholtz, K.A. 1992. Effects of odor control compounds on urease activity in swine manure. *Journal of Animal Science* 70(Suppl. 1):160.

Sweeten, J.M., Parnell, C.B., Shaw, B.W., & Auvermann, B.W. 1998. Particle size distribution of cattle feedlot dust emissions. *Transactions of the American Society of Agricultural Engineers* 41 (5): 477–1481.

Sweeten, J.M., Erickson, L., Woodford, P., Parnell, C.B., Thu, K., Coleman, T., Flocchini, R., Reeder, C., Master, J.R., Hambleton, W., Blume, G. & Tristao, D. 2000. Air quality research and technology transfer programs for concentrated animal feeding operations. Draft report. Presented at United States Agricultural Air Quality Task Force Meeting. July 18 and 19, 2000, Washington, DC.

Takai, H., Moller, F., Iverson, M., Jorsa, S.E., & Bille-Hansen, V. 1993. Dust control in swine buildings by spraying of rapeseed oil. *Livestock Environment IV: 4th International Symposium Coventry, England.* St. Joseph, MI: *American Society of Agricultural Engineers.* p. 726–733.

Takai, H., Jacobson, L.D., & Pedersen, S. 1996. Reduction of dust concentration and exposure in pig buildings by adding animal fat in feed. *Journal of Agricultural Engineering Research* 63(2):113–120.

Takai, H., Pedersen, S., Johnson, J.O., Mertz, H.H.M., Koerkamp, P.W.G.G., Uenk, G.H., Phillips, V. R., Holden, M.R., Sneath, R.W., Short, J.L., White, R.P., Hurtung, J., Seedorf, J., Schroder, M., Linkert, K.H., & Wathes, C.M. 1998. Concentrations and emissions of airborne dust in livestock buildings in Northern Europe. *Journal of Agricultural Engineering Resources* 70: 59–70.

Tamminga, S. 1992. *Gaseous Pollutants Produced by Farm Animal Enterprises. Farm Animals and the Environment* (eds. C. Pludips, D. Piggens) CAB International, Wallingford, UK.

Tate, R.L., III. 1995. *Soil Microbiology.* New York: John Wiley and Sons.

University of Minnesota. 1999. *Generic Environmental Impact Statement on Animal Agriculture: A Summary of the Literature Related to Air Quality and Odor.*

USDA. 1992. *Agricultural Waste Management Field Handbook, National Engineering Handbook, Part 651.* U.S. Department of Agriculture, Natural Resources Conservation Service (NRCS), Washington, DC.

USDA. 1999. Cattle: Final Estimates 1994–1998. Statistical Bulletin 953. U.S. Department of Agriculture, National Agricultural Statistics Service, Washington, DC.

USDA. 2000. Confined Livestock Air Quality Subcommittee, Sweeten, J.M., Chair. U.S. Department of Agriculture, Agricultural Air Quality Task Force Meeting, Washington, DC. Air Quality Research & Technology Transfer Programs for Concentrated Animal Feeding Operations.

USEPA. 1994. *Development and Selection of Ammonia Emission Factors.* Prepared for U.S. Environmental Protection Agency (USEPA), Office of Research and Development by R. Battye, W. Battye, C. Overcash, and S. Fudge of EC/R Incorporated. Durham, NC.

USEPA. 1998. Site visit report to Iowa and Minnesota. Prepared by P. Shriner. U.S. Environmental Protection Agency. May 1998.

USEPA. 2000a. Non-water quality impact estimates for animal feeding operations. Final report. U.S. Environmental Protection Agency, Office of Water, Engineering and Analysis Division, December 15, 2000.

USEPA. 2000b. AgSTAR Digest. U.S. Environmental Protection Agency. Office of Air and Radiation. EPA-430/F-00-012.

USEPA. 2001a. Emissions from animal feeding operations. Research Triangle Park, NC: U.S. Environmental Protection Agency No. 68-D6-0011.

USEPA. 2001b. Inventory of U.S. greenhouse gas emissions and sinks: 1990–1999. EPA 238-R-00-00-1. April, 2001.

Valentine, H. 1964. A study of the effect of different ventilation rates on ammonia concentrations in the atmosphere of broiler houses. *British Journal of Poultry Science* 5:149–159.

Valli, L., Piccinini, S., & Bonazzi, G. 1991. Ammonia emission from two poultry manure drying systems. *Odor and Ammonia Emissions from Livestock Farming.* Elsevier Applied Science, 1991.

Van Der Hoek, K.W. 1998. Summary of the work of the UNECE ammonia expert panel. *Atmospheric Environment* 32: 315–316.

Van Horn, H.H. 1998. Factors affecting manure quantity, quality, and use. Proceedings of the Mid-South Ruminant Nutrition Conference, Dallas-Fort Worth, TX. May 7–8, 1998. Texas Animal Nutrition Council, pp. 9–20.

Wilkie, A.C., et al. 1995. Anaerobic digestion for odor control, in *Nuisance Concerns in Animal Manure Management: Odors and Flies*, Florida Cooperative Extension, University of Florida, Gainesville.

Whitney, M.H., Nicolai, R., & Shurson, G.C. 1999. Effects of feeding low sulfur starter diets on growth performance of early weaned pigs and odor, hydrogen sulfide, and ammonia emissions in nursery rooms. Proceeding from Midwest ASAS/ADSA Annual Meeting, Des Moines, IA.

Witter, E. 1991. Use of CACl2 to decrease ammonia volatilization after application of fresh and anaerobic chicken slurry to soil. *Journal of Soil Science* 42:369–380.

Wright, P. 1997. Survey of manure spreading costs around York, New York. ASAE Paper No. 972040. Presented at the ASAE Annual International Meeting, American Society of Agricultural Engineers, Minneapolis, MN, August 10–14, 1997.

Zhang, Y., Tanaka, A., Barber, E.M., & Feddes, J.J.R. 1996a. Effects of frequency and quantity of sprinkling canola oil on dust reduction in swine buildings. *Transactions of the American Society of Agricultural Engineers.* 39(3):1077–1081.

Zhang, Y., & Gaakeer, W. 1996b. A low cost balloon-type lagoon cover to reduce odour emission. Conference Proceedings: International Conference on Air Pollution from Agricultural Operations, Kansas City, Missouri. Midwest Plan Service. pp. 395–401, Ames, IA.

Zahn, J.A., Tung, A.E., Roberts, B.A. & Hatfield, J.L. 2001. Abatement of ammonia and hydrogen sulfide emissions from a swine lagoon using a polymer biocover. *Journal of the Air and Waste Management Association* 51:562–573.

Zhu, J., Jacobson, L., Schmidt, D. & Nicolai, R. 2000. Daily variations in odor and gas emissions from animal facilities. *American Society of Agricultural Engineers* 16(2):153–158.

8 Beef Cattle Feeding Operations

It's an industry I no longer want to get tangled up in, even at the level of the ninety-nine-cent exchange. Each and every quarter pound of hamburger is handed across the counter after the following production costs, which I've searched out precisely: 100 gallons of water, 1.2 pounds of grain, a cup of gasoline, greenhouse-gas emissions equivalent to those produced by a six-mile drive in your average car, and the loss of 1.25 pounds of topsoil, every inch of which took 500 years for the microbes and earthworms to build. How can all this cost less than a dollar, and who is supposed to pay for the rest of it? If I were a cow, right here is where I'd go mad.

(Kingsolver, 2002, p. 120)

8.1 INTRODUCTION

Although a high-risk enterprise, the United States is the leading beef producer in the world. Almost 30 billion pounds of beef were produced in the United States in 2000, and per capita consumption totaled 78 lb. The beef industry is high risk because the cattle cycle is risky, and currently the cycle is in a declining phase. During some years, an operation may not recover out-of-pocket costs. In the near term, several more years are expected of smaller calf crops, a slight decline in cattle feeding, small decline in slaughter rates, and stable consumption rates. Profitability in the cattle business usually increases as production declines (PSU, 2005).

- *Interesting point:* Traditional feeder-cattle enterprises grow weaned calves (450 to 600 lb) and yearling steers or heifers (550 to 800 lb) to slaughter weights of 1,100 to 1,400 lb.

This chapter discusses beef cattle feeding, confinement, and manure handling operations. This livestock sector includes adult beef cattle (heifers and steers) and calves. Beef cattle may be kept on open pastures or confined to feedlots. In this chapter, we discuss feedlot operations only.

8.2 SIZE AND LOCATION OF INDUSTRY

In 1997, 106,075 beef open feedlots were in operation in the United States, excluding farms where animals graze (USDA, 1999a). These feedlots sold more than 26 million beef cattle in 1997 (USDA, 1999b). Table 8.1 shows the distribution of feedlots by state and estimated capacity. The capacity of a beef feedlot is the maximum number of cattle that can be confined at any one time. The feedlot capacity was derived from annual sales figures (USDA, 1999) by considering the typical number of turnovers of cattle per year and capacity utilization (ERG, 2000).

Table 8.2 shows beef cattle sales by feedlot size in 1997. While most feedlots are small, the majority of production is from larger farms. For example, 2,075 feedlots with capacities greater than 1,000 head accounted for only 2% of all lots but produced 80% of the beef sold in the United States in 1997. Beef feedlots vary in size from feedlots with a confinement capacity of less than 100 head to those in excess of 32,000 head of cattle.

TABLE 8.1
Number of Beef Feedlots by Size in 1997

State	Confinement capacity		
	< 500 head	500–1000 head	> 1000 head
Alabama	921	1	1
Alaska	19	0	0
Arizona	153	2	12
Arkansas	1,039	2	2
California	901	9	41
Colorado	1,400	44	145
Connecticut	151	0	0
Delaware	66	1	1
Florida	549	0	0
Georgia	696	1	2
Hawaii	34	1	3
Idaho	899	8	40
Illinois	7,184	54	51
Indiana	6,001	19	13
Iowa	12,040	233	263
Kansas	2,630	93	298
Kentucky	1,910	6	4
Louisiana	311	0	0
Maine	243	0	0
Maryland	754	1	0
Massachusetts	111	0	0
Michigan	4,455	21	30
Minnesota	8,345	56	58
Mississippi	560	0	0
Missouri	4,392	16	23
Montana	655	14	16
Nebraska	4,855	204	602
Nevada	83	4	4
New Hampshire	79	0	0
New Jersey	335	0	0
New Mexico	321	3	16
New York	1,424	2	3
North Carolina	903	2	3
North Dakota	1,086	9	8
Ohio	7,241	19	11
Oklahoma	1,850	11	35
Oregon	1,864	5	11
Pennsylvania	5,299	16	10
Rhode Island	26	0	0
South Carolina	348	3	1
South Dakota	2,711	65	88
Tennessee	1,965	1	1
Texas	3,574	31	218
Utah	797	5	11

TABLE 8.1 (continued)
Number of Beef Feedlots by Size in 1997

State	Confinement capacity			
	< 500 head	500–1000 head	> 1000 head	
Vermont	158	1	1	
Virginia	1,363	4	3	
Washington	1,170	4	22	
West Virginia	804	0	0	
Wisconsin	7,980	19	10	
Wyoming	345	8	16	
United States	**103,000**	**1,000**	**2,075**	**106,075**

Source: ERG (2000).

TABLE 8.2
Beef Cattle Sold in 1997
(Based on Estimated Maximum Confinement Capacity)

Feedlot size	Number of facilities	Cattle sold	Average cattle sold
< 300 Head	102,000	2,362,000	23
300–500 Head	1,000	600,000	600
500–1,000 Head	1,000	1,088,000	1,088
> 1,000 Head	2,075	22,789,000	10,983
All Operations	106,075	26,839,000	253

Source: USEPA (2001).

Beef cattle are located in all 50 of the United States, but most of the capacity is in the central and western states. Table 8.3 presents information on the total number of animals per state in 1997. The table is divided into heifer (female) population and steer (castrated male) population. The five largest beef-producing states are Colorado, Iowa, Kansas, Nebraska, and Texas. These states account for two-thirds of the steer population and almost 86% of the heifer population on feedlots in the United States.

8.3 BEEF PRODUCTION CYCLES

Three different types of operations are common in the beef industry, each corresponding to a different phase of the animal growth cycle. These operations are referred to as cow–calf operations, backgrounding, and finishing. These operations are typically conducted at separate locations that specialize in each phase of production.

8.3.1 COW–CALF OPERATIONS

Beef cow–calf production is relatively widespread and economically important in most of the United States. According to the USDA *1997 Census of Agriculture*, about a million farms had inventories of cattle and calves that generated $40.5 billion in sales, accounted for 21% of the total market value of agricultural products sold in the United States, and ranked first in sales among all commodities.

TABLE 8.3
Beef Cow Inventory by State in 1997

State	Inventory (1,000 head)	
	Heifers	Steer
Alabama	2	3
Alaska	0	0
Arizona	23	190
Arkansas	6	11
California	68	275
Colorado	410	622
Connecticut	—	0
Delaware	0	1
Florida	3	5
Georgia	2	2
Hawaii	1	1
Idaho	86	161
Illinois	102	140
Indiana	59	123
Iowa	360	554
Kansas	751	1,277
Kentucky	6	12
Louisiana	1	2
Maine	0	1
Maryland	4	6
Massachusetts	0	0
Michigan	31	152
Minnesota	71	190
Mississippi	1	2
Missouri	30	57
Montana	32	45
Nebraska	825	1,203
Nevada	9	14
New Hampshire	0	0
New Jersey	1	5
New Mexico	46	79
New York	14	13
North Carolina	2	7
North Dakota	40	52
Ohio	46	136
Oklahoma	109	256
Oregon	32	41
Pennsylvania	13	56
Rhode Island	0	0
South Carolina	2	3
South Dakota	120	172
Tennessee	7	11
Texas	939	1,463
Utah	16	30
Vermont	0	1

TABLE 8.3 (continued)
Beef Cow Inventory by State in 1997

State	Inventory (1,000 head)		
	Heifers	**Steer**	
Virginia	7	20	
Washington	54	95	
West Virginia	3	4	
Wisconsin	26	111	
Wyoming	33	40	
United States	**4,396**	**7,644**	**12,040**

Source: USDA (1999a).

Cow–calf operations are a source of the heifers and steers (castrated males) fed for slaughter. Cow–calf operations maintain a herd of heifers, brood cows, and breeding bulls, typically on pasture or range land, to produce a yearly crop of calves for eventual sale as feeder cattle. In colder climates and during drought conditions, cow–calf operations using pasture or rangeland provide supplemental feed, primarily hay with some grains and other feedstuffs. Confinement on drylots also is an option used on some cow–calf operations when grazing does not satisfy nutritional needs. Although pasture- or range-based cow–calf operations are most common, operations exclusively using drylots may be encountered. In colder climates, cow–calf operations may have calving barns to reduce calf mortality.

8.3.2 BACKGROUND OPERATIONS

Backgrounding or stocker operations describes a management system where recently weaned calves or yearling cattle are grazed for finishing on high-energy rations to promote rapid weight gain for a period of time before they are placed in the feed yard. Backgrounding operations may be pasture or dry lot–based or some combination thereof. Relatively inexpensive forages, crop residues, and pasture are used as feeds, with the objective of building muscle and bone mass without excessive fat at a relatively low cost. The length of the backgrounding process may be as short as 30 to 60 days or as long as 6 months (Rasby et al., 1996). The duration of the backgrounding process and the size of the animal moving onto the finishing stage of the beef production cycle depend on several factors. High grain prices favor longer periods of backgrounding by reducing feed costs for finishing or fattening, whereas heavier weaning weights shorten the finishing process. Backgrounded beef cattle may be sold to a finishing operation as "feeder cattle," usually at auction or raised under contract with a finishing operation. Large finishing operations commonly have cattle backgrounded under contract to insure a steady supply of animals. In some instances cow–calf and backgrounding operations are combined.

8.3.3 FINISHING OR FEEDLOT OPERATIONS

The final phase of the beef cattle production cycle is called the *finishing* or *feedlot* phase. Beef cattle in the finishing phase are known as "cattle on feed." Finished cattle are "fed cattle." Usually, the finishing phase begins with 6-month-old animals weighing about 400 lb. In between 150 and 180 days, these animals reach the slaughter weights of 1,050 to 1,150 lb for heifers and 1,150 and 1,250 lb for steers, and a new finishing cycle begins. Some feedlot operators start with younger animals weighing about 275 lb, or older or heavier animals initially. This either extends the finishing cycle to about 270 days or shortens it to about 100 days. Accordingly, typical feedlots can have from

1.5 to 3.5 turnovers of cattle herds. On average, most beef feedlots operate at between 80% and 85% of capacity over the course of a year (NCBA, 1999).

8.4   BEEF CONFINEMENT PRACTICES

The cow–calf and backgrounding phases of the beef production cycle are primarily pasture or rangeland based. The underlying rationale for this method of raising cattle is avoidance of the cost of harvesting, transporting, and storing roughages, which is necessary with confinement feeding. Therefore, confinement feeding during these phases of the beef production cycle generally is limited to time periods when grazing can not satisfy nutritional needs.

In the final or finishing phase of the beef cattle production cycle, heifers and steers most typically are fed to slaughter weight in open confinement facilities known as *feedlots* or *feed yards*. The majority of beef feedlots are open feedlots, which may be partially paved. Generally, paving, if present, is limited to a concrete apron typically located along feed bunks and around waterers, because these are areas of heaviest animal traffic and manure accumulation.

Cattle are segregated in pens designed for efficient movement of cattle, optimum drainage, and easy feed truck access. A typical pen holds 150 to 300 head of cattle, but the size can vary substantially. Required pen space may range from 75 to 400 ft^2 of pen space per head, depending on the climate. A varied climate requires 75 ft^2 of pen space per head, whereas a wet climate may require up to 400 ft^2 (Thompson & O'Mary, 1983). Space needs vary with the amount of paved space, soil type, drainage, annual rainfall, and freezing and thawing cycles. These types of operations may use mounds to improve drainage and provide areas that dry quickly, since dry resting areas improve cattle comfort, health, and feed utilization. Typically, pens are constructed to drain as quickly as possible after precipitation events with the resulting runoff conveyed to storage ponds that may be preceded by settling basins to reduce solids entering the ponds. In open feedlots, protection from weather is often limited to a windbreaker near a fence in the winter and sunshade in the summer.

In cold climates and high-rainfall areas, small beef cattle finishing operations may use totally enclosed confinement to reduce the negative impact of cold weather on feed conversion efficiency and rate of weight gain. However, totally enclosed confinement facilities generally are not economically competitive with open feedlots and are relatively few in number.

8.5   FEEDING PRACTICES

Feeding practices in the different phases of the beef production cycle differ, reflecting differences in nutritional requirements for maintenance and growth. As mentioned, cow–calf and backgrounding operations typically depend on grazing, possibly with the feeding of a mineral supplement to satisfy nutritional needs. With feeding in confinement facilities, harvested roughages, hays, and silages are the principal, if not only, feedstuffs.

During the finishing phase of the beef production cycle, a shift occurs from a roughage-based to a grain-based, high-energy ration to produce a rapid weight gain and desirable carcass characteristics. Because beef cattle are ruminant animals, some small level of roughage intake must be maintained to maintain rumen activity. Generally, mixed rations, which are combinations of roughages and concentrates, are fed. However, roughages and concentrates may be fed separately, a practice more common with smaller operations. Roughages have high fiber contents and are relatively dilute sources of energy and protein, whereas concentrates are low-fiber, high-energy feeds, which also may have a high protein content. Feeding practices for beef cattle generally are based on nutrient requirements established by the National Research Council (NRC, 1996).

Key term: The *rumen* is a large, hollow, muscular organ, one of the four stomach compartments in ruminant animals. A fermentation vat, the rumen can hold 160 to 240 L of material and is the site of microbial activity.

- *Interesting point:* Handling moist feeds has a limited potential for particulate emissions, whereas handling dry feeds, such as grain, may be a source of particulate emissions.

While cow–calf and backgrounding operations generally depend on grazing to satisfy nutritional needs, feed must be provided to beef cattle being finished in feedlots. Typically, feed is delivered to feed bunks two or three times per day, with the objective of always having feed available for consumption without the excessive accumulation of uneaten feed to minimize spoilage. Cattle are typically fed using feed bunks located along feed alleys that separate individual pens. Feed is delivered either by self-unloading trucks, tractor-drawn wagons (fence-line feeding), or mechanical feed bunks. Usually, mechanical feed bunks are located between pens, allowing animal access from both sides of the feed bunk. In small feedlots where roughages and concentrates are fed separately, animals may have access to haystacks, self-feeding horizontal silos, or large tubular plastic bags containing roughage. Concentrates are fed separately in portable feed tanks.

Open-front barns and lots with mechanical or fence-line feed bunks are common for feedlots up to 1,000 head, especially in areas with severe winter weather and high rainfall. Portable silage and grain bunks are useful for up to 200 head.

The metabolic requirements for maintenance of an animal typically increases during cold weather, reducing weight gain and increasing feed consumption to provide more energy, thereby increasing the amount of manure that is generated. Feed consumption typically declines under abnormally high temperatures, therefore reducing weight gain. Investigations in California have shown that the effect of climate-related stress could increase feed requirements as much as 33%, resulting in increased manure generation (Thompson & O'Mary, 1983).

8.6 MANURE MANAGEMENT PRACTICES

Beef cattle manure produced in confinement facilities generally is handled as a solid. Runoff from feedlots can be either liquid or slurry. Manure produced in totally enclosed confinement facilities may be handled as slurry or liquid if water is used to move manure. Slurry manure has enough water added to form a mixture capable of being handled by solids handling pumps. Liquid manure usually has less than 8% solids, resulting from significant dilution. It is easier to automate slurry and liquid manure handling, but the large volume of water necessary for dilution increases storage and disposal requirements and equipment costs (USDA, 1992).

Solid manure is scraped or moved by tractors to stockpiles. Runoff from open lots is pumped to solids separation activities to separate the solid and liquid fractions. The liquid fraction is then sent to storage ponds. Both the solid and liquid fractions can be disposed of on land.

8.6.1 MANURE COLLECTION

The following methods are used in feedlots to collect accumulated manure for disposal:

Open lots. Manure most commonly is collected for removal from open lots by scraping using tractor-mounted blades. Very large feedlots commonly use earth-moving equipment such as pan scrapers and front-end loaders. Manure accumulates in areas around feed bunks and water troughs most rapidly, and these areas may be scraped frequently during the finishing cycle. This manure may be removed from the pen immediately or may be moved to another area of the pen and allowed to dry. Usually the entire pen is completely scraped and the manure is removed at the end of finishing, after the animals are shipped for slaughter (Sweeten, 2000).

Totally enclosed confinement. Beef cattle manure accumulations in totally enclosed confinement facilities also are typically collected and removed by scraping using tractor-industry technology. Scrapers also can be used but require a concrete floor. With a concrete floor, use of a flush system for manure collection and removal is also possible. A flush system uses a large volume of water discharged rapidly one or more times per day to transport accumulated manure to an earthen anaerobic lagoon for stabilization and storage. Typically, 100 gal of flush water are used per head twice a day. Frequency of flushing as well as slope and length of the area being flushed determines the amount of flush water required (Loudon et al., 1985). The lagoon usually is the source of the water used for flushing. Because of problems related to freezing, use of flushing in totally enclosed finishing facilities is not common, since totally enclosed confinement operations normally are found only in cold climates.

Slatted floors over deep pits or shallow, flushed alleys have also been used in totally enclosed beef cattle finishing facilities. Most slats are made of reinforced concrete, but they can also be made of wood, plastic, or aluminum. They are designed to support the weight of the slat plus a live load, which includes animals, humans, and mobile equipment. Manure is forced between the slats as the animals walk around the facility, which keeps the floor surface relatively free from accumulated manure. With slatted floors over deep pits, pits typically are emptied at the end of a finishing cycle. Some water may be added to enable pumping, or there may be access room to allow the use of a front-end loader. Because of the cost of slatted floor systems, their use in beef cattle production is rare.

Factors that affect emissions from beef feedlots include the number of animals on the lot and the moisture of the manure. The number of animals influences the amount of manure generated and the amount of dust generated. In well-drained feedlots, emissions of nitrogen oxides are likely to occur because decomposition of manure is aerobic. In wet feedlots, decomposition is anaerobic and emissions of ammonia, hydrogen sulfide, and other odor causing compounds are likely. Additionally, the feedlot is a potential air-release point of particulate matter (PM) and dust from feed and movement of cattle.

8.6.2 Manure Storage, Stabilization, Disposal, and Separation

Manure collected from the feedlot may be stored, stabilized, directly applied to land on-site, or transported off-site for disposal.

8.6.2.1 Storage

If beef cattle manure is handled as a solid, it is stored by stacking it within an area of the feedlot or other open confinement facility or on an adjacent dedicated storage site. Stacking sites are typically uncovered and collection of contaminated runoff is necessary. Manure handled as a slurry or liquid is stored in either earthen storage ponds or anaerobic lagoons. Above-ground tanks are another option for storage of these types of manures but are not commonly used. Storage tanks and ponds are designed to hold the volume of manure and process wastewater generated during the storage period, the depth of normal precipitation minus evaporation, and the depth of the 25-year, 24-hour storm event with a minimum of 1 ft of freeboard remaining at all times. Emissions from storage tanks and ponds include ammonia, hydrogen sulfide, volatile organic compounds (VOCs), and methane. The magnitudes of emissions depend primarily on the length of the storage period and temperature of the manure. Low temperatures inhibit the microbial activity responsible for the creation of these compounds, whereas long storage periods increase the opportunity for emissions.

8.6.2.2 Stabilization

Stabilization is the treatment of manure to reduce odor and volatile solids prior to land application. Because manure is allowed to remain on feedlots for extended periods, a significant degree of decomposition from microbial activity occurs. When stacked for storage, a significant increase in temperature may occur depending on moisture content from microbial heat production. Manure accumulations on feedlots and stored in stacks can be sources of ammonia, hydrogen sulfide, VOCs, and methane, if moisture content is sufficient to promote microbial decomposition. Dry manure is an emission source of nitrous oxide and PM and dust emissions. When beef cattle manure is stored as a slurry or liquid, some decomposition or stabilization also occurs. Anaerobic lagoons, when designed and operated properly, result in a higher degree of stabilization than storage ponds or tanks, which have the single objective of providing storage. In storage ponds and tanks, intermediates in the decomposition process usually accumulate and are sources of odor. Storage tanks and ponds and lagoons can be sources of ammonia, hydrogen sulfide, VOCs, and methane emissions.

8.6.2.3 Land Application

The majority (approximately 83%) of beef feedlots dispose of their manure from storage and stabilization through land application (USDA, 2000). Box-type manure spreaders are used to apply solid manure, whereas flail-type spreaders or tank wagons with or without injectors are used with slurry-type manure. Tank wagons or irrigation systems are used for liquid manure disposal. Beef cattle manure not disposed of by land application may be composted for sale for horticultural and landscaping purposes.

8.6.2.4 Separation

In the beef cattle industry, liquid-solids separation essentially is limited to the removal of solids from runoff collected from feedlots and other open confinement areas using settling basins. However, stationary and mechanical screens also may be used. The objective of these devices is to reduce the organic loading to runoff storage ponds. Although separation also can be used with beef cattle manure handled as a liquid, this form of manure handling is not common in the beef cattle industry, as noted earlier. Emissions from settling basins depend on the hydraulic retention time of the runoff in the basin and frequency of removal of settled solids. If settled solids are allowed to accumulate, ammonia, hydrogen sulfide, VOC, and methane emissions may be significant. Generally, the time spent in separation activities is short (generally, less than 1 day).

8.7 BEEF VIRTUAL FARMS

This section explains a set of virtual farms developed to characterize the beef industry. Virtual farms are hypothetical farms intended to represent the range of design and operating practices that influence emissions from each animal sector. These virtual models can be used to develop emission estimates, control costs, and regulatory assessments.

The virtual farms include four components: confinement areas, solids separation activities, storage and stabilization practices, and land application. Land application includes emissions from the manure application activity and from the soil after manure application. For the virtual farms, emissions from the application of manure are differentiated from emissions from the manure application site (i.e., cropland or other agricultural land) because emission mechanisms are different. Emissions from the application activity occur over a short time, and depend on the methods by which manure is applied. Emissions from the application site occur as substances volatize from the soil over a period of time, as a result of a variety of subsequent microbial and chemical transformations.

Cow–calf and background operations do not typically confine animals and, as such, virtual models were not developed to represent them. Virtual model farms for finishing operations would represent those that do confine cattle.

Two virtual farms were developed to characterize typical beef cattle finishing operations (B1A and B1B). The components of the virtual farms include an open confinement area (feedlot), solids separation for collected surface runoff, manure storage facilities (storage ponds for surface runoff and stockpiles for solids), and land application. In both virtual models, land application includes solid and liquid manure application activities (e.g., irrigations, solid manure spreader) and the manure application site (e.g., emission, released from agricultural soils after the manure is applied). The beef virtual models differ only by presence or absence of solids separation (see Figure 8.1).

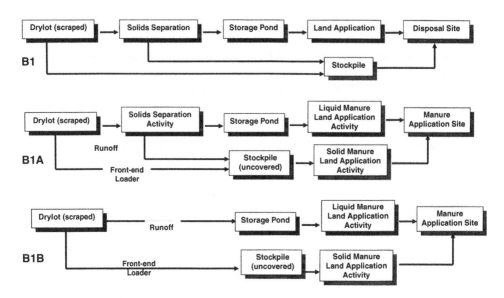

FIGURE 8.1   Beef model farms. (Source: USEPA, 2001a.)

8.7.1 Confinement

Feedlots are the only confinement operation considered for the virtual model farms because most, if not at all, beef operations use feedlots. Industry manure collection information indicates that most of the manure is typically scraped by a tractor scraper or front-end loader and stockpiled for later disposal by land application. Runoff from the feedlot is sent to solids separation processes or directly to storage ponds.

8.7.2 Solids Separation

Runoff from the feedlot is either sent to solids separation activities to remove solids or sent directly to storage ponds. The separated solids are sent to a stockpile and the liquid fraction is sent to a storage pond. Two common types of solids separation were considered in developing the virtual model farms: mechanical screens and gravity-settling basins. After reviewing the emission mechanisms from each type of separation practice, it was determined that emissions should not vary substantially between mechanical screens and settling basins. Additionally, due to the short duration, manure emissions would be relatively small; thus, differences between the separation processes would be insignificant. Therefore, the model virtual farms only represent the option of either having solids separation (B1A) or not (B1B). The virtual models are based on a short manure retention time in solids separation and, therefore, include negligible emissions from this process. The emission differences between the models are from the manure storage following separation.

8.7.3 Storage and Stabilization

The virtual model farms contain storage activities for solid and liquid manure. Two types of solid manure storage activities were considered in developing the virtual model farms. Solid manure could be: (1) stored in an uncovered stockpile or (2) not stored at all and sent directly from the feedlot to be land applied. Review of industry practices indicated that solid manure would generally not be sent directly from the feedlot to be land applied but would have some intermediate storage. Therefore, all the model farms included an uncovered stockpile. The liquid fraction from the runoff or the solids separation process (virtual model B1A) is sent to a storage pond.

8.7.4 LAND APPLICATION

As previously mentioned, land application includes the manure application activity and the manure application site (i.e., cropland or other agricultural land). Solid manure is typically land applied to the manure application site using a solid manure spreader. Three types of land application activities were considered for liquid manure in developing the virtual model farms: land application by (1) liquid surface spreader, (2) liquid injection manure spreader, or (3) irrigation. Review of industry practices indicated that injection is rarely used. The emissions from irrigation and liquid surface spreading were judged to be similar, because of the short duration for each activity and similar emission mechanisms. Therefore, the virtual model farms only refer to liquid manure land application rather than a specific type.

CHAPTER REVIEW QUESTIONS

1. What factors affect beef operation location?
2. What are the three most common types of beef industry operations? What are their functions? Why is the industry configured this way?
3. Describe and discuss typical beef confinement practices. Describe and discuss typical beef feeding practices.
4. Describe and discuss typical beef manure management practices, including methods for collection, open lots, totally enclosed confinement, storage, stabilization, disposal, separation, and land application.
5. What components are included for the two beef virtual farms? Why?
6. Describe and discuss typical beef operations.

THOUGHT-PROVOKING QUESTIONS

1. Read the chapter introduction quote. How *can* a hamburger cost less than $1.00, and who *does* pay for the rest? Why?
2. What makes beef production a high-risk enterprise? Why? How?
3. Research land use and animal per acre ratios for the different regions of the country. Discuss the economics of animal feeding operation and concentrated animal feeding operation success for the different regions.

REFERENCES

ERG. 2000. Facility counts for beef, dairy, veal, and heifer operations. Memorandum from Deb Bartram, Eastern Research Group, Inc. (ERG) to the Feedlots Rulemaking Record. U.S. Environmental Protection Agency (USEPA) Water Docket, W-00-27.

Kingsolver, B. 2002. *Small Wonder.* New York: Perennial: Harper Collins.

Loudon, T.L., Jones, D.D., Petersen, J.B., Backer, L.F., Brugger, M.F., Converse, J.C., Fulhage, C.D., Lindley, J.A., Nelvin, S.W., Person, H.L., Schulte, D.D., & White, R. 1985. *Livestock Waste Facilities Handbook,* 2nd ed., Midwest Plan Service. Ames, IA.

NCBA. 1999. Comments on the draft industry profile. National Cattlemen's Beef Association (NCBA).

NRC. 1996. Nutrient requirements of beef cattle. 7th rev. ed. National Research Council (NRC). Subcommittee on Beef Cattle Nutrition, Committee on Animal Nutrition, Board on Agriculture.

PSU. 2005. Feeding beef cattle. Penn State University. http://agalternatives-aers.psu.edu. Accessed May 5, 2005.

Rasby, R., Rush, I., & Stock, R. 1996. Wintering and backgrounding beef calves. NebGuide. Cooperative Extension, Institute of Agriculture and Natural Resources, University of Nebraska-Lincoln.

Sweeten, J. 2000. Manure management for cattle feedlots. Great Plains Beef Cattle Handbook. Cooperative Extension Service – Great Plans States.

Thompson, G.B., & O'Mary, C.C. 1983. *The Feedlot,* 3rd ed. Philadelphia: Lea & Febiger.

USDA. 1992. *Agricultural Waste Management Field Handbook, National Engineering Handbook,* Part 651. U.S. Department of Agriculture, Natural Resources Conservation Service, Washington, DC.

USDA. 1996. Extension Service. Census of agriculture.

USDA. 1999a. Cattle: Final estimates 1994-1998. Statistical Bulletin 953. U.S. Department of Agriculture, National Agricultural Statistics Service, Washington, DC.

USDA. 1999b. 1997 census of agriculture. U.S. Department of Agriculture, National Agricultural Statistics Service. Washington, DC.

USDA. 2000. *National Animal Health Monitoring System, Part 1: Baseline Reference of Feedlot Management Practices.* U.S. Department of Agriculture, Animal and Plant Health Inspection Service. Fort Collins, CO.

USEPA. 2001a. *Emissions from Animal Feeding Operations.* United States Environmental Protection Agency, Office of Air Quality Planning and Standards. Research Triangle Park, NC.

USEPA. 2001b. Development document for the proposed revisions to the national pollutant discharge elimination system regulation and the effluent guidelines for concentrated animal feeding operations. EPA-821-R-01-003. U.S. Environmental Protection Agency, Office of Water, Washington, DC.

9 Dairy Operations

In the distance ahead he saw a hazy brown cloud and guessed something was on fire. But the smell and immediate choking sensation in his throat as he drove past an enormous feedlot, the cows obscured by the manure dust that loaded the wind and was clearly the source of the cloud, introduced him to the infamous brown days of the Texas panhandle, wind-borne dust he later heard called "Oklahoma rain."

(Proulx, 2002, p. 43)

9.1 INTRODUCTION

In this text, dairy operations are defined as those operations producing milk, raising dairy replacement heifers, or raising calves for veal. Typically, dairy operations combine milk production and the raising of heifers (immature females) as replacements for mature cows that no longer produce milk economically. However, some milk producers obtain some or all replacement heifers (standalone heifer operations). Although some dairies raise veal calves, veal production is typically specialized at operations solely raising veal calves.

9.2 SIZE AND LOCATION OF INDUSTRY

For several decades, the number of milk-producing cows has steadily decreased while the volume of milk produced has continually increased. This increased productivity has been the result of improvements in breeding programs and in feeding and management practices. Concurrently, an ongoing consolidation has occurred in the dairy industry, resulting in fewer but larger farms. Between 1988 and 1997, the number of dairy cows in the United States decreased by 10% and the number of dairy farms decreased by 43% (USDA, 1995, 1999b).

In 1997, approximately 117,000 dairy farms were in operation in the United States (Table 9.1). These farms housed 9,309,000 mature (lactating) cows and 3,829,000 heifers (Table 9.2). Dairy farms vary in size from herds of less than 200 to herds of 3,000 to 5,000 mature cows (Cady, 2000). For this presentation, dairy farm capacity is based on the inventory of mature dairy cows reported to the U.S. Department of Agriculture (USDA).

Table 9.3 shows the number of farms, number of milk cows, and average herd size by size of operation. About 96% of the dairy farms in the United States have herds of 350 or less animals. Farms with 200 or fewer mature cows account for more than 50% of the total number of mature cows in the United States. A typical herd size is 47 head for a small dairy and 1,400 head for a large dairy. Between 1993 and 1997, the number of operations with less than 200 mature cows decreased, but the number of operations with more than 200 mature cows increased by almost 7%. In spite of the trend toward fewer but larger operations, smaller farms still account for a significant percentage of the milk produced in the United States.

Ten states account for 64% of total U.S. milk production capacity. The majority of dairy operations are located in the Midwest, followed by the Mid-Atlantic region. The states with the largest number of dairy operations are Wisconsin (22,576), Pennsylvania (10,920), Minnesota (9,603), and New York (8,732) (Table 9.1). These four states account for almost half the dairy farms in the United States. Although California has only 2,650 dairy farms, it is the largest milk-producing state. Of the large dairies (greater than 700 cows), California has the most operations (46%). Wisconsin has the largest number of mature cows (1,409,000), followed by California (1,379,000), New York (699,000), and Pennsylvania (631,000) (Table 9.2).

TABLE 9.1
Number of Dairy Farms by Herd Size in 1997

State	Capacity		
	< 350 Head	350–700 Head	> 700 Head
Alabama	591	14	3
Alaska	30	0	0
Arizona	163	21	63
Arkansas	1,186	7	0
California	1,440	547	663
Colorado	752	39	23
Connecticut	356	11	3
Delaware	127	4	1
Florida	546	58	62
Georgia	921	45	18
Hawaii	35	4	5
Idaho	1,224	90	90
Illinois	2,220	17	1
Indiana	3,191	21	4
Iowa	4,175	28	4
Kansas	1,449	11	6
Kentucky	3,373	18	2
Louisiana	961	17	4
Maine	673	10	2
Maryland	1,071	16	4
Massachusetts	475	7	1
Michigan	3,887	81	22
Minnesota	9,514	75	14
Mississippi	673	14	1
Missouri	4,154	20	1
Montana	716	5	0
Nebraska	1,336	13	3
Nevada	123	6	9
New Hampshire	323	5	1
New Jersey	293	3	0
New Mexico	406	19	96
New York	8,481	194	57
North Carolina	1,053	31	8
North Dakota	1,164	5	2
Ohio	5,383	38	4
Oklahoma	1,900	15	6
Oregon	992	44	16
Pennsylvania	10,841	71	8
Rhode Island	45	0	0
South Carolina	376	15	2
South Dakota	1,781	17	5
Tennessee	2,060	32	4
Texas	3,828	188	97
Utah	830	47	14
Vermont	1,885	45	10
Virginia	1,632	36	3
Washington	1,100	130	72

TABLE 9.1 (continued)
Number of Dairy Farms by Herd Size in 1997

State	Capacity		
	< 350 Head	350–700 Head	> 700 Head
West Virginia	672	4	0
Wisconsin	22,374	171	31
Wyoming	334	3	0
United States	**113,117**	**2,312**	**1,445**

Source: ERG (2000).

TABLE 9.2
Dairy Cow Inventory by State

State	Inventory (1,000 head)	
	Heifers	Cows
Alabama	10	31
Alaska	0	1
Arizona	20	122
Arkansas	21	54
California	625	1,379
Colorado	42	84
Connecticut	12	29
Delaware	2	10
Florida	38	158
Georgia	32	98
Hawaii	5	10
Idaho	113	268
Illinois	61	135
Indiana	66	140
Iowa	94	235
Kansas	42	81
Kentucky	54	150
Louisiana	17	68
Maine	21	40
Maryland	32	86
Massachusetts	9	27
Michigan	137	312
Minnesota	302	579
Mississippi	18	48
Missouri	71	180
Montana	7	20
Nebraska	24	69
Nevada	9	26
New Hampshire	7	19
New Jersey	6	21

TABLE 9.2 (continued)
Dairy Cow Inventory by State

State	Inventory (1,000 head)		
	Heifers	Cows	
New Mexico	42	197	
New York	288	699	
North Carolina	32	80	
North Dakota	19	60	
Ohio	123	275	
Oklahoma	38	93	
Oregon	47	92	
Pennsylvania	259	631	
Rhode Island	1	2	
South Carolina	9	26	
South Dakota	33	110	
Tennessee	57	115	
Texas	94	390	
Utah	45	90	
Vermont	54	158	
Virginia	61	125	
Washington	91	255	
West Virginia	8	19	
Wisconsin	632	1,409	
Wyoming	1	7	
United States	**3,829**	**9,309**	**13,318**

Source: USDA (1999a).

TABLE 9.3
Total Milk Cows by Size of Operation in 1997

Farm size*	Number of operations	Total number of milk cows	Average milk cow herd size
0–199 head	109,736	5,186,000	47
200–349 head	3,381	795,000	235
350–700 head	2,312	1,064,000	460
> 700 head	1,445	2,050,455	1,419
United States	**116,874**	**9,095,455**	**78**

Source: USEPA (2001b).
* Based on inventory.

The data in Table 9.1 to Table 9.3 do not include standalone heifer operations. While most replacement heifers are raised on dairy farms, an estimated 10% to 15% of dairy cow replacements are obtained from standalone heifer operations (Gardner, 1999; Jordan, 1999). The actual number of standalone heifer operations in the United States is unknown, as is the number raised in total confinement versus pasture-based operations; however, approximately 5,000 such operations are currently estimated in the United States.

9.3 PRODUCTION CYCLES

The primary function of a dairy is the production of milk, which requires a herd of mature, lactating dairy cows. To produce milk, the cows must be bred and give birth. The gestation period is 9 months, and dairy cows are bred again 4 months after calving. Thus, a mature dairy cow produces a calf every 12 to 14 months. Therefore, dairy operations have several types of animal groups present, including calves, heifers, mature cows (lactating and dry cows), veal calves, and bulls.

9.3.1 MATURE COWS (LACTATING AND DRY COWS)

The production cycle in the dairy industry begins with the birth of calves, which causes the onset of lactation (milk production). A period of between 10 and 12 months of milk production is normally followed by a 2-month dry period. The dry period allows for physiological preparation for the next calving (USDA, 1996a). At the time milking normally stops, cows are normally in their seventh to ninth month of pregnancy. A high frequency of calf production is necessary to maintain a cost-effective level of milk production. The rate of milk production peaks shortly after calving and then slowly declines with time. Average U.S. milk production is about 17,000 pounds per cow per year. However, herds with averages of 22,000 to 24,000 pounds of milk per cow per year or higher are not unusual.

About 25% of a milking herd typically is replaced each year, but replacement levels can be as high as 40% for intensively managed herds (USDA, 1996a). Mature cows are replaced or culled for a variety of reasons, including low milk production and diseases such as mastitis, an infection of the udder. Lameness, injury, and belligerence also are reasons for culling. Nearly all culled (i.e., selected) dairy cows, approximately 96%, are slaughtered for beef, used in processed foods, or used in higher-quality pet foods. The remainder is sold to other dairy operations (USDA, 1996a).

- *Important point:* Approximately one-third of all lactating dairy cattle are culled from the herd each year. Animals may be culled due to low milk production, infertility, disease, temperament, lameness, or injury. Most culled animals are destined for slaughter, but some are purchased at auctions and relocated to other farms.

9.3.2 CALVES AND HEIFERS

Shortly after birth, calves are separated from their mothers and are generally kept isolated from older calves or in small groups until they are about 2 months old. After the calves are weaned (at about 3 months of age), they are usually moved from their individual pen or small group into larger groups of calves of similar age.

Because of the continuing need for replenishing the milking herd, approximately 50% of the female calves born are retained as milk cow replacements. Those animals selected as replacements usually are progeny of cows with a record of high milk production. Female calves not raised as replacements are sold for either veal or beef production.

Replacement heifers are either raised on-site or transferred off-site to an operation that specializes in producing dairy cattle replacements (standalone heifer operation). The replacement operation may raise heifers under contract or may purchase calves and sell back the same or other animals at a later date.

- *Important point:* In the dairy industry, both male and female animals are referred to as calves up to an age of about 5 months. From an age of 6 months until the birth of their first calf, females are called *heifers,* with first calving typically occurring at 25 to 28 months of age (USDA, 1996a). Replacements raised off-site may be purchased or returned either as unbred or open (not pregnant) heifers at an age of about 13 months or as bred heifers at an age usually of 22 to 23 months. Dairy farms that raise replacements on-site have three age

TABLE 9.4

Percentage of U.S. Dairies by Housing Type and Animal Group in 1995[a]

Housing type	Unweaned calves	Weaned calves and heifers	Lactating cows	Close-up cows[b]
Drylot	9.1	38.1	47.2	28.9
Freestall	2.5	9.7	24.4	5.6
Hutch	32.5	n/a	n/a	n/a
Individual pens	29.7	6.6	2.3	38.3
Multiple animal areas[c]	40.0	73.9	17.9	26.3
Pasture	7.4	51.4	59.6	41.9
Tie stall/stanchion	10.5	11.5	61.4	26.3

Source: USDA (1996a).

[a] Percentages do not add to 100% because some operations use more than one type of housing.

[b] Cows close to calving.

[c] Superhutches, transition barns, calf barns, and loose housing.

groups of animals present: calves, heifers, and mature lactating and dry (mature nonlactating) cows. Usually, the total number of calves and heifers present is between 50% to 60% of the size of the milking herd.

9.3.3 VEAL CALVES

Roughly 50% of the calves produced by dairy cows are males. Because most dairy cows are bred using artificial insemination, little demand for male calves occurs in the industry. Although some dairy farms have one or more breeding-age bulls for cows that do not conceive by artificial insemination, most male calves are sold either for veal or beef production. Male calves are usually separated from the cows within 3 days of birth. Veal producers typically obtain calves through livestock auctions, although in some cases calves may be taken directly from the dairy farm to the veal operation (Wilson, Stull, & Terosky, 1995).

9.4 CONFINEMENT PRACTICES

How dairy cows are confined depends on the size of operation, age of the animal, and the operator preference. Optimal housing facilities enhance the quality of milk production and allow for the protection of the environment, yet remain cost-effective (Adam et al., 1995). Table 9.4 summarizes the relative percentages of U.S. dairies reporting various types of housing (USDA, 1996a). (Percentages in Table 9.4 do not add to 100% because some operations use more than one type of housing). Information on housing for dry cows was not available, although dry cows are typically housed similarly to lactating cows (Stull, Berry, & DePeters, 1998). Superhutches, transition barns, calf barns, and loose housing may be considered specific types of multiple animal pens. Dairies predominantly use some sort of multiple animal areas for unweaned calves, weaned calves, and heifers.

9.4.1 MATURE COWS—BREEDING CYCLE

The primary objective in housing for cows that are close to calving is to minimize disease and stress to both the cow and calf. Sod pastures are often used in warmer climates or during the summer. Alternatively, cows may be housed in multiple-animal or individual pens prior to calving.

About 2 weeks before a cow is due, she is moved to a "close-up" pen. The cow density in close-up pens is about one-half the density in lactating cow pens to allow the calving cows some space to segregate themselves from other cows if they go into labor, although calving in close-up pens is usually avoided.

When birth is very near, cows are moved to a maturity area for calving. If the climate is sufficiently mild, pastures can be used for a maternity area; otherwise, small individual pens are used. Approximately 45% of all diary farms have maternity housing apart from the housing used for the lactating cows. This feature is more prevalent in larger farms than in smaller farms. Approximately 87% of farms with 200 or more cows have a separated maternity housing (USDA, 1996a).

9.4.2   MATURE COWS—MILKING CENTER

Lactating cows require milking at least twice per day and are either milked in their tie stalls or are led into a separate milking center. Milking centers (also called *parlors*) are separate buildings, apart from the lactating cow confinement. The center is designed to facilitate changing the groups of cows milked and to allow workers access to the cows during milking. A holding area confines cows that are ready for milking. Usually, the holding area is enclosed and is a part of the milking center which, in turn, may be connected to the barn or located in the immediate vicinity of the cow housing.

Cows that are kept in tie stalls may be milked from their stalls. The housing is equipped with a pipeline system that flows around the barn and contains ports in each stall for collecting milk. Approximately 70% of dairy operations reported that they milk the cows from their tie stalls, whereas only 29% reported that they used a milking center. However, more than half of the lactating cow population (approximately 55%) is milked in a milking center (USDA, 1996a, 1996b). Therefore, it can be interpreted that many of the large dairies use milking centers, whereas the smaller dairies typically use tie stalls.

9.4.3   MATURE COWS—LACTATING AND DRY HERD

When not being milked, the herd is confined in freestall barns, drylots, tie stalls and stanchions, pastures, or combinations of these. Dry cows are confined in loose housing or freestalls (Stull et al., 1998). These housing types are described below.

Freestall Barn. The freestall barn is the predominate type of housing system used on larger dairy farms for lactating cows. In a freestall barn, cows are grouped in large pens with free access to feed bunks, waterers, and stalls for resting. Standard freestall barn design include a feed alley in the center of the barn separating two feed bunks on each side. On each side of the barn is an alley between the feed bunk and the first row of freestalls and an alley between the first row of freestalls facing the feed bunk and a second row of freestalls facing the side-wall of the structure. These are the primary areas of manure accumulation, with little manure excreted in the freestalls. There may or may not be access to an outside drylot for exercise or pasture for exercise and grazing. A variety of types of bedding materials are used in freestall barns for animal comfort and to prevent injury. Straw, sawdust, wood shavings, and rubber mats are the most commonly used materials, but bedding materials used also include sand, shredded newspaper, and composted manure solids.

Drylots. In warmer climates, cows simply may be confined in a drylot with unlimited access to feed bunks, water, and usually an open structure to provide shade. Drylot confinement facilities for dairy cattle are similar to beef feedlots described earlier. As with beef feedlots, no bedding materials are used.

Tie stalls and stanchions. Stanchion or tie-stall barns still are common on smaller dairy farms, especially those with older confinement facilities. With this type of housing system, cows are confined in a stall for feeding and, frequently, also milking but have access to a drylot or pasture for exercise. A mechanically cleaned gutter is located behind each row of stalls for manure collection and removal. Usually straw, sawdust, or wood shavings are used as bedding materials in stanchion and tie-stall barns to absorb urine and allow manure to be handled as a solid. Thus, manure produced in stanchion and tie stalls barns contains more bedding than that produced in freestall barns.

Loose housing. Barns, shades, and corrals are defined as loose housing. The design of these facilities depends on the number of cows, climate, and manure-handling techniques.

Pastures. Depending on the farm layout, availability of pastureland, and weather conditions, cows may spend part or most of their day in a pasture. On some farms, cows may be contained outdoors during the day but are housed in a tie stall or freestall overnight.

9.4.4 CALVES

Calves are confined separately from other cattle until they reach 6 months of age. Sickness and mortality rates are highest among calves under 2 months of age; therefore, the housing for this group typically minimizes environmental stress by protecting the calves against heat, wind, and rain. Common calf housing types include individual animal pens and hutches. These housing types are described below.

Individual pens. Individual pens are sized to house animals individually and separate from others (Stull et al., 1998). Individual pens can be used inside a barn to provide isolation for each calf (Bickert et al., 1997).

Hutches. Hutches are portable shelters typically made of wood, fiberglass, or polyethylene, and placed in outdoor areas. One end of a hutch is open and a wire fence may be provided around the hutch to allow the calf to move outside (Bickert et al., 1997).

After calves are weaned, they are typically moved from individual pens or small group pens into housing containing a larger number of calves. Transition housing is used for calves from weaning to about 5 months of age. The most common types of housing used for weaned calves are calf shelters or superhutches and calf barns (Bickert et al., 1997).

9.4.5 HEIFERS

The confinement used for heifers may include the same types used for weaned calves but may also include a pasture in which the herd is allowed to move about freely and to graze. The majority of heifers are on drylots; however, heifers may also be housed in freestall barns.

9.4.6 VEAL CALVES

Veal calves are generally grouped by age in an environmentally controlled building. The majority of veal operations use individual stalls or pens. Floors are constructed of either wood slats or plastic-coated expanded metal. The slotted floors allow for efficient removal of manure. Individual stalls allow regulation of air temperature and humidity through heating and ventilation, effective management and handling of manure, limited cross-contamination of pathogens between calves, individual observation and feeding and, if necessary, examination and medical treatment (Wilson, Stull, & Terosky, 1995).

9.5 FEEDING PRACTICES

Feeding and watering practices vary for each type of animal group at the dairy. Most dairies deliver feed several times each day to the cows and provide a continuous water supply. The type of feed provided varies and is based on the age of the animal and the level of milk production to be achieved.

Feeding requirements of dairy animals influence the physical state of the manure generated, thereby influencing the manure management system. Animals fed liquid diets generate manure that is liquid or slurry in nature, whereas those fed solid diets produce solid manure with different manure management requirements.

Dairy cattle, including calves being raised as replacements after weaning, are fed roughage-based diets. The principal constituents of these diets are corn or grain sorghum silages and legume or grass and legume hays with feed grains and by-product feedstuffs added in varying amounts to satisfy energy, protein, and other nutrient requirements. Because of milk production, lactating cows have higher nutrition requirements than heifers and dry cows and are fed diets containing higher proportions of silages and supplements. Manure that is generated is in solid or semisolid state. To maximize feed intake, lactating cows may be fed several times a day. In contrast, heifers and dry cows usually are fed only twice a day to avoid excessive weight grain. Continuous access to water is critical, especially for lactating cows because milk is about 95% water.

Calves are nursed for 4 to 5 days after birth, until colostrum production ceases and marketable milk production begins. Calves then are fed a milk replacer until weaning, generally at about 8 weeks of age. During this period, a feed-grain-based starter diet is introduced. This starter diet is fed up to about 3 months of age, when rumen development allows a shift to a roughage-based diet. Calves raised for veal only are fed a milk replacer until slaughter. Therefore, manure generated is in a liquid state.

9.6 MANURE MANAGEMENT PRACTICES

Dairy manure management systems are generally designed based on the physical state of the manure being handled. Dairy cattle manure is collected and managed as a liquid, a semisolid or slurry, or a solid. Manure with a total solids or dry matter content of 20% or higher usually can be handled as a solid, while manure with a total solids content of 10% or less can be handled as a liquid. Most dairies have both wet and dry manure management systems (USDA, 1997).

In a slurry or liquid system, manure is flushed from alleys or pits to a storage facility. Typically, effluent from the solids separation system or supernatant from ponds or anaerobic lagoons is used as flush water. The supernatant is the clear liquid overlying the solids that settle below. Dairy manure that is handled and stored as a slurry or liquid may be mixed with dry manure. Liquid systems are usually favored by large dairies for their lower labor cost and because the larger dairies tend to use automatic flushing systems.

9.6.1 DAIRY MANURE COLLECTION AND TRANSPORT

Manure accumulates in confinement areas, such as barns, drylots, and milking centers, and is primarily deposited in areas where the herd is fed and watered. Drylots are used to house calves and heifers. Either drylots or freestall barns are used to house the lactating herd when they are not milked. The milking center houses the lactating herd when they are being milked.

The following methods are used at dairy operations to collect accumulated manure for disposal.

9.6.1.1 Drylots

Manure produced in drylots used for confining dairy cattle, including lactating and dry cows, heifers, and calves being raised as replacements, generally is removed by scraping using a tractor-mounted blade. As with beef feedlots, the rate of manure accumulation in drylots for dairy cattle is highest along feed bunks; this area must be scraped more frequently than other areas of the lot and may be paved. Because of the loss of moisture through evaporation and drainage, drylot manure can either be spread directly after collection or stored in stockpiles for subsequent disposal by land application. Manure scraped from areas along feedbunks usually is stockpiled and spread when the lot is completely scraped. Factors that affect emissions from drylots include the number of animals on the lot and the moisture of the manure. The number of animals influences the amount of manure generated and the amount of dust generated. In well-drained drylots, emissions of nitrogen oxides are likely to occur because decomposition of manure is aerobic. In wet drylots, decomposition is anaerobic and likely has emissions of ammonia, hydrogen sulfide, and other odor-causing compounds. Additionally, the drylot is a potential air-release point of particulate matter (PM) and dust from feed and movement of cattle.

9.6.1.2 Freestall Barns and Milking Centers

Dairy cattle manure accumulations in freestall barns are typically collected and removed by mechanized scraping systems or by using a flush system.

Mechanical/Tractor scraper. Manure and bedding from barns and shade structures are normally collected by tractor or mechanical chain-pulled scrapers. Dairies using scrapers to remove manure from freestall barns are often referred to as *scrape dairies*; 85% of operations with more than 200 milking cows use a mechanical or tractor scraper (USDA, 1996b). Tractor scraping is more common since the same equipment can be used to clean outside lots as well as freestalls and loose housing. A mechanical alley scraper consists of one or more blades wide enough to scrape the entire alley in one pass. A timer can be set so that the scraper runs two to four times a day or continuously in colder conditions to prevent the blade from freezing to the floor. Scrapers reduce daily labor requirements but have a higher maintenance cost because of corrosion and deterioration.

Flush systems. Manure can be collected from areas with concrete flooring by using a flushing system. A large volume of water is introduced at the head of a paved area, and the cascading water removes the manure. Flush water can be introduced from storage tanks or high-volume pumps. The required volume of flush water varies with the size of the area to be flushed and slope of the area. Recycling from the supernatant of a storage pond or anaerobic lagoon can minimize the total amount of flush water introduced; however, only fresh water can be used to clean the milking parlor area.

Gutter cleaner/Gravity gutters. Gutter cleaners or gravity gutters are frequently used in confined stall dairy barns. The gutters are usually 16 to 24 in. wide, 12 to 16 in. deep, and flat on the bottom. Either shuttle-stroke or chain-and-flight gutter cleaners are typically used to clean the gutters. About three-fourths (74%) of U.S. dairy operations with less than 100 milking cows and approximately one-third of U.S. dairy operations with 100 to 1999 milking cows use gutter cleaners (USDA, 1996b).

Slatted floors/Slotted floors. Freestall dairy barns also may have slatted floors located over a storage tank. As the animals move about the barn, manure is forced through the openings between the slats, which are manufactured using reinforced concrete. The cost of slatted floors has limited their use in the dairy industry. Generally, some water must be added to allow removal of manure from storage tanks under slatted floors by pumping.

Most dairies can be grouped into one of three categories, depending on the method of removing manure from the freestall barn: flush dairy, scrape dairy, or flushed alley dairy. Flushing systems are the only method of manure removal from the milking center. Some dairy operations use flush water in freestall barns but only in areas where animals are fed (i.e., the feed alleys). Mechanical scrapers are used in the rest of the barn. Dairies using this type of manure removal method are referred to as *flushed alley dairies*. Flushing systems are predominantly used in freestall barns by large dairies with 200 or more head (approximately 27%). These systems are much less common in freestall barns at dairies with less than 200 head (less than 5% reported using this system) (USDA, 1996b). These systems are also more common at dairies located in warmer climates. A farm type of dairy, a feedlot dairy, confines animals in a drylot, similar to beef cattle, and does not use a freestall barn. This type of confinement and manure management system is common in California.

The method used to transport manure from confinement depends largely on the consistency of the manure. Liquids and slurries from milking centers, freestall barns that are flushed, and runoff from drylots can be transferred through open channels, through pipes, and in liquid tank wagons. Pumps can be used to transfer liquid and slurry manure as needed; however, the higher the solids content of the manure, the more difficult it is to pump.

Solid and semisolid manure from drylots can be transferred by mechanical conveyance or in solid manure spreaders. Slurries can be transferred in large pipes by using gravity, piston pumps, or air pressure. Gravity systems are preferred because of their low operating costs.

Emissions from freestall barns and milking centers are influenced by the frequency of manure removal (i.e., flush frequency or scrape frequency). The longer the manure is present, the more emissions occur from the confinement area. Because of the wet nature of manure in these areas, decomposition is anaerobic and emissions of ammonia, hydrogen sulfide, and other odor-causing compounds occurs. These areas may also be a source of PM emissions from feeding systems.

9.6.2 MANURE STORAGE, STABILIZATION, AND SEPARATION

Manure collected from the confinement facilities may be transferred directly to storage or undergo solids separation or stabilization prior to storage and land application.

9.6.2.1 Storage

Solid manure (from the feedlot and from scraped freestall barns) is typically stored in uncovered storage stockpiles. Because open piles are subjected to rain, they exhibit emission profiles of both aerobic and anaerobic conditions over time. When wet, the stockpiles are potential sources of ammonia, hydrogen sulfide, nitrous oxide, and odor-causing compounds from anaerobic decomposition. When dry, they are emission sources of nitrous oxide from aerobic decomposition and PM.

Manure handled as a slurry or liquid is stored in either earthen storage ponds or anaerobic lagoons. Aboveground tanks are another option for storage of these types of manures but are not commonly used. Storage tanks and ponds are designed to hold the volume of manure and process wastewater generated during the storage period, the depth of normal precipitation minus evaporation, and the depth of the 25-year, 24-hour storm event with a minimum of 1 ft of freeboard remaining at all times. Emissions from storage tanks and ponds include ammonia, hydrogen sulfide, volatile organic compounds (VOCs), and methane. The magnitude of emissions depends primarily on the length of the storage period and temperature of the manure. Low temperatures inhibit the microbial activity responsible for the creation of hydrogen sulfide and methane but may increase VOC emissions and odors. Long storage periods increase the opportunity for emissions of VOCs, hydrogen sulfide, and ammonia.

9.6.2.2 Stabilization

Stabilization is the treatment of manure to reduce odor and volatile solids prior to land application. Runoff from drylots and liquid manure from flush alleys are often stabilized in anaerobic lagoons, which use bacterial digestion to decompose organic carbon into methane, carbon dioxide, water, and residual solids. Single-cell systems combine both stabilization and storage in one earthen structure, whereas two-cell systems separate stabilization and storage (i.e., anaerobic lagoon followed by a storage pond).

Emissions from anaerobic lagoons depend on the loading rate, hydraulic retention time, and temperature. The loading rate determines the size of the lagoon and how much manure can be stored; the more manure stored, the higher the emissions potential. The hydraulic retention time refers to the length of time that liquids are stored. The longer the retention time, the more likely that compounds will volatize from the lagoon. Emissions also increase with higher temperatures. Another factor influencing emissions is proper design and maintenance. A properly operated system should have little or no VOC emissions or odors. Anaerobic lagoons at dairies emit methane, hydrogen sulfide, and ammonia.

If manure is allowed to remain on drylots for extended periods, a significant degree of decomposition from microbial activity occurs. When stacked for storage, a significant increase in temperature may occur, depending on moisture content from microbial heat production. Manure accumulations on drylots and stored in stacks can be sources of ammonia, hydrogen sulfide, VOCs, and methane if moisture content is sufficient to promote microbial decomposition. Dry manure is a source of nitrous oxide and PM and dust.

9.6.2.3 Solids Separation

In the dairy industry, liquid-solids separation may be used to remove solids from runoff collected from drylots and flushed manure from freestall barns and milking centers. The liquid from solids separation is sent to a storage pond or anaerobic lagoon; the solid is stored in piles. Solids separation is necessary to reduce organic loading to storage ponds and lagoons so they do not overflow. Mechanical separators (stationary screens, vibrating screens, presses, and centrifuges) or gravity settling basins may be used for this purpose. Emissions from separation and mechanical separation systems where wastes are held longer, emissions of ammonia, hydrogen sulfide, VOCs, and methane emissions may be significant. Generally, the time spent in separation activities is short (i.e., less than 1 day).

9.7 DAIRY AND VEAL VIRTUAL FARMS

In this section, we explain a set of virtual model farms developed to characterize the dairy and veal industries. As mentioned in Chapter 8, virtual farms are hypothetical farms intended to represent the range of design and operating practices that influence emissions from each animal sector. These virtual models can be used to develop emission estimates, control costs, and regulatory assessments. Cow–calf and standalone heifer operations using drylots for confinement are similar to beef feedlots and are assumed to be adequately represented by the beef virtual farms. Separate virtual model farms were developed for veal because of the differences in manure characteristics and handling operations from dairies.

Key term: A veal calf is a calf fed a liquid diet at an age of up to 8 weeks and a live weight of up to 190 lb.

The virtual model farms include four components: confinement areas, solids separation activities, storage and stabilization practices, and land application. Land application includes emissions from the manure application activity and from agricultural soils after manure application. For the virtual farms, the manure land application activity was differentiated from the manure application site (i.e., cropland or other agricultural land) because emission mechanisms are different. Emissions from the application activity occur over a short period and depend on the method by which manure is applied. Emissions from the application site occur as substances volatize from the soil over a period of time as a result of a variety of chemical and biological transformations in the soil.

9.7.1 Dairy Virtual Model Farms

Eight virtual farms were developed to represent typical dairy operations. The common components of the dairy virtual models include confinement areas (freestall barn, drylot, and milking centers), solids separation, manure storage and stabilization (anaerobic lagoons or storage ponds for liquid manure and stockpiles for solids), and land application.

All dairies have milking centers and drylots to confine animals. Most dairies also have a freestall barn. Those dairies using flush water to remove manure in the freestall barn are referred to as *flush dairies* (D1) (see Figure 9.1). Those using flush water to remove manure from only the freestall barn alleys are referred to as *flushed alley dairies* (D4) (see Figure 9.4). Those using scraping to remove manure are referred to as *scrape dairies* (D2) (see Figure 9.2). Dairies not having a freestall barn at all are referred to as *feedlot dairies* (D3) (see Figure 9.3). Within each of the four basic virtual models, two variations were developed with and without solids separation activities (D1A and D1B; D2A and D2B; D3A and D3B; and D4A and D4B).

9.7.1.1 Confinement

In a dairy, cows are mostly kept in drylots, freestall barns, flushed alley freestall barns, or milking centers. In the virtual models, freestall barns and flushed alley freestall barns are used for mature

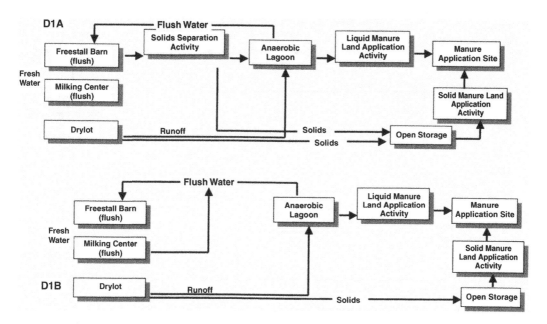

FIGURE 9.1 Flush dairy. (Source: USEPA, 2001a.)

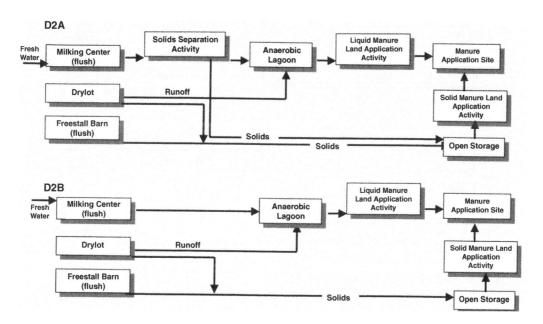

FIGURE 9.2 Scrape dairy. (Source: USEPA, 2001a.)

cows when they are not being milked. Heifers and dry cows are kept on drylots. Where there is no freestall barn or flushed alley barn, lactating cows are kept in drylots except during milking.

In all virtual models, manure is collected from milking centers by flushing with fresh water. Manure is collected from drylots by a tractor scraper or front-end loader. The method used to collect manure from freestall barns varies among the models and has been discussed earlier.

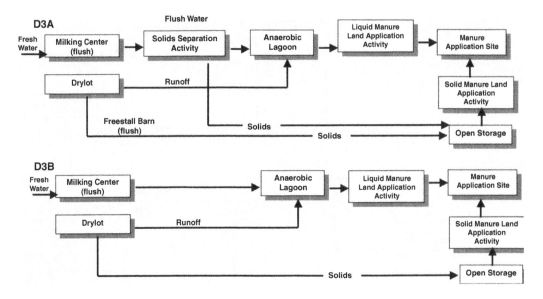

FIGURE 9.3   Feedlot dairy. (Source: USEPA, 2001a.)

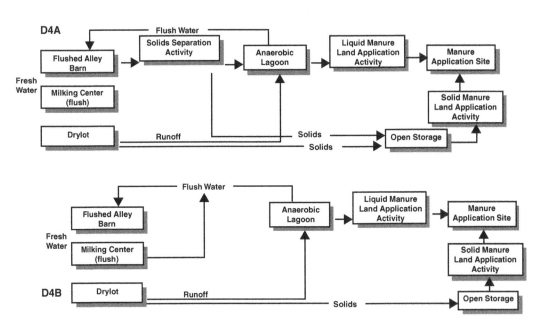

FIGURE 9.4   Flushed alley dairy. (Source: USEPA, 2001a.)

The flushed manure from the freestall barns and milking centers is combined as it is removed and then sent to solids separation. Manure from the drylot is transported to an uncovered stockpile. Runoff from the drylot is sent to solids separation.

- *Important point:* "Flushing dairy manure is an alternative to blade scraping of freestalls or holding pens. It offers the advantage of labor reduction with automated systems, limited scraping requirements, lower operating cost, drier floors, and cleaner facilities. Disadvantages may include additional water requirements per cow, initial fixed cost and/or handling of sand laden manure" (Harner et al., 2005, p. 1).

9.7.1.2 Solids Separation

Two virtual model variations were developed for solids separation at each of the four types of virtual farms. In one variation, runoff from the drylot and flushed manure from the milking center and freestall barn is sent to solids separation processes prior to storage. In the other variation, manure is sent directly to storage and treatment lagoons.

In the virtual models that used solids separation, the separated solids are sent to a stockpile and the liquid fraction is sent to a storage and stabilization lagoon. Two common types of solids separation activities were considered in developing the virtual farms: mechanical screens and gravity settling basins. Review of the emission mechanisms from each type of separation practice indicated that emissions would not substantially vary between mechanical screens and settling basins. Additionally, because of the short duration manure would be present in these activities, emissions are expected to be relatively small, and thus differences between the separation processes would be insignificant. Therefore, the virtual farms do not distinguish the methods of solids separation. Manure retention is expected to be short and, therefore, no emissions were estimated from solid separation activities.

9.7.1.3 Storage and Stabilization

All virtual model dairy farms contain storage activities for solid and liquid manure. Two types of solid manure storage activities were considered in developing the virtual farms. Solid manure could be: (1) stored in an uncovered stockpile or (2) not stored at all and sent directly from the drylot to be land applied. Review of industry practices indicated that solid manure would generally not be sent directly from the drylot to be land applied but would have some intermediate storage. Therefore, all the virtual farms included an uncovered stockpile.

Stabilization is the treatment of manure for reducing volatile solids and controlling odor prior to application to agricultural lands. The use of the word "stabilization" rather than "treatment" is intended to avoid the implication that treated animal manures can be discharged to surface water or groundwater.

Two types of storage and stabilization processes were considered to handle the liquid fraction from the drylot runoff and the solids separation process (if used): (1) an anaerobic lagoon (sometimes referred to as a *combined lagoon and storage pond* or a *one-cell lagoon*) or (2) an anaerobic lagoon followed by a spare storage pond (i.e., two-cell lagoon). A review of industry practices indicated that two-cell lagoons were not commonly used. Therefore, they were not considered in developing the virtual farms.

9.7.1.4 Land Application

As mentioned earlier, land application includes the manure application activity and the manure application site (i.e., cropland or other agricultural land). Solid manure is assumed to be land-applied to the manure application site using a solid manure spreader. Three types of liquid manure land application activities were considered in developing the virtual farms: land application by (1) liquid surface spreader, (2) liquid injection manure spreader, or (3) irrigation. Review of industry practices indicated that injection is rarely used. The emissions from irrigation and liquid surface spreading are expected to be similar because of the short duration of each activity and similar emission mechanisms. Therefore, the virtual farms only refer to liquid manure land application rather than a specific type.

9.7.2 VEAL VIRTUAL FARMS

Two virtual model farms were developed for veal (V1 and V2). The components of the virtual farms include confinement areas (enclosed housing), manure storage and stabilization facilities (anaerobic

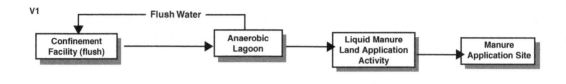

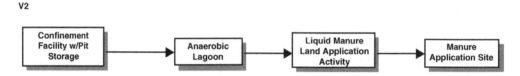

FIGURE 9.5 Confinement facility. (Source: USEPA, 2001a.)

lagoons or storage pits), and land application. The two differ only by the method of manure collection and storage.

9.7.2.1 Confinement

Because of the liquid nature of veal manure, it is flushed or stored in a pit. In virtual farm V1, veal are kept in a confinement facility, and their manure is flushed to an anaerobic lagoon. In virtual farm V2, veal are kept in a confinement facility with a pit underneath to store manure. The manure is then pumped to land application devices. Both methods are used in the veal industry (see Figure 9.5).

9.7.2.2 Storage and Stabilization

In virtual model farm V1, flushed manure is sent to stabilization and storage. Two types of storage and stabilization processes were considered in developing the virtual model farms: (1) an anaerobic lagoon (one-cell) or (2) an anaerobic lagoon followed by a separate storage pond (two-cell). Review of industry practices indicated that only anaerobic lagoons (one-cell) were commonly used. Additionally, a review of emission mechanisms and existing emissions data indicated that total emissions would not be substantially different between the one-cell and two-cell systems. Therefore, the virtual farms only include an anaerobic lagoon. The supernatant from the anaerobic lagoons is used as flush water.

Virtual model farm V2 does not have an anaerobic lagoon. Instead, manure is directly transported from the confinement area (i.e., pit storage) to the land application device.

9.7.2.3 Land Application

Land application includes the manure application activity and the manure application site (i.e., cropland or other agricultural land). In virtual farms V1 and V2, the manure from the storage and stabilization system is land-applied in a liquid form. Three types of land application activities were considered for liquid manure in developing the virtual farms: land application by (1) liquid surface spreader, (2) liquid injection manure spreader, or (3) irrigation. Review of industry practices indicated that injection is rarely used. The emissions from irrigation and liquid surface spreading are expected to be similar because of the short duration for each activity and similar emission mechanisms.

CHAPTER REVIEW QUESTIONS

1. What are the recent changes and trends in dairy operations? What factors affect this?
2. Describe and discuss dairy operation production cycles for mature cows, calves and heifers, and veal calves.
3. Describe and discuss confinement practices for dairy operations for mature cows (breeding cycle), mature cows (milking center), mature cows (lactating and dry herd), calves, heifers, and veal calves.
4. Describe and discuss feeding practices for dairy operations.
5. Describe and discuss dairy operation manure management practices. Include collection and transportation, drylots, freestall barns and milking centers, and storage, stabilization, and separation.
6. What components are included for the eight virtual dairy farms? Why?
7. Describe and discuss typical dairy operations.
8. What advantages are presented by flushing dairy manure?
9. What components are included for the two typical veal operations? Why?
10. Describe and discuss the typical veal operations.

THOUGHT-PROVOKING QUESTIONS

1. Why are dairy operations increasing in size and reducing in overall number? What are the industry issues that drive this trend? Is this trend common only to dairy operations? How does it affect other sectors of the industry?

REFERENCES

Adams, R.S., Comerford, J.W., Ford, S.A., Graves, R.E., Heald, C.W., Heinriche, A.J., Henning, W.R., Hutchinson, L.G., Ishler, V.A., Keyser, R.B., O'Connor, M.L., Specht, L.W., Spencer, S.B., Varga, G.A., & Yonkers, R.D. 1995. *Dairy Reference Manual,* 3rd Ed. NRAES-63. Natural Resource, Agricultural, and Engineering Service, Cooperative Extension.

Bickert, W.G., Bodman, G.R., Holmes, B.J., Janni, K.A., Kammel, D.W., Zulovich, J.M., & Stowell, R. 1997. *Dairy Freestall Housing and Equipment,* 6th Ed. MWPS-7.Midwest Plan Service. Ames, IA.

Cady, R. 2000. Monsanto Company and Founder of the Professional Dairy Heifer Growers Association. Personal communication with Eastern Research Group, Inc., February 18, 2000.

ERG. 2000. Facility counts for beef, dairy, veal, and heifer operations. Memorandum from Deb Bartram, Eastern Research Group, Inc. to the Feedlots Rulemaking Record. U.S. Environmental Protection Agency Water Docket, W-00-27.

Gardner, R. 1999. South East District Director for the Professional Diary Heifer Growers Association. Personal communication with Eastern Research Group, Inc., December 9, 1999.

Harner, J.P., Brouk, M.J., Smith, J.F., & Murphy, J.P. 2005. *Flush vs. Scrape: Nutrient Management Considerations.* Manhattan, KS: Kansas State University.

Jordan, L. 1999. South East Regional Director for the Professional Dairy Heifer Growers Association. Personal communication with Eastern Research Group, Inc., December 8, 1999.

Proulx, A. 2002. *That Old Ace in the Hole.* New York: Scribner.

Stull, C., Berry, S., & DePeters, E. 1998. *Animal Care Series: Dairy Care Practices,* 2nd ed. Dairy Workgroup, University of California Cooperative Extension. University of California Publishing. Davis, California.

USDA. 1995. Milk: Final estimates 1988–1992. Statistical Bulletin 909. U.S. Department of Agriculture, National Agricultural Statistics Service. Washington, DC.

USDA. 1996a. *National Animal Health Monitoring System, Part I: Reference of 1996 Dairy Management Practices.* U.S. Department of Agriculture, Animal and Plant Health Inspection Service. Fort Collins, CO.

USDA. 1996b. *National Animal Health Monitoring System Part III; Reference of 1996 Dairy Health and Health Management.* U.S. Department of Agriculture, Animal and Plant Health Inspection Service. Fort Collins, CO.

USDA. 1997. *A Guide to Dairy Calf Feeding and Management.* U.S. Department of Agriculture, Bovine Alliance on Management & Nutrition. Fort Collins, CO.

USDA. 1999a. Cattle: Final estimates 1994–1998. Statistical Bulletin 953. U.S. Department of Agriculture, National Agricultural Statistics Service, Washington, DC.

USDA. 1999b. Milking cows and production: Final estimates 1993–1997. Statistical Bulletin 952. U.S. Department of Agriculture, National Agricultural Statistics Service. Washington, DC.

USEPA. 2001a. *Emission from Animal Feeding Operations.* United States Environmental Protection Agency, Office of Air Quality Planning and Standards. Research Triangle Park, NC.

USEPA. 2001b. Development document for the proposed revisions to the national pollutant discharge elimination system regulation and the effluent guidelines for concentrated animal feeding operations. EPA-821-R-01-003. U.S. Environmental Protection Agency, Office of Water, Washington, DC.

Wilson, L.L., Stull, C., & Terosky, T.L. 1995. Scientific advancements and legislation addressing veal calves in North America. Veal Perspectives to the Year 2000, Internal Symposium. September 12 and 13, 1995, LeMans, France.

10 Swine Feeding Operations

This is your panhandle hog farm cough. I been workin over at Murphy Farms haulin waste. It's occupational. But I got laid off last week so it's getting better. A bad cough goes with a corporate hog farm.

(Proulx, 2002, p. 112)

10.1 INTRODUCTION

Currently, two prevailing views on swine feeding operations are popular in the United States. According to the U.S. Environmental Protection Agency (2005), the swine industry faces growing scrutiny of its environmental stewardship. The potential impact of an individual operation on the environment varies with animal concentration, weather, terrain, soils, production and waste management strategies, and numerous other conditions. On the other hand, according to the U.S. National Pork Producers Council (2005), "Concentrated pork production operations are subject to comprehensive regulations at the federal, state, and local level. Pork producers have taken the lead in working with the USEPA, state regulators, and environmental organizations in developing additional, science-based options for regulatory programs" (p. 1).

- *Important point:* A typical finishing pig produces an average of 1.2 gal of manure each day, a total of 438 gal of manure each year, at an average rate of 0.05 lb of nitrogen. In total, pork production accounts for 12% to 15% of all livestock manure in the United States.

The U.S. swine industry has undergone major consolidation over the past several decades. The number of hog operations, which approached 3 million in the 1950s, had declined to about 110,000 by 1997 (USDA, 1999a). The rate of consolidation has increased dramatically in the last decade, during which the number of swine operations decreased by more than 50% (USDA, 1999b). This trend toward consolidation appears to be continuing today.

- *Important point:* The total of amount of collectible manure from swine confinement operations is about 100 lb for every person in the United States.

While the number of operations has decreased, annual hog production has risen. The domestic hog industry is increasingly dominated by large totally enclosed confinement operations capable of handling 5,000 hogs or more at a time (USDA, 1999a). These operations typically produce no other livestock or crop commodities.

- *Important point:* For every pound of nitrogen in swine manure produced in the United States and used as crop fertilizer, 2 lb of nitrogen is piped directly into rivers and streams by municipal and industrial wastewater treatment facilities, and 4 lb are released into the atmosphere, primarily from industrial fuel combustion and transportation.

Another trend in the industry is an increasing degree of vertical integration that has accompanied consolidation. Hogs are raised by independent producers under contract with integrators who slaughter and market the hogs produced. The integrator provides the animals, feed, required vaccines and other drugs, and management guidance. The grower provides the labor and facilities, and is responsible for manure and carcass disposal. In return, each grower receives a fixed payment, adjusted for production efficiency.

These changes at both the industry and farm levels represent a significant departure from earlier eras, when hogs were produced primarily on relatively small but integrated farms where crop production and other livestock production activities occurred and where animals spent their complete life cycle at one location.

- *Important point:* Manure from pigs in a concentrated feeding operation must be totally contained at the site and then land-applied to crops in a way that does not enter surface water or groundwater. A municipal wastewater lagoon treats human sewage, but does not remove all nutrients or other contaminants, and then legally discharges it into a nearby stream or other surface water. Changes in wastewater treatment facilities in recent years have changed the form of nitrogen in sewage effluent from ammonia to nitrate; this does not change the total amount of nitrogen in the effluent and may not resolve the problem of eutrophication.

10.2   SIZE AND LOCATION OF SWINE INDUSTRY

In 1997, 109,754 swine operations were functioning in the United States. These operations produced 142.6 million pigs (USDA, 1999b). Farms vary in size from operations with a few hundred pigs to some newer operations that house hundreds of thousands of animals at one time. Table 10.1 shows the distribution of farms by size (based on 1997 inventory) and state. Table 10.2 shows the 1997 animal population by farm size. These data show the increasing dominance by large operations. In

TABLE 10.1
Number of Swine Operations by Size in 1997

State	Inventory		
	< 2,000 head	2,000–4,999 head	>5,000 head
Alabama	909	15	8
Alaska	53	0	0
Arizona	201	4	1
Arkansas	1,115	89	43
California	1,579	4	10
Colorado	1,202	9	14
Connecticut	210	0	0
Delaware	127	4	1
Florida	1,429	2	0
Georgia	1,706	39	19
Hawaii	247	1	0
Idaho	711	3	0
Illinois	6,673	381	114
Indiana	6,003	326	113
Iowa	15,711	1,224	308
Kansas	2,719	76	36
Kentucky	1,826	38	17
Louisiana	631	1	1
Maine	341	0	0
Maryland	574	10	0

TABLE 10.1 (continued)
Number of Swine Operations by Size in 1997

State	Inventory		
	< 2,000 head	2,000–4,999 head	>5,000 head
Massachusetts	382	1	0
Michigan	2,729	91	33
Minnesota	6,873	463	176
Mississippi	627	23	12
Missouri	5,192	165	62
Montana	597	23	7
Nebraska	5,753	189	75
Nevada	112	0	1
New Hampshire	249	0	0
New Jersey	428	2	1
New Mexico	346	0	0
New York	1,498	9	1
North Carolina	1,756	648	582
North Dakota	782	10	5
Ohio	5,801	125	26
Oklahoma	2,936	36	30
Oregon	1,382	1	0
Pennsylvania	3,305	115	36
Rhode Island	60	0	0
South Carolina	1,184	27	15
South Dakota	2,775	68	56
Tennessee	2,019	18	6
Texas	5,410	5	13
Utah	499	3	9
Vermont	238	0	0
Virginia	1,140	20	10
Washington	974	4	0
West Virginia	645	0	0
Wisconsin	3,629	51	6
Wyoming	292	0	4
United States	**103,580**	**4,323**	**1,851**	**109,754**

Source: USDA (1999a).

TABLE 10.2
U.S. Swine Operations and Inventory by Farm Size in 1997

Farm size*	Percent of operations	Percent of national inventory
< 1,999 head	94.4	39.3
2,000–4,999 head	3.9	20.8
> 5,000 head	1.7	40.2

Source: USEPA (2001).
* Based on inventory.

1997, 94% of the farms had a capacity of 2,000 pigs or less. These smaller operations confined 40% of the total inventory of pigs. In contrast, larger operations, which represent 6% of the number of farms, confined 60% of the inventory. The largest 2% of farms (> 5000 head) confined 40% of the inventory (USEPA, 2001). Table 10.3 shows the total inventory by state of breeding sows and hogs raised for market.

Swine production historically has been centered in the Midwest, with Iowa being the largest hog producing state in the country. Although the Midwest continues to be the nation's leading hog producer (five of the top seven producing states are still in the Midwest), significant growth has taken place in other areas. Perhaps the most dramatic growth has occurred in the Mid-Atlantic region, specifically in North Carolina. From 1987 to 1997, North Carolina advanced from being the 12th largest pork producer in the nation to second, behind only Iowa. The idea of locating production phases of different sites was developed in North Carolina. The state also has a much higher per-farm-average inventory than any of the states in the Midwest. Whereas Iowa had an average of fewer than 850 head per farm, North Carolina had an average of more than 3,200 head per farm in 1997 (USEPA, 2001).

- *Important point:* All swine manure, solid and liquid, applied at agronomic rates would supply about one-eighth of the nitrogen needs of the nation's corn crop.

Growth has occurred elsewhere as well. Significant growth has occurred in recent years in the panhandle area of Texas, Oklahoma, Colorado, Utah, and Wyoming. Some of the very large new operations have been constructed in these states.

TABLE 10.3
Swine Inventory by State in 1997

State	Inventory (1,000 head)	
	Breeding	Market
Alabama	20	170
Alaska	1	2
Arizona	15	130
Arkansas	113	768
California	27	183
Colorado	160	630
Connecticut	1	4
Delaware	4	26
Florida	10	45
Georgia	70	496
Hawaii	5	24
Idaho	4	26
Illinois	545	3,993
Indiana	448	3,265
Iowa	1,295	11,980
Kansas	196	1,296
Kentucky	71	499
Louisiana	5	27
Maine	1	5
Maryland	11	74
Massachusetts	3	16

TABLE 10.3 (continued)
Swine Inventory by State in 1997

State	Inventory (1,000 head)	
	Breeding	**Market**
Michigan	130	895
Minnesota	625	4,800
Mississippi	28	192
Missouri	445	3,016
Montana	20	160
Nebraska	440	3,085
Nevada	1	7
New Hampshire	1	4
New Jersey	3	20
New Mexico	1	5
New York	11	68
North Carolina	1,000	8,675
North Dakota	24	176
Ohio	203	1,335
Oklahoma	211	1,319
Oregon	5	30
Pennsylvania	119	941
Rhode Island	1	2
South Carolina	35	270
South Dakota	161	1,069
Tennessee	45	295
Texas	75	505
Utah	55	240
Vermont	1	2
Virginia	43	357
Washington	6	33
West Virginia	3	13
Wisconsin	126	639
Wyoming	19	76
United States	**6,810**	**51,697**

Source: USDA (1999b).

10.3 SWINE PRODUCTION CYCLES

The production cycle for hogs has three phases: farrowing, nursing, and finishing. Some farms specialize in a single phase of the growth cycle, whereas other farms may handle two or all three phases.

The first phase begins with breeding and gestation over a 114-day period, followed by farrowing (giving birth). After farrowing, the newly born pigs or piglets normally are nursed for a period of 3 to 4 weeks until they reach a weight of 10 to 15 lb. Typically, litters range from 9 to 11 pigs per litter, with a practical range of 6 to 13. The average number of pigs weaned per litter in 1997 was 8.7. Sows can be bred again within a week after a litter is weaned. Sows normally produce five to six litters before they are sold for slaughter, at a weight of 400 to 460 lb. After weaning, pigs are relocated to a nursery.

Nursery operations receive weaned pigs and grow them to a weight of 40 to 60 lb (feeder pigs). Weaned pigs are fed a starter ration until they reach a weight of 50 to 60 lb. At this point, they are 8 to 10 weeks of age. The third phase of swine production is the growing-finishing phase, where the gilts (young females) and young castrated boars (males) not retained for breeding are fed until they reach a market weight, typically between 240 and 280 lb. In this phase of swine production, a growing ration is fed to a weight of 120 lb, followed by a finishing ration. Growing-finishing usually takes between 15 and 18 weeks. Hogs normally are slaughtered at about 26 weeks of age. After weaning, swine typically are fed a corn-soybean, meal-based diet that may include small grains such as wheat and barley and other ingredients; they are fed this diet until they are slaughtered.

- *Important point:* hogs are produced in three types of specialized enterprises:
  - Farrow-to-finish operations raise hogs from birth to slaughter weight, about 240 to 280 lb.
  - Feeder pig producers raise pigs from birth to about 10 to 60 lb and then generally sell them for finishing.
  - Feeder pig finishers buy feeder pigs and grow them to slaughter weight (ERS, 2005).

The most common operation type is the farrow-to-finish operation that encompasses all three phases of swine production. Another common production mode is the combination of the farrowing and nursing phases, which provide feeder pigs for standalone grow-finish operations. Although not as common, some newer farms operate only the farrowing phase or only the nursery phase.

The annual production capacity of a farrowing operation is determined by the number of sows that can be confined and the number of litters of pigs produced per sow each year. Because the gestation period for pigs is 114 days, more than one litter of pigs can be produced per sow each year.

The annual production capacity of a farrow-to-finish or grow-finish operation is determined by capacity of the confinement facility, the duration of the growing period, and the time required for cleaning out and disinfecting the confinement facility between herds. The latter two factors determine the number of groups of pigs (turnovers) per year. The grow-finish production phase usually takes between 15 and 18 weeks. The length of the grow-finish cycle depends on the finished weight specified by the processor. Extremely hot or cold weather can reduce the rate of weight gain and also lengthen the grow-finish period. The duration of the clean-out period between groups of feeder pigs may be only a few or several weeks, depending on market conditions. A typical range for a grow-finish operation is 2.4 to 3.4 turnovers per year.

Turnovers affect the amount of manure generation. A grow-finish operation with a confinement capacity of 1,000 pigs and 2.4 turnovers per year produces approximately 2,400 pigs for slaughter per year, whereas the same operation with 3.4 turnovers per year produces 3,400 pigs per year. Assuming the same initial and final weights and the same rate of weight gain, this difference translates into one-third more manure production per year.

Production practices tend to vary regionally, depending on climate conditions, historical patterns, and local marketing and business practices. Table 10.4 presents the frequency of farrowing, nursing, and finishing operations in the three major hog production regions. Based on survey results in 1995, 61.9% of respondents were farrow-to-finish operations and 24.3% were grow-finish operations (USDA, 1995). Although many large operations are farrow-to-finish operations, this no longer is the norm. New operations commonly specialize in feeder pig production, nursery, or grow-finish phases of the production cycle. These operations may be linked by common ownership or separately owned but all under contract with a single integrator. Thus, pigs may begin their life-cycle in a sow herd on one site, move to a nursery on another, and then move again to a finishing facility. Specialized operations can take advantage of skilled labor, expertise, advanced technology, streamlined management, and disease control.

TABLE 10.4
Frequency of Production Phase in 1995 (Percent of Farms)*

Production Phase	Size	USDA APHIS Region**		
		Midwest	North	Southeast
Farrowing		76.6	68.6	69.3
Nursery	< 5,000 hogs marketed	20.1	51	57.8
Finishing		78.8	79.7	93.4
Farrowing		44.8	80.4	89
Nursery	> 5,000 hogs marketed	75	67.1	97.4
Finishing		45.8	69.7	62.8

Source: USDA (1995).

* Totals do not add to 100% because many operations combine production phases.

** Midwest = SD, NE, MN, IA, IL; North = WI, MI, IN, OH, PA; Southeast = MO, KY, TN, NC, GA.

TABLE 10.5
Typical Swine Housing Confinement Facilities

Facility Type*	Description	Applicability
Total confinement	Pigs are raised in pens or stalls in environmentally controlled building.	Most commonly used in nursery and farrowing operations and all phases of very large operations. Particularly common in the Southeast.
Open building with no outside access	Pigs are raised in pens or stalls but are exposed to natural climate conditions	Relatively uncommon, but used by operations of all sizes
Open building with outside access	Pigs are raised in pens or stalls but may be moved to outdoors	Relatively uncommon, but used by some small to mid-sized operations
Lot with hut or no building	Pigs are raised on cement or soil lot and are not confined to pens or stalls	Used by small to mid-sized operations
Pasture with hut or no building	Pigs are raised on natural pasture land and are not confined to pens or stalls	Traditional method of raising hogs. Currently used only at small operations

* These are the main facility configurations contained in the Swine '95 Survey conducted by USDA, 1995.

10.4 SWINE CONFINEMENT PRACTICES

Table 10.5 summarizes the five major housing configurations used by domestic swine producers. Although many operations still raise pigs outdoors, the trend in the swine industry is toward larger operations where pigs are raised in totally or partially enclosed confinement facilities. Typically, the gestation and farrowing, nursery, and grow-finish phases of the production cycle occur in separate, specially designed facilities.

Farrowing operations require intense management to reduce piglet mortality. Houses have farrowing pens (typically, 5 ft by 7 ft), and the piglets are provided a protected area of about 8 ft^2. Nursery systems are typically designed to provide a clean, warm, dry, and draft-free environment in which animal stress is minimized to promote rapid growth and reduce injury and mortality. Nursery buildings are cleaned and disinfected thoroughly between groups of pigs to prevent transmission of disease from one herd to another. Finishing pigs require less intensive management and can tolerate greater variations in environmental conditions without incurring health problems. Finishing operations allow about 6 ft^2 per pig.

TABLE 10.6
Housing Frequency in 1995 (Percent of Farms)

Swine production phase	Size	Housing	USDA APHIS Region*		
			Midwest	North	Southeast
Farrowing	< 5,000 hogs marketed	Total confinement	22.6	53.1	56
		Open building, no outside access	13.1	8.0	8.8
		Open building, outside access	25.7	33.8	31.2
		Lot	16.2	3.2	1.1
		Pasture	22.4	1.9	2.8
	> 5,000 hogs marketed	Total confinement	98.3	100	100
Nursery	< 5,000 hogs marketed	Total confinement	52.3	55.4	62
		Open building, no outside access	9.1	11.5	8.8
		Open building, outside access	27.7	33.8	31.2
		Lot	7.0	Not available	3.7
	> 5,000 hogs marketed	Total confinement	99	100	96.4
Finishing	< 5,000 hogs marketed	Total confinement	19.9	36.5	23.4
		Open building, no outside access	15.4	14.1	9.5
		Open building, outside access	24.5	42.1	55.9
		Lot	17.1	4.6	9.3
		Pasture	23.0	2.5	1.9
	> 5,000 hogs marketed	Total confinement	96.8	95.5	83.9

Source: USDA (1995).
* Midwest = SD, NE, MN, IA, IL; North = WI, MI, IN, OH, PA; Southeast = MO, KY, TN, NC, GA.

A typical confinement building is 40 ft by 300 to 500 ft. The buildings are either totally enclosed or open-side with curtains. Totally enclosed facilities are mechanically ventilated throughout the year. Open-sided buildings are naturally ventilated during warm weather and mechanically ventilated during cold weather, when curtains are closed. Swine houses have an integrated manure collection system as described in the next section. As shown in Table 10.6, smaller facilities tend to use open buildings.

10.5   SWINE MANURE MANAGEMENT PRACTICES

Although use of open lots for swine production still occurs, this method of confinement generally is limited to small operations. Swine manure produce in open lots is handled as a solid in methods similar to those at beef cattle feedlots and dairy cattle drylots. In enclosed confinement facilities, swine manure is handled as either slurry or liquid.

Four principal types of waste management systems are commonly used with total and partially enclosed confinement housing in the swine industry: deep pit, pull-plug pit, pit recharge, and flush systems. The deep pit, pull-plug pit, and pit recharge systems are used with slatted floors; flush systems can be used with either solid or slatted floors. We present brief descriptions of these management systems below. These practices do not represent all of the practices in use today; however, they are the predominant practices currently used by swine operations.

10.5.1   COLLECTION PRACTICES

Flush Systems. Flush systems use either fresh water or, more commonly, supernatant from an anaerobic lagoon to transport accumulated wastes to an anaerobic lagoon. Flush frequency can be daily, or as frequently as every 2 hours. Frequency depends on flushed channel length and slope and volume of water used per flush. Because pigs defecate as far away as possible from their feeding and resting areas, facilities with solid floors usually have a flush channel formed in that area. With slatted floors, usually a series of parallel flush channels are formed in the shallow pit under the slats. Methane emissions from flushed swine confinement facilities are low, but ammonia, hydrogen sulfide, and volatile organic compound (VOC) emissions may be higher than from pit recharge and pull-plug pit systems because of turbulence during flushing.

Pit Recharge. Pit recharge systems use relatively shallow pits drained periodically by gravity to an anaerobic lagoon. The frequency of draining varies, but between 4 and 7 days is standard. Pit recharge systems generally use 16- to 18-in. deep pits located under slatted floors. Previously, 24-in. deep pits were preferred, but now shallower pits are used. Following draining, the empty pit is partially refilled with water, typically with supernatant from the anaerobic lagoon. Generally, about 6 to 8 in. of water is added. With pit recharge systems, emissions of ammonia, hydrogen sulfide, methane, and VOCs from the confinement facility are lower than those with deep pits. However, if the manure is sent to an anaerobic lagoon, facility-wide emissions of ammonia, hydrogen sulfide, and methane from pit recharge may be greater than those from deep pits.

Pull-Plug Pits. Pull-plug pits are similar to pit recharge in that pit contents are drained by gravity to a storage or stabilization system. Pits are drained about every 1 to 2 weeks. However, water is not added back into the pit. The system relies on the natural moisture in the manure. Manure drained from pull-plug pits may be discharged to a manure storage tank, earthen storage pond, or an anaerobic lagoon for stabilization and storage. Gaseous emissions from confinement facilities with pull-plug pits are similar in magnitude to those with pit recharge systems.

Deep Pit Storage. Deep pits normally are sized to collect and store 6 months of waste in a pit located directly under a slatted flooring system. Accumulated manure is emptied by pumping. The accumulated manure may be directly applied to land or transferred either to storage tanks or earthen storage ponds for land application later. Because of the relatively high total solids (dry matter) concentration in swine manure collected and stored in deep pits, irrigation is not an option for disposal. To reduce odor, ammonia, and hydrogen sulfide concentrations in confinement facilities with deep pits, ventilation air may flow through the animal confinement area, down through the slatted floor, and over the accumulated manure before discharge from the building. Alternatively, deep pits may be ventilated separately. In either case, emissions of ammonia, hydrogen sulfide, methane, and VOCs from confinement facilities with deep pits at least theoretically should be higher than from facilities with other types of manure collection and storage systems.

10.5.2   SWINE MANURE STORAGE AND STABILIZATION

Most large hog farms have from 90 to 365 days of manure storage capacity (NPPC, 1996). Storage is in either an anaerobic lagoon or a storage facility. Typical storage facilities include deep pits, tanks, and earthen ponds. Anaerobic lagoons provide both manure stabilization and storage. The use of storage tanks and ponds generally is limited to operations with deep pits and pull-plug pits where manure is handled as slurry. Pit recharge and flush systems typically use anaerobic lagoons, because of the need for supernatant for use as recharge or flush water. Anaerobic lagoons emit less VOCs and noxious odors than storage facilities but emit more methane.

Storage facilities and anaerobic lagoons are operated differently. Storage facilities hold manure until the vessel is full and then are fully emptied at the next available opportunity. To maintain proper microbial balance, lagoons are never fully emptied, are sized for a design manure acceptance rate, and are emptied on a schedule. In this section, we describe the types of lagoons and storage facilities used and the factors affecting their design.

10.5.2.1   Anaerobic Lagoons

The anaerobic lagoon has emerged as the overwhelmingly predominant method used for the stabilization and storage of liquid swine manure. Methods of aerobic stabilization (e.g., oxidation ditches or aerated lagoons) were abandoned many years ago because of high electricity costs and operational problems such as foaming.

Several factors have contributed to the use of anaerobic lagoons for swine waste management. One is the ability to handle the manure as a liquid and use irrigation for land application. A second is the potential to reduce noxious odors by maximizing the complete reduction of complex organic compounds to methane and carbon dioxide, which are odorless gases. Finally, the use of anaerobic lagoons in the swine industry was driven, in part, by the potential to maximize nitrogen losses through ammonia volatilization, thereby reducing land requirements for ultimate disposal. With the shift to phosphorus as the basis for determining acceptable land application rates for animal manures, maximizing nitrogen loss is ceasing to be an advantage.

The design and operation of anaerobic lagoons for swine and other animal manure have the objective of maintaining stable populations of the microorganisms responsible for the reduction of complex organic compounds to methane and carbon dioxide. As mentioned earlier in the text, the microbial reduction of complex organic compounds to methane and carbon dioxide is a two-step process, in which a variety of VOCs are formed as intermediates. Many of these VOCs, such as butyric acid, are sources of noxious odors when not reduced further to methane. Methanogenic microorganisms have slower growth rates than the microbes responsible for the formation of VOCs; anaerobic lagoons must be designed and operated to maintain a balance between the populations of these microorganisms and methanogens to avoid accumulations of VOCs and releases of associated noxious odors.

Emissions of methane and VOCs from anaerobic lagoons vary seasonally. Since reaction rates of all microbial processes are temperature dependent, microbial activity decreases as the temperature approaches freezing; emissions can be very low during winter. Where significant seasonal variation occurs in lagoon water temperature, an imbalance in the microorganisms occurs in late spring and early summer, leading to high VOC emissions and associated odors. This variation is unavoidable, and the severity depends on seasonal temperature extremes.

10.5.2.2   Storage Facilities

Storage facilities include deep pits (beneath confinement buildings), in-ground tanks, above-ground tanks, and earthen ponds. Most storage facilities are open to the atmosphere.

Manure storage tanks and earthen ponds not only must have adequate capacity to store the manure produced during the storage period but also any process wastewaters or runoff that require storage. In addition, provision for storage of the volume of settled solids that accumulate for the period between solids removal is necessary. Because of the storage size required for liquid and slurry manures, completely mixing and emptying these facilities during draw down at the end of each storage period is difficult. Thus, an accumulation of settled solids periodically occurs, requiring a complete clean out of the facility. Estimates of rates of settled solids accumulation for various manures can be found in the Agricultural Waste Management Field Handbook (USDA, 1992).

The microbial processes responsible for methane and VOC formation also occur in storage tanks and ponds. However, the necessary balance in microbial populations for the complete reduction of organic carbon to methane and carbon dioxide never is established, because of higher organic loading rates and accumulations of high concentrations of VOCs, which inhibit methane formation. Thus, emissions of methane from manure storage tanks and ponds are lower than at anaerobic lagoons, and emissions of VOC are higher. Rates of formation of ammonia and hydrogen sulfide do not differ, but emission rates may differ depending on hydraulic retention time, pH, and the area of the liquid-atmosphere interface. The pH of storage facilities normally is acidic because of the accumulation of organic acid, which reduces the rate of ammonia emission but increases the rate of hydrogen sulfide emission. The reverse is true for anaerobic lagoons, which have pH values that typically are slightly above neutral. However, time and surface area probably are the more significant variables controlling the masses of ammonia and hydrogen sulfide emitted.

10.5.2.3 Anaerobic Lagoon Design

Both single-cell and two-cell systems are used for the stabilization and storage of swine manure. In single-cell systems, stabilization and storage are combined. In two-cell systems, the first cell has a constant volume and provides stabilization, while the second cell provides storage. With two-cell systems, water for pit recharge or flushing is withdrawn from the second cell. In climates with low precipitation and high evaporation rates, one or more additional cells may be added for the ultimate disposal of excess liquid by evaporation. Anaerobic lagoons use bacterial digestion to decompose organic carbon into methane, carbon dioxide, water, and residual solids. Periodic removal of settled solids is necessary. Typically, lagoons are dredged every 10 to 15 years, and the sludge is applied to land.

The design of lagoon treatment cells is similar to storage ponds, with one exception: lagoons are never completely emptied, except when accumulated solids are removed. Lagoons require permanent retention of what is known as the *minimum treatment volume* that should be reflected in design. Thus, lagoons must be larger in total volume than ponds that provide storage for the same volume of manure.

Determination of minimum treatment volume for lagoons is based on Natural Resources Conservation Service's (NRCS's) recommended total volatile solids (TVS) loading rates and the daily TVS loading to the lagoon. For anaerobic lagoons, recommended rates range from 3 lb TVS per 1,000 ft^3 per day in northern parts of Montana and North Dakota to 12 lb TVS per 1,000 ft^3 per day in Puerto Rico and Hawaii. This is a reflection of the effect of temperature on the rate of microbial activity. The calculation of minimum treatment volume is simply the daily TVS loading to the lagoon divided by the recommended TVS loading rate for the geographical location of the lagoon (USDA, 1992).

With open manure storage tanks, ponds, and lagoons, provision also is necessary to store the accumulation of normal precipitation directly falling into the structure, less evaporation during the storage period. The storage requirement for normal precipitation less evaporation varies geographically. In addition, provisions for storage of precipitation from a 25-year, 24-hour storm event must be in place, which also varies geographically, with a minimum of 1 ft of free board remaining. Design values used for the accumulation of normal precipitation, less evaporation, are based on mean monthly precipitation values for the location of the storage facility, obtained from the National Oceanic and Atmospheric Administration.

In some situations, manure storage ponds and lagoons also may be used for the storage of runoff captured from open confinement areas. In these situations, provision for storage of runoff collected from normal precipitation during the storage period as well as from a 25-year, 24-hour storm event must be included in the design storage capacity of the pond. Expected annual and monthly runoff values for the continental United States, expressed as percentages of normal precipitation, for paved and unpaved open lots can be found in the *Agricultural Waste Management Field Handbook* (USDA, 1992).

10.5.2.4 Regional Differences in Manure Management Systems

Regional differences occur in methods of swine manure management, driven primarily by climate but also influenced by the size of an operation. For example, small operations with less than 500 head of confinement capacity commonly use drylots that are scraped periodically for manure removal. Manure storage is rare, but runoff collection and storage ponds may also be used for storage of any confinement facility wash water. Operations with greater than 500 head of confinement capacity typically use one of the management systems previously described. As confinement capacity increases, the probability that either a pull-plug pit or flush system with an anaerobic lagoon will be used also increases.

However, other regional differences still occur, even among operations with greater than 1,000-head confinement capacity. For example, use of flushing generally is limited to the Central and Southern regions of the United States because freezing of flush water is not a problem, and use of deep pits generally is limited to the Mid-Atlantic, Midwest, and Pacific regions (Table 10.7). In contrast, pH

TABLE 10.7
Frequency (in percent) of Operations in 1995 that Used Certain Manure Storage Systems for Operations that Marketed 5,000 or More Hogs in a 12-Month Period (Percent of Farms)

	USDA APHIS region*		
Manure storage system	Midwest	North	Southeast
Deep pit storage	21.5	28.5	85.7
Above ground storage	NA	NA	27.2
Below ground storage	NA	NA	43.3
Anaerobic lagoon	91.2	4.8	33.3
Aerated lagoon	NA	**	NA
Solids separated from liquids	NA	NA	14.4

Source: USDA, 1995.

*   Midwest = SD, NE, MN, IA, IL; North = WI, MI, IN, OH, PA; Southeast = MO, KY, TN, NC, GA.

** Aerated lagoons were reported on 70% of the operations. The standard error of the data as reported by NAHMS exceeds 21% and therefore was determined by NAHMS not to be statistically valid.

recharge systems are used in all regions. The database used to create Table 10.7 did not include frequency of use of pull-plug pits. However, pull-plug pits generally are used primarily in climates where winter temperatures severely impact anaerobic lagoon performance.

10.5.3   SWINE MANURE LAND APPLICATION

Essentially all swine manure is disposed of by application to cropland. Manure from deep pits and pull-plug pits typically is surface-applied and may be incorporated by disking or plowing. Subsurface injection also may be used but is a less common practice. Incorporation following application and injection are used most commonly when odors from land application sites are a concern. Irrigation is the most common method of disposal of supernatant from anaerobic lagoons. In arid areas, evaporation is another option for disposal of lagoon liquids. Methods of swine manure disposal by USDA region are summarized in Table 10.8.

10.5.4   SWINE MORTALITY

A variety of methods are used for the disposal of mortalities in the swine industry (Table 10.9). Commonly used methods for disposal of young pig carcasses are burial, composting, and incineration. However, burial is becoming less common because of water quality concerns. It is being replaced primarily by composting. Incineration is more expensive because of equipment and fuel costs, but requires less labor. Carcass composting is a mixed aerobic and anaerobic process and therefore is a source of those gaseous compound emissions associated with aerobic and anaerobic microbial decomposition of organic matter. Land application is used for the disposal of composted carcasses. Larger animals usually are disposed of off-site by rendering, although they also may be buried or composted.

10.6   SWINE VIRTUAL FARMS

Four basic virtual model farms were identified for swine. These virtual models represent grow-finish operations. The components of the virtual model farms include the confinement houses, manure

TABLE 10.8
Method of Manure Application on Land in 1995

Variable	Size	USDA APHIS region*		
		Midwest	North	Southeast
Irrigation	< 5,000 hogs	47.6	11.2	2.9
Broadcast	marketed	18.4	57.8	69.0
Slurry-surface		33.0	55.7	46.6
Slurry-subsurface		NA	26.6	22.9
Irrigation	> 5,000 hogs	100	74.8	16.4
Broadcast	marketed	NA	NA	39.4
Slurry-surface		NA	6.3	68.1
Slurry-subsurface		NA	23.6	72.1

Source: USDA (1995).

Note: Swine farms use more than one method of disposal; totals will add to more than 100%.

* Midwest = SD, NE, MN, IA, IL; North = WI, MI, IN, OH, PA; Southeast = MO, KY, TN, NC, GA.

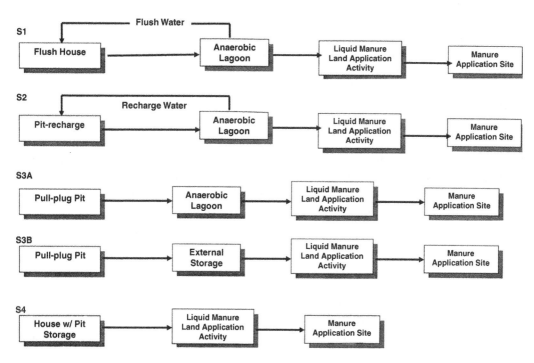

FIGURE 10.1 Swine models. (Source: USEPA, 2001a.)

storage facilities (anaerobic lagoons, external storages, or pit storages), and land application. The four virtual models represent the most common manure collection methods: flush, pit-recharge, pull-plug pit, and pit storage (S1, S2, S3, and S4) (see Figure 10.1). For the pull-plug pit virtual model, two variations were developed to account for different manure storage practices (S3A and S3B). The four swine virtual model farms differ in the type of manure management systems in the confinement area and the method of storage (see Figure 10.1).

10.6.1 Confinement

Swine are kept in confinement buildings, usually with slatted floors to separate the manure from the animals. The manure falls through the slats where it is stored for a period of time. Periodically, manure is removed to a storage and stabilization site. The time that the manure is stored in the confinement house depends on the type of manure management system. For storage pits, the storage time varies from several days to several months. For flush systems, manure is removed several times a day. The virtual model swine farms are differentiated by their manure management systems, which are flush house (S1), pit recharge (S2), pull-plug pit (S3A and S3B), and pit storage (S4). The virtual models with pit storage are sources of emissions of ammonia, hydrogen sulfide, methane, and VOCs. The flush house virtual model emits ammonia and hydrogen sulfide. All models emit particulate matter from feed and swine dander.

10.6.2 Storage and Stabilization

In virtual model farms S1 and S2, manure is sent to an anaerobic lagoon. Two types of lagoon systems were considered: (1) an anaerobic lagoon (sometimes referred to as a *combined lagoon and storage pond* or *one-cell lagoon*) or (2) an anaerobic lagoon followed by a separate storage pond (two-cell lagoon). Review of industry practices indicated that the one-cell anaerobic lagoon was the most commonly used method. Additionally, a review of emission mechanisms and existing emission data indicated that total emissions would not be substantially different between the one-cell and two-cell systems. Therefore, the virtual model farms only include an anaerobic lagoon. The supernatant from the lagoon is used as flush water or pit recharge water.

In the pull-plug pit virtual model farms, the manure is either sent to an anaerobic lagoon (S3A) or to external storage (S3B). For the pit storage model (S4) manure is sent directly from the confinement facility (i.e., pit storage) to be land-applied.

10.6.3 Land Application

Land application includes the manure application activity and the manure application site (i.e., cropland or other agricultural land). All manure from the swine virtual model farm is land applied in a liquid form. Three types of liquid land application activities were considered in developing the virtual model farms: land application by (1) liquid surface spreader, (2) liquid injection manure spreader, or (3) irrigation. Information was not available to estimate or differentiate emissions from the three activities. Therefore, the virtual model farms do not distinguish among methods of liquid land application.

CHAPTER REVIEW QUESTIONS

1. Where are the majority of swine operations located? Why?
2. Discuss and describe typical swine operation production cycles.
3. Discuss and describe the three types of specialized swine enterprises.
4. Describe and discuss typical swine confinement practices.
5. Describe and discuss typical swine manure management practices. Include collection, storage, and stabilization.
6. How are storage practices changed to fit regional conditions? Include land application practices and swine mortality.
7. What components are included for the four virtual swine operations? Why?
8. Describe and discuss typical swine operations.

THOUGHT-PROVOKING QUESTIONS

1. Two prevailing views on the swine industry are presented—one from the USEPA and one from the U.S. National Pork Producers Council. Take each point of view and discuss the negative and positive aspects of the industry.
2. What is the current trend in swine production operations? Why?
3. Discuss the problems related to swine manure production.
4. Research and discuss vertical integration in the U.S. swine industry. What are the advantages and disadvantages to the growers and owners?
5. What effect do state regulations have on swine operation population? Why is the Mid-Atlantic region such a high growth area?

REFERENCES

ERS. 2005. Briefing room hogs: Background. Economic Research Service. United States Department of Agriculture. http://www.ers.usda.gov/Briefing/Hogs/Background.htm. Accessed May 27, 2005.

NPPC. 1996. *Swine Care Handbook*. National Pork Producers Council, Des Moines, IA.

Proulx, A. 2002. *That Old Ace in the Hole*. New York: Scribner.

USDA. 1992. *Agricultural Waste Management Field Handbook, National Engineering Handbook, Part 651*. U.S. Department of Agriculture, Natural Resources Conservation Service, Washington, DC.

USDA. 1995. *Swine '95 Part 1: Reference of 1995 Swine Management Practices*. U.S. Department of Agricultural, Animal and Plant Health Inspection Service, Fort Collins, CO.

USDA. 1999a. 1977 Census of Agriculture. U.S. Department of Agriculture, National Agricultural Statistics Service, Washington, DC.

USDA. 1999b. Hogs and pigs: Final estimates 1993–1997, Statistical Bulletin 951. U.S. Department of Agriculture, National Agricultural Statistics Service. Washington, DC.

USEPA. 2001a. Emission from animal feeding operations. United States Environmental Protection Agency, Office of Air Quality Planning and Standards. Research Triangle Pack, NC.

USEPA. 2001b. Development document for the proposed revisions to the national pollutant discharge elimination system regulation and the effluent guidelines for concentrated animal feeding operations. EPA-821-R-01-003. U.S. Environmental Protection Agency, Office of Water, Washington, DC.

USEPA. 2005. *Swine Production and Environmental Stewardship*. http://www.epa.gov/agriculture/anafoidx.html. Accessed May 26, 2005.

U.S. National Park Producers Council. (2005). 8 pork industry environmental impact "facts." http://www.vortexcombustion.com/pork_facts.htm. Accessed May 26, 2005.

11 Poultry Feeding Operations

Poultry litter, liquid poultry manure, and dead bird residues are nutrient-rich, bacteria-laden materials that can be a threat to groundwater if not properly managed. These waste materials can benefit the farmstead if they are properly stored, handled, and applied to land.... The manner in which poultry waste is collected, stored, treated, and applied to land can make a big difference in its fertilizer value. Stored poultry waste and compost residue material should be sampled and tested to determine how much nitrogen, phosphorus, and potassium they contain. The litter, liquid poultry manure, and compost nutrient information will help determine fertilizer needs, along with waste application rates for a pasture or crop.

(Oklahoma State University, 2005)

11.1 INTRODUCTION

Broilers, layers, turkeys, ducks, geese, and game fowl encompass the subsectors of the poultry industry. This chapter focuses only on broilers, layers, and turkeys, which accounted for 99% of the annual farm receipts from the sale of poultry in 1997 (USDA, 1998a). Broilers accounted for approximately 65% of poultry sales, with sales of eggs and turkeys accounting for 21% and 13%, respectively (USDA, 1998b).

Up until the 1950s, most of the nation's poultry was produced on small family farms in the Midwestern United States. Midwestern states provided favorable climatic conditions for seasonal production of poultry and close proximity to major sources of grain feed. With the advent of controlled environment housing facilities, poultry production ceased to be a seasonal activity. With improvement of transportation and distribution systems, the poultry industry eventually expanded from the Midwest to other regions. By 1997, the value of poultry production exceeded $21.6 billion, and much of the poultry output was generated on large facilities with confinement capacities in excess of 100,000 birds (USDA, 1998a).

Poultry production (especially broiler production) is a highly vertically integrated industry, and as a result, management strategies at the facility level tend to be more uniform than in other sectors of animal feeding operations (AFOs). Growers working under contract with integrators produce more than 90% of all chickens raised for human consumption in the United States. Under contract, the integrators provide the growers with birds, feed, medicines, transportation, and technical help. The contract growers provide the labor and the production facilities to grow the birds from hatchlings to market age and receive a minimum guaranteed price for the birds moved for slaughter. The contract growers are responsible for disposal of manure and animal carcasses.

11.2 BROILERS

Broiler production refers to raising a meat-type chicken typically slaughtered at about 7 weeks of age at a live weight of about 5 lb. A broiler is a young chicken of either sex, characterized as having tender meat, flexible breastbone cartilage and soft, pliable, smooth-textured skin.

Key terms:

- Compost: Organic residues that have been collected and allowed to decompose.
- Composting: A controlled process of decomposing organic matter by microorganisms.
- Cost sharing: A program in which the Consolidated Farm Service Agency (formerly the Agricultural Stabilization and Conservation Service) pays a percentage of the costs of a project, facility, or effort.
- Decompose: The breakdown of organic materials.
- Leaching: The removal of soluble substance from soils or other material by water.
- Mortality: Birds that died during production.
- Nutrient: Those elements necessary for plant growth.
- Nutrient management plan: A specific plan designed to manage animal manures and mortalities so that the most benefit is obtained and the environment is protected.
- Stacking shed: A structure designed and built for the storage of poultry manure.

11.2.1 SIZE AND LOCATION OF THE BROILER INDUSTRY

In 1997, 23,937 broiler operations produced 6.7 billion broilers. The consolidation of the broiler industry from small, family-run to large operations began earlier than other poultry and livestock sectors and was well entrenched by the 1970s. Table 11.1 illustrates the trend. Between 1982 and 1992, more than 6,000 broiler operations (20% of the industry's producers), went out of business. During this period, total broiler production increased by 50%, with new, larger operations becoming more predominant. Between 1992 and 1997, the number of operations stabilized, but production increased 24%, from 5.4 billion broilers to 6.7 billion broilers.

Larger operations dominate broiler production. In 1997, most operations had a confinement capacity of 90,900 birds or less, as shown in Table 11.2. The confinement capacity was estimated

TABLE 11.1

Broiler Operations and Production in the United States

Year	Operations	Production
1982	30,100	3,516,095,408
1987	27,645	4,361,198,301
1992	23,949	5,427,532,921
1997	23,937	6,741,476,153

TABLE 11.2

Number of Broiler Operations by Size in 1997

State	Confinement Capacity		
	< 10,900 birds	10,900–90,900 birds	> 90,900 birds
Alabama	90	1885	502
Alaska	9	0	0
Arizona	19	1	0
Arkansas	262	2974	414
California	137	36	67
Colorado	74	0	0
Connecticut	28	2	0
Delaware	37	687	81
Florida	55	191	75

TABLE 11.2 (continued)
Number of Broiler Operations by Size in 1997

State	Confinement Capacity		
	< 10,900 birds	10,900–90,900 birds	> 90,900 birds
Georgia	82	1,475	688
Hawaii	6	3	0
Idaho	55	0	0
Illinois	114	1	0
Indiana	153	48	3
Iowa	490	28	1
Kansas	93	0	0
Kentucky	44	119	80
Louisiana	29	215	75
Maine	71	2	0
Maryland	117	777	103
Massachusetts	40	1	0
Michigan	334	2	0
Minnesota	520	95	6
Mississippi	66	928	399
Missouri	132	180	139
Montana	61	0	0
Nebraska	224	0	1
Nevada	6	0	0
New Hampshire	33	2	0
New Jersey	79	0	0
New Mexico	11	0	0
New York	165	7	0
North Carolina	141	1670	275
North Dakota	82	1	0
Ohio	306	178	10
Oklahoma	93	476	63
Oregon	109	33	14
Pennsylvania	421	374	50
Rhode Island	4	1	0
South Carolina	32	218	116
South Dakota	91	1	0
Tennessee	91	402	55
Texas	259	483	258
Utah	19	0	0
Vermont	57	0	0
Virginia	59	467	145
Washington	104	36	22
West Virginia	37	85	64
Wisconsin	529	38	20
Wyoming	17	0	0
United States	**6,089**	**14,122**	**3,726**

Total 23,937

Source: USDA (1999a).

TABLE 11.3
Broiler Inventory by State

State	Inventory (1,000 birds)	State	Inventory (1,000 birds)
Alabama	164,764	Nebraska	291
Alaska	—	Nevada	—
Arizona	—	New Hampshire	—
Arkansas	211,745	New Jersey	—
California	43,145	New Mexico	—
Colorado	—	New York	255
Connecticut	—	North Carolina	120,909
Delaware	46,709	North Dakota	—
Florida	24,073	Ohio	8,327
Georgia	215,055	Oklahoma	35,891
Hawaii	182	Oregon	3,945
Idaho	—	Pennsylvania	24,582
Illinois	—	Rhode Island	—
Indiana	—	South Carolina	33,236
Iowa	—	South Dakota	—
Kansas	—	Tennessee	25,200
Kentucky	20,100	Texas	82,745
Louisiana	—	Utah	—
Maine	—	Vermont	—
Maryland	53,691	Virginia	47,164
Massachusetts	—	Washington	7,055
Michigan	116	West Virginia	16,509
Minnesota	8,418	Wisconsin	5,982
Mississippi	130,964	Wyoming	—
Missouri	45,455	Other	35,156
Montana	—	**United States**	**1,411,673**

Source: USDA (1998b).

from 1997 sales, assuming 5.5 flock turnovers per year. Operations with more than 90,900 birds of confinement capacity represented only 11% of the total number of broiler operations, but accounted for nearly half the annual production. Smaller operations with fewer than 10,900 bird confinement capacity accounted for nearly 78% of the broiler operations but less than 30% of the annual production (USDA, 1999a).

In addition to being dominated by large producers, the broiler industry is concentrated in several states. Georgia, Arkansas, and Alabama are some of the largest broiler producing states, followed by Mississippi, North Carolina, and Texas. Table 11.3 shows the broiler population by state.

11.2.2 BROILER PRODUCTION CYCLES

The production cycle of broilers is divided into two phases: brooding and grow-out. The brooding phase begins when day-old chicks are placed in a heated section of a broiler house known as the *brood chamber*. The brood chamber is maintained at over 100°F when the birds are a day or two old. During the birds' first few weeks of growth, the temperature of the brood chamber is gradually decreased. Once the birds need floor space, the remainder of the house is opened and chicks fed out to market weight.

The length of the grow-out phase ranges from 28 to 63 days, depending on the size of the bird desired. Broilers are produced to meet specific customer requirements, which can be a retail grocery store, fast-food chain, or institutional buyer. For broilers, the typical grow-out period is 49 days, resulting in an average weight of 4.5 to 5.5 lb. The grow-out period may be as short as about 28 days to produce a 2.25 to 2.5 lb bird, commonly referred to as a Cornish game hen. For producing roasters weighing 6 to 8 lb, the grow-out period can be up to 63 days. Broiler houses are operated on an "all in–all out" basis and require time for cleaning and repair between flocks. For broilers, five to six flocks per house per year is typical. The number of flocks per year is lower for roasters and higher for Cornish hens. When roasters are produced, females usually are harvested at 49 days of age to provide more floor space per bird to accommodate the added weight gain by the males that remain.

Female broilers grown to lay eggs for replacement stock are called *broiler breeders* and are usually raised on separate farms. These farms produce only eggs for broiler replacements. A typical laying cycle for hens is about 1 year, after which the hens are sold for slaughter.

11.2.3 BROILER CONFINEMENT

The most common type of housing for broilers, roasters, and breeding stock is enclosed housing with a compacted soil floor covered with dry bedding. Dry bedding (litter) can be sawdust, wood shavings, rice hulls, chopped straw, peanut hulls, or other products, depending on availability and cost. Manure as excreted by birds has high water content. The litter absorbs moisture excreted by the birds.

- *Important point:* Litter should always be stored in a roofed area that is well protected from rain and wind. A stacking shed (roofed structure with a concrete floor) is the safest method for temporary storage of litter.

Mechanical ventilation is typically provided using a negative-pressure system, with exhaust fans drawing air out of the house and fresh air returning through ducts around the perimeter of the roof. The ventilation system uses exhaust fans to remove moisture and noxious gases during the winter season and excess heat during the summer. Advanced systems use thermostats and timers to control exhaust fans. Many houses have side curtains that are opened in warm weather for natural ventilation.

Broilers and Roasters. Houses for broilers and roasters are usually 40 ft wide and 400 to 500 ft long and typically designed for 25,000 to 30,000 broilers per flock.

Broiler Breeders. Houses are usually 40 to 45 ft wide and 300 to 600 ft long. Most of the breeder houses contain wooden slats elevated 18 to 24 in. and laid across supports for the birds to roost. The slats are spaced 1 in. apart, which allows most of the manure produced by the birds to fall beneath the slat area, keeping the area accessible to the birds cleaner. Drinkers, mechanical feeders, and nests are placed over the slats. The slats cover two-thirds of the area of the house, running along the outside walls, with the center corridor containing bedding litter. The center corridor is covered with 2 to 6 in. of bedding before young breeder layers are placed in the breeder house. Equipment can access the center section of the house to aid in clean-out between flocks.

11.2.4 BROILER MANURE MANAGEMENT

A typical broiler house with capacity for 22,000 birds at a time produces 120 tons of litter per year (NCC, 1999). Two kinds of manure are removed from broiler houses: litter and cake. Litter is a mixture of bedding and manure. Cake is a compacted and concentrated mixture of manure and litter that usually builds up on the surface of the litter around waterers and feeders, where much of the manure is deposited.

- *Important point:* Most broiler operations produce 1.0 to 1.4 tons of litter per 1,000 birds. For a flock of 15,000 birds, this is between 15 and 21 tons of litter per flock.

- *Important point:* Assuming total nitrogen (N) content of litter is 3%, this would result in the litter containing one-half to two-thirds ton of total N for each 15,000-bird flock. As much as 25% of the total nitrogen contained in fresh litter could reach groundwater.

11.2.4.1   Broiler Manure Collection

Broiler houses are partially cleaned between flocks to remove cake and are fully cleaned out less often. The remaining litter may be "top dressed" with an inch or so of new bedding material. The litter (bedding and manure) is typically completely cleaned out annually, although there is a trend toward performing complete clean-outs less often. When the broiler house is completely cleaned out, the litter is typically removed with a front-end loader.

A broiler breeder house is cleaned after the hens have finished the lay cycle, which is typically about 1 year. When the house is cleaned, the equipment (including slats) is removed from the house to allow a front-end loader to push all of the manure to the center litter section of the house. Then the front-end loader places the mixture of manure and litter into a spreader for land application. A thorough cleaning after each flock removes pathogens that could be transferred to the next flock. After removal of all organic matter, the house is disinfected.

Factors that affect emissions from broiler houses include the moisture content of the manure, the time the manure is present in the broiler house, and the ventilation rate. The moisture content affects the volatilization of compounds that are soluble in water, such as ammonia, hydrogen sulfide, and volatile organic compounds (VOCs). The more moisture present, the more likely these compounds are to be emitted. Manure as excreted by birds has a high water content, most of which evaporates, emitting ammonia as the manure dries out. Since broiler manure storage is integrated with the broiler house, ammonia emissions continue throughout the year. The ventilation rate affects the amount of ammonia and particulate matter (PM) carried out of the broiler house. During the growth of the flock, continuous airflow removes ammonia and other gases, reducing the moisture content of the litter over that of freshly excreted manure. Another result of continuous airflow is lower nitrogen content of the litter (manure and bedding).

11.2.4.2   Broiler Manure Storage

Once broiler manure has been collected, it is either immediately applied to cropland or stored for later land application. Because cake removal occurs after each grow-out cycle, cake storage is a necessity. Traditionally, cake from broiler production facilities was stored in uncovered stockpiles until conditions permitted land application. However, water quality concerns have led to the increased use of storage structures known as *litter sheds* for cake storage. Litter sheds typically are partially enclosed pole-type structures. Water quality concerns have also led to the recommendation that cake not stored in litter sheds be placed in well-drained areas and covered to prevent contaminated runoff and leaching. However, covering of stockpiles of cake is rare. Because of the larger volume involved, broiler manure and litter from a total facility clean-out is usually stored in open or covered stockpiles if immediate land application is not possible. Because of cost, litter sheds generally are sized only to provide capacity for cake storage. To avoid long-term storage of broiler manure and litter in stockpiles, the timing of total facility clean-outs gradually is shifting to early and mid-spring.

Factors that affect emissions from broiler litter storage are moisture content and length of storage. High moisture content leads to the development of anaerobic conditions and the production of hydrogen sulfide and other reduced sulfur compounds, VOCs, and methane and facilitates the further mineralization of organic nitrogen to ammonia. As the time of storage increases, the opportunity for the generation and emission of these compounds increases. Open stockpiles of litter can be intermittent sources of PM emissions if the surface layer of the stored litter is sufficiently dry. Thus, frequencies of precipitation events, evaporation rates, and wind speed are important variables. In litter sheds, protection from precipitation increases the probability of particulate matter emission, and partial protection from wind decreases the probability.

11.2.5 MORTALITY MANAGEMENT

With broilers, the highest rate of mortality normally occurs during the first 2 weeks of the grow-out cycle but continues at a lesser rate throughout the rest of the cycle. Typically, about 4% to 5% of the birds housed die during the grow-out cycle. To prevent the possible spread of disease, dead birds must be removed at least daily, if not more frequently. Several options are available for dead bird disposal. Of these options, composting is one of the more desirable approaches and has been promoted heavily by the broiler industry. As an alternative to composting or burial, at least one integrator has been distributing freezers to preserve carcasses for subsequent disposal by rendering.

Carcass composting is an aerobic process using oxygen, bacteria, and heat to reduce the volume and weight of bird carcasses. The birds are placed in the composting bins, piles, or elongated piles called *windrows* within 24 hours of death and covered with appropriate composting material. The mixture generates heat and rapidly decays the dead birds into a product suitable for land application. Carcass composting is very popular in areas where birds cannot be taken to rendering. The finished compost is suitable for disposal by land application without attracting scavengers and other vermin.

Catastrophic losses of broiler chickens also occur, especially during periods of extremely hot weather but also because of weather events such as hurricanes, tornadoes, and snow or ice storms. Catastrophic losses of broilers from excessive heat usually are more severe with older birds. Several options for disposal of catastrophic losses are available, with burial being the most commonly used practice. Large-scale composting is another, and probably more desirable, option from a water quality perspective.

- *Important point:* Composting of dead birds is showing real promise as an on-farm disposal method. Composting stabilizes nutrients, eliminates odors, and disinfects the material. As long as the composter is self-contained with installed odor control, properly sized, properly sited, and sheltered from rainfall, there are no environmental hazards from composting.

11.3 LAYING HENS

Laying hens or layers are sexually mature female chickens maintained for the production of eggs, primarily for human consumption. These eggs are known as *table eggs* and may be sold as shell eggs (table eggs) or may be used in the production of liquid, frozen, or dehydrated eggs. Fertile eggs also are produced for hatching to provide broiler and laying hen chicks, but such production occurs in a relatively small number of specialized operations that are not discussed in this presentation.

11.3.1 SIZE AND LOCATION OF THE TABLE EGG INDUSTRY

Trends in the egg industry have paralleled those in other livestock industries, increasing overall production on fewer and larger farms. Table 11.4 illustrates the degree of consolidation that has occurred in the industry in the last 15 years. In 1982, there were 212,000 operations with mature hens in the United States. Between 1982 and 1997, the number of operations dropped by 69%, while the number of hens increased slightly from 310 to 313 million. Overall, table egg production has not increased as rapidly as has broiler production.

Table 11.5 shows the number of layer operations by size in 1997. The size distribution is based on the inventory of layers that are 20 weeks or older (i.e., excluding immature birds) and excludes

TABLE 11.4
Layer Operations and Production in the United States

Year	Number of operations	Inventory (number of layers 20 weeks and older)
1982	212,608	310,515,367
1987	141,880	316,503,065
1992	86,245	301,467,288
1997	69,761	313,851,480

Source: USDA (1999a).

TABLE 11.5
Number of Layer Operations by Size in 1997

	Inventory (layers 20 weeks or older)		
State	< 20,000 birds	20,000–100,000 birds	> 100,000 birds
Alabama	1,022	108	16
Alaska	63	0	0
Arizona	367	0	1
Arkansas	1,455	182	6
California	2,541	62	67
Colorado	1,568	4	5
Connecticut	363	10	4
Delaware	83	3	0
Florida	1,104	37	16
Georgia	903	191	28
Hawaii	129	5	2
Idaho	862	1	2
Illinois	1,671	9	7
Indiana	1,688	59	38
Iowa	1,753	37	41
Kansas	1,948	13	3
Kentucky	1,855	23	4
Louisiana	813	13	2
Maine	516	13	3
Maryland	601	8	9
Massachusetts	491	4	2
Michigan	2,182	10	13
Minnesota	1,833	44	15
Mississippi	826	57	2
Missouri	3,507	43	9
Montana	1,001	0	0
Nebraska	1,458	10	8
Nevada	200	0	0
New Hampshire	390	4	0
New Jersey	808	3	2

TABLE 11.5 (continued)
Number of Layer Operations by Size in 1997

	Inventory (layers 20 weeks or older)		
State	**< 20,000 birds**	**20,000–100,000 birds**	**> 100,000 birds**
New Mexico	647	2	2	
New York	1,812	20	10	
North Carolina	1,409	146	11	
North Dakota	534	3	0	
Ohio	2,958	58	49	
Oklahoma	3,138	28	3	
Oregon	2,193	1	5	
Pennsylvania	2,960	117	70	
Rhode Island	91	1	0	
South Carolina	644	37	14	
South Dakota	717	4	4	
Tennessee	2,504	20	1	
Texas	6,090	138	31	
Utah	521	1	5	
Vermont	499	2	1	
Virginia	1,491	22	6	
Washington	1,482	6	16	
West Virginia	1,073	12	0	
Wisconsin	2,438	12	7	
Wyoming	436	0	0	
United States	**67,638**	**1,583**	**540**	**69,761**

Source: USDA (1999a).

farms that raise only pullets. 98% of the table egg operations in 1997 housed less than 20,000 birds. Although the majority of operations are in the small size category, large operations are responsible for a continually increasing larger share of total egg production. Between 1982 and 1992, the average number of hens and pullets on poultry farms increased from 1,460 birds per farm to 3,495 per farm. The 326 largest operations represent less than 0.5% of the total number of operations (70,857) but confined over 55% of the laying hens (Abt, 1998).

Laying operations, although primarily performed in 10 states, are much less geographically concentrated than the broiler industry. States in the Midwest accounted for the largest number of operations, and the large production facilities are fairly evenly spaced throughout the country. Table 11.6 presents the 1997 inventory of layers by state.

11.3.2 LAYER PRODUCTION CYCLES

Laying hens reach sexual maturity and begin laying eggs between 16 and 20 weeks of age, depending on breed. Before the onset of egg production, these birds are referred to as *pullets*. Pullets that are about to start egg production are known as *starter pullets*. Some table egg producers raise their own starter pullets; others purchase birds from starter pullet operations. Starter pullet operations may raise birds in facilities like those used for broiler production or in cages like those used for egg-producing hens.

TABLE 11.6
Layer Inventory by State

State	Inventory (number of layers 20 weeks and older)		State	Inventory (number of layers 20 weeks and older)	
	Hens > 1 year	Pullets		Hens > 1 year	Pullets
Alabama	4,292	6,200	Nebraska	6,011	3,979
Alaska	0	0	Nevada	0	0
Arizona	0	0	New Hampshire	53	106
Arkansas	6,070	8,351	New Jersey	931	1,023
California	15,270	11,010	New Mexico	636	536
Colorado	1,910	1,760	New York	1,070	2,400
Connecticut	1,141	2,318	North Carolina	4,307	7,306
Delaware	150	250	North Dakota	100	140
Florida	6,216	4,522	Ohio	10,863	16,195
Georgia	6,680	13,840	Oklahoma	1,909	2,166
Hawaii	465	263	Oregon	1,800	1,200
Idaho	546	385	Pennsylvania	9,400	13,605
Illinois	1,534	1,929	Rhode Island	10	61
Indiana	10,238	12,076	South Carolina	2,205	2,424
Iowa	11,655	10,130	South Dakota	800	1,370
Kansas	505	843	Tennessee	316	922
Kentucky	1,450	1,650	Texas	5,630	11,545
Louisiana	940	963	Utah	939	759
Maine	2,256	2,523	Vermont	9	186
Maryland	1,518	1,644	Virginia	704	2,759
Massachusetts	72	473	Washington	2,815	2,156
Michigan	2,343	2,817	West Virginia	285	760
Minnesota	6,740	5,215	Wisconsin	1,994	1,989
Mississippi	2,487	4,424	Wyoming	8	4
Missouri	3,490	3,605	Other	168	132
Montana	35	255	**United States**	**140,966**	**171,171**

Source: USDA (1998c).

- *Important point:* Slightly more than 10% of all layer farms have pullet-raising facilities on the farm. Pullet houses are similar in construction to broiler houses.

Usually, laying hens are replaced after about 12 months of egg production, when the natural decreasing rate of egg production becomes inadequate to cover feed costs. At this point, laying hens become spent hens and may be slaughtered or rendered to recover any remaining value. Although a second egg production cycle can be obtained from a flock of laying hens following a resting period, this practice is rarely used.

- *Important point:* Spent hens may be slaughtered for meat for human or pet foods or disposed of by rendering. More than three-fourths of layer farms molt their birds, followed by a second period of egg production. Routine molting by withholding or restricting feed is the most common method. Placement and removal of birds are on an "all in–all out" basis. Typically, laying hens are also fed corn- and soybean-based diets, which may also include various cereal grains such as wheat and barley and a variety of other ingredients.

11.3.3 LAYER CONFINEMENT PRACTICES

Laying hens maintained for table egg production are almost exclusively confined in cages, which allow automation of feed distribution and egg collection. Most confinement facilities for laying hens are mechanically ventilated to remove moisture and carbon dioxide produced by respiration. Exhaust fans draw air into the building through slots located along the perimeter of the roof under the eaves. Several types of cage systems are common, including full and modified stair-step systems. With modified stair-step cage systems, upper cages are partially offset with a baffle diverting manure from upper cages away from lower cages. Cage systems are also available that stack cages without any offset to maximize the number of birds per unit floor area.

Both one- and two-story buildings are used to house laying hens. Two-story buildings are known as high-rise houses. In a high-rise house, full or modified stair-step cage systems are located in the upper story with manure collected and stored in the lower story of the building. Ventilation fans are located in a sidewall of the manure collection and storage area with airflow passing down through the cages and over the accumulated manure to remove moisture evaporating from the manure. With proper design and management, including prevention of watering system leakage, laying hen manure moisture content can be reduced from 75% to as low as 25% to 30%.

In single-story buildings, full or modified stair systems are located over shallow manure collection pits that may be cleaned either by scraping or flushing. With stacked cage systems, a belt system under the cages collects and removes manure.

When high-rise houses are properly designed and operated, emissions of PM are higher than from single story houses, because of manure drying. Emissions of ammonia are also higher, because of an increased rate of volatilization as moisture evaporates. However, emissions of hydrogen sulfide, VOCs, and methane are lower, because of the predominately aerobic microbial environment created by drying. Emissions from scraped and flushed manure collection pits are similar to deep pit and flush systems for swine with emission factors depending on frequency of scraping or flushing. The frequency of operation of the belt system also affects emission factors for ammonia, hydrogen sulfide, VOCs, and methane. Because laying hen manure in single-story houses is handled without any drying, manure particles are an insignificant component of PM emissions.

- *Important point:* Unlike broilers, laying hens are confined in cages and no litter or bedding material is used.

11.3.4 LAYER MANURE MANAGEMENT

Manure produced by laying hens is handled both as a liquid or slurry and as a solid, with handling as a liquid or slurry being more common in older production facilities. As older production facilities are replaced, however, handling laying hen manure as a liquid or slurry is becoming less common every year. When laying hen manure is handled as a liquid or slurry, flushing or scraping is used to remove laying hen manure from the production facility. With scraping systems, a tank or an earthen structure is usually used for storage if the manure is not applied directly to cropland, while flush systems use anaerobic lagoons for stabilization and storage. Typically, these lagoons are the source of the water used for flushing, although fresh water may be used in rare instances.

Accumulated manure in high-rise houses normally is removed annually during the period between flocks of birds when the house is cleaned and disinfected in preparation for new birds. However, manure can be stored for 2 or possibly 3 years. Manure removed from high-rise-type houses is directly applied to cropland for disposal.

Scraped pits typically are cleaned at least weekly, with the manure either directly applied to cropland or stored in a tank or an earthen pond. With belt systems, manure may be removed as frequently as daily and applied directly to cropland or stored for application later. However, removal may be less frequent if partial drying is desired.

TABLE 11.7
Primary Manure Handling Method by Region (Percent of Farms)

Primary manure handling method	Great Lakes	Southeast	Central	West	All farms
High rise	63.0	31.4	48.1	7.8	39.7
Deep pit below ground	0.0	0.0	6.4	7.3	2.9
Shallow pit (pit at ground level with raised cages)	23.4	19.9	1.6	24.1	18.9
Flush system to anaerobic lagoon	0.0	41.0	0.0	12.0	12.5
Belt system	13.6	4.3	20.2	5.2	10.6
Scrape system	0.0	2.5	23.7	43.6	15.4
Total	100	100	100	100	100

Source: USDA (2000).
Note: Great Lakes: IN, OH, PA; Southeast: AL, FL, GA, NC; Central: AR, IO, MN, MO, NE; West: CA, TX, WA.

As with flush systems for swine and dairy cattle manure, anaerobic lagoons are used for the stabilization of flushed laying hen manure, and supernatant from the lagoon serves as the source of flush water. Both single-cell and two-cell lagoons are used.

As shown in Table 11.7, significant differences occur regionally in methods of handling laying hen manure. Nationally, the high-rise house is the most commonly used method of handling laying hen manure. The use of flush systems with anaerobic lagoons is limited to the Southeast and West.

- *Important point:* The majority of eggs marketed commercially in the United States are washed using automatic washers. Cleaning compounds, such as sodium carbonate, sodium metasilicate, or trisodium phosphate, together with small amounts of other additives, are commonly used in these systems. Wash water is contaminated with shell, egg solids, dirt, manure, and bacteria washed from the egg surface into the recycled water. Eggs may be washed either on-farm or off-farm. Over three-fourths of layer farms process eggs off-farm, and one-third of the largest farms are likely to wash eggs off-farm. Operations that wash their eggs on-farm may do so in-line or off-line. Larger operations commonly collect and store egg wash water on site in large tanks or lagoons for treatment and storage.

11.3.5 MORTALITY MANAGEMENT

About 1% of the started pullets housed are expected to die each month through the laying cycle. To prevent the possible spread of disease, dead birds must be removed from cages daily, if not more frequently. As mentioned earlier, disposal of dead birds is the responsibility of the grower. Several options are available for dead bird disposal. Of these options, composting is one of the more desirable approaches.

Catastrophic losses of laying hens also occur. Loss of power and mechanical ventilation during periods of extremely hot weather is the most common cause of loss. Weather events such as hurricanes and tornadoes can also cause catastrophic losses. Several options are available for the disposal of catastrophic losses, with burial being the most common. (Note that burial is prohibited or highly regulated in some states.) Large-scale composting is another, and probably more desirable, option from a water quality perspective.

- *Important point:* Composting of dead birds is showing real promise as an on-farm disposal method. Composting stabilizes nutrients, eliminates odors, and disinfects the material. As long as the composter is sheltered from rainfall, there are no environmental hazards from composting. Several different versions of composters are available.

TABLE 11.8

Turkey Operations and Production in the United States*

Year	Operations	Production
1982	7,498	172,035,000
1987	7,347	243,336,000
1992	6,257	279,230,000
1997	6,031	307,587,000

Source: USDA (1998c).

* Total operations that sold turkeys for slaughter.

11.4 TURKEYS

Turkey production is very similar to broiler production. The principal differences between turkey and broiler production are the size of bird produced and the length of the grow-out cycle. The grow-out period for female or hen turkeys is usually about 14 to 16 weeks, resulting in a live weight at slaughter of between 13 and 20 lb. However, the usual grow-out period for toms or male turkeys is longer, ranging from 17 to 21 weeks, resulting in a live weight at slaughter of between 30 and 37 lb. Typically, two flocks of turkeys are produced annually because of the longer grow-out cycle and the somewhat seasonal demand for turkey. Turkeys are primarily fed corn- and soybean-based diets, which may also include various cereal grains and a variety of other ingredients.

11.4.1 SIZE AND LOCATION OF TURKEY INDUSTRY

In 1997, 6,031 turkey operations sold 307 million turkeys for wholesale distribution. In total, the USDA reports more than 12,000 operations, including breeding operations, poult-raising operations, small retail operations, and facilities that specialize in a first stage of growing. Turkey production has increased steadily over the past two decades, with a shift in production to fewer but larger operations. Table 11.8 illustrates how the number of turkey operations dropped while production nearly doubled from 1982 to 1997. Between 1982 and 1992, almost 21% of the turkey operations went out of business, while production rose by almost 80% (USDA, 1998b).

Table 11.9 shows the size distribution of turkey operations based on sales in 1997. Although most turkey operations are relatively small, most of the production comes from larger operations. These larger operations can have an average confinement capacity of more than 130,000 birds. In 1997, the 369 largest operations (2.7% by number) confined 43.6% of the turkey population (USDA, NASS, 1997).

State-level data from the 1997 Census of Agriculture (USDA, 1999) indicate that the north-central and southeast areas of the United States account for approximately half of all turkey farms. The key production states (determined by number of turkeys produced) are North Carolina, Minnesota, Virginia, Arkansas, California, and Missouri. Other states with significant production include Indiana, South Carolina, Texas, Pennsylvania, and Iowa. Table 11.10 shows the turkey populations by state in 1997.

11.4.2 TURKEY PRODUCTION CYCLE

The growth of a turkey is commonly divided into two phases: brooding and grow-out. The brooding phase of a poult (young turkey) is from 1 day old to about 6 to 8 weeks. During this time, poults need supplemental heat. Brooder heaters are used to keep the ambient temperature at 90 to 95°F when the poults arrive. Thereafter, the producer decreases the temperature by 5°F for the next 3 weeks until

TABLE 11.9
Number of Turkey Operations by Size in 1997

State	Annual sales (birds)		
	< 30,000 birds	30,000–60,000 birds	> 60,000 birds
Alabama	26	0	0
Alaska	4	0	0
Arizona	13	0	0
Arkansas	69	66	154
California	108	8	95
Colorado	78	0	1
Connecticut	35	0	0
Delaware	6	0	0
Florida	52	0	0
Georgia	22	3	0
Hawaii	0	0	0
Idaho	40	0	0
Illinois	80	11	18
Indiana	119	60	80
Iowa	142	20	44
Kansas	46	3	18
Kentucky	31	0	0
Louisiana	13	0	0
Maine	99	0	0
Maryland	42	2	5
Massachusetts	70	0	0
Michigan	206	5	30
Minnesota	157	42	160
Mississippi	11	0	0
Missouri	122	135	145
Montana	46	0	0
Nebraska	47	1	13
Nevada	11	0	0
New Hampshire	55	0	0
New Jersey	58	1	0
New Mexico	20	0	0
New York	146	0	1
North Carolina	150	268	355
North Dakota	16	2	11
Ohio	187	56	38
Oklahoma	41	11	13
Oregon	97	0	0
Pennsylvania	177	63	64
Rhode Island	11	0	0
South Carolina	28	51	89
South Dakota	21	2	28
Tennessee	41	0	0
Texas	153	8	54
Utah	41	31	25
Vermont	77	0	0

TABLE 11.9 (continued)
Number of Turkey Operations by Size in 1997

State	Annual sales (birds)			
	< 30,000 birds	30,000–60,000 birds	> 60,000 birds	
Virginia	104	100	185	
Washington	62	1	0	
West Virginia	31	13	36	
Wisconsin	160	19	9	
Wyoming	12	0	0	
Total	3,378	982	1,671	6,031

Source: USDA (1999a).

TABLE 11.10
Turkey Inventory by State

State	Inventory (1,000 birds)	State	Inventory (1,000 birds)
Alabama	—	Nebraska	—
Alaska	—	Nevada	—
Arizona	—	New Hampshire	5
Arkansas	10,465	New Jersey	26
California	7,326	New Mexico	—
Colorado	1,360	New York	478
Connecticut	2	North Carolina	18,663
Delaware	—	North Dakota	907
Florida	—	Ohio	2,337
Georgia	61	Oklahoma	—
Hawaii	—	Oregon	—
Idaho	—	Pennsylvania	4,047
Illinois	1,221	Rhode Island	—
Indiana	5,068	South Carolina	3,907
Iowa	2,442	South Dakota	1,256
Kansas	663	Tennessee	—
Kentucky	—	Texas	—
Louisiana	—	Utah	—
Maine	—	Vermont	14
Maryland	258	Virginia	9,070
Massachusetts	29	Washington	—
Michigan	—	West Virginia	1,570
Minnesota	15,872	Wisconsin	—
Mississippi	—	Wyoming	—
Missouri	7,326	Other	11,027
Montana	—	**United States**	**105,088**

Source: USDA (1998b).

the temperature reaches 75°F. Brooding can occur either in a partitioned area of the house called the *brooding chamber* or in an entirely separate house. Separate poult housing is more prevalent in larger operations for purposes of disease control.

The grow-out phase starts after the brooding phase. Depending on the sex of the birds, the grow-out phase typically lasts up to 21 weeks, resulting in a live slaughter weight of between 30 to 37 lb. At the end of the production cycle, the house is completely cleaned out.

Typically, two flocks of turkeys are produced annually because of the longer grow-out cycle and the somewhat seasonal demand for turkey. As the demand for turkey has increased and become somewhat less seasonal, a third flock may be started with grow-out completed in the following year. Turkeys are fed primarily corn- and soybean-based diets, which also may include various cereal grains and a variety of other ingredients.

11.4.3 Turkey Confinement Practices

Confinement facilities for turkeys are similar to those used for broilers, typically being 40 ft wide but usually only 300 to 400 ft in length. They also may be totally enclosed or partially enclosed with partially open, screened sidewalls that can be closed using curtains. Size of sidewall opening depends on climate and may be as much as 4 to 5 ft high in warm climates. Partially enclosed facilities are more common in warmer climates such as the South and Southeast; totally enclosed facilities are more common in the North. As with broilers and laying hens, totally enclosed facilities generally have automatic delivery and mechanical ventilation. Negative pressure ventilation is the principal method of ventilation use.

Like broiler production, essentially all turkey production occurs in partially or totally enclosed facilities that are divided into two or three chambers. Initially, only one chamber, also known as the *brood chamber,* is used; this is the area where the newly hatched turkeys, known as *poults,* are placed. Like broiler chicks, poults are unable to maintain a constant body temperature until about 6 to 8 weeks of age and thus require supplemental heat. Brood chambers for turkeys, therefore, are also heated at the beginning of the grow-out cycle. As with broiler chickens, the second or the second and third chambers are opened to provide more floor space per bird as the birds grow. In cold weather, some heat may be provided throughout the grow-out cycle.

- *Important point:* Separate poult housing is more prevalent in larger operations for purposes of disease control.

Some turkey producers use separate brood and growing houses (brooding and grow-out) and move the birds from the brooding house to the growing house after about 6 to 8 weeks. Another production practice is to use the brood chamber in a house exclusively for brooding and use the remainder of the house for grow-out after the birds reach the age of 6 to 8 weeks. These management systems are known as *two-age management systems.* Such systems produce more flocks each year than single-age farms.

11.4.4 Turkey Manure Management

As with broiler chickens, turkeys are raised unconfined in the production facility on litter, typically sawdust or wood shavings. Total clean-out of brood chambers and brood houses after each flock is common, as is total clean-out of growing chambers or houses annually. Crust removal between flocks followed by top dressing with new litter also occurs in the production of turkeys.

- *Important point:* In the turkey sector, the use of litter sheds to store crust and total clean-outs from brood chambers or brood houses is also emerging. When land is not available for disposal, storage of these materials in uncovered piles is common.

11.4.5 Mortality Management

Typically, about 5% to 6% of hens and 9% to 12% of toms die during the grow-out cycle, with the highest rate of loss occurring during the initial weeks. As with broilers and layer hens, dead birds must be removed daily, if not more frequently, with dead bird disposal being the responsibility of the grower. Again, several options for dead bird disposal are available; composting is one of the more desirable approaches from a water quality perspective.

Catastrophic losses of turkeys occur during periods of extremely hot weather, but they may also be caused by weather events such as hurricanes, tornadoes, and snow or ice storms. Older turkeys, like older broilers, are more susceptible to catastrophic losses during periods of extremely hot weather. Several options are available for disposal of catastrophic losses, with burial being the most common practice.

11.5 POULTRY VIRTUAL MODEL FARMS

Four basic virtual model farms were identified for poultry based on current practices: broiler house, caged layer high-rise house, caged layer flush house, and turkey house (see Figure 11.1). Broiler houses and turkey houses are similar; therefore, the virtual model farms for broilers (C1, C2) and turkeys (T1, T2) follow the same confinement, storage and stabilization, and land application phases. In the broiler and turkey virtual model farms, operators either store litter or directly apply it to land. The caged layer house differs because the manure is not mixed with bedding, and in some caged layer houses, manure is removed by flushing to an anaerobic lagoon. None of the virtual model farms has solids separation activities.

11.5.1 Confinement

Virtual model farms C1 and T1 represent broiler chickens and turkeys kept in enclosed housing with bedding derived from wood shavings, rice hulls, chopped straw, peanut hulls, or other materials. The litter (bedding and manure) is removed using a front-end loader every 1 to 3 years. Cake is removed using specially designed equipment after each flock is cycled.

Virtual model farm C2 reflects caged layers kept in a high-rise house without bedding. Virtual model farm C3 represents a caged layer flush house. In this virtual model, cages are suspended over shallow pits with water used to flush manure to storage and stabilization systems.

The confinement facility is a source of PM (from the litter, feather particles, and feed), ammonia, and hydrogen sulfide. For this analysis, it was assumed that emissions during solids transport (i.e., front-end loader) would have negligible air impacts because of the short duration that the manure would spend in transport.

11.5.2 Storage and Stabilization

The dry manure from broiler and turkey houses is either stored or directly applied to land. In all cases, the virtual models assume that cake is stored separately in a covered shed. Manure from total clean-out of barns can either be stored in an open storage pile and then applied to land (C1A and T1A) or directly applied to land (C1B and T1B).

The caged layer high-rise house (C2) does not have a separate manure storage facility. Manure is sent directly from the confinement facility to be land applied.

Two types of storage and stabilization processes were considered for caged layer flush houses (C3): (1) an anaerobic lagoon (also referred to as a *combined lagoon and storage pond* or *one-cell lagoon*) or (2) a separate storage pond following a stabilization lagoon (two-cell lagoon). Review of industry practices indicated that the anaerobic lagoon was the most commonly used method. Additionally, a review of emission mechanisms and existing emission data indicated that total emissions would not be substantially different between the one-cell and two-cell systems. Therefore,

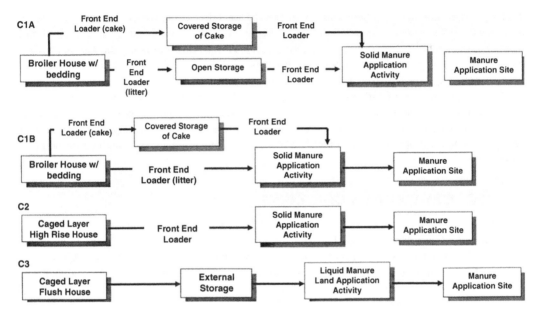

FIGURE 11.1A Broilers. (Source: USEPA, 2001a.)

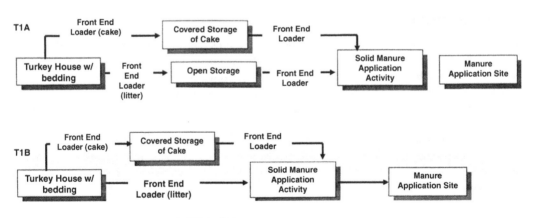

FIGURE 11.1B Turkeys. (Source: USEPA, 2001a.)

the virtual model farms only include an anaerobic lagoon. The supernatant (clear liquid overlying material deposited by settling) from the lagoon is used as flush water.

11.5.3 LAND APPLICATION

Land application includes the manure application activity and the manure application site (i.e., cropland or other agricultural land). In model farms C1, C2, and T1, the dry manure is assumed to be land applied to the manure application site using a solid manure spreader. Three types of land application activities were considered for liquid manure in developing the model farms: land application by (1) liquid surface spreader, (2) liquid injection manure spreader, or (3) irrigation.

CHAPTER REVIEW QUESTIONS

1. Describe and discuss typical broiler operation types, production cycles, confinement practices, and manure management. Include discussions of manure collection and storage and mortality management.
2. Describe and discuss typical egg-laying hens operation types, production cycles, confinement practices, and manure management. Include discussions of manure collection and storage and mortality management.
3. Describe and discuss typical chicken and turkey industry operations, including production cycle, confinement practices, manure management, and mortality management.
4. What components are included for the four virtual poultry operations? Why?
5. Describe and discuss the typical poultry operations.

THOUGHT-PROVOKING QUESTIONS

1. Research, describe, and discuss recent poultry industry trends.
2. Research and discuss vertical integration in the poultry industry.

REFERENCES

Abt. 1998. *Preliminary Study of the Livestock and Poultry Industry: Appendices.* Abt Associates, Inc., Cambridge, MA.

NCC. 1999. National Chicken Council comments from John Thorne to Janet Goodwin and Paul Shriner, U.S. Environmental Protection Agency (USEPA). September 8, 1999.

Oklahoma State University. 2005. Farm & Ranch *A* Syst: Fact Sheet 8. Oklahoma Cooperative Extension Service, Oklahoma State University.

USDA. 1998a. Agricultural statistics. U.S. Department of Agriculture, National Agricultural Statistics Service. Washington, DC.

USDA. 1998b. Poultry production and value, final estimates 1994–97. Statistical Bulletin 958. U.S. Department of Agriculture, National Agricultural Statistics Service. Washington, DC.

USDA. 1998c. Chickens and eggs, final estimates 1994–97. Statistical Bulletin 944. U.S. Department of Agriculture (USDA), National Agricultural Statistics Service. Washington, DC.

USDA. 1999a. 1997 census of agriculture. U.S. Department of Agriculture, National Agricultural Statistics Service. Washington, DC.

USDA. 2000. Part II: Reference of 1999 table egg layer management in the United States (Layer '99). U.S. Department of Agriculture, Animal and Plant Health Inspection Service (APHIS). Fort Collins, CO.

USEPA. 2001a. Emissions from animal feeding operations. United States Environmental Protection Agency, Office of Air Quality Planning and Standards. Research Triangle Park, NC.

USEPA. 2000lb. Development document for the proposed revisions to the national pollutant discharge elimination system regulation and the effluent guidelines for concentrated animal feeding operations. EPA-821-R-01-003. U.S. Environmental Protection Agency, Office of Water, Washington, DC.

12 CAFO Component Design*

And if you think the lagoons and exhaust fans make a stink, wait until you get a whiff of a field fresh spread with hog poop. The ammonia will burn your eyes out of your head. Your hair will fall out. They could make the stink better by coverin over the waste ponds or aeratin, but that costs money. Cheaper just to let it sit there. And the state don't care.

(Proulx, 2002, p. 113)

12.1 INTRODUCTION (USDA, 1996)

A CAFO can function for any one or all of the following purposes: production, collection, storage, treatment, transfer, and ultilization. *Production* is the function of the amount and nature of agricultural waste generated by an agricultural enterprise. *Collection* refers to the initial capture and gathering of waste from the point of origin or deposition to a collection point. *Storage* is the temporary containment of the waste. *Treatment* is any function designed to reduce the pollution potential of waste, including physical, biological, and chemical treatment. It includes activities that are sometimes considered pretreatment, such as the separation of solids. *Transfer* refers to the movement and transportation of waste throughout the system and includes the transfer of waste from the collection point to the storage facility, to the treatment facility, and to the utilization site. *Utilization* includes recycling reusable waste products and reintroducing nonreusable waste products into the environment. These functions are carried out by planning, applying, and operating individual components.

A *component* can be a piece of equipment, such as a pump; a structure, such as a waste storage tank; or an operation, such as composting. The combination of the components should allow the flexibility needed to efficiently handle all forms of waste generated for a given enterprise. In addition, the components must be compatible and integrated with each other. All components should be designed to be simple, manageable, and durable, and they should require low maintenance. In this chapter, we discuss these components under section headings that describe the function that they are to accomplish.

12.2 PRODUCTION

The production function includes components that affect the volume and consistency of the agricultural waste produced. Roof gutters and downspouts and diversion to exclude clean water from areas of waste are examples of components that reduce the volume of waste material that needs management. Fences and walls that facilitate collection of waste confine the cattle, thus increasing the volume of waste collected.

12.2.1 ROOF RUNOFF MANAGEMENT

Roof runoff should be diverted from feedlots and manure storage areas unless it is needed for some use, such as dilution water for waste storage ponds or treatment lagoons. As illustrated in

---

* This chapter is a modified version of NRCS/USDA (1992), *National Engineering Handbook, Part 651. Agricultural Waste Management Field Handbook,* chap. 10.

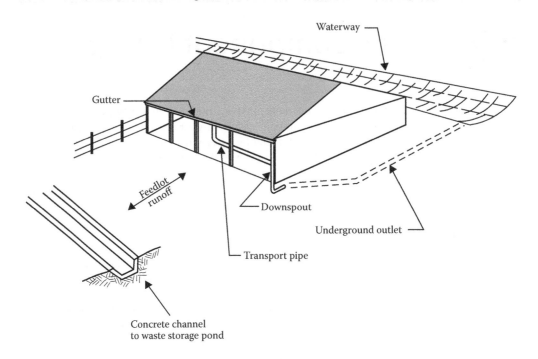

FIGURE 12.1   Roof gutter and downspout. (Source: NRCS/USDA, 1992, p. 10-1.)

Figure 12.1, roof gutters and downspouts with underground connections to open channel outlets can accomplish this. Gutters and downspouts may not be needed if the roof drainage does not come into contact with areas accessible to livestock.

The area of a roof that can be served by a gutter and downspout system is controlled by either the flow capacity of the gutter (channel flow) or by the capacity of the downspout (orifice flow). The gutter's capacity may be computed using Manning's equation. Design of a gutter and downspout system is based on the runoff from a 10-year frequency, 5-minute rainfall, except that a 25-year frequency, 5-minute rainfall is used for exclusion of roof runoff from waste treatment lagoons, waste storage ponds, or similar practices.

- *Important point:* The common design equation for fluid flow in open channels (of any configuration) is Manning's formula.

A procedure for the design of roof gutters and downspouts follows:

Step 1—Compute the capacity of the selected gutter size. This may be computed using Manning's equation. Using the recommended gutter gradient of $1/16$ in./ft and a Manning's roughness coefficient of 0.012, this equation can be expressed as:

$$q_g = 0.01184 \times A_g \times r^{0.67}$$ (12.1)

where:
q_g = Capacity of gutter, ft³/sec.
A_g = Cross sectional area of gutter, in.².
r = A_g/wp, inches.
wp = Wetted perimeter of gutter, inches.

Step 2—Compute capacity of downspout. Using an orifice discharge coefficient of 0.65, the orifice equation may be expressed as:

$$q_d = 0.010457 \times A_d \times h^{0.5} \tag{12.2}$$

where:
q_d = Capacity of downspout, ft³/sec.
A_d = Cross sectional area of downspout, in.².
h = Head, inches (generally the depth of the gutter minus 0.5 in.).

Step 3—Determine whether the system is controlled by the gutter capacity or downspout capacity and adjust number of downspouts if desired.

$$Nd = qg/qd \tag{12.3}$$

where:
N_d = Number of downspouts.

If N_d is less than 1, the system is gutter-capacity controlled. If it is equal to or greater than 1, the system is downspout capacity controlled unless the number of downspouts is equal to or exceeds N_d.

Step 4—Determine the roof area that can be served, based on the following equation:

$$Ar = q \times 3,600/P \tag{12.4}$$

where:
A_r = Area of roof served, ft².
q = Capacity of system, either q_g or q_d whichever is smallest, ft³/sec.
P = 5-minute precipitation for appropriate storm event, inches.

- *Important point:* The above procedure is a trial and error process. Different sizes of gutters and downspouts should be evaluated, as well as multiple downspouts, to determine the best gutter and downspout system to serve the roof area involved.

12.2.2 RUNOFF CONTROL

All livestock facilities in which the animals are housed in open lots or the manure is stored in the open must deal with runoff. As illustrated in Figure 12.2, "clean" runoff from land surrounding livestock facilities should be diverted from barns, open animal concentration areas, and waste storage or treatment facilities.

- *Important point:* Diversions are to be designed according to National Resource Conservation Service (NRCS) Conservation Practice Standard, Diversion, Code 362 (USDA 1985). Diversion channels must be maintained to remain effective. If vegetation is allowed to grow tall, the roughness increases and the channel velocity decreases, causing possible channel overflow. Therefore, vegetation should be periodically mowed. Earth removed by erosion from earthen channels should be replaced. Unvegetated, earthen channels should not be used in regions of high precipitation because of potential erosion.

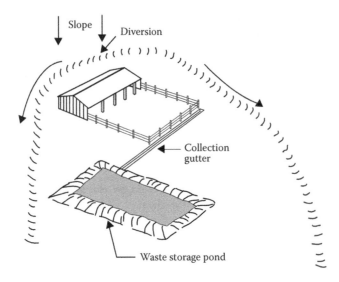

FIGURE 12.2   Diversion of "clean" water around feedlot.

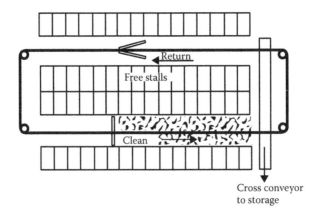

FIGURE 12.3   Scrape alley used in dairy barns. (Source: NRCS/USDA, 1992, p. 10-4.)

12.3   COLLECTION

Livestock and poultry manure collection often depends on the degree of freedom that is allowed the animal. If animals are allowed freedom of movement within a given space, the manure produced will be deposited randomly. Components that provide efficient collection of animal waste include paved alleys, gutters, and slatted floors with associated mechanical and hydraulic equipment as described in the sections below.

12.3.1   ALLEYS

Alleys are paved areas where animals walk. They generally are arranged in straight lines between animal feeding and bedding areas. On slatted floors, animal hoofs work the manure through the slats into the alleys below, and the manure is collected by flushing or scraping the alleys.

12.3.1.1   Scrape Alleys and Open Areas

Figure 12.3 illustrates a scrape alley used in dairy barns. Two kinds of manure scrapers are used to clean alleys: mechanical and tractor scrapers. A mechanical scraper is dedicated to a given alley and propelled using electrical drives attached by cables or chains. The drive units are often used to power two mechanical scrapers that are traveling in opposite directions in parallel alleys in an oscillating manner. Some mechanical scrapers are in alleys under slatted floors.

A tractor scraper can be used in irregularly shaped alleys and open areas where mechanical scrapers cannot function properly. It can be a blade attached to either the front or rear of a tractor or a skid-steer tractor with a front mounted bucket.

The width of the alleys depends on the desires of the producer and the width of available equipment. Scrape alley widths typically vary from 8 to 14 ft for dairy and beef cattle and from 3 to 8 ft for swine and poultry.

12.3.1.2 Flush Alley

Flushing is also used to clean alleys. Grade is critical and can vary between 1.25% and 5%. It may change for long flush alleys. The alley should be level perpendicular to the centerline. The amount of water used for flushing is also critical. An initial flow depth of 3 in for underslat gutters and 4 to 6 in for open alleys is necessary.

The length and width of the flush alley are also factors. Most flush alleys should be less than 200 ft long. The width generally varies from 3 to 10 ft, depending on animal type. For underslat gutters and alleys, channel width should not exceed 4 ft. The width of open flush alleys for cattle is frequently 8 to 10 ft.

- *Important point:* Flush alleys and gutters should be cleaned at least twice per day. For pump flushing, each flushing event should have a minimum duration of 3 to 5 minutes.

Table 12.1 and Table 12.2 indicate general recommendations for the amount of flush volume. Table 12.3 gives the minimum slope required for flush alleys and gutters. Figure 12.4 and Figure 12.5 illustrate flush alleys.

Several mechanisms are used for flushing alleys. The most common rapidly empties large tanks of water or uses high-volume pumps. Several kinds of flush tanks are used (Figure 12.6). One known as a *tipping tank* pivots on a shaft as the water level increases. At a certain design volume, the tank tips, emptying the entire amount in a few seconds, which causes a wave that runs the length of the alley.

Some flush tanks have manually opened gates. These tanks are emptied by opening a valve, a standpipe, a pipe plug, or a flush gate. Float switches can be used to control flushing devices.

TABLE 12.1
Recommended Total Daily Flush Volumes

Animal type	gal/head
Swine	
Sow and litter	35
Prenursery pig	2
Nursery pig	4
Growing pig	10
Finishing pig	15
Gestating sow	25
Dairy cow	100
Beef feeder	100

Source: MWPS (1985).

TABLE 12.2
Flush Tank Volumes and Discharge Rates

Initial flow depth, in	Tank volume, gal/ft of gutter width	Tank discharge rate, gpm/ft of gutter width	Pump discharge gpm/ft of gutter width
1.5	30	112	55
2.0	40	150	75
2.5	45	195	95
3.0	55	255	110
4.0	75	615	150
5.0	100	985	175
6.0	120	1,440	200

Source: MWPS (1985).

TABLE 12.3
Minimum Slope for Flush Alleys

	Underslat alley	Open alley narrow width (< 4 ft)			Open alley wide width (> 4 ft)		
Initial flow depth, in	3.0	1.5	2.0	2.5	4.0	5.0	6.0
Slope, %	1.25	2.0	1.5	1.25	5.0	4.0	3.0

Source: MWPS (1985).

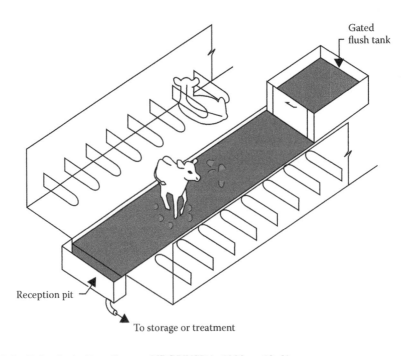

FIGURE 12.4 Dairy flush alley. (Source: NRCS/USDA, 1992, p. 10-6.)

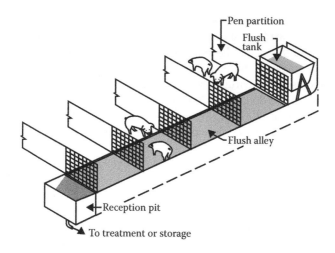

FIGURE 12.5   Swine flush alley. (Source: NRCS/USDA, 1992, p. 10-6.)

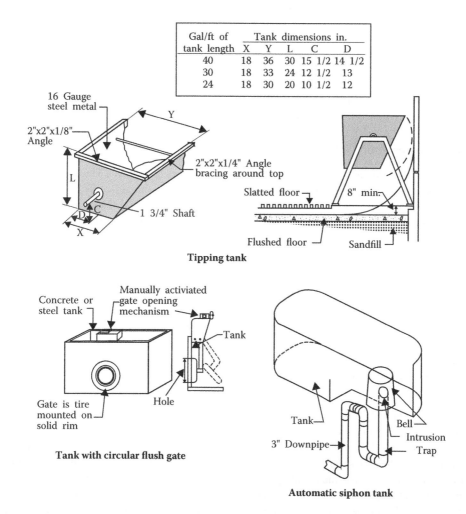

Gal/ft of	Tank dimensions in.				
tank length	X	Y	L	C	D
40	18	36	30	15 1/2	14 1/2
30	18	33	24	12 1/2	13
24	18	30	20	10 1/2	12

FIGURE 12.6   Flush tanks. (Source: NRCS/USDA, 1992, p. 10-7.)

Another kind of flush tank uses the principle of a siphon. In this tank, the water level increases to a given point where the head pressure of the liquid overcomes the pressure of the air trapped in the siphon mechanism. At this point, the tank rapidly empties, causing the desired flushing effect.

Most flush systems use pumps to recharge the flush tanks or to supply the necessary flow if the pump flush technique is used. Centrifugal pumps typically are used. The pumps should be designed for the work that they will be doing. Low volume pumps (10 to 150 gpm) may be used for flush tanks, but high volume pumps (200 to 1,000 gpm) are needed for alley flushing. Pumps should be the proper size to produce the desired flow rate. Flush systems may rely on recycled lagoon water for the flushing liquid.

In some parts of the country where wastewater is recycled from lagoons for flush water, salt crystals (struvite) may form inside pipes and pumps and cause decreased flow. Use of plastic pipes and fittings and pumps that have plastic impellers can reduce the frequency between cleaning or replacing pipes and pumps. If struvite formation is anticipated, recycle systems should be designed for periodic cleanout of pumps and pipes. A mild acid, such as dilute hydrochloric acid (1 part 20 mole hydrochloric acid to 12 parts water), can be used. A separate pipe may be needed to accomplish acid recycling. The acid solution should be circulated throughout the pumping system until normal flow rates are restored. The acid solution should then be removed.

Caution! Spent acid solution must be disposed in accordance with applicable environmental and safety regulations.

12.3.2 GUTTERS

Gutters are narrow trenches used to collect animal waste. They are often employed in confined stall or stanchion dairy barns and in some swine facilities.

12.3.2.1 Gravity Drain Gutters

Deep, narrow gutters can be used in swine finishing buildings (Figure 12.7). These gutters are at the lowest elevation of the pen. The animal traffic moves the waste to the gutter. The gutter fills and is periodically emptied. Gutters with Y, U, V, or rectangular cross-sectional shapes are used in farrowing and nursery swine facilities. These gutters can be gravity drained periodically.

12.3.2.2 Step-Dam Gutters

Step-dam gutters (i.e., gravity gutters or gravity flow channels) provide a simple alternative for collecting dairy manure (Figure 12.8). A 6-in. high dam holds back a lubricating layer of manure in a level, flat-bottomed channel. Manure drops through a floor grate or slats and flows down the gutter under its own weight. The gutter is about 30 in wide and steps down to a deeper cross channel below the dam.

12.3.2.3 Scrape Gutters

Scrape gutters are frequently used in confined stall dairy barns. The gutters are 16 to 24 in. wide, 12 to 16 in. deep, and generally do not have any bottom slope. They are cleaned using either shuttle-stroke or chain and flight gutter cleaners (Figure 12.9 and Figure 12.10). Electric motor drive shuttle stroke gutter cleaners have paddles that pivot on a drive rod. The drive rod travels alternately forward for a short distance and then backward for the same distance. The paddles are designed to move manure forward on the forward stroke and to collapse on the drive rod on the return stroke. This action forces the manure down the gutter. Shuttle stroke gutters can only be used on straight gutters.

Chain and flight scrapers are powered by electric motors and are used in continuous loops to service one or more rows of stalls.

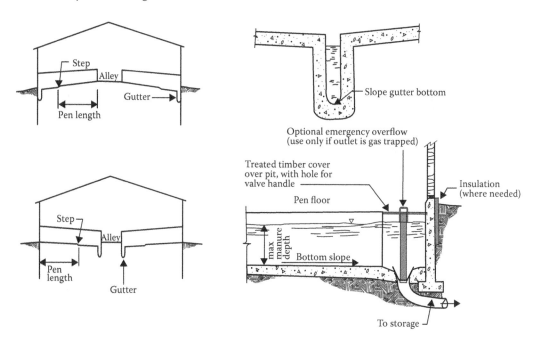

FIGURE 12.7   Flush and gravity flow gutters for swine manure. (Source: NRCS/USDA, 1992, p. 10-8.)

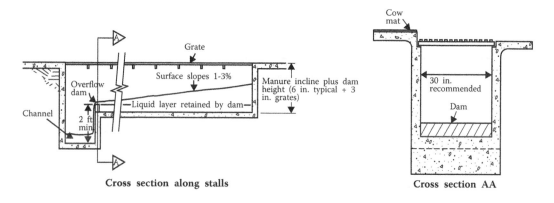

FIGURE 12.8   Gravity gutter for dairy manure. (Source: NRCS/USDA, 1992, p. 10-9.)

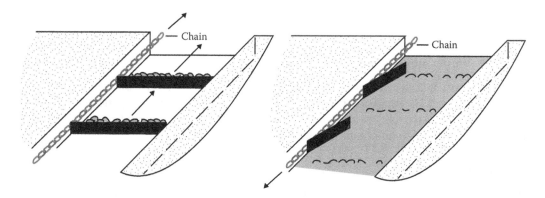

FIGURE 12.9   Shuttle-stroke gutter cleaner. (Source: NRCS/USDA, 1992, p. 10-9.)

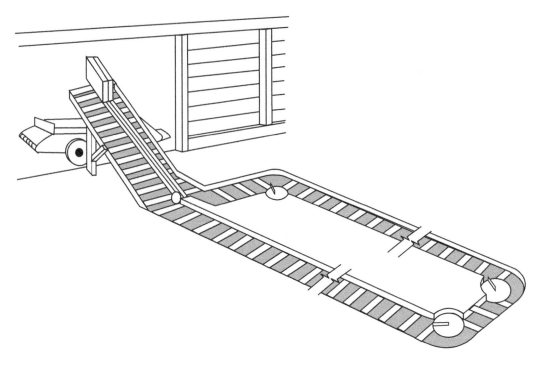

FIGURE 12.10   Chain and flight gutter cleaner. (Source: NRCS/USDA, 1992, p. 10-10.)

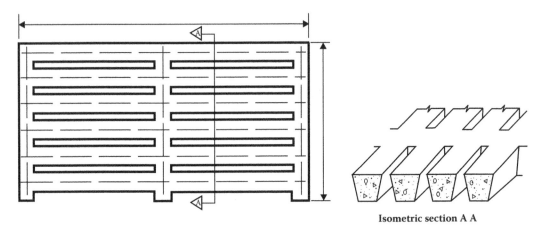

Isometric section A A

FIGURE 12.11   Concrete gang slats. (Source: NRCS/USDA, 1992, p. 10-10.)

12.3.2.4 Flush Gutters

Narrow gutters can also be cleaned by flushing. Flush gutters are usually a minimum of 2 ft deep on the shallow end. The depth may be constant or increase as the length of the gutter increases. The bottom grade can vary from 0% to 5% depending on storage requirements and cleanout technique. Flushing tanks or high volume pumps may be used to clean flush gutters.

12.3.3 SLATTED FLOORS

Waste materials are worked through the slats by the animal traffic into a storage tank or alley below. As shown in Figure 12.11, most slats are constructed of reinforced concrete; however, some are made of wood, plastic, or aluminum. They are manufactured either as individual units or as gangs of several slats. Common slat openings range from ⅜ in. to 1¾ in., depending on animal type. For swine, openings between ⅜ and ¾ in. are not recommended.

Slats are designed to support the weight of the slats plus the live loads (animal, humans, and mobile equipment) expected for the particular facility. Reinforcing steel is required in concrete slats to provide needed strength.

12.4 STORAGE

Waste generally must be stored so that it can be used when conditions are appropriate. Storage facilities for wastes of all consistencies must be designed to meet the requirements of a given enterprise.

Determining the storage period for a storage facility is crucial to the proper management of an agricultural waste management system. If too short a period is selected, the facility may fill before the waste can be used in an environmentally sound manner. Too long a period may result in an unjustified expenditure for the facility.

Many factors are involved in determining the storage period. They include the wealth, crop, growing season, equipment availability, soil, soil condition, labor requirements, and management flexibility. Generally, when waste utilization is by land application, a storage facility must be sized so that it can store the waste during the nongrowing season. A storage facility with a longer storage period generally allows more flexibility in managing waste because it can accommodate weather variability, equipment availability, equipment breakdown, and overall operation management.

12.4.1 WASTE STORAGE FACILITIES FOR SOLIDS

Storage facilities for solid manure include waste storage ponds and waste storage structures. Waste storage ponds are earthen impoundments used to retain manure, bedding, and runoff liquid. Solid and semisolid manure placed into a storage pond most likely has to be removed as a liquid unless precipitation is low or a means of draining the liquid is available. The pond bottom and entrance ramps should be paved if emptying equipment will enter the pond.

Waste storage structures can be used for manure that stacks and can be handled by solid manure handling equipment. These structures must be accessible for loading and hauling equipment. They can be open or covered. Roofed structures are used to prevent or reduce excess moisture content. Open stacks can be used in either an arid or a humid climate. Seepage and runoff must be managed. Structures for open and covered stacks often have wooden, reinforced concrete or concrete block sidewalls. The amount of bedding material often dictates whether the manure can be handled as a solid.

In some instances, manure must be stored in open stacks in fields. Runoff and seepage from these stacks must be managed to prevent movement into streams or other surface water or groundwater. Figure 12.12 and Figure 12.13 show various solid manure storage facilities.

12.4.1.1 Design Considerations

Solid waste storage ponds and structures must be designed correctly to ensure desired performance and safety. Considerations include materials selection, control of runoff and seepage, necessary storage capacity, and proper design of structural components, such as sidewalls, floors, and roofs.

The primary materials used in constructing timber structures for solids storage are pressure-treated or rot-resistant wood and reinforced concrete. These materials are suitable for long-term exposure to animal waste without rapid deterioration. Structural grade steel is also used, but it corrodes and must be protected against corrosion or be periodically replaced. Similarly, high quality and protected metal fasteners must be used with timber structures to reduce corrosion problems.

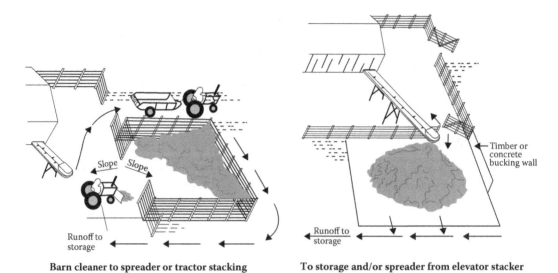

FIGURE 12.12   Solid manure stacking facilities. (Source: NRCS/USDA, 1992, p. 10-12.)

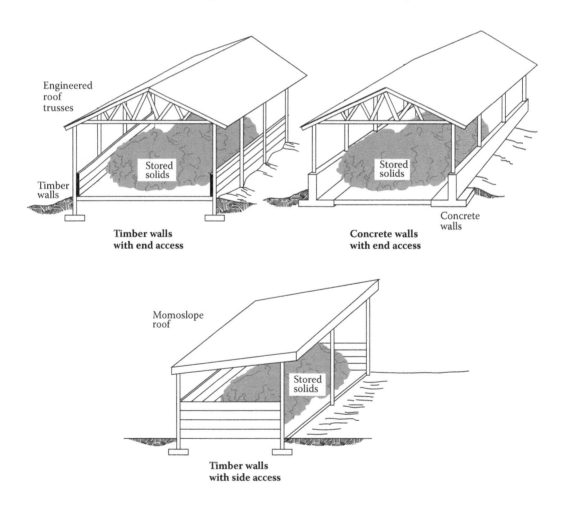

FIGURE 12.13   Roofed solid manure storage. (Source: NRCS/USDA, 1992, p. 10-13.)

Seepage and runoff, which frequently occur from manure stacks, must be controlled to prevent access into surface water and groundwater. One method of control is to channel any seepage into a storage pond. At the same time, uncontaminated runoff, such as that from the roof and outside the animal housing and lot area, should be diverted around the site.

Concrete ramps are used to gain access to solid manure storage areas. Ramps and floors of solid manure storage structures must be designed so that handling equipment can be safely operated. Ramp slopes of 8 to 1 (horizontal to vertical) or flatter are considered safe. Slopes steeper than this are difficult to negotiate. Concrete pavement for tamps and storage units should be rough finished to aid in traction. Ramps must be wide enough that equipment can be safely backed and maneuvered.

Factors to consider in the design of storage facilities for solids include type, number and size of animals, number of days storage desired, and the amount of bedding that will be added to the manure. Equation 12.5 can be used to calculate the manure storage volume:

$$VMD = AU \times DVM \times D \tag{12.5}$$

where:
VMD = Volume of manure production for animal type for storage, ft^3.
AU = Number of 1,000-pound animal units by animal type.
DVM = Daily volume of manure production for animal type, ft^3/AU/day.
D = Number of days in storage period.

The bedding volume to be stored can be computed using:

$$BV = FR \times WB \times AU \times D / BUM \tag{12.6}$$

where:
FR = Volumetric void ratio (ASAE 1982) (values range from 0.3 to 0.5).
WB = Weight of bedding used for animal type, lb/AU/day.
BUW = Bedding unit weight, lb/ft.^{3}

Using the recommended volumetric void ratio of 0.5, the equation becomes:

$$BV = 0.5 \times WB \times AU \times D / BUM \tag{12.7}$$

Earlier we discussed and described characteristics of manure and bedding. Other values may be available locally or from the farmer or rancher.

- *Important point:* Allowance must be made for the accumulation of precipitation that may fall directly into the storage. Contaminated runoff should be handled separately from a solid manure storage facility. Uncontaminated runoff should be diverted from the storage unit.

12.4.2 LIQUID AND SLURRY WASTE STORAGE

Liquid and slurry manure can be stored in waste storage ponds or in aboveground or below-ground tanks. Solids separation of manure and bedding is a problem that must be considered in planning and design. Solids generally can be resuspended with agitation before unloading, but this involves costs for time, labor, and energy. Another option allows solids to accumulate if the bottom is occasionally cleaned. This requires a paved working surface for equipment.

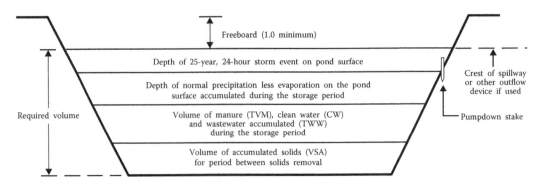

FIGURE 12.14   Cross section of waste storage pond without a watershed. (Source: NRCS/USDA, 1992., p. 10-16)

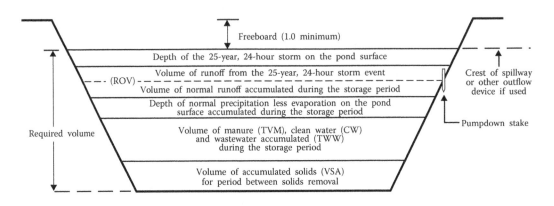

*or other outflow device

FIGURE 12.15   Cross section of waste storage pond with a watershed. (Source: NRCS/USDA, 1992, p. 10-17.)

Earthen storage is frequently the least expensive type of storage; certain restrictions, such as limited space availability, high precipitation, water table, permeable soils, or shallow bedrock, can limit the types of storage considered.

Storage ponds are earthen basins designed to store wastewater and manure (Figure 12.14, Figure 12.15, and Figure 12.16). They generally are rectangular but may be circular or any other shape that is practical for operation maintenance. The inside slopes range from 1.5 to 1 (horizontal to vertical) to 3 to 1. The combined slopes (inside plus outside) should not be less than 5 to 1 for embankments. The soil, safety, and operation and maintenance must be considered in designing the slopes. The minimum top width of embankments should be 8 ft; however, greater widths should be provided for operation of tractors, spreaders, and portable pumps.

- *Important point:* Storage ponds should provide capacity for normal precipitation and runoff (less evaporation) during the storage period. A minimum of 1 ft of freeboard is provided.

Inlets to storage ponds can be of any permanent material designed to resist erosion, plugging or, if freezing is a problem, damage by ice. Typical loading methods are pipes and ramps. Flow of wastes away from the inlet should be considered in selecting the location of the inlet.

Gravity pipes, pumping platforms, and ramps are used to unload storage ponds. A method for removing solids should be designed for the storage pond. If the wastes will be pumped, adequate access must be provided to thoroughly agitate the contents of the pond. A ramp should have a slope of 8 to 1 or flatter and be wide enough to provide maneuvering room for unloading equipment.

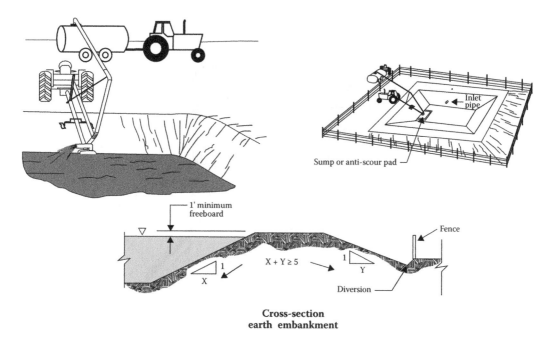

FIGURE 12.16 Waste storage ponds. (Source: NRCS/USDA, 1992, p. 10-17.)

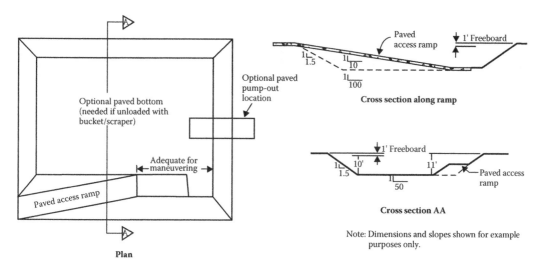

FIGURE 12.17 Layout of waste storage ponds. (Source: NRCS/USDA, 1992, p. 10-18.)

Pond liners are used in many cases to compensate for site conditions or improve operation of the pond. Concrete, geomembrane, and clay linings reduce permeability and can make an otherwise unsuitable site acceptable. Concrete also provides a wear surface if unloading equipment enters the pond.

- *Important point:* Figures 12.17, 12.18, and 12.19 represent various kinds of storage ponds and tanks.

Liquid manure can be stored in aboveground (Figure 12.18) or belowground (Figure 12.19) tanks. Liquid manure storage tanks can be constructed of metal, concrete, or wood. Belowground tanks

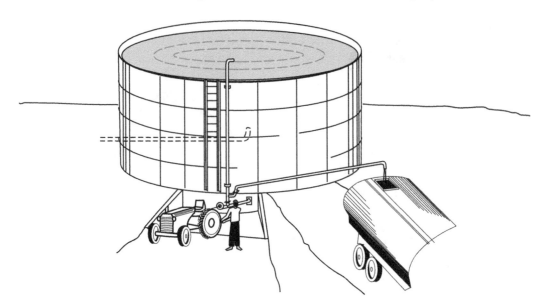

FIGURE 12.18   Aboveground waste storage tank. (Source: NRCS/USDA, 1992, p. 10-19.)

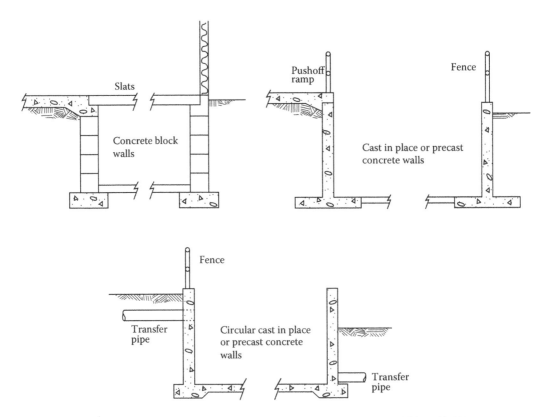

FIGURE 12.19   Belowground waste storage structure. (Source: NRCS/USDA, 1992, p. 10-19.)

can be loaded using slatted floors, push-off ramps, gravity pipes or gutters, or pumps. Aboveground tanks are typically loaded by a pump moving the manure from a reception pit. Tank loading can be from the top or bottom of the tank, depending on such factors as desired agitation, minimized pumping head, weather conditions, and system management.

Storage volume requirements for tanks are the same as those for ponds except that provisions are normally made to exclude outside runoff from waste storage tanks because of the relatively high cost of storage. Of course, if plans include storage of outside runoff, accommodation for its storage must be included in the tank's volume.

Tanks located beneath slatted floors can sometimes be used for temporary storage, with subsequent discharge into lagoons or other storage facilities. Recycled lagoon effluent is added to a depth of 6 to 12 in in underslat pits to reduce tendency for manure solids to stick to the pit floor. Wastes are allowed to collect for several days, typically 1 to 2 weeks, before the pits are gravity drained.

12.4.2.1 Design Considerations

Tank material types—The primary materials used to construct manure tanks are reinforced concrete, metal, and wood. Such tanks must be designed by a professional engineer and constructed by experienced contractors. A variety of manufactured, modular, and cast-in-place tanks are available from commercial suppliers. The NRCS concurs in the standard detail drawings for these structures based on a review and approval of the drawings and supporting design calculations. A determination must be made that the site conditions are compatible with the design assumptions upon which the design is based. Structures can also be designed on an individual site-specific basis.

Cast-in-place reinforced concrete, the principal material used in belowground tanks, can be used in aboveground tanks as well. Tanks can also be constructed of precast concrete panels that are bolted together. Circular tank panels are held in place with metal hoops. The panels are positioned on a concrete foundation or have footings cast as an integral part of the panel. Tank floors are cast-in-place slabs.

Other aboveground tanks are constructed of metal. Glass-fused steel panels are widely used. Such tanks are manufactured commercially and must be constructed by trained crews. Other kinds of metal panels are also used.

At least one company offers a wooden aboveground tank for liquid storage. The preservative-treated boards have tongue-and-groove edges and are held in place using metal hoops similar to those used for concrete panel tanks. All manure tanks should meet the standards identified in the section on solid manure storage.

Sizing—Liquid waste storage ponds and structures should be sized to hold all of the manure, bedding, wastewater from milkhouse, flushing, and contaminated runoff that can be expected during the storage period. Equation 12.8 can be used to compute the waste volume:

$$WV = TVM = TWM = TBV \tag{12.8}$$

where:
WV = Waste volume for storage period, ft³.
TVM = Total volume of manure for storage period, ft³ (see Equation 12.5).
TWW = Total wastewater volume for storage period, ft³.
TBV = Total bedding volume for storage period, ft³ (see Equation 12.6).

In addition to the waste volume, waste storage tanks must, if uncovered, provide a depth to accommodate precipitation less evaporation on the storage surface during the most critical storage period, generally the consecutive months that represent the storage period that gives the greatest depth of precipitation less evaporation. Waste storage tanks must also provide a depth of 0.5 ft for material not required during emptying. A depth for freeboard of 0.5 ft is also recommended.

Waste storage ponds must also provide a depth to accommodate precipitation less evaporation during the most critical storage period. If the pond does not have a watershed, the depth of the 25-year, 24-hour precipitation on the pond surface must be included. Frequently, waste storage ponds are designed to include outside runoff from watersheds. For these, the runoff volume of the 25-year, 24-hour storm must be included in the storage volume.

12.4.2.2   Design of Sidewalls and Floors

The information on the design of sidewalls and floors discussed earlier on solid manure storage materials is applicable to these items used for liquid manure storage. All possible influences, such as internal and external hydrostatic pressure, flotation and drainage, live loads from equipment and animals, and dead loads from covers and supports, must be considered in the design.

Pond sealing—Waste storage ponds must not allow excess seepage. The soil in which the pond is located must be evaluated and, if needed, tested during planning and design to determine need for an appropriate line.

12.5   TREATMENT

In many situations, agricultural waste must be treated before final utilization. The purpose of treatment is to reduce the pollution potential of the waste through biological, physical, and chemical processes using such components as lagoons, oxidation ditches, and composting. These types of components reduce nutrients, destroy pathogens, and reduce total solids. Composting also reduces the volume of the waste. Treatment also includes any step that might be considered pretreatment, such as solids separation, drying, and dilution that prepares the waste for facilitating another function. By their nature, treatment facilities require a higher level of management than that of storage facilities.

12.5.1   Anaerobic Lagoons

Anaerobic lagoons are widely accepted in the United States for the treatment of animal waste. Anaerobic treatment of animal waste helps to protect water quality by reducing much of the organic concentration quality, the organic concentration biological oxygen demand, and the carbonaceous oxygen demand of the waste. Anaerobic lagoons also reduce the nitrogen content of the waste through ammonia volatilization and effectively reduce animal waste odors if the lagoon is managed properly.

12.5.1.1   Design

The maximum operating level of an anaerobic lagoon is a volume requirement plus a depth requirement. The volume requirement is the sum of the following volumes:

- Minimum treatment volume, ft³ (MTV)
- Manure volume, wastewater volume, and clean water, ft³ (WV)
- Sludge volume, ft³ (SV)

- *Important point:* The depth requirement is the normal precipitation less evaporation on the lagoon surface.

Polluted runoff from a watershed must not be included in a lagoon unless a defensible estimate of the volatile solid (VS) loading can be made. Runoff from a watershed, such as a feedlot, is not included in a lagoon because loading would only result during storm events and because the magnitude of the loading would be difficult, if not impossible, to estimate. As a result, the lagoon would be shocked with an overload of VS.

If an automatic outflow device, pipe, or spillway is used, it must be placed at a height above the maximum operating level to accommodate the 25-year, 24-hour storm precipitation on the lagoon

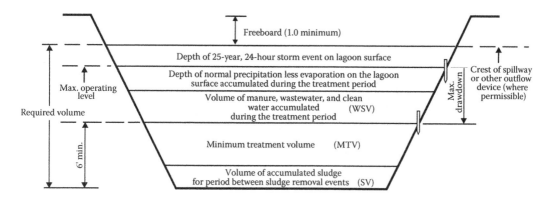

Note: The minimum treatment volume for an
anaerobic waste treatment lagoon is based
on volatile solids.

FIGURE 12.20 Anaerobic lagoon cross section. (Source: NRCS/USDA, 1992, p. 10-28.)

surface. This depth added to the maximum operating level of the lagoon establishes the level of the required volume or the outflow device, pipe, or spillway. A minimum of 1 ft of freeboard is provided above the outflow and establishes the top of the embankment. Should state regulations preclude the use of an outflow device, pipe, or spillway, or if for some other reason the lagoon does not have these, the minimum freeboard is 1 ft above the top of the required volume.

The combination of these volumes and depths is illustrated in Figure 12.20. The terms and derivation are explained in the flowing paragraphs.

Anaerobic waste treatment lagoons are designed on the basis of VS loading rate (VSLR) per 1,000 ft³. VS represents the amount of solid material in wastes that will decompose as opposed to the mineral (inert) fraction. The rate of solids decomposition in anaerobic lagoons is a function of temperature; therefore, the acceptable VSLR varies from one location to another. Figure 12.21 indicates the maximum VSLRs for the United States. If odors need to be minimized, VSLR should be reduced by 25% to 50%.

The MTV represents the volume needed to maintain sustainable biological activity; the minimum treatment volume for VS can be determined using Equation 12.9.

$$MTV = TVS/VSLR \tag{12.9}$$

where:
MTV = Minimum treatment volume, ft³.
TVS = Total daily volatile solids loading (from all sources), lb/day.
$VSLR$ = Volatile solids loading rate, lb/1,000 ft³/day (from Figure 12.21).

- *Important point:* If feed spillage exceeds 5%, VSP should be increased by 4% for each additional 1% spillage.

WV should reflect the actual volume of manure, wastewater, and flush water that will not be recycled and clean dilution water added to the lagoon during the treatment period (Equation 12.10). The treatment period is either the detention time required to obtain the desired reduction of pollution potential of the waste or the time between land application events, whichever is longer. State regulations may govern the minimum detention time. Generally, the maximum time between land application events determines the treatment period because this time generally exceeds the detention time required.

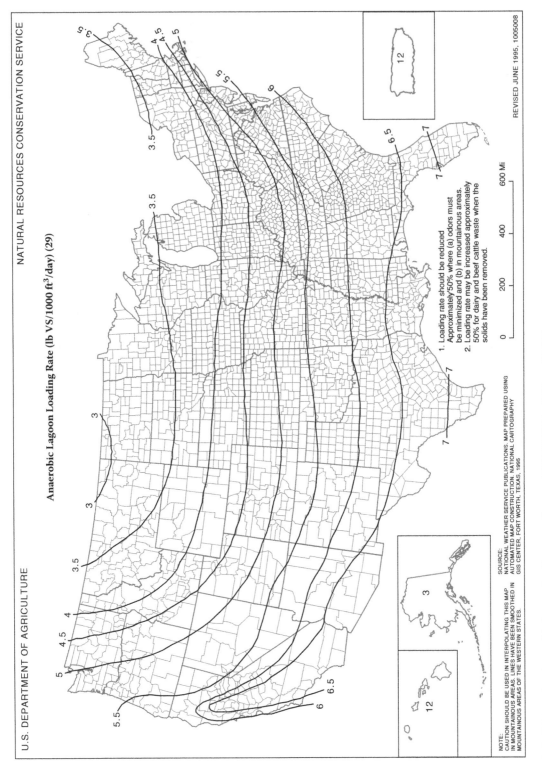

FIGURE 12.21 Anaerobic lagoon loading rate. (Source: NRCS/USDA, 1992, p. 10-29.)

$$MTV = TVM + TWW + CW \qquad (12.10)$$

where:
WV = Waste volume for treatment period, ft³.
TVM = Total volume of manure for treatment period, ft³.
TWW = Total volume of wastewater for treatment period, ft³.
CW = Clean water added during treatment period, ft³.

As the manure decomposes in the anaerobic lagoon, only part of the total solids (TS) is reduced. Some of the TS is mineral material that does not decompose, and some of the VS require a long time to decompose. These materials, referred to as *sludge*, gradually accumulate in the lagoon. To maintain the MTV, the volume of sludge accumulation over the period between sludge removal must be considered. Lagoons are commonly designed for a 15- to 20-year sludge accumulation period. The SV can be determined using Equation 12.11.

$$SV = 365 \times AU \times TS \times SAR \times T \qquad (12.11)$$

where:
SV = Sludge volume (ft³).
AU = Number of 1,000-pound animal units.
T = Sludge accumulation time (years).
TS = Total solids production per animal unit per day (lb/AU/day).
SAR = Sludge accumulation ratio (ft³/lb TS).

TS can be obtained from the applicable tables. Sludge accumulation ratios should be taken from Table 12.4. An SAR is not available for beef, but it can be assumed to be similar to that for dairy cattle.

The lagoon volume requirements are for accommodation of the MTV, the SV and the WV for the treatment period, as expressed in Equation 12.12.

$$LV = MTV + SV + TWV \qquad (12.12)$$

where:
LV = Lagoon volume requirement, ft³.
MTV = Minimum treatment volume, ft³ (see Equation 12.9).
SV = Sludge volume accumulation for period between sludge removal events, ft³ (see Equation 12.11).
WV = Waste volume for treatment period, ft³ (see Equation 12.10).

TABLE 12.4
Sludge Accumulation Ratios

Animal Type	SAR
Poultry	
Layers	0.0295
Pullets	0.0455
Swine	0.0483
Dairy cattle	0.0729

Source: Adaptation from Barth (1985).

TABLE 12.5

Minimum Top Width for Lagoon Embankments

Minimum height of embankment, ft	Top width, ft
10 or less	6
11–14	8
15–19	10
20–24	12
25–34	14
35 or more	15

Source: USDA (1984).

In addition to the lagoon volume requirement (LV), a provision must be made for depth to accommodate the normal precipitation less evaporation on the lagoon surface; the 25-year, 24-hour storm precipitation; the depth required to operate the emergency outflow; and freeboard. Normal precipitation on the lagoon surface is based on the critical treatment period that produces the maximum depth. This depth can be offset to some degree by evaporation losses on the lagoon surface. This offset varies, according to the climate of the region, from a partial amount of the precipitation to an amount in excess of the precipitation. Precipitation and evaporation can be determined from local climate data.

- *Important point:* The minimum acceptable depth for anaerobic lagoons is 6 ft but, in colder climates, at least 10 ft is recommended to assure proper operation and odor control.

The design height of an embankment for a lagoon should be increased by the amount needed to ensure that the design elevation is maintained after settlement. This increase should not be less than 5% of the design fill height. The minimum top width of the lagoon should be as shown in Table 12.5, although a width of 8 ft or less is difficult to construct.

The combined side slopes of the settled embankment should not be less than 5 to 1 (horizontal to vertical). The inside slopes can vary from 1 to 1 for excavated slopes to 3 to 1 or flatter where embankments are used. Construction technique and soil type must also be considered. In some situations a steep slope may be used below the design liquid level, while a flatter slope is used above the liquid level to facilitate maintenance and bank stabilization. The minimum elevation of the top of the settled embankment should be 1 ft above the maximum design water surface in the lagoon.

A lagoon should be constructed to avoid seepage and potential groundwater pollution. Care in site selection, soils investigation, and design can minimize the potential for these problems. Figure 12.22 shows a two-lagoon system.

If overtopping can cause embankment failure, an emergency spillway to overflow pipe should be provided. A lagoon can have an overflow to maintain a constant liquid level if the overflow liquid is stored in a waste storage pond or otherwise properly managed. The inlet to a lagoon should be protected from freezing. This can be accomplished by using an open channel that can be cleaned out or by locating the inlet pipe below the freezing level in the lagoon. Because of possible blockages, access to the inlet pipe is needed. Venting inlet pipes prevents backflow of lagoon gases into the animal production facilities.

Sludge removal is an important consideration in the design. This can be accomplished by agitating the lagoon and pumping out the mixed sludge or by using a dragline for removing floating or settled sludge. Some pumps can remove sludge, but not deposited rocks, sand, or grit. The sludge removal technique should be considered when determining lagoon surface dimensions. Many

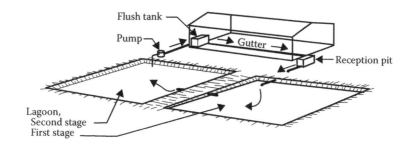

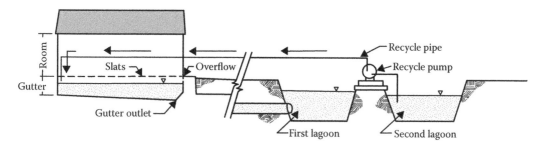

FIGURE 12.22 Anaerobic lagoon recycle systems. (Source: NRCS/USDA, 1992, p. 10-31.)

agitation pumps have an effective radius of 75 to 100 ft. Draglines may only reach 30 to 50 ft into the lagoon.

12.5.1.2 Management

Anaerobic lagoons must be managed properly if they are to function as designed. Specific instructions about lagoon operation and maintenance must be included in the overall waste management plan supplied to the decisionmaker. Normally an anaerobic lagoon is managed so that the liquid level is maintained at or below the maximum operating level, as shown in Figure 12.23. The liquid level is lowered to the minimum treatment level at the end of the treatment period. It is good practice to install markers at the minimum treatment and maximum operating levels.

The minimum liquid level in an anaerobic lagoon before wastes are added should coincide with the MTV. If possible, a lagoon should be put into service during the summer to allow adequate development of bacterial populations. A lagoon operates more effectively and has fewer problems if loading is by small, frequent (daily) inflow, rather than large, infrequent slug loads.

The pH should be measured frequently. Many problems associated with lagoons are related to pH in some manner. The optimum pH is about 6.5. When pH falls below this level, methane bacteria are inhibited by the free hydrogen ion concentration. The most frequent cause of low pH in anaerobic digestion is the shock loading of organic material that stimulates the facultative acid-producing bacteria. Add hydrated lime or lye if pH is below 8.5. Add 1 lb per 1,000 ft^2 daily until pH reaches 7.

Lagoons are designed based on a given loading rate. If an increase in the number of animals is anticipated, sufficient capacity to handle the entire expected waste load should be available. The most common problem in using lagoons is overloading, which can lead to odors, malfunctioning, and complaints. When liquid removal is needed, the liquid level should not be dropped below the MTV plus SV levels. If evaporation exceeds rainfall in a series of dry years, the lagoon should be partly drawn down and refilled to dilute excess concentrations of nutrients, minerals, and toxics.

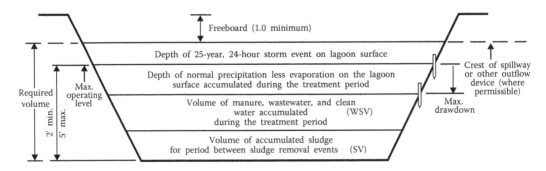

FIGURE 12.23   Aerobic lagoon cross section. (Source: NRCS/USDA, 1992, p. 10-35.)

Lagoons are typically designed for 15 to 20 years of sludge accumulation. After this time, the sludge must be cleaned out before adding additional waste.

Sometimes operators want to use lagoon effluent as flush water. To polish and store water for this purpose, waste storage ponds can be constructed in series with the anaerobic lagoon. The capacity of the waste storage pond should be sized for the desired storage volume. A minimum capacity of the waste storage pond is the volume for rainfall (RFV), runoff (ROV), and emergency storm storage (ESV). By limiting the depth to less than 6 ft, the pond functions more nearly like an aerobic lagoon. Odors and the level of ammonia, ammonium, and nitrate are more effectively reduced.

12.5.2   Aerobic Lagoons

Aerobic lagoons can be used if minimizing odors is critical (see Figure 12.23). These lagoons are operated within a depth range of 2 to 5 ft to allow for the oxygen entrainment necessary for the aerobic bacteria.

The design of aerobic lagoons is based on the amount of BOD added per day. If local data are not available, use the BOD values from applicable tables. Figure 12.24 shows the acceptable aerobic loading rates for the United States in pounds of BOD per acre per day (lb-BOD/acre/day). The lagoon surface area at the average operating depth is sized so that the acceptable loading rate is not exceeded.

Even though an aerobic lagoon is designed on the basis of surface area, it must have enough capacity to accommodate the WV and SV. In addition, depth must be provided to accommodate the normal precipitation less evaporation on the lagoon surface; the 25-year, 24-hour storm precipitation on the lagoon surface; and freeboard. Should state regulations not permit an emergency outflow or for some other reason one is not used, the minimum freeboard is 1 ft above the top of the required volume. Figure 12.23 demonstrates these volume depth requirements.

Aerobic lagoons need to be managed similarly to anaerobic lagoons in that they should never be overloaded with oxygen-demanding material. The lagoon should be filled to the minimum operating level, generally 2 ft, before being loaded with waste. The maximum liquid level should not exceed 5 ft. The water level must be maintained within the designed operating range. Sludge should be removed when it exceeds the designed sludge storage capacity. Aerobic lagoons should also be enclosed in fences and marked with warning signs.

12.5.3   Mechanically Aerated Lagoons

Note: Much of this material was adapted from technical notes on the design of mechanically aerated lagoons for odor control in *USDA Soil Conservation Service Technical Notes* (Moffitt, 1980).

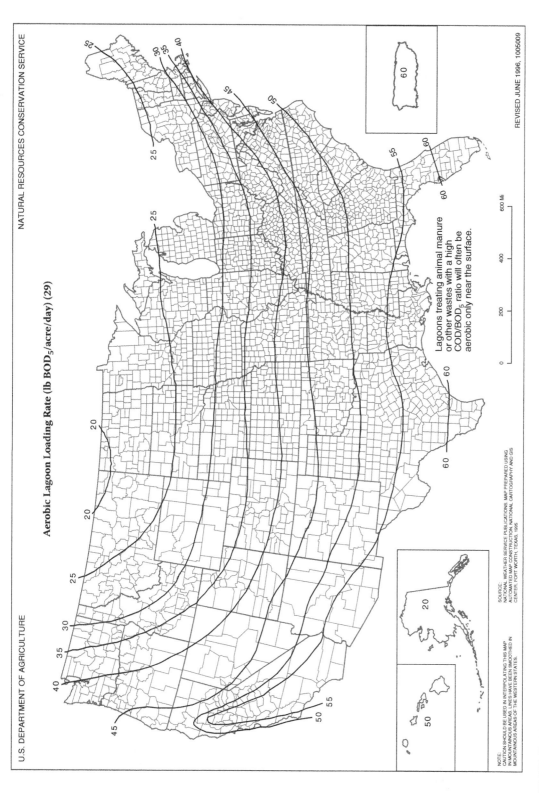

FIGURE 12.24 Aerobic lagoon loading rate. (Source: NRCS/USDA, 1992, p. 10-36.)

Aerated lagoons operate aerobically and are dependent on mechanical aeration to supply the oxygen needed to treat waste and minimize odors. This type of design is used to convert an anaerobic lagoon to an aerobic condition or, as an alternative, to a naturally aerated lagoon that would otherwise need to be much larger. Mechanically aerated lagoons combine the small surface area feature of anaerobic lagoons with the relatively odor-free operation of an aerobic lagoon. The main disadvantages of this type of lagoon are the energy requirements to operate the mechanical aerators and the high level of management required.

The typical design includes 1 lb of oxygen transferred to the lagoon liquid for each pound of BOD added. The TS content in aerated lagoons should be maintained between 1% and 3% with dilution water. The depth of aerated lagoons depends on the type of aerator used. Agitation of settled sludge must be avoided. As with naturally aerobic lagoons, consideration is required for storage of manure and rainfall.

Two kinds of mechanical aerators are used: the surface pump and the diffused air system. The surface pump floats on the surface of the lagoon, lifting water into the air, thus assuring an air-water mixture. The diffused air system pumps air through water but is generally less economical to operate than the surface pump.

12.5.3.1   Lagoon Loading

Lagoon loading should be based on BOD or COD. NRCS designs on the basis of BOD. Applicable tables show recommended BOD production rates, but local data should be used where available.

12.5.3.2   Aerator Design

Aerators are designed primarily on their ability to transfer oxygen (O_2) to the lagoon liquid. Of secondary importance is the ability of the aerator to mix or disperse O_2 throughout the lagoon. Where the aerator is intended for minimizing odors, complete mixing is not a consideration except as it relates to the surface area.

For the purpose of minimizing odors, aerators should transfer between 1 and 2 lb of oxygen per pound of BOD. Even a limited amount of oxygen transfer (as little as ⅓ lb O_2 per lb BOD) reduces the release of volatile acids and accompanying gases. For design purposes, use 1 lb of oxygen per pound of BOD unless local research indicates a higher value is needed.

Aerators are tested and rated according to their clean water transfer rate (SWTR) or laboratory transfer rate (LTR), whichever term is preferred. The resulting value is given for transfer at standard atmospheric pressure (14.7 psi), dissolved oxygen equal to 0%, and water at 20°C. The actual transfer rate expected in field operation can be determined by using Equation 12.13.

$$FTR = CWTR \times (B \times C_{dc}) - DO/C_{sc} \times C^{t-20} \times a \qquad (12.13)$$

where:

FTR = Pound of O_2 per horsepower-hour transferred under field conditions.

$CWTR$ = Clean water transfer rate in pound per horsepower-hour transferred under standard laboratory conditions.

B = Salinity-surface tension factor, the ratio of the saturated in the wastewater to that of clean water. Values range from 0.95 to 1.0.

C_{dc} = O_2 saturation concentration at design conditions of altitude and temperature (mg/L) from Figure 12.25 and Figure 12.26.

DO = Average operating O_2 concentration (mg/L). The recommended value of DO can vary from 1 to 3 depending on the reference material. A value of 1.5 should be considered a minimum. For areas where minimizing odors is particularly critical, a DO of 2 or more should be used.

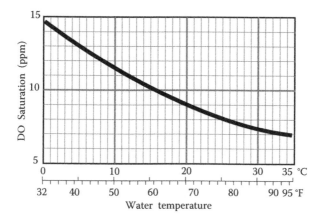

FIGURE 12.25 Relation of dissolved oxygen (DO) saturation to water temperature (clean water at 20°C ad sea level). (Source: NRCS/USDA, 1992, p. 10-40.)

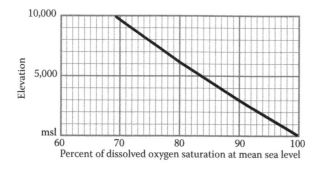

FIGURE 12.26 Relation of dissolved oxygen saturation to elevation above mean sea level. (Source: NRCS/ USDA, 1992, p. 10-40.)

T	= Design temperature (°C).
O	= Temperature correction factor; values range from 1.024 to 1.035.
A	= The ratio of the rate of O_2 transfer in the wastewater to that of clean water. Generally taken as 0.75 for animal waste.
C_{sc}	= Saturation concentration of O_2 in clean water, 20°C and sea level (9.17 mg/L).

- *Important point:* Unless local information supports using other values, the following values for calculating field transfer rates should be used: B = 1.0, DO = 1.5, O = 1.024, a = 0.75, and C_{sc} = 9.17.

Figure 12.27 provides a quick solution to the term O^{t-20}, where O is equal to 1.024. Designs for both summer and winter temperatures are often necessary to determine the controlling (least) transfer rate.

Having calculated FTR, the next step is to determine horsepower requirements of aeration based on loading rates and FTR as calculated above. Horsepower requirements can be estimated using Equation 12.14.

$$HP = BOD/FTR \times HO \qquad (12.14)$$

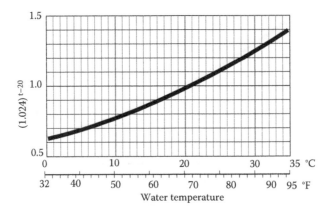

FIGURE 12.27 Numeral values for 0^{t-20} at different temperatures where $0 = 1.024$. (Source: NRCS/USDA, 1992, p. 10-40.)

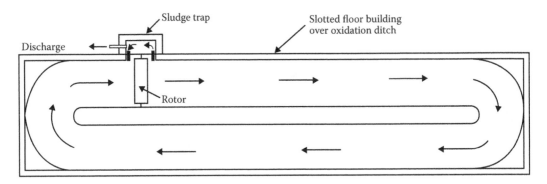

FIGURE 12.28 Schematic of an oxidation ditch. (Source: NRCS/USDA, 1992, p. 10-41.)

where:
HP = Horsepower.
BOD = 5-day biochemical oxygen demand loading of waste, lb/day.
HO = Hours of operation per day.

- *Important point:* Most lagoon systems should be designed on the basis of continual aerator operations.

The actual selection of aerator(s) is a subjective process and often depends on the availability of models in the particular area. In general, multiple small units are preferred to one large unit. The multiple units provide better coverage of the surface area and permit flexibility for the real possibility of equipment failure and reduced aeration.

12.5.4 Oxidation Ditch

In some situations sufficient space is not available for a lagoon for treating animal waste, and odor control is critical. One option for treating animal waste under these circumstances is an oxidation ditch (see Figure 12.28). The shallow, continuous ditch generally is in an oval layout, with a special aerator spanning the channel. The action of the aerator moves the liquid waste around the channel and keeps the solids in suspension. Because of the need for continuous aeration, this process can be expensive to operate. Oxidation ditches should only be designed by a professional engineer familiar with the process.

The range of loading for an oxidation ditch is 1 lb of BOD per 30 to 100 ft³ of volume. This provides for a retention time of 30 to 70 days. Solids accumulate over time and must be removed by settling. The TS concentration is maintained in the 2% to 6% range, and dilution water must be added periodically.

If oxidation ditches are not overloaded, they work well for minimizing odors. The degree of management required, however, may be more than desired by some operators. Daily attention is often necessary, and equipment failure can lead to toxic gas generation soon after the aerators are stopped. If the ditches are properly managed, they can be effective in reducing nitrogen to N_2 through cyclic aerobic/anaerobic periods, which allows nitrification and then denitrification.

12.5.5 Drying/Dewatering

If the water is removed from freshly excreted manure, the volume to handle is reduced. The process of removing water is referred to as *dewatering*. In arid regions of the United States, most manure is dewatered (dried) by evaporation from sun and wind. Some nutrients may be lost in the drying process.

Dried or dewatered manure solids are often sold as a soil conditioner or garden fertilizer. These solids may also be used as fertilizer on agricultural land. They are high in organic matter and can be expected to produce odors if moisture is added and the material is not redried or composted. Because the water is removed, the concentrations of some nutrients and salts change. Dried manure should be analyzed to determine the nutrient concentrations before land application.

In humid climates, dewatering is accomplished by adding energy to drive off the desired amount of moisture. Processes have been developed for drying manure in greenhouse-type facilities; however, the drying rate is dependent on the temperature and relative humidity. The cost of energy often makes the drying process unattractive.

12.5.6 Composting

Composting is the aerobic biological decomposition of organic matter, a natural process that is enhanced and accelerated by the mixing of organic waste with other ingredients in a prescribed manner for optimum microbial growth.

Composting converts an organic waste material into a stable organic product by converting nitrogen from the unstable ammonia form to a more stable organic form. The end result is a product that is safer to use than raw organic material and one that improves soil fertility, tilth (i.e., the physical condition of soil as related to its ease of tillage, fitness as a seedbed, and its impedance to seedling emergence and root penetration), and water holding capacity. In addition, composting reduces the bulk of organic material to be spread; improves its handling properties; reduces load, fly, and other vector problems; and can destroy weed seeds and pathogens.

12.5.6.1 Composting Methods

Three basic methods of composting—windrow, static pile, and in-vessel—are described below.

Windrow method—The windrow method involves the arrangement of compost mix in long, narrow piles or windrows (see Figure 12.29). To maintain an aerobic condition, the compost mixture must be periodically turned. This exposes the decomposing material to the air and keeps temperatures from getting too high (> 170°F). The minimum turning frequency varies from 2 to 10 days, depending on the type of mix, volume, and the ambient air temperature. As the compost ages, turning frequency can be reduced.

Only the type of turning equipment used limits the width and depth of the windrows. Turning equipment can range from a front-end loader to an automatic mechanical turner. Windrows generally are 4 to 6 ft deep and 6 to 10 ft wide.

Some advantages and disadvantages of the windrow method include:

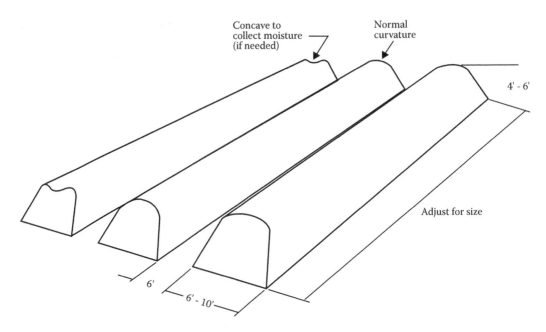

FIGURE 12.29   Windrow schematic. (Source: NRCS/USDA, 1992, p. 10-42.)

Advantages:

- Rapid drying with elevated temperatures
- Drier product, resulting in easier product handling
- Ability to handle high volumes of material
- Good product stabilization
- Low capital investment

Disadvantages:

- Not space efficient
- High operational costs
- Piles should be turned to maintain aerobic conditions
- Turning equipment may be required
- Vulnerable to climate changes
- Odors released on turning of compost
- Large volume of bulking agent might be required

Static pile method—The static pile method consists of mixing the compost material, then stacking the mix on perforated plastic pipe or tubing though which air is drawn or forced. Forcing air through the compost pile may not be necessary with small compost piles that are highly porous or with a mix that is stacked in layers with highly porous material. The exterior of the pile generally is insulated with finished compost or other material. In nonlayered operations, the materials to be composted must be thoroughly blended before pile placement.

The dimensions of the static pile are limited by the amount of aeration that can be supplied by the blowers and the stacking characteristics of the waste. The compost mixture height generally ranges from 8 to 15 ft; the width is usually twice the depth. Individual piles generally are spaced about a half the distance of the height.

With forced air systems, air movement through the pile occurs by suction (vacuum) or by positive pressure (forced) through perforated pipes or tubing. A filter pile or material is normally used to absorb odor if air is sucked through the pile (see Figure 12.30).

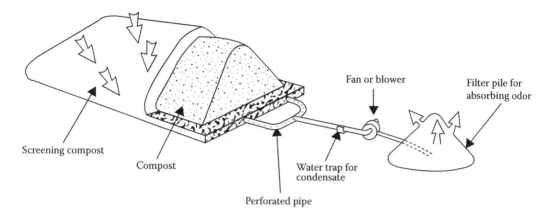

FIGURE 12.30 Static pile composting schematic. (Source: NRCS/USDA, 1992, p. 10-43.)

Some advantages and disadvantages of the static pile method include:

Advantages:

- Low capital cost
- High degree of pathogen destruction
- Good odor control
- Good product stabilization

Disadvantages:

- Not space efficient
- Vulnerable to climate impacts
- Difficult to work around perforated pipe unless recessed
- Operating cost and maintenance on blowers

In-vessel method—The in-vessel method involves the mixing of manure or other organic waste with a bulking agent in a reactor, building, container, or vessel (see Figure 12.31) and may involve the addition of a controlled amount of air over a specific detention time. This method has the potential to provide a high level of process control because moisture, aeration, and temperature can be maintained with some of the more sophisticated units.

Some of the advantages and disadvantages of the in-vessel method include:

Advantages:

- Space efficient
- Good process control because of self-containment
- Protection from adverse climate conditions
- Good odor control because of self-containment and process control
- Potential for heat recovery dependent on system design
- Can be designed as a continuous process rather than a hatch process

Disadvantages:

- High capital cost for sophisticated units
- Lack of operating data, particularly for large systems
- Careful management required

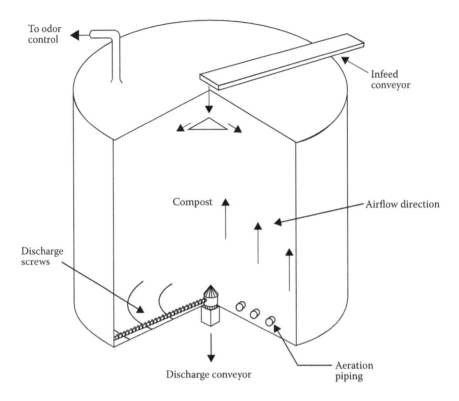

FIGURE 12.31   In-vessel composting schematic. (Source: NRCS/USDA, 1992, p. 10-44.)

- Dependent on specialized mechanical and electrical equipment
- Potential for incomplete stabilization
- Mechanical mixing needs to be provided
- Less flexibility in operation mode than with other methods

12.5.6.2 Method Selection

The composting method must fit the individual farm operation. Highly sophisticated and expensive composting operations are not likely to be a viable option for small farming operations. Some factors to consider when selecting the particular method of composting include:

Operator management capability—The management capability of the operator is an important consideration when selecting the right composting method. Even simple composting methods require that the operator spend additional time in monitoring and material handling. The operator should fully understand the level of management required. The windrow method generally is the simplest method to manage but requires additional labor for periodically turning the compost mix. The static pile is generally next in complexity because of having to maintain blowers and work around perforated pipe. In-vessel composting can be the simplest or the most difficult to mange, depending on the sophistication of the system.

Equipment and labor availability—Consider what equipment is available for loading, unloading, turning, mixing, and hauling. The windrow method requires extra equipment and labor to periodically turn rows.

Site features—If a limited amount of space is available, the static pile or in-vessel method may be the only viable composting alternatives. Proximity to neighbors and the appearance of the compost operation may make the windrow and static pile methods unattractive alternatives. If the only composting site has limited accessibility, then the static pile or in-vessel method should be considered because of less mixing requirements.

Compost utilization—If the compost is to be marketed commercially, then a composting method that produces a predictable, uniform product should be considered. Because of varying climatic conditions, the windrow method may not produce a predictable end product. Sophisticated in-vessel methods provide the most process control; therefore, they produce the most uniform and predictable product.

Climate—In extremely wet climates, the static pile and aerated composting methods may become too wet to compost properly unless measures are taken to protect the compost from the weather. In very cold climates, the composting process may slow in the winter. Sheltering the compost pile from the wind helps to prevent a slowdown in the composting process. The windrow and static pile methods are the most vulnerable to freezing temperatures because they are exposed to the elements. All methods may perform unsatisfactorily if the organic waste and amendments are initially mixed in a frozen state.

Cost—Composting capital and operating costs vary considerably depending on the degree of sophistication. The windrow method generally has the least capital cost but has the highest operational costs. The in-vessel method usually has the highest initial capital cost but the lowest operational cost.

12.5.6.3 Siting and Area Considerations

The location of the composting facility is a very important factor in a successful compost operation. To minimize material handling, the composting facility should be located as close as possible to the source of organic waste. If land application is the preferred method of utilization, the facility should also be located with convenient access to the land application sites. Several other important considerations when locating a compost facility include:

- **Wind direction**—Improperly managed compost facilities may generate offensive odors until corrective actions are taken. Wind direction and proximity to neighbors should be considered when locating a composting facility.
- **Topography**—Avoid locating composting facilities on steep slopes, where runoff may be a problem, and in areas where the composting facility is subject to inundation.
- **Groundwater protection**—The composting facility should be located downgradient and at a safe distance from any wellhead. A roofed compost facility that is properly managed should not generate leachate that could contaminate groundwater. If a compost facility is not protected from the weather, it should be sited to minimize the risk to groundwater.
- **Area requirements**—The area requirements for each composting method vary. The windrow method requires the most land area. The static pile method requires less land area than the windrow method but more than the in-vessel method. The pile dimensions also affect the amount of land area necessary for composting. A large pile with a low surface area to total volume ratio requires less composting area for a given volume of manure but is also harder to manage. The size and type of equipment used to mix, load, and turn the compost should also be considered when sizing a compost area. Enough room must be provided in and around the composting facility to operate equipment. In addition, a buffer area around the compost site should be considered if a visual barrier is needed or desired. In general, given the pile dimensions, a compost bulk density of 35 to 45 lb/ft^3 can be used to estimate the surface area necessary for stacking the initial compost mix. To this area, add the amount of area necessary for equipment operation, pile turning, and buffer.
- **Existing areas**—To reduce the initial capital cost, existing roofed, concrete, paved, or gravel areas should be used if possible as a composting site.

12.5.6.4   Compost Utilization

Finished compost is used in a variety of ways but is primarily used as a fertilizer supplement and soil conditioner. Compost improves soil structure and soil fertility, but it generally contains too low a quantity of nitrogen to be considered the only source of crop nitrogen. Nutrients in finished compost are slowly released over a period of years, thus minimizing the risk of nitrate leaching and high nutrient concentrations in surface runoff.

Good quality compost can result in a product that can be marketed to home gardeners, landscapers, vegetable farmers, garden centers, nurseries, greenhouses, turf growers, golf courses, and ornamental crop producers. Generally, the marketing of compost from agricultural operations has not provided enough income to completely cover the cost of composting. If agricultural operations do not have sufficient land to spread the waste, marketing may still be an attractive alternative compared to hauling the waste to another location for land spreading. Often, compost operators generate additional income by charging municipalities and other local governments for composting urban yard waste with the waste products of the agricultural operations.

Finished compost has also been successfully used as a bedding material for livestock. Because composting generates high temperatures that dry out and sterilize the compost, the finished product is generally acceptable as a clean, dry, bedding material. Refeeding of poultry compost as a food supplement is currently being tested and may prove to be an acceptable use of poultry compost.

12.5.6.5   Compost Mix Design

Composting of organic waste requires the mixing of an organic waste with amendments or bulking agents in the proper proportions to promote aerobic microbial activity and growth and to achieve optimum temperatures. The following elements must be provided in the initial compost mix and maintained during the composting process:

- a source of energy (carbon) and nutrients (primarily nitrogen)
- sufficient moisture
- sufficient oxygen for an aerobic environment
- a pH in the range of 6 to 8

- *Important point:* The proper proportion of waste, amendments, and bulking agents is commonly called the "recipe."

A composting amendment is any item added to the compost mixture that alters the moisture content, C:N ratio, or pH. Many materials are suitable for use as a composting amendment. Crop residue, leaves, grass, straw, hay, and peanut hulls are just some of the examples that may be available. Others, such as sawdust, wood chips, or shredded paper and cardboard, may be available inexpensively from outside sources. Table 12.6 shows typical C:N ratios of common composting amendments. The C:N ratio is highly variable, and local information or laboratory values should be used whenever possible.

A bulking agent is used primarily to improve the ability of the compost to be self-supporting (structure) and to increase porosity to allow internal air movement. Wood chips and shredded tires are examples of bulking agents. Some bulking agents, such as large wood chips, may also alter the moisture content and C:N ratio, in which case they would be both a bulking agent and a compost amendment.

Compost design parameters—To determine the recipe, the characteristics of the waste and the amendments and bulking agents must be known. Characteristics most important in determining the recipe are moisture content (wet basis), carbon content, nitrogen content, and the C:N ratio. If any two of the last three components are known, the remaining one can be calculated.

TABLE 12.6
Typical Carbon to Nitrogen Ratios of Common Composting Amendments

Material	C:N ratios	Material	C:N ratios
Alfalfa (broom stage)	20	Pig manure	5–8
Alfalfa hay	12–18	Pine needles	225–1,000
Asparagus	70	Potato tops	75
Austrian pea straw	59	Poultry manure (fresh)	6–10
Austrian peas (green manure)	18	Poultry manure (henhouse litter)	12–18
Bark	100–130	Reeds	20–50
Bell pepper	30	Residue of mushroom culture	40
Breading crumbs	28	Rice straw	48–115
Cantaloupe	20	Rotted manure	20
Cardboard	200–500	Rye straw	60–350
Cattle manure (with straw)	25–30	Sawdust	300–723
Cattle manure (liquid)	8–13	Sawdust (beech)	100
Clover	12–23	Sawdust (fir)	230
Clover (sweet and young)	12	Sawdust (old)	500
Corn and sorghum stover	6–100	Seaweed	19
Cucumber	20	Shredded tires	95
Dairy manure	10–18	Soil organic matter	10–24
Garden wastes	20–60	Soybean residues	20–40
Grain rice	36	Straw	40–80
Grass clippings	12–25	Sugar cane (trash)	50
Green leaves	30–60	Timothy	80
Green rye	36	Tomato leaves	13
Horse manure (peat litter)	30–60	Tomatoes	25–30
Leaves (freshly fallen)	40–80	Watermelon	20
Newspaper	400–500	Water hyacinth	20–30
Oat straw	48–83	Weeds	19
Paper	173	Wood straw	60–373
Pine vines (native)	29	Wood (pine)	723
Peat (brown or light)	30–50	Wood chips	100–441

Source: USDA (1996).

Carbon to nitrogen (C:N) ratio—The balance between carbon and nitrogen in the compost mixture is a critical factor for optimum microbial activity. After the organic waste and the compost ingredients are mixed together, microorganisms multiply rapidly and consume carbon as a food source and nutrients to metabolize and build proteins. The C:N ratio of the compost mix should be maintained for most compost operations between 25 and 40 to 1. If the C:N ratio is low, a loss of nitrogen generally occurs through rapid decomposition and volatilization of ammonia. If it is high, the composting time increases because the nitrogen becomes the limiting nutrient for growth.

Moisture—Microorganisms need moisture to convert the carbon source to energy. Bacteria generally can tolerate moisture content as low as 12% to 13%; however, with less than 40% moisture, the rate of decomposition is slow. At greater than 60% moisture, the process turns from one that is aerobic to one that is anaerobic. Anaerobic composting is less desirable because it decomposes more slowly and produces putrid odors. The finished product should result in a material with a low moisture content.

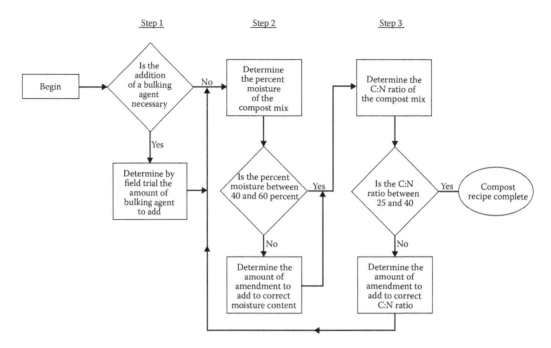

FIGURE 12.32 Compost mixture design flowchart. (Source: NRCS/USDA, 1992, p. 10-48.)

pH—Generally, pH is self-regulating and not a concern when composting agricultural waste. Bacterial growth generally occurs within the range of pH 6.0 to 7.5, and fungi growth usually occurs within the range of 5.5 to 8.5. The pH varies through the compost mixture and during the various phases of the composting process. The pH in the compost mixture is difficult to regulate once decomposition is started. Optimum pH control can be accomplished by adding alkaline or acidic materials to the initial mixture.

Compost mix design process—The determination of the compost mix design (recipe) is normally an iterative process of adjusting the C:N ratio and moisture content by the addition of amendments. If the C:N ratio is out of the acceptable range, amendments are added to adjust it. If this results in high or low moisture content, more amendments are added to adjust the moisture content. The C:N ratio is again checked, and the process may be repeated. After a couple of iterations, the mixture is normally acceptable. Figure 12.32 is a mixture design process flow chart that outlines the iterative procedure necessary in determining the compost recipe.

The iterative process of the compost mix design can be summarized to a series of steps to determine the compost mix design. These steps follow the mixture design process flowchart shown in Figure 12.32.

Step 1: Determine the amount of bulking agent to add. The process normally begins with determining whether a bulking agent is needed. The addition of a bulking agent is necessary if the raw waste cannot support itself or if it does not have sufficient porosity to allow internal air movement. A small field trial is the best method to determine the amount of bulking agent required. To do this, a small amount of raw waste would be weighed, and incremental quantities of bulking would be added and mixed until the mix has the structure and porosity desired. Wood chips, bark, and shredded tires are examples of commonly used bulking agents.

Step 2: Calculate the moisture content of the compost mix. After the need for and quantity of bulking agent have been determined, the moisture content of the mixture or raw waste should be calculated. Because water is often added as result of spillage from waterers and in the cleaning processes, raw waste to be composted may have significantly higher moisture content than that of "as excreted" manure. If the amount of water added to the manure can be determined, the moisture content can be calculated using Equation 12.16, ignoring the inappropriate terms.

In addition to extra water, feed spillage and bedding material can constitute a major part of the raw waste to be composted. The moisture content for each additive can be determined individually and used to determine the moisture content of the entire mix (Equation 12.16). A sample of the raw waste (including the bedding, wasted feed, and water) can also be taken, weighed, dried, and weighed again to determine the moisture content of the mix. Using this procedure, the moisture content can be calculated using Equation 12.15.

$$M_t = Wet\ weight - Dry\ weight/Wet\ weight \times 100 \qquad (12.15)$$

where:
M_t = Percent moisture content (wet basis).

Note: To avoid confusion and repetition, the combination of "as excreted" manure, bedding, water, and bulking agent will be referred to as the "compost mix."

The general equation for the moisture content of the compost mix is:

$$M_M \frac{\dfrac{(W_w \times M_w) + (W_b \times M_b) + (W_a \times M_a)}{100} + H_2O}{W_m} \qquad (12.16)$$

where:
M_m = Percent moisture of the compost mixture (wet basis), Equation 12.15.
W_w = Wet weight of waste (lb).
M_w = Percent moisture content of waste (wet basis), Equation 12.15.
W_b = Wet weight of bulking agent (lb).
M_b = Percent moisture content of bulking agent (wet basis), Equation 12.15.
W_a = Wet weight of amendment (lb).
M_a = Moisture content of amendment (wet basis).
H_2O = Weight of water added (lb) = G × 8.36, where G = gallons of water.
W_m = Weight of the compost mix (lb) including wet weight of waste, bulking agent, amendments, and added water.

(*Note:* This equation may contain variables that are not needed in every calculation.)

Step 2 (continued): Determine the amount of amendment to add, if any, to the compost mix to result in final moisture content between 40% and 60%. If the moisture content of the compost mix is less than 40%, adding an amendment is necessary to raise the moisture content to an acceptable level. Water is the amendment that is generally added to raise the moisture content, but an amendment with a higher moisture content than the desired moisture content of the compost mix is acceptable. It is generally best to begin the composting process when the moisture content is closer to 60%, because the process of composting elevates the temperature and reduces moisture.

- *Important point:* If the moisture content of the compost mix is above 60%, the addition of an amendment is necessary to lower the moisture content at or below 60%. Straw, sawdust, wood chips, and leaves are commonly used amendments.

Equation 12.17 can be used to determine the amount of amendment to add to lower or raise the moisture content of the compost mix.

$$W_{aa} \frac{(W_{mb} \times (M_{mb} - M_d))}{M_d - M_{aa}} \tag{12.17}$$

where:
W_{aa} = Wet weight of amendment to be added.
W_{mb} = Wet weight of mix before adding in amendment.
M_{mb} = Percent moisture of mix before adding amendment.
M_d = Desired 12.12 can be used for the addition of water by using: $M_{aa} = 100\%$ for water.

- *Important point:* Equation 12.17 can be used for the addition of water by using:

Step 3: Calculate the C:N ratio. The C:N ratio for the compost mix is calculated from the C:N ratios of the waste-bulking agents and amendments. The C:N ratios for various waste products and amendments are also shown in Table 12.6. The C:N ratios not reported in the table can be estimated from the amount of fixed solids (amount of ash left after organic matter is burned off) or the VS and the nitrogen content. Equations 12.13 and 12.14 are used to estimate the C:N ratio from fixed solids or VS.

$$W_{aa} =$$

$$\%C = 100 - \%FS/1.8$$

$$Wc = VS/1.8$$

$$C : N = \%/\%N = W_c/W_n$$

where:
C = Percent carbon (dry basis).
$\%FS$ = Percent fixed solids (dry basis).
W_c = Dry weight of carbon.
VS = Weight of volatile solids.
$C:N$ = Carbon to nitrogen ratio.
$\%N$ = Percent total nitrogen (dry basis).
W_n = Dry weight of nitrogen.

The C:N ratio and nitrogen content of manure and of other amendments are highly variable. Using local values for C:N ratios and nitrogen or testing of the compost constituents is highly recommended. The general equation for estimating the C:N ratio of the compost mix is given by Equation 12.18.

$$R_m = \frac{W_{cw} + W_{cb} + W_c a}{W_{nw} + W_{nb} + W_{na}} \tag{12.18}$$

where:
R_m = C:N ratio of compost mix.
W_{cw} = Weight of carbon in waste (lb).
W_{cb} = Weight of carbon in bulking agent (lb).
W_{ca} = Weight of carbon in amendment (lb).
W_{nw} = Weight of nitrogen in waste (lb).

W_{nb} = Weight of nitrogen in bulking agent (lb).
W_{na} = Weight of nitrogen in amendment (lb).

The weight of carbon and nitrogen in each ingredient can be estimated using the following equations:

$$W_w = \%N \times W_{dry}$$

$$W_n = W_c / C:N$$

$$W_c = \%C \times W_{dry}$$

$$W_c = C:N \times W_n$$

where:
W_{dry} = Dry weight of material in question

The dry weight of material can be calculated using Equation 12.19.

$$W_{dry} = W_{wet} \times 100 - M_{wet}/100 \qquad (12.19)$$

where:
W_{wet} = Wet weight of material in question.
M_{wet} = Percent moisture content of material (wet basis).

Step 3 (continued): Determine the amount of amendment, if any, to add to the compost mix to result in an initial C:N ratio between 25 and 40. If the C:N ratio calculated in Step 3a is less than 25 or more than 40, the type and amount of amendment to add to the compost mix must be determined. For a compost mix with a C:N ratio below 25, an amendment should be added with a C:N ratio higher than the desired C:N ratio. For a compost mix with a C:N ratio of more than 40, an amendment must be added with a C:N ratio less than the desired C:N ratio.

Equation 12.20 or 12.21 can be used to calculate the weight of amendment to add to achieve a desired C:N ratio.

$$W_{aa} = \frac{W_{nm} \times (R_d - R_{mb}) + 10,000}{N_{aa} \times (100 - M_{aa}) \times (R_{aa} - R_d)} \qquad (12.20)$$

$$W_{aa} = \frac{N_m W_{nm} \times (100 - M_{mb}) + (R_d - R_{mb})}{N_{aa} \times (100 - M_{aa}) \times (R_{aa} - R_d)} \qquad (12.21)$$

where:
W_{nm} = Weight of nitrogen in compost mix (lb).
R_d = Desired C:N ratio.
R_m = C:N ratio of the compost mix before adding amendment.
N_{aa} = Percent nitrogen in amendment to be added (dry basis).
R_{aa} = C:N ratio of compost amendment to be added.
N_m = Percent nitrogen in compost mix (dry basis).
M_{mb} = Percent moisture of compost mix before adding amendment (wet basis), Equation 12.10.

For a compost mix with a C:N ratio of more than 40, a carbonless amendment such as fertilizer can be added to lower the C:N ratio to within the acceptable range. In this special case, Equation 12.22 can be used to estimate the dry weight of nitrogen to add to the mix:

$$W_{nd} = \frac{W_{cw} + W_{cb} + W_{ca})}{R_d} - (W_{nw} + Wn_b + W_{na})$$ (12.22)

where:
W_{nd} = Dry weight of nitrogen to add to mix.

After the amount of an amendment to add to correct the C:N ratio has been determined, the design process then returns to step 2. If no change is necessary in steps 2 and 3, the compost mix design process is complete.

12.5.6.6 Composting Operational Considerations

The landowner or operator should be provided a written set of instructions as a part of the waste management plan. These instructions should detail the operation and maintenance requirements necessary for successful composting operation. They should include the compost mix design (recipe), method or schedule of turning or aerating, and instructions on monitoring the compost process and on long-term storage compost. The final use of the compost should be detailed in the waste utilization plan.

Composting time—One of the primary composting considerations is the amount of time it takes to perform the composting operation. Composting time varies with C:N ratio, moisture content, climate, type of operation, management, and the types of wastes and amendments being composted. For a well-managed windrow or static pile composting operation, the composting time during the summer months ranges from 14 days to a month. Sophisticated in-vessel methods may take as little as 7 days to complete the composting operation. In addition to the actual composting time, the amount of time necessary for compost curing and storage should be considered.

Temperature—Consideration should be given to how the compost temperature is going to be monitored. The temperature probe should be long enough to penetrate a third of the distance from the outside of the pile to the center of the mass. The compost temperature should be monitored on a daily basis if possible. The temperature is an indicator of the level of microbial activity within the compost. Failure to achieve the desired temperatures may result in the incomplete destruction of pathogens and weed seeds and can cause fly and odor problems.

- *Important point:* Initially, the compost mass is at ambient temperature; however, as the microorganisms multiply, the temperature rises.

The composting process is commonly grouped into three phases based on the prominent type of bacteria present in the compost mix. Figure 12.33 illustrates the relationship between time, temperature, and compost phase. If the temperature is less than 50°F, the compost is said to be in the *psychrophillic stage*. If it is in the range of 50°F to 105°F, the compost is in the *mesophillic stage*. If the compost temperature exceeds 105°F, the compost is in the thermophillic stage. For complete pathogen destruction, the compost temperature must exceed 135°F.

The compost temperature declines if moisture or oxygen is insufficient or if the food source is exhausted. In compost methods where turning is the method of aerating, a temperature rhythm often develops with the turning of the compost pile (see Figure 12.34).

Moisture—The moisture content of the compost mixture should be monitored periodically during the process. Low or high moisture content can slow or stop the compost process. High moisture content generally results in the process turning anaerobic and foul odors developing. A high

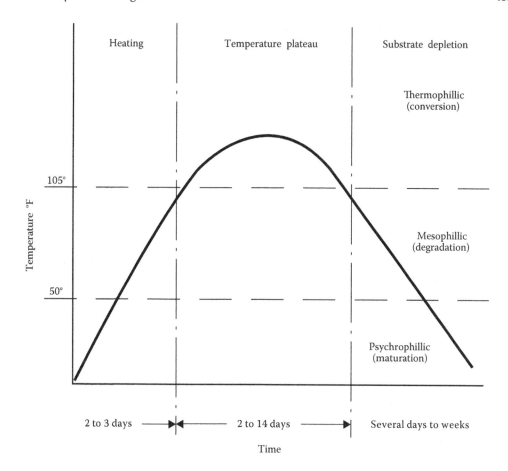

FIGURE 12.33 Composting temperature. (Source: NRCS/USDA, 1992, p. 10-55.)

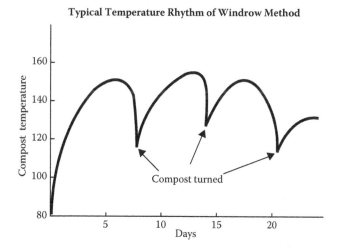

FIGURE 12.34 Typical temperature rhythm of windrow method. (Source: NRCS/USDA, 1992, p. 10-56.)

temperature drives off significant amounts of moisture, and the compost mix may become too dry, resulting in a need to add water.

Odor—The odor given off by the composting operations is a good indicator of how the compost operation is proceeding. Foul odors may mean that the process has turned from aerobic to anaerobic. Anaerobic conditions are the result of insufficient oxygen in the compost. This may be caused by excessive moisture in the compost or the need for turning or aerating of the compost.

12.5.6.7   Compost Process Steps

The composting operation generally follows these steps (see Figure 12.35):

Preconditioning of materials (as needed)—Grinding or shredding of the raw material may be necessary to increase the exposed surface area of the compost mixture to enhance decomposition by microorganisms.

Mixing of the waste with a bulking agent or amendment—A typical agricultural composting operation involves mixing the raw waste with a bulking agent, amendment, or both, according to a prescribed mix or design. The prescribed mix should detail the quantities of raw waste, amendments, and bulking agents to be mixed. The mixing operation is generally done with a front-end loader on a tractor, but other, more sophisticated methods can be used.

Aeration by forced air or mechanical turning—Once the materials are mixed, the composting process begins. Bacteria begin to multiply and consume carbon and free oxygen. To sustain microbial activity, air must be added to the mix to resupply the oxygen to the compost pile. Simply remixing or turning the compost pile can add air. With more sophisticated methods, such as an aerated static pile, air is forced or sucked through the compost mix using a blower. The pounds of air per pound of volatile matter per day generally range from 5 to 9. Given in percentage, the optimum oxygen concentration of the compost mixture ranges from 5% to 15%, by volume. An increase of oxygen beyond 15% generally results in a decrease in temperature because of greater air flow. Low oxygen concentration generally results in anaerobic conditions and slow processing times. Inadequate aeration results in anaerobic conditions and increased odors. Odor is an excellent indicator of when to turn and aerate a compost pile.

Moisture adjustment (as needed)—Water should be added with caution because too much moisture can easily be added. A compost mix that has excessive moisture problems does not compost properly, appears soggy and compacted, and is not loose and friable. Leachate from the compost mixture is another sign of excessive moisture conditions.

Curing (optional)—Once the compost operation is completed, it can be applied directly to the field or stored and allowed to cure for a period of months. During the curing process, the compost temperature returns to ambient conditions and the biological activity slows down. During the curing phase, the compost nutrients are further stabilized. Typical curing time ranges from 30 to 90 days, depending on the type of raw material and end use.

Drying (optional)—Further drying of the compost to reduce weight may be necessary if the finished compost is to be marketed, hauled long distances, or used as bedding. Drying can be accomplished by spreading the compost out in warm, dry weather or under a roofed structure until a sufficient quantity of moisture evaporates.

Bulking agent recovery (as needed or required)—If such bulking agents as shredded tires or large wood chips are used in the compost mixture, they can be recovered from the finished compost by screening. The recovered bulking agents are then reused in the next compost mix.

Storage (as needed)—Finished compost may need to be stored for a period of time during frozen or snow-covered conditions or until the compost product can be marketed. If possible, finished compost should be covered to prevent leaching or runoff.

12.5.6.8   Dead Animal Composting

The disposal of dead animals is a major environmental concern. Composting can be an economical and environmentally acceptable method of handling dead animals. This process produces little

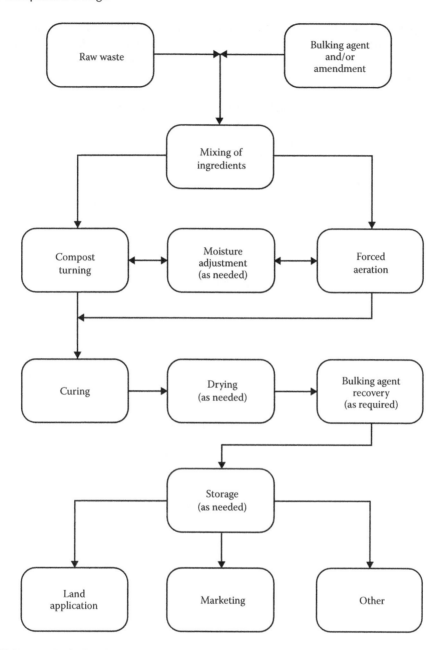

FIGURE 12.35 Agricultural composting process flow. (Source: NRCS/USDA, 1992, p. 10-57.)

odor and destroys harmful pathogens. Although composting of dead poultry is the most common process, the process does apply equally well to other animals. Some operators have composted dead animals weighing as much as 100 lb by grinding or cutting them into smaller pieces.

Composting of dead animals should be considered when:

- a preferred use, such as rendering, is not available
- the mortality rate as a result of normal animal production is predictable
- sufficient land is available for nutrient utilization
- state or local regulations permit dead animal composting
- other disposal methods are not permitted or desired
- marketing of finished compost is feasible

Special planning considerations—Because composting of dead animals is similar in may ways to other methods of composting, the same siting and planning considerations apply. These considerations, therefore, are not be repeated here. Composting of dead animals does, however, have unique problems that require special attention.

- *Important point:* Many states and localities regulate the disposal of dead animals. A construction permit may be required before installation of the facility begins, and an operating permit may be necessary to operate the facility. The animal producer is responsible for procuring all necessary permits to install and operate the facility.

- *Important point:* The size of the animals to be composted should be considered when planning a compost facility. Larger animals require additional equipment, labor, and handling to cut the animals into smaller pieces to facilitate rapid composting.

- *Important point:* Dead animal composting facilities should be roofed to prevent rainfall from interfering with the compost operations. Dead animal composting must reach a temperature in excess of 130°F to destroy pathogens. The addition of rainfall can elevate the moisture content and result in an anaerobic compost mix. Anaerobic composting takes much longer and creates odor problems.

Sizing dead animal composting facilities—A typical dead animal composting facility consists of two stages. The first stage, also called the *primary composter,* is made up of equally size bins in which the dead animals and amendments are initially added and allowed to compost. The mixture is moved from the first stage to the second stage, or secondary digester, when the compost temperature begins to decline. The second stage can also consist of a number of bins, but it is most often one bin or concrete area or alley that allows compost to be stacked with a volume equal to or greater than the sum of the first stage bins.

The design volume for each stage should be based on peak disposal requirements for the animal operation. The peak disposal period normally occurs when the animals are close to their market weight. The volume for each stage is calculated by multiplying the weight of dead animals at maturity times a volume factor. The volume factor (VF) can vary from 1.0 to 2.5 ft³/lb, depending on the type of composter, local conditions, and experiences. Equation 12.23 can be used to calculate the volume for each stage in the compost facility.

$$Vol = B \times M/T \times W \times VF/100 \tag{12.23}$$

where:
Vol = Volume required for each stage (ft³)
B = Number of animals
M = Normal mortality of animals for the entire life cycle expressed as a percent
T = Number of days for animal to reach market weight (days)
W = Market weight of animals (lb)
VF = Volume factor.

Note: M/T is used to estimate the percentage of dead animals to be composted at maturity. Other estimators or field experience may be more accurate.

The number of bins required for the first and second stages can be estimated to the nearest whole number by dividing the total volume required by the volume of each bin (Equation 12.24).

$$\# \text{ Bins} = \text{Total 1}^{\text{st}} \text{ stage volume (ft}^3)/\text{volume of single bin (ft}^3) \tag{12.24}$$

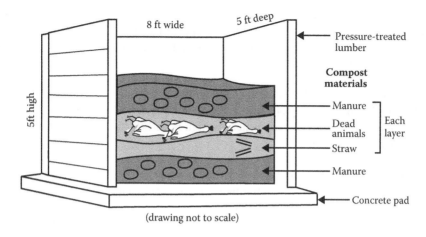

FIGURE 12.36 Dead animal composting bin. (Source: NRCS/USDA, 1992, p. 10-59.)

TABLE 12.7
Poultry Mortality Rates

Poultry type	Loss rate %	Flock life (days)	Cycles/year	Market weight (lb)
Broiler	4.5–5.5	42–49	5.5–6.0	4.2
Roaster females	3	42	4	4.0
Roaster males	8	70	4	7.5
Laying hens	14	440	0.9	4.5
Breeding hens	10–12	440	0.9	7–8
Breeder males	20–25	300	1.1	10–12
Turkey females	5–6	95	3	14
Turkey males	9	112	3	24
Turkey feather producers	12	126	2.5	30

Source: USDA (1996).

Bins are typically 5 ft high, 5 ft deep, and 8 ft across the front. The width across the front should be sized to accommodate the equipment used to load and unload the facility. To prevent spontaneous combustion and to allow for ease of monitoring, a bin height of no more than 6 ft is recommended. The depth should also be sized to accommodate the equipment used.

A high volume to surface area ratio is important to insulate the compost and allow the internal temperature to rise. The bin height and depth should be no less than one-half the width. Shallow bins are easier to unload and load; therefore, the bin depth should be no more than the width. Figure 12.36 provides an example of a dead animal composting bin.

Mortality rates vary considerably because of climate and among varieties, species, and types of operation. Information provided by the animal producer or operator should be used whenever possible. Table 12.7 gives typical mortality rates, flock life, and market weights for poultry.

Mix requirements—Rapid composting of dead animals occurs when the C:N ratio of the compost mix is maintained between 10 and 20. This is considerably lower than what is normally recommended for other types of composting. Much of the nitrogen in dead animal mass is not exposed on the surface; therefore, a lower C:N ratio is necessary to ensure rapid composting with elevated temperatures. If the dead animals are shredded or ground up, a C:N ratio of 25:1 would be more appropriate. The initial compost mix should have a C:N ratio that is between 1 and 15. As composting proceeds, nitrogen, carbon, and moisture are lost. Once composting is complete, the C:N ratio should be between 20 and 25. A C:N ratio of more than 30 in the initial compost mixture is not recommended because excessive composting time and failure to achieve the temperature necessary to destroy pathogens may result.

- *Important point:* For the C:N ratio, a general guideline is: materials that are still green and moist are high in nitrogen (e.g., fresh grass clippings, green weeds, most vegetative kitchen wastes); materials that are dry are generally high in carbon (for example, sawdust, paper, dried grass, dried leaves).

The moisture content of the initial compost mixture should be between 45% and 55% by weight to facilitate rapid decomposition. An initial moisture content of more than 60% would be excessively moist and would retard the compost process. The most common problem in dead animal composting is the addition of too much water. Depending on the mass of dead animals and the moisture content of the amendments, water may not need to be added to the initial mix. Because water is relatively dense compared to the compost mix, the addition of a little water can raise the moisture content significantly. Although water may not need to be added to the initial mix, a source of water should be provided at the compost site for temperature control.

Composting of dead animals should remain aerobic at all times throughout the process. Anaerobic conditions result in putrid odors and may not achieve temperatures necessary to destroy pathogens. Foul odor during the compost process indicates that the compost process has turned anaerobic and that corrective action is needed. These actions will be discussed later. To prevent the compost process from going anaerobic, the initial mix should have enough porosity to allow air movement into and out of the compost mix. This can be accomplished by layering dead animals and amendments in the mix. For example, a dead poultry compost mix would be layered with straw, dead birds, and manure of waste cake from the poultry houses. Layers of such high porosity material as straw, wood chips, peanut hulls, and bark allow lateral movement of air in the compost mix. Figure 12.37 is an example of commonly recommended layering of manure, straw, and dead poultry.

Table 12.8 is a typical recipe for composting dead birds. The ingredients are presented by volume as well as weight.

- *Important point:* Research and evaluation on composting dead animals other than poultry is limited. The differences between livestock and poultry as related to composting are insignificant except for the size of the animal to be composted and the density of skeletal material. Large birds, such as turkeys, have been successfully composted. If large animals are to be composted, they should be cut into no larger than 15-lb pieces and be cut in a manner to maximize surface exposure. Large animal composting is a promising technology but is not well documented. Caution is advised.

Operational considerations—Efficient and rapid composting requires careful control of the C:N ratio, percent moisture and aerobic conditions, and the internal temperature of the compost mix. A deficiency in any of these three areas retards and possibly inhibits the composting process from achieving temperatures too low for pathogen destruction. Careful planning and monitoring is required to ensure that the process is proceeding as expected.

The landowner or operator should be provided a written set of instructions as a part of the waste management plan that details the operation and maintenance requirements necessary for successful dead animal composting. The instructions should include compost mix design (recipe), method or schedule of when to unload the primary digester (first stage) and load the secondary digester (second stage), methods to monitor the compost process, and information on long-term compost storage. The final use of the compost should be detailed in the waste utilization plan.

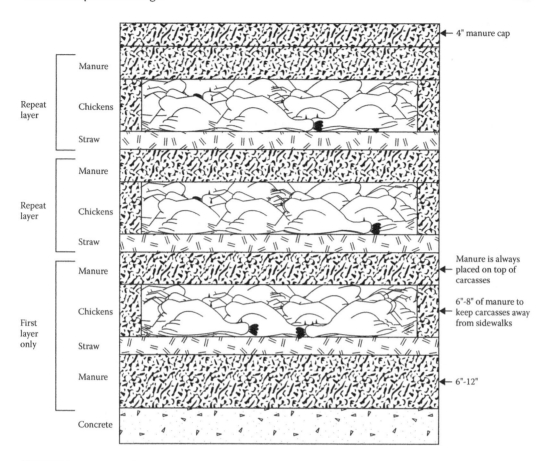

FIGURE 12.37 Recommended layering for dead bird composting. (Source: NRCS/USDA, 1992, p. 10-61.)

TABLE 12.8
Broiler Compost Mix

Ingredient	Volumes (parts)	Weights (parts)
Straw	1.0	0.1
Broiler	2.0	1.0
Manure	2.0	1.5
Water*	0.5	0.75

Source: USDA (1996).

* More or less water may be necessary depending on
   the moisture content of the straw and manure.

Temperature is an important gauge of the progress of the composting operation. After initial loading into the first stage, the compost temperature should peak between 130 and 140 degrees in 5 to 7 days. The same is true for when the compost is moved and stacked in the second stage. Elevated temperatures are necessary to destroy the fly larvae, pathogenic bacteria, and viruses. The two-stage process maximizes the destruction of these elements.

When the compost is initially loaded into the compost bin, the internal temperature begins to rise as a result of bacterial activity. Maximum internal temperatures within the first stage should exceed 130°F within a few days. Although internal compost temperatures rise to a level necessary for the destruction of pathogenic organisms and fly larvae, the temperatures near the edge of the compost pile is not sufficient to destroy these elements. The edge of the compost stack in the first stage may remain an incubation area for fly larvae and allow the survival of the more heat-resistant pathogens.

Removing the compost from the first stage and restacking in the second stage mixes and aerates the compost. The compost that was on the edge of the compost pile is mixed with the internal compost material and subsequently is exposed to temperatures in excess of 130°F in the second stage stack.

The internal temperature of the compost in the first and second stages should be monitored on a daily basis. The compost should be moved from the first stage to the second stage when the internal temperature of the first stage compost begins to decline. This generally occurs after 5 to 7 days.

If internal temperatures fail to exceed 130°F in the first or second stages of the composter, the compost material should immediately be incorporated if land applied or remixed and composted a second time.

Excessively high temperatures are also a danger in dead animal composting because spontaneous combustion of the compost material can occur when the compost temperature exceeds 170°F. If the temperature exceeds 170°F, the compost should be removed from the bin and spread out in a uniform layer no more than 6 in deep. Water should be used, if necessary, to further cool the compost. Once the temperature has fallen to a safe level, the compost can be restacked. Adding moisture to the compost should retard the biological growth and reduce the temperature. Excessive application of water stops the process and can cause anaerobic conditions to develop. The compost mix should be rehydrated to a moisture content of 55% to 65%, by weight, to reduce excessive temperatures.

Anaerobic conditions may develop if the initial porosity or the compost mix is too low, excessive amounts of water are added to the mix, or the C:N ratio is excessively low. Odor generally is a good indicator of anaerobic conditions. If foul odors develop, the reason for the odor problem must be identified before corrective action can be taken. Anaerobic conditions may be the result of any one or a combination of factors, including excessive moisture, low porosity, or low C:N ratio.

12.5.7 Mechanical Separation

Animal manure contains material that can often be reclaimed. Much of the partly digested feed grain can be recovered from manure of poultry and livestock fed high-grain rations. This material can be used as a feed ingredient for other animals. Solids in dairy manure from animals fed a high roughage diet can be removed and processed for use as good quality bedding. Some form of separation must be used to recover these solids. Typically, a mechanical separator is employed. Separators are also used to reduce solids content and required storage volumes.

Separators also facilitate handling of manure. For example, solid separation can allow the use of conventional irrigation equipment for land application of the liquids. Separation eliminates many of the problems associated with the introduction of solids into waste storage ponds and treatment lagoons. For example, it eliminates the accelerated filling of storage volumes with solids and also minimizes agitation requirements.

Several kinds of mechanical separators can be used to remove by-products from manure (see Figure 12.38). One commonly used kind is a screen. Screens are statically inclined or in continuous motion to aid separation. The most common type of continuous motion screen is a vibrating screen. The TS concentration of manure to be processed by a screen should be reduced to less than 5%. Higher TS concentrations reduce the effectiveness of the separator.

A centrifuge separator uses centrifugal force to remove solids, which are eliminated from the machine at a different point than liquids. In addition, various types of presses can be used to force the liquid part of the waste from the solid part.

Flat belt separator

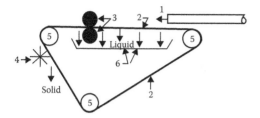

Roller-press separator

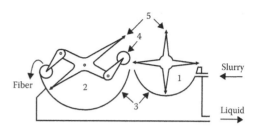

1 Slurry input
2 Polyester mesh belt
3 Press rollers
4 Rotary brush
5 Belt guide rollers
6 Liquid collection trough

1 Screening stage
2 Roller pressing stage
3 Screens
4 Spring loaded press roller
5 Brushes

Vibrating screen separator

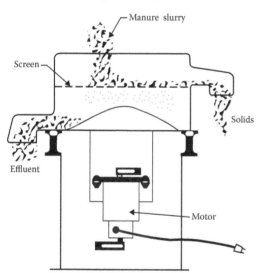

Stationary inclined screen separator

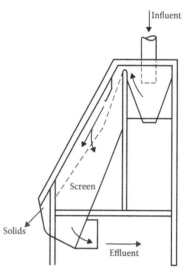

FIGURE 12.38 Schematic for mechanical solid-liquid separators. (Source: NRCS/USDA, 1992, p. 10-63.)

Several design factors should be considered when selecting a mechanical separator. One factor is the amount of liquid waste that the machine can process in a given amount of time. This is referred to as the *throughput* of the unit. Some units have a relatively low throughput and must be operated for a long time. Another very important factor is the TS content required by the given machine. Centrifuges and presses can operate at a higher TS level than can static screens.

Consideration should be given to handling the separated materials. Liquid can be collected in a reception pit and later pumped to storage or treatment. The separated solids have a TS concentration of 15% to 40%. While a substantial amount of nutrients are removed with the solids, the majority of the nutrients and salt remain in the liquid fraction. In many cases, water drains freely from piles of separated solids. This liquid needs to be transferred to storage to reduce odors and fly breeding.

Typically, solids must still be processed before they can be used. If they are intended for bedding, the material should be composted or dried. If the solids are intended for animal feed, they may need to be mixed with other feed ingredients and ensiled before feeding to prevent bacteriological disease transmission. A feed ration using manure must be proportioned by an animal nutritionist so that it is both nutritious and palatable.

TABLE 12.9
Operational Data for Solid and Liquid Separators

Waste type	Separator	TS concentration (N)			% retained in separated solids				
		Raw waste	Separated liquids	Separated solids	TS	VS	COD	N	P
Dairy	Vibrating screen								
	16 mesh	5.8	5.2	12.1	56	—	—	—	—
	24 mesh	1.9	1.5	7.5	70	—	—	—	—
	Decanter centrifuge								
	16–30 gpm	6–8	4.9–6.5	13–33	35–40	—	—	—	—
	Static inclined screen								
	12 mesh	4–6	1.6	12.2	49	—	—	—	—
	32 mesh	2.8	1.1	6.0	68	—	—	—	—
Beef	Static inclined screen	4.4	3.8	13.3	15	—	—	—	—
	Vibrating screen	1–2	—	—	40–50	—	—	—	—
Swine	Decanter centrifuge								
	3 gpm	7.6	2.6	37	14	—	—	—	—
	Vibrating screen 22 gpm/ft²								
	18 mesh	4.6	3.6	10.6	35	39	39	22	26
	30 mesh	5.4	3.5	9.5	52	56	49	33	34

Source: USDA (1996).

TABLE 12.10
Characteristics of Solid and Liquid Separators

Characteristic	Decanter centrifuge	Vibrating screen	Stationary inclined screen
Typical screen opening	—	20 mesh	10–20 mesh
Maximum waste TS concentration	8%	5%	5%
Separated solids TS concentration	To 35%	To 15%	To 10%
TS reduction*	To 45%	To 30%	To 30%
COD reduction*	To 70%	To 25%	To 45%
N reduction*	To 20%	To 15%	To 30%
P reduction*	To 25%	—	—
Throughout (gpm)	To 0	To 300	To 1,000

Source: Adapted from Barker (1986).
* Removed in separated solids

A planner or designer needs to know the performance characteristics of the separator being considered for the type of waste to be separated. The best data, if available, should be that provided by the separator manufacturer. If data are not available, the manufacturer or supplier may agree to demonstrate the separator with waste material to be separated. This can also provide insight as to the effectiveness of the equipment.

If specific data on the separator are not available, Table 12.9 and Table 12.10 can be used to estimate performance characteristics. Table 12.9 gives data for separating different wastes using different separators, and Table 12.10 presents general operational characteristics of mechanical separators.

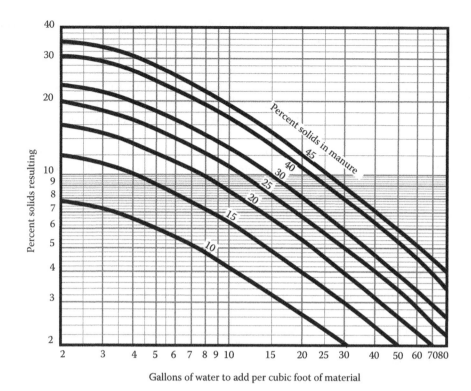

Gallons of water to add per cubic foot of material

FIGURE 12.39 Design aid to determine quantity of water to add to achieve a desired TS concentration. (Source: NRCS/USDA, 1992, p. 10-66.)

12.5.8 SETTLING BASINS

In many situations, removing manure solids, soil, and other material from runoff from livestock operations is beneficial. The most common device to accomplish this is the settling or solids separation basin. A settling basin used in association with livestock operations is a shallow basin or pond designed for low velocities and accumulation of settled materials. It is positioned between the waste source and the waste storage or treatment facilities. Most readily settleable solids settle from the flow if the velocity of the liquid is below 1.5 ft/sec.

- *Important point:* The basins should be planned and designed in accordance with Soil Conservation Service Conservation Practice Standard, Sediment Basin, Code 350 (USDA, 1978). Settling basins should have access ramps that facilitate removal.

12.5.9 DILUTION

Dilution is often used to prepare the waste to facilitate another function. This involves adding clean water or another waste with less TS to the waste, resulting in a waste with the desired percentage of TS. A common use of dilution is to prepare the waste to facilitate use by land application using a sprinkler system. Figure 12.39 is a design aid for determining the amount of clean dilution water required to lower the TS concentration.

12.5.10 VEGETATIVE FILTERS

A vegetative filter can be a shallow channel or a wide, flat area of vegetation used for removing suspended solids and nutrients from concentrated livestock area runoff and other liquid wastes. The filters are designed with adequate length and limited flow velocities to promote filtration, deposition, infiltration, absorption, adsorption, decomposition, and volatilization of contaminants. Consideration must be given to hydraulic as well as contaminant loading.

Vegetative filters rely on infiltration to remove nitrates and microorganisms in solution, because these waste constituents are very mobile in water. Provision for rest periods between loadings is recommended. In cases where a large volume of solids is expected, settling basins are needed above the filter area or channel. "Clean" water must be diverted from the filter. Installation and maintenance are critical.

Vegetation filters are planned and designed according to Conservation Practice Standard, Filter Strip, Code 393 (USDA, 1982), which gives more detailed planning considerations and design criteria. If state or local government has restrictions on the use of vegetative filters, the requirements must be met before design and construction. This is especially true if the outflow from the vegetative filter will flow into a stream or waterway. Unless permitted by state regulations, wastewater treatment by vegetative filters is not sufficient to allow discharge to surface water.

12.6   TRANSFER

Manure collected from within a barn or confinement area must be transferred to the storage or treatment facility. In the simplest system, the transfer component is an extension of the collection method. More typically, the transfer method must be designed to overcome distance and elevation changes between the collection and storage facilities. In some cases, gravity can be used to move the manure. In many cases, however, mechanical equipment is needed. Transfer also involves movement of the waste from storage or treatment to the point of utilization. This may involve pumps, pipelines, and tank wagons.

12.6.1   RECEPTION PITS

Slurry and liquid manure collected by scraping, gravity flow, or flushing are often accumulated in a reception pit (see Figure 12.40). Feedlot runoff can also be accumulated. These pits can be sized to hold all the waste produced for several days to improve pump efficiency or to add flexibility in management. Additional capacity might be needed for extra liquids, such as milk parlor water or

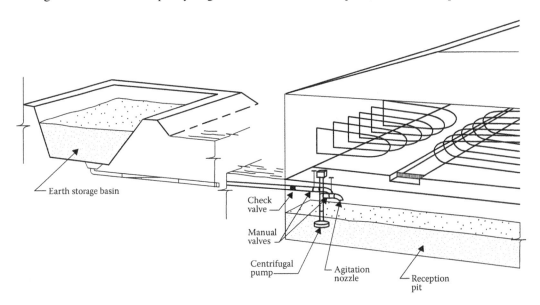

FIGURE 12.40   Reception pit for dairy freestall barn. (Source: NRCS/USDA, 1992, p. 10-67.)

runoff from precipitation. For example, if the daily production of manure and parlor cleanup water for a dairy is estimated at 2,500 gal and 7 days of storage is desired, a reception pit with a capacity of 17,500 gal (2,500 gal/day × days) is the minimum required. Additional volume should be allowed for freeboard emergency storage.

- *Important point:* Reception pits are rectangular or circular and are often constructed of cast-in-place reinforced concrete or reinforced concrete block. Reinforcing steel must be added so the walls can withstand internal and external loads.

Waste can be removed with pumps or by gravity. Centrifugal pumps can be used for agitating and mixing the manure before transferring the material. Both submersible pumps and vertical shaft pumps with the motor located above the manure can be used. Diluted manure can be pumped using submersible pumps, often operated with float switches. The entrance to reception pits should be restricted by guard rails or covers.

Debris, such as pieces of metal, wood, and rocks, must sometimes be removed from the bottom of a reception pit. Most debris must be removed manually, but if possible, this should be done remotely from outside the pit. The pit should be well ventilated before entering. If waste is in the pit, a self-contained breathing apparatus must be used. Short baffles spaced around the pump intake can effectively guard against debris clogging the pump.

In cold climates, reception pits should be protected from freezing. Covering or enclosing them in a building can accomplish this. Adequate ventilation must be provided in all installations. In some installations, hoppers and either piston pumps or compressed air pumps are used instead of reception pits and centrifugal pumps. These systems are used with semisolid manure that does not flow readily or cannot be handled using centrifugal pumps.

12.6.2 GRAVITY FLOW PIPES

Liquid and slurry manure can be moved by gravity if sufficient elevation differences are available or can be established. For slurry manure, a minimum of 4 ft of elevation head should exist between the top of the collection pit or hopper and the surface of the material in storage when storage is at maximum design depth.

Gravity flow slurry manure systems typically use 18- to 36-in. diameter pipe. In some parts of the country, 4- to 8-in. diameter pipe is used for the gravity transport of low (< 3%) TS concentration waste. The planner or designer should exercise caution when specifying the 4- to 8-in. pipe. Smooth steel, plastic, concrete, and corrugated metal pipe are used. Metal pipes should be coated with asphalt or plastic to retard corrosion, depending on the type of metal. All joints must be sealed so that the pipe is water tight.

Gravity flow pipes should be designed to minimize changes in grade or direction over the entire length. Pipe slopes that range from 4% to 15% should work satisfactorily, but a 7% to 8% slope is preferable. Excessive slopes allow separation of liquids and solids and increase the chance of plugging. The type and quantity of bedding and the amount of milkhouse waste and wash water added have an effect on the flow characteristics and the slope needed in a particular situation. Straw bedding should be discouraged, especially if it is not chopped. Smooth, rounded transition from reception pit to pipe and the inclusion of an air vent in the pipeline aid the flow and prevent plugging.

Figure 12.41 illustrates the use of gravity flow for manure transfer. At least two valves should be located in an unloading pipe. Proper construction and operation of gravity unloading waste storage structures are extremely important. Containment berms should be considered if the contamination risk is high downslope of the unloading facility.

12.6.3 PUSH-OFF RAMPS

Manure that is scraped from open lots can be loaded into manure spreaders or storage and treatment facilities using push-off ramps (see Figure 12.42) or docks. A ramp is a paved structure leading to

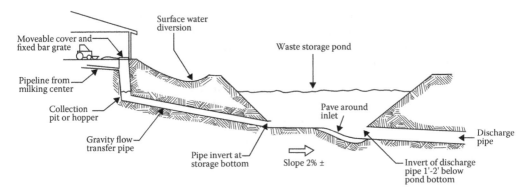

Gravity flow transfer

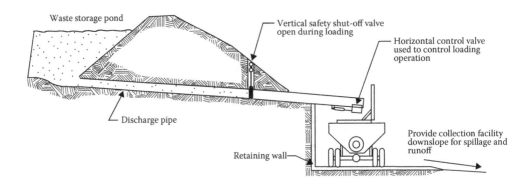

Gravity flow from storage

FIGURE 12.41 Examples of gravity flow transfer. (Source: NRCS/USDA, 1992, p. 10-69.)

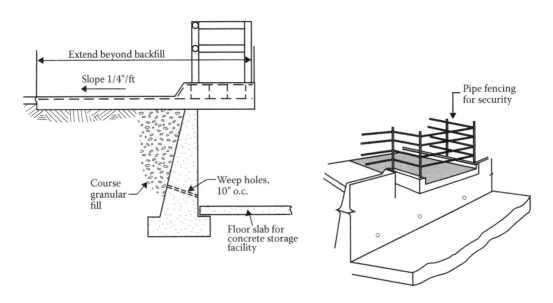

FIGURE 12.42 Push-off ramp. (Source: NRCS/USDA, 1992, p. 10-70.)

a manure storage facility. It can be level or inclined and usually includes a retaining wall. A dock is a level ramp that projects into the storage or treatment facility. Runoff should be directed away from ramps and docks unless it is needed for waste dilution. Ramp slopes should not exceed 5%. Push-off ramps and docks should have restraints at each end to prevent the scraping tractors from accidentally going off the end.

12.6.4 PICKET DAMS

Manure with considerable bedding added can be stored as a solid or semisolid. If the manure is stored uncovered, precipitation can accumulate in the storage area. Picket dams can be used to drain runoff from the storage area while retaining the solid manure and bedding within the storage area. Any water drained should be channeled to a waste storage pond. The amount of water that drains from the manure depends on the amount of precipitation and the amount of bedding in the manure. Water will not drain from manure once the manure and water are thoroughly mixed. Picket dams will not dewater liquid manures.

The picket dam should be near the unloading ramp to collect runoff and keep the access as dry as possible. It should also be on the side of the storage area opposite the loading ramp. Water should always have a clear drainage path from the face (leading edge) of the manure pile to the picket dam.

The floor of the storage area using a picket dam should have slope of no more than 2% toward the dam. Picket dams should be made of pressure-treated timbers with corrosion-resistant fasteners. The openings in the dam should be about 0.75-in. wide vertical slots. Figure 12.43 shows different aspects of picket dam design.

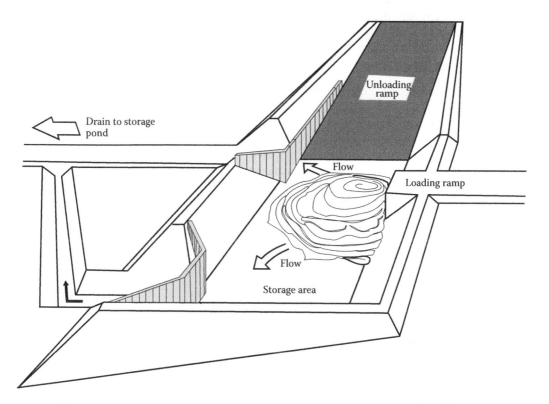

FIGURE 12.43 Solid manure storage with picket dam. (Source: NRCS/USDA, 1992, p. 10-71.)

12.6.5 PUMPS

Most liquid manure handling systems require one or more pumps to either transport or agitate manure. Pumps are in two broad classifications—displacement and centrifugal. The displacement group are piston, air pressure transfer, diaphragm, and progressive cavity pumps. The first two are used only for transferring manure; however, diaphragm and progressive cavity pumps can be used for transferring, agitating, and irrigating manure.

The centrifugal group is vertical shaft, horizontal shaft, and submersible pumps. They can be used for agitation and transfer of liquid manure; however, only vertical and horizontal shaft pumps are used for irrigation because of the head that they can develop.

Pump selection is based on the consistency of the material to be handled, the total head to overcome, and the desired capacity (pumping rate). Pump manufacturers and suppliers can provide rating curves for a variety of pumps.

- *Important point:* Other equipment used in the transfer of agricultural wastes include a variety of pumps, including chopper/agitator, centrifugal, ram, and screw types. Elevators, pipelines, and hauling equipment are also used.

12.7   UTILIZATION

Utilization is a function in a waste management system employed for a beneficial purpose. The typical method is to apply the waste to the land as a source of nutrients for plant growth and of organic matter to improve soil tilth and water holding capacity and to help control erosion. The vast majority of animal waste produced in the United States is applied to cropland, pasture, and hayland. Manure properly managed and applied at the appropriate rates and times can significantly reduce the amount of commercial fertilizer needed for crop production. An anaerobic digester used for bio-gas production is considered a utilization function component because the waste is being managed for use, even though further management of the digester effluent is required.

12.7.1 NUTRIENT MANAGEMENT

Manure should be applied at rates where the nutrient requirements of the crop to be grown are met. Concentration of nutrients in the manure should be known, and records on manure application rates should be maintained.

Between the time of manure production and the time of application, nutrient concentrations can vary widely because of storage, dilution, volatilization, settling, drying, or treatment. To accurately use manure, representative examples of the material to be land applied should be analyzed for nutrient content. Before application can be computed, the soil in the fields where the manure will be applied should be analyzed and nutrient recommendations obtained. This information should indicate the amount of nutrients to be applied for a given crop yield.

Scheduling land application of wastes is critical. Several factors must be considered, including:

- Amount of available manure storage
- Major agronomic activities, such as planting and harvesting
- Weather and soil conditions
- Availability of land and equipment
- Stage of crop growth

- *Important point:* A schedule of manure application should be prepared in advance and should consider the most likely periods when application is not possible. This can help in determining the amount of storage, equipment, and labor needed to make applications at desired times.

12.7.2 Land Application Equipment

Animal waste is land applied using a variety of equipment. The kind of equipment used depends on the TS concentration of the waste. If the manure handles as a solid, a box spreader or flail spreader is used. Solids spreaders are used for manure from solid manure structures and for the settled solids in sediment basins.

Slurry wastes are applied using tank wagons or flail spreaders. Some tank wagons can be used to inject the waste directly into the soil. Slurry spreaders are typically used for waste stored in aboveground or belowground storage structures, earthen storage structures, and sometimes lagoons.

Waste with a TS concentration of less than 5% can be applied using tank wagons, or it can be irrigated using large diameter nozzles. Irrigation is used primarily for land application of liquids from lagoons, storage ponds, and tanks. Irrigation systems must be designed on a hydraulic loading rate as well as on nutrient utilization.

- *Important point:* Custom hauling and application of manure are becoming popular in some locations. This method of utilization reduces the amount of specialized equipment needed by the owner or operator.

12.7.3 Land Application of Municipal Sludge (Biosolids)

Municipalities in the United States treat wastewater biologically using either anaerobic or aerobic processes. These processes generate biosolids that has agronomic value as a nutrient source and soil amendment. Land application of biosolids is currently recognized as acceptable technology; however, strict regulations and practices must be followed.

12.7.4 Biogas Production

Some of this material was adapted directly from *Tentative Guidelines for Methane Production by Anaerobic Digestion of Manure* (Fogg, 1981) and *Handbook of Water and Wastewater Treatment Plant Operations* (Spellman, 2003).

Liquid manure confined in an air-tight vessel decomposes and produces methane, carbon dioxide, hydrogen sulfide, and water vapor as gaseous by-products. This process is known as *anaerobic digestion.* Many municipalities use this technique to treat biosolids generated in wastewater treatment.

Anaerobic digestion is the traditional method of biosolids stabilization. It involves using bacteria that thrive in the absence of oxygen and is slower than aerobic digestion but has the advantage that only a small percentage of the wastes are converted into new bacterial cells. Instead, most of the organics are converted into carbon dioxide and methane gas.

Note: In an anaerobic digester, the entrance of air should be prevented because of the potential for air mixed with the gas produced in the digester that could create an explosive mixture.

Equipment used in anaerobic digestion includes a sealed digestion tank with either a fixed or a floating cover, heating and mixing equipment, gas storage tanks, solids and supernatant withdrawal equipment, and safety equipment (e.g., vacuum relief, pressure relief, flame traps, explosion proof electrical equipment).

In operation, process residual (thickened or unthickened biosolids) is pumped into the sealed digester. The organic matter digests anaerobically by a two-stage process. Sugars, starches, and carbohydrates are converted to volatile acids, carbon dioxide, and hydrogen sulfide. The volatile acids are then converted to methane gas. This operation can occur in a single tank (single stage) or in two tanks (two stages). In a single-stage system, supernatant and digested solids must be removed whenever flow is added. In a two-stage operation, solids and liquids from the first stage flow into

TABLE 12.11
Anaerobic Digester—Biosolids Parameters

Raw sludge solids	Impact
< 4% solids	Loss of alkalinity
Decreased sludge retention time	
Increased heating requirements	
Decreased volatile acid:alk ratio	
4 –8% solids	Normal operation
> 8% solids	Poor mixing
Organic overloading	
Decreased volatile acid:alk ratio	

Source: Spellman (2003).

the second stage each time fresh solids are added. Supernatant is withdrawn from the second stage to provide additional treatment space. Periodically, solids are withdrawn for dewatering or disposal. The methane gas produced in the process may be used for many treatment plant activities.

Note: The primary purpose of a secondary digester is to allow for solids separation.

Various performance factors affect the operation of the anaerobic digester. For example, percent volatile matter in raw sludge, digester temperature, mixing, volatile acids–alkalinity ratio, feed rate, percent solids in raw sludge, and pH are all important operational parameters that the operator must monitor.

Along with being able to recognize normal and abnormal anaerobic digester performance parameters, operators must also know and understand normal operating procedures. Normal operating procedures include sludge additions, supernatant withdrawal, biosolids withdrawal, pH control, temperature control, mixing, and safety requirements. Important performance parameters are listed in Table 12.11.

- *Important point:* Many livestock and poultry producers have become interested in the anaerobic digestion process because of the potential for onsite energy production.

Biogas, the product of anaerobic digestion, is typically made up of 55% to 65% methane (CH_4), 35% to 45% carbon dioxide (CO_2), and traces of ammonia (NH_3) and hydrogen sulfide (H_2S). Pure methane is a highly combustible gas with an approximate heating value of 994 BTU/ft³. Biogas can be burned in boilers to produce hot water, in engines to power electrical generators, and in absorption coolers to produce refrigeration.

The most frequent problem with anaerobic digestion systems is related to the economical use of the biogas. The biogas production rate from a biologically stable anaerobic digester is reasonably constant; however, most on-farm energy use rates vary substantially. Because compression and storage of biogas is expensive, economical use of biogas as an on-farm energy source requires that farm use must closely match the energy production from the anaerobic digester.

Because of the presence of hydrogen sulfide, biogas may have an odor similar to that of rotten eggs. Hydrogen sulfide mixed with water vapor can form sulfuric acid, which is highly corrosive. It can be removed from biogas by passing the gas through a column of iron-impregnated wood chips. Condensers or condensate traps can remove water vapor. Carbon dioxide can be removed by passing biogas through limewater under high pressure.

Biogas can be used to heat the slurry manure in the digester. From 25% to 50% of the biogas is required to maintain a working digester temperature of 95°F, depending on the climate and the amount of insulation used. Belowground digesters require less insulation than those above ground. Engines can burn biogas directly from digesters; however, removal of hydrogen sulfide and water vapor is recommended.

- *Important point:* If digested solids are separated from digester effluent and dried, they make an excellent bedding material. A brief period of composting may be necessary before it is used.

Anaerobic digestion in itself is not a pollution control practice. Digester effluent must be managed similarly to undigested manure by storing in waste storage ponds or treating in lagoons. Initial start-up of a digester is critical. The digester should be partly filled with water (50% to 75% full) and brought to temperature using an auxiliary heater. Feeding of the digester with manure should increase over a period of 3 to 6 weeks, starting with a feeding rate of about 25% of full feed (normal operation).

- *Important point:* Biogas production rates can be measured using specially designed corrosion resistant gas meters. These rates and carbon dioxide levels are good indicators of digester health during start-up. Several simple tests can be used in the field to determine carbon dioxide.

12.7.4.1 Design Procedure

Because of the safety issues and economic and operational complexities involved, SCS assistance on biogas production is generally limited to planning and feasibility. The information presented here is intended for that type of assistance. Interested farmers and ranchers should be advised to obtain other assistance in the detailed design of the facility.

The guidelines present here are based on digestion of manure in the mesophillic temperature range (about 95°F) and may be subject to change as a result of additional research and experience. They provide a basis for considering biogas production facilities based on current knowledge as part of a waste management system.

Several digester types are used (see Figures 12.44, 12.45, and 12.46). The mixed tank is a concrete or metal cylindrical vessel constructed aboveground. If the manure is highly liquid (low TS),

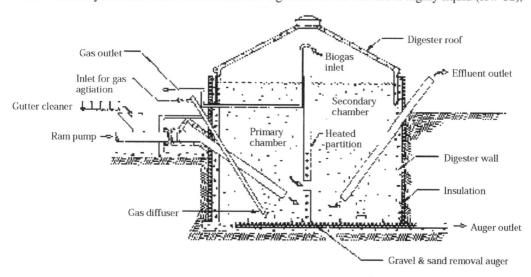

FIGURE 12.44 Two-storage, mixed tank anaerobic digester. (Source: NRCS/USDA, 1992, p. 10-73.)

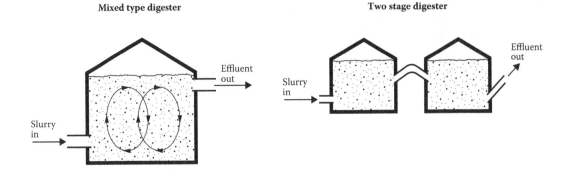

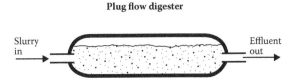

FIGURE 12.45 Typical anaerobic digester system. (Source: NRCS/USDA, 1992, p. 10-74.)

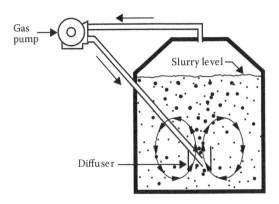

FIGURE 12.46 Gas agitation in an anaerobic digester. (Source: NRCS/USDA, 1992, p. 10-75.)

the digester must be periodically mixed to get good digestion. This can be done mechanically using a mechanical mixer, recirculating digestion liquid, or pumping biogas into the bottom biosolids to remix the contents of the digester.

Another digester, known as the *plug flow,* is used for relatively thick manure (12% to 14% TS), such as dairy manure. The manure is introduced at one end and theoretically moves as a "plug" to the other end. However, if the TS content of the influent manure is too low, the manure "channel," the actual retention time is reduced, and the biogas yield diminishes.

For any digester, the influent must be managed for consistency in frequency of feeding as well as in the VS concentration. For this to happen, the rations fed and manure management must be consistent. Some manure requires preprocessing before it enters the digester. For example, poultry manure must be diluted to about 6% TS to allow grit to settle before the manure is pumped into the digester. Grit material is very difficult to remove from digesters. All digesters must be periodically cleaned. The frequency of cleaning can vary from 1 to 4 years.

Determine manure production—Manure production can be based on applicable tables or on reliable local data. The following data is needed:

Volume of manure produced	= ———————— ft³/day
Wet weight of manure produced	= ———————— lb/day
Total solids (TS)	= ———————— lb/day
Volatile solids (VS)	= ———————— lb/day
Percent solids (TS/wet weight)	= ———————— percent

- *Important point:* Fresh manure is desirable for digestion. Characteristics of beef feed lot manure must be determined for each operation.

Establish TS concentration for digester feed—TS concentrations considered desirable as input to the digester can range from about 6% to 12%. The following are guidelines:

Dairy manure	10 to 12 %
Confined beef manure	10 to 12 %
Beef feedlot manure (after settling grit)	8 to 10 %
Swine manure	8 to 10 %
Chicken manure	7 to 9 %

These percentages may need to be adjusted to eliminated scum formation and promote natural mixing by the gas produced within the mass. If scum forms, a small increase in percent solids may be desirable. This increase may be limited by pumping characteristics and should seldom go above 12% solids.

Determine effective digester volume—A hydraulic detention time of 20 days is suggested. This time appears to be about optimum for efficient biogas production. The daily digester inflow in cubic feet per day can be determined using Equation 12.25.

$$DMI = TMTS \times 100/DDFSC \times 62.4 \tag{12.25}$$

where:
DMI = Daily manure inflow, ft³.
$TMTS$ = Total manure TS production, lb/day.
$DDSFC$ = Desired digester input TS concentration, %.

The necessary digester volume in cubic feet can be determined using Equation 12.26.

$$DEV = DMI \times 20 \tag{12.26}$$

where:
DEV = Digester effective volume, ft³.
20 = Recommended detention time, days.

Select digester dimensions—Optimum dimensions of the liquid part of the digester volume have not been established. The digester should be longer than it is wide to allow raw manure to enter one end and digested slurry to be withdrawn at the other. An effectively operating digester has much mixing by heat convection and gas bubbles. True plug flow does not occur.

Sufficient depth should be provided to preclude excessive delay at start-up because of the oxygen interchange at the surface. A combination of width equal to about two times the depth and length equal to about four times the depth is a realistic approach. Other proportions of width and length should work well. For the purpose of discussion assume:

$$H = (DEV/8)^{033} \qquad (12.27)$$

$$WI = 2 \times H$$

$$L = 2 \times H$$

where:
H = Height, ft.
WI = Width, ft.
L = Length, ft.

- *Important point:* Dimensions should be adjusted to round numbers to fit the site and provide economical construction.

Estimate biogas production—Biogas production is dependent on VS destruction within the digester. An efficient digester with a 20-day retention should reduce VS by 50%. Some research indicates a reduction of 55% of VS in swine manure and 50% to 65% in poultry manure. Biogas production from poultry manure may vary significantly from the estimates presented below. Animals fed a high roughage ration produce less biogas than those fed a high concentrate ration. Estimated VS reductions are:

Dairy and beef   50%
Swine   55%
Poultry   60%

Estimated daily biogas production rates are:

Dairy   12 ft³/lb VS destroyed
Beef   10 ft³/lb VS destroyed
Swine   13 ft³/lb VS destroyed
Poultry   13 ft³/lb VS destroyed

Biogas production per day is estimated by multiplying the percent VS reduction times the estimated daily biogas production rate times the daily VS input. Biogas production in cubic feet per day would be:

Dairy   6 × daily VS input
Beef   5 × daily VS input
Swine   7.2 × daily VS input
Poultry   7.8 × daily VS input

Initial start-up of a digester requires a period of time for anaerobic bacteria to become acclimated and multiply to the level required for optimum methane production. If available, biosolids from a municipal anaerobic digester or another anaerobic manure digester can be introduced to speedup the start-up process.

- *Important point:* The digester contents must be maintained at about 95°F for continuous and uniform biogas production. Hot water tubes within the digester can serve this purpose.

12.7.4.2   Other Considerations

Biogas is difficult to store because it cannot be compressed at normal pressures and temperatures. Storage pressures above 250 psi are rarely used. Because of these reasons, biogas usage is generally planned to match production and thus eliminate the need for storage.

The most common use of biogas is the production of electricity using an engine-generator set. The thermal conversion efficiency is about 25% for this type of equipment. The remainder of the energy is lost as heat. Heat exchangers can be used to capture as much as 50% of the initial thermal energy of the biogas from the engine exhaust gases and the engine cooling water. This captured heat can sometimes be used onsite for heating. Some of it must be used to maintain the digester temperature.

Effluent from anaerobic digesters has essentially the same amount of nutrients as the influent. Some of the organic nitrogen is converted to ammonia, making it more plant available but more susceptible to volatilization unless the liquid is injected. Only a little volume is lost by processing the manure through an anaerobic digester. For manure requiring dilution before digestion, the amount of liquid to be stored and handled is actually greater than the original amount of manure.

12.8 ANCILLARY COMPONENTS

12.8.1 FENCES

Fences are an important component in some agricultural waste management systems. They are planned and designed in accordance with Conservation Practice Standard, Fencing, Code 382 (USDA, 1980a). As they apply to agricultural waste management, fences are used to:

- Confine livestock so that manure can be more efficiently collected
- Exclude livestock from surface water to prevent direct contamination
- Provide the necessary distance between the fence and surface water to be protected for the interception of lot runoff in a channel, basin, or other collection or storage facility located above the lot
- Reduce the lot area and thus reduce the volume of lot runoff to be collected or stored
- Exclude livestock from hazardous areas, such as waste storage ponds
- Allow management of livestock for waste utilization purposes
- Protect vegetative filters from degradation by livestock

12.8.2 DEAD ANIMAL DISPOSITION

Every livestock and poultry facility experiences loss of animals by death. Regardless of the method used, the disposition of dead animals should be accomplished in a sanitary manner and in accordance with all state and local laws.

Use of the energy contained in the dead animals should be given first consideration. Rendering and composting of dead animals both result in useful by-products. If use is not viable, consideration can be given to disposal by incineration or burial. Incineration can cause odor problems unless an afterburner or excess air system issued.

A common method for onsite dead animal disposal is burial. The burial sites need to be at least 150 ft downgradient from any groundwater supply source. Sites with highly permeable soils, fractured or cavernous bedrock, and a seasonal high water table are not suitable and should be avoided. In no case should the bottom of the burial pit be closer than 5 ft from the groundwater table. Surface water should be diverted from the pit.

For large animals (cattle and mature swine), individual pits should be opened for each occasion of burial. The pits should be closed and marked after burial. For small animals (poultry and small pigs), pits can be constructed for use over a period of time.

Typical pit sizes for small animals are 4 to 6 ft wide, 4 to 12 ft long, and 4 to 6 ft deep. The sides of the pit should be constructed of concrete block, treated timber, or precast concrete. The side walls must have some openings to allow for pressure equalization. The bottom of small animal pits is not lined. The top should be airtight with a single-capped opening to allow for adding dead animals. Figure 12.47 illustrates one possible disposal pit configuration.

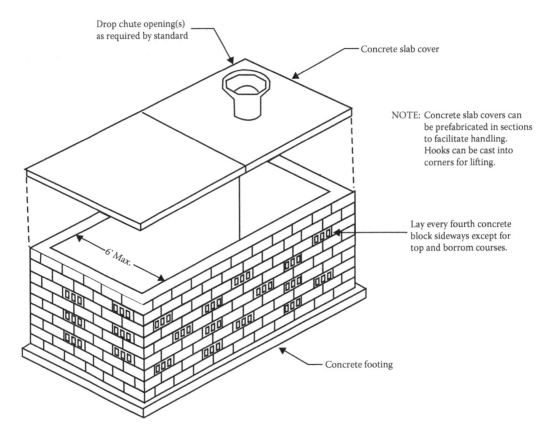

Drop chute opening(s) as required by standard

Concrete slab cover

NOTE: Concrete slab covers can be prefabricated in sections to facilitate handling. Hooks can be cast into corners for lifting.

Lay every fourth concrete block sideways except for top and borrom courses.

6' Max.

Concrete footing

FIGURE 12.47   Poultry and suckling pig disposal pit constructed with 8" × 8" × 16" concrete blocks. (Source: NRCS/USDA, 1992, p. 10-78.)

- *Important point:* Disposal pits should have adequate capacity. The recommended capacity for broilers is 100 ft³ per 10,000 broilers. For small pigs, the capacity is 1 ft³ per sow.

12.9   SAFETY

Much of the material in this section was taken from *Handbook for Water/Wastewater Treatment Plant Operations* (Spellman, 2003) or adapted from the publication *Safety and Liquid Manure Handling* (White & Young, 1980).

Safety must be a primary consideration in managing animal waste. It must be considered during planning and designing of waste management system components as well as during the actual operation of handling waters. The operator must be made aware of safety aspects of any waste management system components under consideration. Accidents involving waste management may be the result of:

- Poor design or construction
- Lack of knowledge or training about components and their characteristics
- Poor judgment, carelessness, or lack of maintenance
- Lack of adequate safety devices, such as shields, guard rails, fences, or warning signs

The potential for an accident with waste management components is always present. However, accidents do not have to happen if components are properly designed, constructed, and maintained and if all persons involved with the components are adequately trained and supervised.

First-aid equipment should be near storage units and lagoons. A special, easily accessible area should be provided for storing the equipment. The area should be inspected periodically to ensure that all equipment is available and in proper working condition. The telephone numbers of the local fire department and rescue squad should be posted near the safety equipment and near all telephones.

12.9.1   Confined Spaces

Manure gases can accumulate when manure is stored in environments without adequate ventilation, such as underground covered waste storage tanks. These gases can reach toxic concentrations, and displace oxygen.

- *Caution:* Anaerobic digesters are inherently dangerous, and several catastrophic failures have been recorded. To prevent such failures, safety equipment such as pressure relief and vacuum relief valves, flame traps, condensate traps, and gas collection safety devices are installed. These critical safety devices must be checked and maintained for proper operation.

Note: Because of the inherent danger involved with working inside anaerobic digesters, they are automatically classified as permit-required confined spaces. Therefore, all operations involving internal entry must be made in accordance with the Occupational Safety and Health Administration's Confined Space Entry Standard.

12.9.2   Aboveground Tanks

Aboveground tanks can be dangerous if access is not restricted. Uncontrolled access can lead to injury or death from falls from ladders and to death from drowning if someone falls into the storage tank. These rules should be enforced:

- Permanent ladders on the outside of aboveground tanks should have entry guards locked in place or the ladder should be terminated above the reach of individuals.
- Ladders must never be left standing against aboveground tanks.

12.9.3   Lagoons, Ponds, and Liquid Storage Structures

Lagoons, ponds, and liquid storage structures present the potential for drowning of animals if access is not restricted. Floating crusts can appear capable of supporting a person's weight and provide a false sense of security. Tractors and equipment can fall or slide into storage ponds or lagoons if they are operated too close to them. The following rules should be obeyed:

1. Strong rails should be built along all walkways or ramps of open manure storage structures.
2. Place fences around storage ponds and lagoons, and post signs such as "Caution Manure Storage (or Lagoon)." The fence keeps livestock and children away from the structure. Additional precautions include a minimum of one lifesaving station equipped with a reaching pole and a ring buoy on a line.
3. Place a barrier strong enough to stop a slow-moving tractor on all push-off platforms or ramps.
4. If manure storage is outside the livestock building, use a water trap or other device to prevent gases in the storage structure from entering the building, especially during agitation.

12.9.4 EQUIPMENT

All equipment associated with waste management, such as spreaders, pumps, conveyors, and tractors, can be dangerous if improperly maintained or operated. Operators should be thoroughly familiar with the operator's manual for each piece of equipment. Equipment should be inspected frequently and service as required. All guards and safety shields must be kept in place on pumps, around pump hoppers, and on manure spreaders, tank wagons, and power units.

12.10 SUMMARY

The ultimate success of an animal feeding operation (AFO) or concentrated animal feeding operation (CAFO) rests firmly on effective, efficient, practical component design. While, of course, many other factors affect the bottom line, successful operation begins with careful planning, long before the first animal enters the facility. Ultimate successful operation includes minimal negative environmental impact as well.

CHAPTER REVIEW QUESTIONS

1. Define "component" as it relates to CAFO/AFO design.
2. What should ideal component design convey?
3. Describe and discuss roof runoff management.
4. How are gutter size capacities computed?
5. How are downspout capacities computed?
6. How are roof area needs computed?
7. Describe and discuss the importance of and necessary considerations for runoff control and diversions.
8. Discuss "freedom of movement" and how it affects waste collection.
9. Describe and discuss common collection systems. Include alleys, gutters, and slatted floors.
10. Describe and discuss storage component factors for sizing requirements.
11. Describe and discuss storage facility requirements for manure solids.
12. How is manure storage volume calculated?
13. How is bedding storage computed?
14. Describe and discuss liquid and slurry waste storage planning and requirements.
15. How are storage ponds structure needs calculated?
16. How are storage ponds volume needs calculated?
17. What functions do common treatment components provide?
18. What management considerations are necessary for treatment components?
19. Describe and discuss common treatment component systems.
20. How is MTV calculated?
21. How is WV determined?
22. How is SV determined?
23. What factors are used to determine LV requirements?
24. Describe and discuss anaerobic lagoon management.
25. Describe and discuss aerobic lagoon management.
26. Describe and discuss mechanically aerated lagoon management.
27. How is transfer rate determined for aeration?
28. How are horsepower requirements calculated?
29. Describe and discuss the use of oxidation ditches. What are the design requirements? Why?
30. What purpose does dewatering serve?
31. Describe and discuss commonly used composting methods. What are their advantages and disadvantages?

32. What are the critical factors to consider when selecting a composting method?
33. What siting and area factors must be considered for composting components? How do they differ, method to method?
34. What are the financial advantages that efficient composting can offer?
35. What are the requirements for the composting mix?
36. What amendments are in common use? How are they chosen? What are their advantages and disadvantages?
37. Discuss the C:N ratio for composting.
38. Discuss moisture's role in composting.
39. Discuss pH and composting.
40. Describe the iterative process of compost mix design.
41. How is amount of needed amendment determined?
42. How is C:N ratio calculated?
43. How does an operator determine the amount of amendment needed for C:N ratio adjustment?
44. What variables affect compost time? Why is this important?
45. How is temperature optimally monitored?
46. Describe and discuss the three phases of the composting process.
47. Describe the relationship between moisture and odor in the composting process.
48. Discuss the common operational steps for composting.
49. Why is composting of dead animals an important process for certain CAFOs/AFOs?
50. When should dead animal composting be considered?
51. What are the special considerations for siting?
52. What are the special considerations for handling?
53. How is design volume calculated?
54. How are bin size and number calculated?
55. Describe and discuss the optimum C:N ratio for dead animal composting.
56. Describe and discuss the optimum moisture content for dead animal composting.
57. Describe and discuss the optimum amendments for dead animal composting.
58. Describe and discuss the optimum temperature range for dead animal composting.
59. What operational considerations should be examined?
60. What are the advantages of mechanical separation?
61. What are the critical design factors involved?
62. What functions do settling basins provide? Why are they important?
63. How is dilution used?
64. What are the benefits and drawbacks for vegetative filters?
65. What is involved in waste transfer?
66. What advantages and problems are related to use of receptor pits?
67. What advantages and problems are related to the use of gravity flow pipes?
68. What advantages and problems are related to the use of pushoff ramps?
69. What advantages and problems are related to the use of picket dams?
70. What advantages and problems are related to the use of pumps?
71. Discuss waste utilization.
72. What scheduling factors should be considered in terms of nutrient management?
73. How are different consistencies of animal wastes land applied?
74. What are the advantages of anaerobic digestion?
75. What equipment is needed?
76. What does biogas consist of? How can it be used? What can be done with the digested solids?
77. Discuss the design procedure for biogas production.
78. How is digester volume calculated?

79. How are digester dimensions calculated?
80. How is biogas production estimated?
81. What are biogas storage considerations?
82. How are fences used?
83. What considerations for dead animal disposition should be made?
84. Describe and discuss the safety concerns related to animal waste management.

THOUGHT-PROVOKING QUESTIONS

1. Research and discuss the costs versus benefits of regulation on the CAFO and AFO industry. How much of the regulatory requirements would be performed voluntarily by CAFO owner management? By CAFO on-site management?
2. Do the CAFO regulations a particular state has in place affect whether CAFO investors site there?
3. Research and discuss: If all applicable CAFO operations' water intakes were required to be placed directly below the operations' outputs, would this make the industry self-regulating on water pollution issues?
4. What role does component design play in CAFO management? How can good or poor component design affect profitability, neighborhood relations, odor control, and regulatory compliance?
5. What are your state's regulations for composting dead animals?

REFERENCES/RECOMMENDED READING

Allen, M.W. 1986. Roof runoff management—gutter selection size. NENTC Tech. Notes, Ag. Engr. No. 1. U.S. Dep. Agric., Soil Conserv. Serv., Chester, PA.

Alpert, J.E. 1987. Composting process and operations. Univ. of MA, On-Farm Composting Conference, E & A Environmental Consultants, Sloughton, MA.

Alpert, J.E. 1987. Windrow and static pile composting. Univ. of MA, On-Farm Composting Conference, E & A Environmental Consultants, Sloughton, MA.

American Society of Agricultural Engineers. 1982. Solid and liquid manure storages. Engineering Practice 393. *Agricultural Engineering Yearbook*, ASAE, St. Joseph, MI. pp. 303–305.

Barker, J.C. 1986. Course notes from agricultural waste management class. North Carolina State University.

Barth, C.L. 1985. The rational design standard for anaerobic livestock lagoons. Proceedings of the Fifth International Symposium on Agricultural Wastes. ASAE, St. Joseph, MI. pp. 638–646.

Barth, C.L. 1985. Livestock waste characterization-a new approach. Proceedings of the Fifth Internal Symposium on Agricultural Wastes. ASAE, St. Joseph, MI. pp. 286–294.

Bartlett, A. (n.d.). Windrow composting of poultry and horse manure with hay. White Oak Farm, Belchertown, MA.

Bos, R.E. 1974. Dewatering bovine manure. M.S. Thesis, Agric. Eng. Dep., The PA State Univ., State College, PA.

Brinton, W. (n.d.). Agricultural and horticultural applications of compost. Woods End Lab., Mt. Vernon, ME.

Brinton, W. F., Jr. 1990. Agricultural waste management and composting. Woods End Res. Lab., Amer. Soc. Argon., NE An. Meet., Univ. NH.

Brinton, W.R., & Seekins, M.D. 1988. Composting fish by-products—A feasibility study.

Cassidy, J. M. (n.d.). Agricultural regulations for Massachusetts compost products. Bur. Farm Prod. MA. Dep. Food and Agric.

Cathcart, T.P, Lipton, D.W., Wheaton, F.W., Brinsfield, R.B., Swartz, D.G., & Strand, L.E. (n.d.). Composting of blue crab waste. Univ. MD, College Park, Pub. No. UM-SG Ts-84-01.

Cathcart, T.P., Wheaton, F.W., & Brinsfield, R.B. 1986. Optimizing variables affecting composting of blue crab scrap. Univ. MD, College Park.

Cheresminoff, P.N & Young, R.A. 1975. *Pollution Engineering Practice Handbook*. Ann Arbor Sci Pub., Inc. pp. 788–792.

Clark, J.W. Viessman, W., Jr., & Hammer, MJ. 1971. *Water Supply and Pollution Control.* Int. Textbook Co., pp. 579–584.

Commonwealth Marketing and Development. 1988. Midcoast compost project market study. Portland, MA.

Costa, C.A. 1987. Introduction: Why consider composting. MA Dep. Food and Agric., On-Farm Composting Conf., Univ. MA.

Eberhardt, D. L., & Pipes, W.O. 1972. Composting applications for Illinois. IIEQ Doc. No. 73-5, IL Inst. Environ Qual.

Eccles, C., & Stentiford, E.I. 1987. Microcomputers monitor static pile performance. *Biocycle*, pp. 42–45.

Ely, J.F., & Spencer, E.I. 1978. The composting alternative waste disposal in remote locations. Res. Dep., Appalachian MT.

Fairbank, C.W. 1974. Energy values. *Agricultural Engineering.* ASAE, St. Joseph, MI. Sept. Issue.

Fogg, C.E. 1981. Tentative guidelines for methane production by anaerobic digestion of manure. National Bulletin No. 210-1-13. U.S. Dep. Agric., Soil Conserv. Serv., Wash., DC.

Fulford, B. (n.d.). Composting dairy manure with newspaper and cardboard. The New Alchemy Inst., Falmouth, MA.

Geiger, J.S. (n.d.). Composting with dairy and horse manure and fish wastes. Appleton farms, Ipswich, MA.

Glerum, J.C., Klamp, G., & Poelma, H.R. 1971. The separation of solid and liquid parts of pig slurry. In *Livestock Waste Management and Pollution Abatement,* ASAE, St. Joseph, MI, pp. 345–347.

Graves, R.E., & Clayton, J.T. 1972. Stationary sloping screen to separate solids from dairy cattle manure slurries. ASAE Paper 72-951, St Joseph, MI.

Graves, R.E. et al. 1986. Manure management for environmental protection. PA Dep. Env. Resource, Harrisburg, PA.

Guest, R.W. 1984. Gravity manure handling. NRAES/NDPC 27.10, Agric. Eng. Dep., Cornell Univ., Ithaca, NY.

Hansen, R.B., & Mancl, K.M. Modern composting, a natural way to recycle wastes. The Ohio State Univ.

Haug, R.T., & Tortorici, L.D. 1986. Composting process design criteria. *Biocycle,* Nov/Dec issue.

Hegg, R.O., Larson, R.E., & Moore, J.A. 1981. Mechanical liquid-solid separation in beef, dairy, and swine waste slurries. *Transactions* of ASAE 24(1):159–163.

Higgins, A. J. 1990. Engineering parameters for the selection of compost bulking agents. ASAE, NAR 83-207.

Holmberg, R.D., Hill, D.T., Prince, T.J., & Van Dyke, N.J. 1982. Solid-liquid separation effect on physical properties of flushed swine waste. ASAE Paper 82-4081, St. Joseph, MI.

Jones, D.D., Day, D.L., & Dale, A.C. 1971. Aerobic treatment of livestock wastes. Univ. of IL. Exp. Sta. Bull. 737, Urbana-Champaign, IL.

Kilmer, V.J. 1982. *Handbook of Soils and Climate in Agriculture.* Boca Raton, FL: CRC Press.

Kuter, G. (n.d.). Commercial in-vessel composting of agricultural wastes. Int. Process Sys., Lebanon, CT.

Laliberty, L. (n.d.). Composting for a cash crop. Farm Res. Ctr., Putnam, CT.

MacLean, A.J., & Hore, F.R. 1974. Manures and compost. Canada Dep. Agric. Pub. 868.

McKinney, R. E. 1962. *Microbiology for Sanitary Engineers.* New York: McGraw-Hill, Inc., pp. 260–265.

Mezitt, R. (n.d.). Composting for the nursery industry. Weston Nurs., Inc.

Midwest Plan Service. 1985. Biological treatment. Livest. waste fac. Handb. MWPS-18., IA State Univ., ch. 7.

Millar, C.G., Turk, L. M. & Foth, H.D. 1965. *Fundamentals of Soil Science.* New York: John Wiley & Sons, Inc.

Moffitt, D.C. 1980. Design of mechanically aerated lagoons for odor control. U.S. Dep. Agric., Soil Conserv. Serv. Tech. Notes, WTSC, Portland, OR.

Murphy, D.W. (n.d.). Composting of dead birds. Dep. Poult. Sci., Univ. MD, Princess Anne.

Nessen, F. (n.d.). Present and future composting regulation in Massachusetts. MA Dep. Environ. Qual. & Eng.

New Hampshire Department of Agriculture. 1989. Good neighbor guide for horse-keeping: Manure management. USDA Ext. Serv. Pub. No. 89-EWQI-1-9186.

Ngoddy, P.O., Hayser, J.P., Collins, R.K., Wells, G.D., & Heider, F.A. 1971. Closed system waste management for livestock water pollution control. *Research Series,* USEPA, Project 13040DKP.

North Carolina State University. 1980. Earthen liquid manure storage basin with access ramp. Agric. Ext. Serv., Raleigh, NC.

Parsons, R.A. 1984. On-farm biogas production. NRAES-20. NE Reg. Agric. Eng. Serv., Cornell Univ.

Pennsylvania Department of Environmental Resources. 1990. Assessment of field manure nutrient management with regards to surface water and groundwater quality.

Plovanich, C.J. (n.d.). In-vessel composing of cattle and swine manure. Revere Copper and Brass, Inc., Rome, NY.

Poincelot, R.P. 1975. The biochemistry and methodology of composting. CT Agric. Exp. Stat.

Proulx, A. 2002. *That Old Ace in the Hole*. New York: Scribner

Rynk, R. 1987. On farm composting: The opportunities, benefits and drawbacks. Eng. Notes, Univ. MA.

Rynk, R. 1988. On farm composting: The process and methods. Eng. Notes, Univ. MA.

Rynk, R. 1989. Composting as a dairy manure management technique. Dairy Manure Mgt. Symp., Syracuse, NY, NRAES-31.

Rodale, J.I. 1975. *The Complete Book of Composting*. Emmaus, PA: Rodale Books, Inc.

Rodale, J.I. 1968. *The Encyclopedia of Organic Gardening*. Emmaus, PA: Rodale Books, Inc. pp. 151–153.

Safley, L.M., Jr., & Fairbank, W.C. 1983. Separation and utilization of manure solids. Proceed. Sec. Natl. Dairy Housing Conf. ASAE, St. Joseph, MI. pp. 77–91.

Salvato, J.A., Jr. 1982. *Environmental Engineering and Sanitation*. New York: Wiley Interscience, pp. 403–406.

Shutt, J.W. White, R.K. Taiganides, E.P., & Mote, C.R. 1975. Evaluation of solids separation devices. In *Managing Livestock Wastes,* ASAE, St. Joseph, MI. pp. 463–467.

Simpson, M. (n.d.). Economics of agricultural composting. Bur. Solid Waste Disposal, MA.

Singley, M.E. (n.d.). Preparing organic materials for composting. ASAE Paper NAP 83-208.

Spellman, F.R. 2003. *Handbook of Water and Wastewater Treatment Plant Operations*. Boca Raton, FL: CRC Press.

Sweeten, J.M. (n.d.). Composting manure and sludge. TX Agric. Exp. Sta. Pub. No. 1-2289.

USDA. 1965. Rule of thumb estimates of the relationships of carbon to nitrogen in crop residues. USDA, Soil Conserv. Serv., Portland, Or., Tech. Notes.

USDA. 1978. Sediment basin. SCS Conservation Practice Standard, Code 350. Soil Conserv. Serv., Wash., DC.

USDA. 1979. Waste management system. SCS Conservation Practice Standard, Code 312. Soil Conserv. Serv., Wash., DC.

USDA. 1979. Waste storage pond. SCS Conservation Practice Standard, Code 425. Soil Conserv. Serv., Wash., DC.

USDA. 1980. Fencing. SCS Conservation Practice Standard, Code 382. Soil Conserv. Serv., Wash., DC.

USDA. 1980. Waste storage structure. SCS Conservation Practice Standard, Code 313. Soil Conserv. Serv., Wash., DC.

USDA. 1982. Filter strip. SCS Conservation Practice Standard, Code 393. Soil Conserv. Serv., Wash., DC.

USDA. 1984. Pond sealing or lining. SCS Conservation Practice Standard, Code 521. Soil Conserv. Serv., Wash., DC.

USDA. 1984. Waste treatment lagoon. SCS Conservation Practice Standard, Code 359. Soil Conserv. Serv., Wash., DC.

USDA. 1985. Diversion. SCS Conservation Practice, Code 382. Soil Conserv. Serv., Wash., DC.

USDA. 1987. Guide on design, operation and management of anaerobic lagoons. Tech. Note Ser. 711. SNTC, Soil Conserv. Serv., Wash., DC.

USDA. 1987. Interim engineering standard for poultry and swine disposal pit. SNTC, Soil Conserv. Serv., Wash., DC.

USDA. 1989. Interim standard for dead poultry composting. Soil Conserv. Serv., Auburn, AL.

USDA. 1990. Design and construction guidelines for considering seepage from agricultural waste storage ponds and treatment lagoons. Tech. Note Ser. 716, SNTC, Soil Conserv. Serv., Wash., DC.

USDA. 1996. Agricultural Waste Management System Component Design. AWMFH-Chapter 10. http://www.wcc.nrcs.usda.gove/awm/awmfh.htim. Accessed June 2005.

USEPA. 1977. Municipal sludge management: Environmental factors. EPA 430/9-77-004, USEPA, Wash., DC.

USEPA. 1988. Design manual—Constructed wetlands and aquatic plant systems for municipal wastewater treatment. EPA/625/1-88/022.

USEPA. 1989. In-vessel composting of municipal wastewater sludge. USEPA Summary Rep. No. EPA/625/8-89/016.

University of Arkansas. (n.d.). Cooperative Extension Service composting poultry carcasses. Mp 317.

University of Delaware. 1989. Dead poultry disposal. College of Agricultural Sciences. Agricultural Experiment Station.

University of Tennessee. 1977. Covered dry stack manure storage. Agric. Ext. Serv., Plan No. T4029, Univ. TN, Knoxville, TN.

White, R.K., & Forster, D.L. 1978. Evaluation and economic analysis of livestock waste management systems. EPA/2-78-102. USEPA, ADA, OK.

White, R.K., & Young, C.W. 1980. Safety and liquid manure handling. OH Coop. Ext. Serv., AEX 703. OH State Univ., Columbus, OH.

Wright, R.E. Assoc., Inc. 1990. Assessment of field manure nutrient management with regards to surface and ground water quality. PA dep. Environ. Resource., Harrisburg, PA.

Index

Printed and bound by CPI Group (UK) Ltd, Croydon, CR0 4YY

23/10/2024

01778249-0016